ELECTROMYOGRAPHY

ELECTROMYOGRAPHY
Physiology, Engineering, and Noninvasive Applications

EDITED BY

Roberto Merletti
Philip Parker
IEEE Press Engineering in Medicine and Biology Society, *Sponsor*

IEEE PRESS

A JOHN WILEY & SONS, INC., PUBLICATION

Cover: The background image represents a multichannel surface EMG signal, four classified motor units and their firing rates during a voluntary contraction. From the top to the bottom is a survey of electrodes for research and clinical applications: linear dry arrays (LISiN, Politecnico di Torino, Italy); electrode matrixes (Helmholtz Institute for Biomedical Engineering, Aachen, Germany); adhesive arrays (LISiN, Politecnico di Torino, Italy and Spes Medica, Salerno, Italy); a 128 electrode matrix (University Medical Center, Nijmegen, The Netherlands). Bottom picture: glass slide image of a motor end plate.

Published by John Wiley & Sons, Inc., Hoboken, New Jersey.
Published simultaneously in Canada.

For general information on our other products and services please contact our Customer Care Department within the U.S. at 877-762-2974, outside the U.S. at 317-572-3993 or fax 317-572-4002.

Wiley also publishes its books in a variety of electronic formats. Some content that appears in print, however, may not be available in electronic format.

Library of Congress Cataloging-in-Publication Data is available.

ISBN 0-471-67580-6

Printed in the United States of America

10 9 8 7 6 5 4 3 2 1

CONTENTS

3 DECOMPOSITION OF INTRAMUSCULAR EMG SIGNALS   47

D. W. Stashuk, D. Farina, K. Søgaard

4 BIOPHYSICS OF THE GENERATION OF EMG SIGNALS   81

D. Farina, R. Merletti, D. F. Stegeman

5 DETECTION AND CONDITIONING OF THE SURFACE EMG SIGNAL   107
R. Merletti, H. Hermens

6 SINGLE-CHANNEL TECHNIQUES FOR INFORMATION EXTRACTION FROM THE SURFACE EMG SIGNAL   133
E. A. Clancy, D. Farina, G. Filligoi

9 MYOELECTRIC MANIFESTATIONS OF MUSCLE FATIGUE   233

R. Merletti, A. Rainoldi, D. Farina

INTRODUCTION

Interpretation, decomposition, and application of biological signals, including the electro-cardiogram (ECG), the electromyogram (EMG), the electroencephalogram (EEG), and other forms of bioelectrical manifestations of physiopathological events, have fascinated many researchers. Decoding and extracting the information contained in these signals is a tempting task that many engineers and physiologists undertake with pleasure and determination—a challenge that is difficult to resist.

This book is dedicated to EMG, the signal generated by skeletal muscles; the motors that allow us to move. The mechanism of operation of these motors is a masterpiece of engineering. It incorporates central control strategies, signal transmission along nerve fibers and across neuromuscular junctions, electrical activation of the muscle fibers organized in elementary motors (the motor units) and, through a chain of complex biochemical events, the production of forces acting on the tendons of the agonist and/or antagonist muscles and moving the bones. It also incorporates a number of feedback circuits relaying back to the spinal cord and the brain information concerning the length and velocity of shortening of the muscles and the forces acting on the tendons. Through these mechanisms we can hold a flower without crushing it, play a violin or a piano, lift a weight, climb a mountain, or hit a competitor in a boxing match. The EMG signal provides a window on the motor as well as on its controller [30,44]. If this book will induce in its readers a sense of marvel and curiosity, a feeling of awe and respect toward the beauty and elegance of the design of the system, a desire to understand it by breaking its codes and unravel its rules, then our goal will be completely fulfilled.

HISTORICAL NOTES

H. Piper is considered to be the first investigator to study EMG signals. He did this in Germany in 1912 using a string galvanometer [33]. In 1924 Gasser and Erlanger [11] did similar investigations using an oscilloscope. Four years later Proebster observed signals generated by denervated muscles and opened the field of clinical EMG [34]. The concentric needle electrode, developed by Adrian and Bronk in 1929 [1] provided a powerful tool, still widely used today, for EMG studies. Vacuum tube amplifiers [27] and, later, solid state circuits were used in the following decades and allowed fundamental contributions by Kugelberg, Petersen, Buchthal, Guld, Gydikov, Kosarov, Pinelli, Rosenfalck, and Stalberg [3,4,13,14,15,19,20,21,38,39] who are the founding fathers of the methodology and introduced quantitative analysis of the Motor Unit Action Potential (MUAP). Among the work of many others it is worth mentioning that of Denny-Brown, who discussed the "Interpretation of the Electromyogram" in 1949 [8], Uchizono [41], who described the propagation of signals in toad sartorious muscle, and Willison, who introduced the turns and amplitude analysis of EMG in 1964 [43]. A milestone was the first edition of *Muscles Alive* by J. V. Basmajian published in 1962 and now in its fifth edition

[2]. The availability of powerful computers in the late 1970s and early 1980s made it possible to take up the enormous task of decomposition of needle EMG into the constituent MUAP trains and thereby "see," after hours of computer processing, the control mechanisms of individual motor units. Today this task requires a few minutes. Pioneers in this challenge were LeFever and De Luca, Guiheneuc, McGill, and others [12,23,24,28]. Physiology and mathematics of the myoelectric (EMG) signal were first jointly discussed by De Luca in a classic paper published in 1979 providing the first structured approach to the issue of EMG information content and extraction [6].

The availability of computers also allowed the development of models and simulation approaches in the study of EMG. In the 1970s and 1980s many groups took up this challenge and the number of researchers, and published papers, in the field grew steadily. Pioneers in the modeling field were Dimitrova and Lindstrom, [9,25]. These tools greatly contributed to the understanding of EMG biophysics and therefore to the unraveling of the information contained in the signal. In addition, models are powerful teaching instruments and are fundamental in the training of experts in the field, developments of new applications, and progress in current applications. An early and natural area of application of EMG signals is that of driving input for the control of powered upper limb prostheses. This application, which came to be known as myoelectric control, was first demonstrated in the 1940s [37], and developed rapidly throughout the 1960s, 1970s, and 1980s [17,26,40,42]. EMG pattern recognition based controllers for multifunction powered prostheses with potential for simultaneous control are at the forefront of current research and development in this area [10,22].

The International Society of Electromyography and Kinesiology (ISEK) was conceived at a meeting in 1965 of J. Basmajian, S. Carlsöö, B. Johnson, M. MacConaill, J. Pauly, and L. Scheving, and formally founded in 1966. This society, through its activities of conferences, meetings, and simposia, has been central to the development and progress of the field of electromyography. A proposal for a journal in which to publish the results of EMG research was approved by the ISEK council in 1988. The *Journal of Electromyography and Kinesiology* published its first issue in 1991.

Needle EMG techniques can detect MUAPs in a small volume near the needle tip and provide very localized information concerning either superficial or deep muscle structures. Surface techniques can detect MUAPs in a large volume and provide global information that is dominated by the most superficial motor units. Recent surface techniques based on linear electrode arrays or two dimensional electrode grids allow the implementation of spatial filters and the detailed observation of properties of individual superficial motor units.

Because of the blurring introduced by the tissue interposed between the sources and the electrodes, information is lost. For this reason the interpretation of surface EMG is much more challenging than needle EMG; however, the noninvasiveness and the ease of use of the technique, often abused by nonexpert and nonmedical operators, contributed to its large diffusion in many applications ranging from biofeedback, to movement analysis, to fatigue assessment. This fact has not always led to positive results and has generated confusion in the field because of the lack of guidelines and standards. Many hundreds of papers on surface EMG have been published in the last 20 years, often contributing to the confusion with contradictory findings. Surface EMG became a treacherous territory, since it is too easy to use and consequently too easy to abuse [7]. It is indeed very easy to apply an electrode pair on a muscle, display a signal and draw some "conclusions" concerning the pattern, modalities, timing, and intensity of muscle activation. Rather negative reviews appeared in the literature [16,35,36] and triggered a discussion in the field that partially appeared in the Correspondence section of *Neurology* (2001, **56**, 1421–1422).

In an attempt to correct this situation, in 1995 a group of researchers proposed that the European Community (EC) sponsor a concerted action on *Surface EMG for Noninvasive Assessment of Muscles (SENIAM)* which was funded in 1996, for three years [29]. The aim of SENIAM was to enhance international cooperation and reach an acceptable level of consensus among European laboratories active in the field. A second concerted action on *Prevention of Neuromuscular Disorders in the Use of Computer Input Devices (PROCID)* was funded by the EC for the period 1998 to 2001 and is largely based on EMG techniques. A three-year shared cost project on *Neuromuscular Assessment of the Elderly Worker* (NEW, 2001–2004) applied noninvasive electromyographic and mechanomyographic techniques to the study of work- and age-related neuromuscular problems. An additional EC project is applying surface EMG techniques to the study of innervation and control asymmetry of the anal and urethral sphincters in relation to incontinence (OASIS).

The fact that the EC decided to fund these projects, which were largely focused on proposition of recommendations, guidelines, and standards, is an important acknowledgment of the maturity and potential benefits of surface EMG. The journal *Medical Engineering and Physics* published a special issue on Intelligent Data Analysis in Electromyography and Electroneurography (nos. 6 and 7, 1999) [32]. The October 2000 special issue of the *Journal of Electromyography and Kinesiology* was devoted to SEMG [31]. The November–December 2001 issue of the *IEEE Engineering in Medicine and Biology Magazine* was devoted to dynamic EMG. A special issue of the European *Journal of Applied Physiology* was devoted to project PROCID. A special issue of *Medical and Biological Engineering and Computing* and of the European *Journal of Applied Physiology* are entirely devoted to the results of project NEW.

Today needle and surface EMG techniques are complementary instruments and integrate each other. Both are important tools for physiological investigations. The first is more suitable and widely accepted for diagnostic applications; the second has major applications in the fields of biofeedback, prosthesis control, ergonomics, occupational and sport medicine, movement analysis, and allows frequent and painless assessment and evaluation of neuromuscular functions. Applications of surface EMG techniques are less known and often neglected in academic curricula. For this reason this book is mostly focused on these noninvasive applications.

Many books exist on needle EMG and a few on surface EMG. The reference book *Muscles Alive: Their Function Revealed by Electromyography* by J. Basmajan and C. J. De Luca [2] remains by far the most important, with a strong interdisciplinary character. Some textbooks deal primarily with EMG biofeedback techniques and describe these specific applications. The books by J. Cram and G. Kasman [5,18] are more oriented toward clinical applications and aimed at the allied health professionals.

OVERVIEW OF THE BOOK

This textbook is aimed at graduate students in biomedical engineering, life sciences, movement sciences, exercise physiology, neurophysiology and neurology, rehabilitation, sport and occupational medicine, and related fields. A general background in biomedical instrumentation, signal processing, and mathematical modeling is assumed. This background is more easily found among engineers but is not uncommon, and is becoming more and more frequent among physiologists and life scientists. The general approach is quantitative and analytical, and provides a basis for clinical applications, which, however, are not the main issue in this book.

Chapter 1 provides the basic physiological background, and Chapter 2 summarizes the well-established needle EMG methodologies. Chapter 3 focuses on decomposition of needle EMG, a great engineering challenge not yet satisfactorily solved. Chapter 4 addresses the issue of surface EMG biophysics, while the critical problems related to signal detection and conditioning are described in Chapter 5. Chapters 6 and 7 are devoted to information extraction using signal processing techniques that are available today. Chapter 8 deals with the simulation and modeling approaches to EMG interpretation. Myoelectric manifestations of muscle fatigue are addressed in Chapter 9. New advanced tools for information extraction are described in the rather technical Chapter 10: they open up possibilities of great interest with relevant clinical applications in the near future. Interesting possibilities emerge also from the combined analysis of the electromyogram and mechanomyogram, discussed in Chapter 11. Applications of surface EMG are described in the following seven chapters. They range from applications in neurology and ergonomics to exercise physiology, from rehabilitation medicine to biofeedback and prosthesis control.

The contributors to this work do not include all members of the community of EMG researchers. However, the contributors are well representative of the community, providing a sampling of senior and junior members. Some investigators of great experience and competence were unable to contribute; however, they were mentors of many of the contributors who carry on their teachings and bring innovative points of view. Among those with major early and continuing contributions to this field, of particular note are Drs. John Basmajian, Carlo DeLuca, Nonna Dimtrova, and Kevin McGill.

We tried to merge as smoothly as possible physiology, engineering, and some important applications by providing suggestions and recommendations to the authors. We believe that they did an excellent job in producing a harmonious result that can be appreciated by readers of different backgrounds. Any errors or failings of this work are certainly not attributable to the contributors but are strictly the responsibility of the editors.

ACKNOWLEDGMENT

The Editors had the privilege of coordinating an excellent team of contributors and are greatly indebted with them for their efforts and results. They have devoted, without compensation, a considerable portion of their time to this endeavor that was originally envisioned by Prof. Metin Akay. To Prof. Metin Akay and to Prof. Lars Arendt-Nielsen, early reviewers of this book, we express our sincere thanks for the many suggestions and positive criticism they provided. The book reports and disseminates knowledge that was in large part acquired within the European Concerted Actions—Surface EMG for Noninvasive Assessment of Muscles (SENIAM), Prevention of Neuromuscular Disorders in the Use of Computer Input Devices (PROCID), and the shared cost project Neuromuscular Assessment of the Elderly Worker (NEW).

R. Merletti
P. A. Parker

REFERENCES

1. Adrian, E. D., and D. W. Bronk, "The discharge of impulses in motor nerve fibers II: The frequency of discharge in reflex and voluntary contractions," *J Physiol* **67**, 119–151 (1929).

2. Basmajian, J., and C. J. De Luca, *Muscles Alive: Their function revealed by electromyography*, 5th ed., Williams and Wilkins, Baltimore, 1985.

3. Buchthal, F., P. Pinelli, and P. Rosenfalck, "Action potential parameters in normal human muscle and their physiological determination," *Acta Physiol Stand* **32**, 219–229 (1954).

4. Buchthal, F., C. Guld, and P. Rosenfalck, "Action potential parameters in normal human muscle and their dependence on physical variables," *Acta Physiol Stand* **32**, 200–215 (1954).

5. Cram, J., G. Kasman, and J. Holtz, *Introduction to surface electromyography*, Aspen Publisher, Gaithersburg, Maryland, 1998.

6. De Luca, C. J., "Physiology and mathematics of myoelectric signals," *IEEE Trans BME* **26**, 313–325 (1979).

7. De Luca, C. J., "The use of surface electromyography in biomechanics," *J Appl Biomech* **13**, 135–163 (1997).

8. Denny-Brown, D., "Interpretation of the electromyogram," *Arch Neur Psychia* **61**, 99 (1949).

9. Dimitrova, N., "Model of the extracellular potential field of a single striated muscle fiber," *Electromyogr Clin Neurophysiol* **14**, 53–66 (1974).

10. Englehart, K., B. Hudgins, and P. A. Parker, "A wavelet based continuous classification scheme for multifunction myoelectric control," *IEEE Trans BME* **48**, 302–311 (2001).

11. Gasser, H. S., and J. Erlanger, "The compound nature of the action current of nerve as disclosed by the cathode ray oscillograph," *Am J Physiol* **70**, 624–666 (1924).

12. Guiheneuc, P., J. Calamel, C. Doncarli, D. Gitton, and C. Michel. "Automatic detection and pattern recognition of single motor unit potentials in needle EMG," in J. E. Desmedt, ed., *Computer-aided electromyograph*, Basel Press, Karger, vol 10, 1983.

13. Gydikov, A., and D. Kosarov, "Volume conduction of the potential from separate motor units in human muscle," *Electromyography* **12**, 127–147 (1972).

14. Gydikov, A., and D. Kosarov, "Influence of various factors on the length of the summated depolarized area of the muscle fibers in voluntary activating of motor units and in electrical stimulation," *Electromyog Clin Neurophysiol* **14**, 79–85 (1974).

15. Gydikov, A., K. Kostov, A. Kossev, and D. Kosarov, "Estimations of the spreading velocity and of parameters of the muscle potentials by averaging of the summated electromyogram," *Electromyog Clin Neurophysiol* **24**, 191–195 (1984)

16. Haig, A. J., J. B. Geblum, J. J. Rechtien, and A. J. Gitter, "Technology assessment: the use of surface EMG in the diagnosis and treatment of nerve and muscle disorders," *Muscle Nerve* **19**, 392–395 (1996).

17. Jacobson, S., D. Knutti, R. Johnson, and H. Sears, "Development of the Utah arm.," *IEEE Trans BME* **29**, 249–269 (1982).

18. Kasman, G., J. Cram, S. Wolf, and L.Barton, *Clinical applications in surface electromyography*, Aspen Publisher, Gaithersburg, Maryland, 1998.

19. Kosarov, D., A. Gydikov, N. Tankov, and N. Raditcheva, "Examining the work of motor units of various sizes in the muscles of dogs," *Agressologie* **12**, 29–39 (1971).

20. Kugelberg, E., "Electromyogram in muscular disorders," *J Neurol Neurosur Psychiat* **60**, 140 (1947).

21. Kugelberg, E., and I. Petersen, "Insertion activity in electromyography," *J Neurol Neurosur Psychiat* **12**, 268 (1949).

22. Kyberd, P., M. Evans, and S. Winkel, "An intelligent anthropomorphic hand with automatic grasp," *Robotica* **16**, 531–536 (1998).

23. LeFever, R. S., and C. J. DeLuca, "A procedure for decomposing the myoelectric signal into its constituent action potentials: I. Technique, theory and implementation," *IEEE Trans BME* **29** (3), 149–157 (1982).

24. LeFever, R. S., and C. J. DeLuca, "A procedure for decomposing the myoelectric signal into its constituent action potentials: II execution and test of accuracy," *IEEE Trans BME* **29** (3), 158–164 (1982).

25. Lindstrom, L., and R. Magnusson, "Interpretation of myoelectric power spectra: A model and its applications," *Proc IEEE* **65**, 653–662 (1977).

26. Mann, R. W., "Force and position proprioception for prostheses, In P. Herberts, R. Kadefors, R. Magnusson, I. Petersen, eds., *Control of upper extremity prostheses and orthoses*, Thomas, Springfield, I., pp. 202–209, 1974.

27. Matthews, B. H. C., "A special purpose amplifier," *J Physiol (Lond)* **81**, 28–29 (1934).

28. McGill, K. C., K. L. Cummins, and L. J. Dorfman, "Automatic decomposition of the clinical electromyogram," *IEEE Trans BME* **32** (7), 470–477 (1985).

29. Merletti, R. "Surface electromyography: The SENIAM project"; *Europa Medicophysica* **36**, 167–169 (2000).

30. Merletti, R., D. Farina, M. Gazzoni, A. Merlo, P. Ossola, and A. Rainoldi, "Surface electromyography: A window on the muscle, a glimpse on the central nervous system," *Europa Medicophysica* **37**, 57–68 (2001).

31. Merletti, R, and H. Hermens, "Introduction to the special issue on the SENIAM European Concerted Action," *J Electrom Kinesiol* **10**, 283–286 (2000).

32. Pattichis C. S., I. Schofield, R. Merletti, P. A. Parker, and L. T. Middleton. Introduction to the special issue on intelligent data analysis in electromyography and electroneurography, *Med Eng Phys* **21**, 379–388 (1999).

33. Piper, H., *Elektrophysiologie Menschlicher Muskeln*, Berlin: Springer Verlag, 1912.

34. Proebster R., "Uber Muskelationsstrome am gesunder und Kranken Menschem," *Zeitschr Orthop Clin* **50**, 1 (1928).

35. Pullman, S. L., D. S. Goodin, A. I. Marquinez, S. Tabbal, and M. Rubin, "Clinical utility of surface EMG," *Neurology* **55**, 171–177 (2000).

36. Rechtien, J., B. Gelblum, A. Haig., and A. Gitter, "Technology review: dynamic electromyography in gait and motion analysis," *Muscle Nerve* **22** (Suppl), S233–S238 (1999).

37. Reiter, R., "Eine neu elecktrokunstand," *Grenzgebeite der Medicin* **1**, 133–135 (1948).

38. Rosenfalck P., "Intra and extracellular potential field of active nerve and muscle fibre," *Acta Physiolog Scand* **321** (Suppl), 1–168 (1969).

39. Stalberg E, and J. V. Trontelj, *Single fibre elecromyography*, Mirvalle Press, UK 1979.

40. Scott, R. N., and D. S. Dorcas, "A three-state myoelectric control system," *Med Biol Eng* **4**, 367 (1966).

41. Uchizono, K., "Experimental measurement of the length of excitatory wave in nerve and muscle," *Jap J Physiol* **4**, 59–64 (1953).

42. Williams, T. W., *Clinical application of the improved Boston arm, Proc.* Conf. Energy Devices in Rehab., Boston, 1976.

43. Willison, R. G., "Analysis of electrical activity in healthy and dystrophic muscle in man," *J Neurol Neurosurg Psychiat* **27**, 386–394 (1964).

44. Wolf, W., "The EMG as a window to the brain: signal processing tools to enhance the view," in I. Gath, and G. Inbar, eds., *Advances in processing and pattern analysis of biological signals*, Plenum Press, New York, 1996, pp. 339–356.

CONTRIBUTORS

R. Casale Department of Clinical Neurophysiology, Rehabilitation Institute of Montescano, IRCCS, S. Maugeri Foundation, Montescano, Pavia, Italy

E. A. Clancy Electrical and Computer Engineering Department, Biomedical Engineering Department, Worcester Polytechnic Institute, Worcester, MA

J. R. Cram Sierra Health Institute, Nevada City, Ca

C. Disselhorst-Klug Helmholtz-Institute for Biomedical Engineering, Aachen, Germany

C. Doncarli IRCCyN, Ecole Centrale de Nantes, Nantes, France

K. B. Englehart Department of Electrical and Computer Engineering and Institute of Biomedical Engineering, University of New Brunswick, Fredericton, New Brunswick, Canada

D. Farina Laboratory for Engineering of the Neuromuscular System, Department of Electronics, Politecnico di Torino, Italy

F. Felici Exercise Physiology Laboratory, Department of Human Movement and Sport Sciences, University Institute of Motor Sciences (IUSM), Faculty of Motor Sciences, Roma, Italy

G. Filligoi Department INFOCOM, School of Engineering, CISB, Centro Interdipartimentale Sistemi Biomedici, Università degli Studi "La Sapienza," Roma

C. Frigo Department of Bioengineering and Laboratory of Biomedical Technologies, Polytechnic of Milano, Italy

G. M. Hägg Department for Work and Health, National Institute for Working Life, Stockholm, Sweden

H. J. Hermens Roessingh Research and Development Enschede, The Netherlands

P. Hodges Department of Physiotherapy, The University of Queensland, Brisbane, QLD, Australia

B. S. Hudgins Department of Electrical and Computer Engineering and Institute of Biomedical Engineering, University of New Brunswick, Fredericton, New Brunswick, Canada

J. Jabre Department of Neurology, Boston University, Acting Chief, Harvard-BU Neurology Service, Boston VA Medical Center, Boston

G. Jull Cervical Spine and Whiplash Research Unit, Department of Physiotherapy, The University of Queensland, Brisbane, QLD, Australia

R. Kadefors Industry and Human Resources, National Institute for Working Life, Göteborg, Sweden

S. Karlsson University Hospital, Department of Biomedical Engineering and Informatics, Umea, Sweden

B. Melin Department of Work and Health, National Institute for Working Life, Stockholm, Sweden

R. Merletti Laboratory for Engineering of the Neuromuscular System, Department of Electronics, Politecnico di Torino Italy

M. Mihelin Department of Biomedical Engineering, Institute of Clinical Neurophysiology, University Medical Center, Ljubljana, Slovenia

T. Moritani Laboratory of Applied Physiology, The Graduate School of Human and Environmental Studies, Kyoto University, Kyoto, Japan

C. Orizio Department of Biomedical Sciences and Biotechnologies, Section of Human Physiology, University of Brescia, Brescia, Italy

P. A. Parker Department of Electrical and Computer Engineering and Institute of Biomedical Engineering, University of New Brunswick, Fredericton, New Brunswick, Canada

A. Rainoldi Laboratory for Engineering of the Neuromuscular System, Department of Electronics, Politecnico di Torino, Italy, University of Tor Vergata, Roma, Italy

R. Shiavi Departments of Biomedical Engineering and Electrical Engineering, Vanderbilt University, Nashville, TN

K. Søgaard Department of Physiology, National Institute of Occupational Health, Copenhagen, Denmark

D. W. Stashuk Department of Systems Design Engineering, University of Waterloo, Ontario, Canada

D. F. Stegeman Department of Clinical Neurophysiology, University Medical Center, Nijmegen, Interuniversity Institute for Fundamental and Clinical Human Movement Sciences (IFKB), Amsterdam, Nijmegen, The Netherlands

J. V. Trontelj Department of Neurology, Institute of Clinical Neurophysiology, University Medical Center of Ljubljana, Slovenia

J. G. van Dijk Department of Neurology and Clinical Neurophysiology, Leiden University Medical Centre, Leiden, The Netherlands

D. Zazula Faculty of Electrical Engineering and Computer Science, University of Maribor, Maribor, Slovenia

M. J. Zwarts Department of Clinical Neurophysiology, Institute of Neurology, University Medical Center, Nijmegen, Interuniversity Institute of Fundamental, and Clinical Human Movement Sciences (IFKB), The Netherlands

1

BASIC PHYSIOLOGY AND BIOPHYSICS OF EMG SIGNAL GENERATION

T. Moritani

Laboratory of Applied Physiology
The Graduate School of Human and Environmental Studies
Kyoto University, Kyoto, Japan

D. Stegeman

Department of Clinical Neurophysiology
University Medical Center, Nijmegen
Interuniversity Institute for Fundamental and Clinical Human Movement Sciences (IFKB)
Amsterdam, Nijmegen, The Netherlands

R. Merletti

Laboratory for Engineering of the Neuromuscular System
Department of Electronics
Politecnico di Torino, Italy

1.1 INTRODUCTION

Understanding EMG signals implies the understanding of muscles and the way they generate bioelectrical signals. It also implies the understanding of the "forward problem," that is, how specific mechanisms and phenomena influence the signals, as well as the more difficult "inverse problem", that is, how the signals reflect certain mechanisms and phenomena and allow their identification and description. The concept of forward and inverse problem is familiar to physiologists and engineers and is strictly associated to the concept

Electromyography: Physiology, Engineering, and Noninvasive Applications, edited by Roberto Merletti and Philip Parker.
ISBN 0-471-67580-6   Copyright © 2004 Institute for Electrical and Electronics Engineers, Inc.

of a system as a set of inputs, transfer functions and outputs, and of a model, as a set of descriptions and relations associating, under certain conditions and assumptions, the inputs to the outputs.

In this chapter we provide a basic description of the physiological system whose output is the needle or surface detected EMG signal. We summarize the large number of factors and phenomena that contribute to such signals and provide a basis of knowledge for the signal analysis approaches that will be addressed in the subsequent chapters. In Section 1.2 we introduce basic concepts and mechanisms of muscle physiology and motor control. In Section 1.3 we consider the basic electrophysiology of the muscle membrane. We assume that some of the concepts described in Section 1.2 (action potential, power spectrum, etc.) are known to the reader. They are discussed in greater detail in Section 1.3 and in other chapters of this book.

1.2 BASIC PHYSIOLOGY OF MOTOR CONTROL AND MUSCLE CONTRACTION

1.2.1 Motor Unit

The human motor system must cope with a great diversity of internal and external demands and constraints. These include the regulation of force output for precise and powerful movements, upright posture, locomotion, and even our repertoire of gestures. As it is impossible to describe all the specific control features of the various motor systems in isolation, we will attempt to delineate the basic principles of motor control, with special attention to the skeletomotor system, which plays the major role in the control of force and movements in humans.

Simplified schematic diagrams of the central motor system and the concept of the motor unit (MU are presented in Figure 1.1). The central nervous system is organized in a hierarchical fashion. Motor programming takes place in the premotor cortex, the supplementary motor area, and other associated areas of the cortex. Inputs from these areas, from the cerebellum and, to some extent, from the basal ganglia converge to the primary motor cortex and excite or inhibit the various neurons of the primary motor cortex. The outputs from the primary motor cortex have a powerful influence on interneurons and motoneurons of the brain stem and of the spinal cord. There exists a link between the corticospinal tract and alpha (α)-motoneurons, providing direct cortical control of muscle activity, as indicated in Figure 1.1.

A motor unit (MU) consists of an α-motoneuron in the spinal cord and the muscle fibers it innervates (Fig. 1.1). The α-motoneuron is the final point of summation for all the descending and reflex input. The net membrane current induced in this motoneuron by the various synaptic innervation sites determines the discharge (firing) pattern of the motor unit and thus the activity of the MU. The number of MUs per muscle in humans may range from about 100 for a small hand muscle to 1000 or more for large limb muscles [33]. It has also been shown that different MUs vary greatly in force generating capacity, with a 100-fold or more difference in twitch force [27,97].

The wide variation in the morphological and electrophysiological properties of the individual motoneurons comprising a motoneuron pool is matched by an equally wide range in the physiological properties of the muscle units they innervate. Interestingly the muscle fibers that are innervated by a particular motoneuron manifest nearly identical bio-

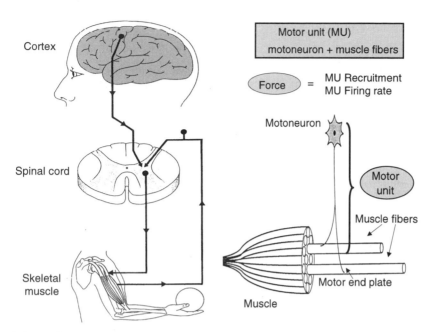

Figure 1.1. A schematic representation of basic motor control mechanisms and of the motor unit and its components. (Modified from [93] with permission)

chemical, histochemical, and contractile characteristics, together defining the typing of the specific MU. Earlier studies [10] identified three types of motor units based on physiological properties such as speed of contraction and fatigability (sensitivity to fatigue): (1) fast-twitch, fatigable (FF or type IIb); (2) fast-twitch, fatigue-resistant (FR or type IIa); and (3) slow-twitch (S or type I), which is most resistant to fatigue. The FF type motor units are predominantly found in pale muscles (high ATPase enzyme for anaerobic energy utilization), low capillarization, less hemoglobin, myoglobin, and mitochondria for oxidative energy supply), while red muscles (low ATPase, high capillarization, abundant hemoglobin, myoglobin and mitochondria for oxidative energy supply) such as the soleus are predominantly composed of type S motor units.

Figure 1.2 shows typical contractile properties of predominantly fast-twitch (extensor digitorum longus, EDL) and slow-twitch (soleus, SOL) fibers obtained from an isolated rat muscle. Note the large differences in contractile force, contraction time (CT), electromechanical delay time (EMD), and maximal rate of force development (dF/dt) and relaxation.

In humans a classification of motor units based on their physiological properties is difficult to achieve. Therefore an identification of muscle fiber populations in the muscle cross section based on histochemical criteria has been commonly adopted after obtaining a small sample of muscle tissue by a needle biopsy technique. Type I muscle fibers have high levels of ATPase activity and low levels of succinic dehydrogenase (SDH, one of the major enzymes for aerobic energy production), and type II fibers demonstrate the reverse pattern of enzyme activity. Type II fibers are subdivided in two subgroups type IIa and type IIb with different properties. Figure 1.3 shows histochemical fiber typing in human skeletal muscle demonstrating different myofibrillar ATPase reactions after preincubation at pH 4.6. In this preparation, type I (slow-twitch) fibers stain dark, type IIa fibers remain unstained, and type IIb fibers moderately stained (see Fig. 1.3).

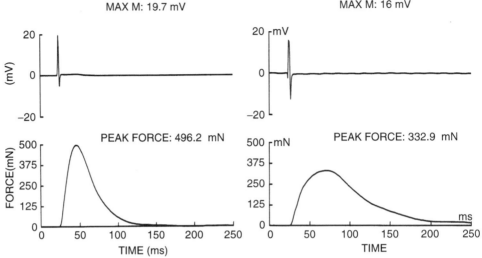

COMMENT:EDL CONTROL (#6)

PEAK FORCE: 496 mN
MAX dF/dt: 37.2 mN/ms
RELAXATION dF/dt: 13.3 mN/ms
EMD: 4.5 CT: 20 HALF RELAX: 24.5 ms

MAX M: 19.7 mV

COMMENT:SOL CONTROL (#6)

PEAK FORCE: 333 mN
MAX dF/dt: 16.5 mN/ms
RELAXATION dF/dt: 4.91 mN/ms
EMD: 5 CT: 43.5 HALF RELAX: 49 ms

MAX M: 16 mV

PEAK FORCE: 496.2 mN

PEAK FORCE: 332.9 mN

Figure 1.2. Contractile characteristics of typical fast-twitch (*left*: rat EDL) and slow-twitch (*right*: soleus) muscle fibers. EMD: Electormechanical delay (time delay between the onset of EMG and the onset of force). CT: Contraction time (time from onset to peak of force).

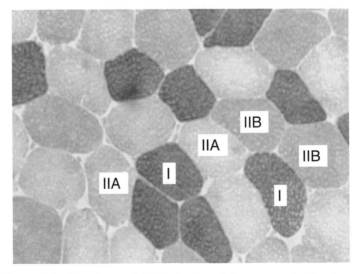

Figure 1.3. Histochemical determination of human muscle fiber types.

During aerobic work glycolytic energy metabolism furnishes pyruvate, which is then transferred to the mitochondria where its carbon skeleton is entirely degraded to CO_2 through oxidative phosphorylation. This process of full oxidation of glucose in mitochondria yields 36 ATP molecules for each glucose molecule degraded. Note that anaerobic glycolysis of glucose to pyruvate only yields 2 ATP with a subsequent formation of lactic acid, which may affect muscle contractile activity. Thus the net ATP production differs by a factor 18 between aerobic and anaerobic energy metabolism. Consequently type I fibers are fatigue resistant due to their high oxidative metabolism and their higher energy efficiency.

Type II (fast-twitch) fibers stain weaker for succinic dehydrogenase than type I fibers, but stain stronger for the enzymes necessary for anaerobic metabolism. Type II fibers therefore generate the ATP for muscular contraction mainly through anaerobic glycolysis, which results in the production of lactic acids and other metabolic by-products. They possess small amounts of mitochondria, and their power output during repetitive activation cannot be achieved through ATP production by oxidative process in their mitochondria. Thus type II fibers are prone to fatigue quickly because they accumulate lactic acids (up to 30-fold the concentration in resting muscle). The low pH associated with this lactate accumulation, as well as the corresponding increases in free phosphate and other metabolic by-products, inhibits the chemical reactions including the myosin ATPase, slowing contraction speed or stopping active contraction entirely (see Fig 1.4). The different metabolic pathways are activated depending on the speed, the intensity, and the duration of muscular contraction.

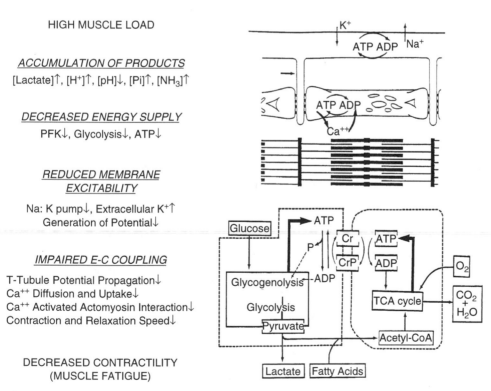

Figure 1.4. Possible metabolic and electrophysiologic consequences of muscular activity that lead to fatigue.

TABLE 1.1. Summary of Different Motor Units and Their Physiological Properties

Motor Unit Type	Histochemical and Metabolic Properties	Mechanical Properties	Electrical Properties	Others
S or SO or type I	Oxidative, do not work well in ischemic or low oxygen conditions	Slow twitch, small forces, fatigue resistant, smaller fiber diameter and MU size	Lower nerve conduction velocity	Small axons recruited at low force levels
FR or FOG or type IIa	Oxidative glycolytic	Fast twitch, fatigue resistant	Intermediate nerve conduction velocity	Intermediate axons recruited at moderate force levels
FF or FG or type IIb	Glycolytic, work well in ischemic or low oxygen conditions	Fast twitch, large forces, fatigable, larger fiber diameter and MU size	High nerve conduction velocity	Large axons recruited at high force levels

In addition the MU type is not only reflected in mechanical and histological differences but also in the single-fiber action potential and in the MU action potential features. Wallinga et al. [102] investigated the action potential of individual muscle fibers of the rat soleus (type I) and the extensor digitorum longus (EDL, predominantly type II). They found that in comparison to type I fibers, type II fibers have more negative resting potential, larger peak excursion, faster rate of depolarization and repolarization and shorter action potential duration. Furthermore type I and types IIa and IIb muscle fibers appear to be randomly distributed across the muscle cross section. Depending on the muscle function, the percentage of the two fiber types may be different. Antigravity muscles (e.g., soleus) tend to be predominantly type I, while muscles suitable for rapid movements have similar proportions of the two fiber types. Table 1.1 summarizes different MU properties.

1.2.2 Motor Unit Recruitment and Firing Frequency (Rate Coding)

In voluntary contractions, force is modulated by a combination of MU recruitment and changes in MU activation frequency (rate coding) [54,69,72]. The greater the number of MUs recruited and their discharge frequency, the greater the force will be. During full MU recruitment the muscle force, when activated at any constant discharge frequency, is approximately 2 to 5kg/cm^2, and in general, this is relatively independent of species, gender, age, and training status [1,41].

Our current understanding of motor unit recruitment is based on the pioneer work of Henneman and colleagues in the 1960s, who proposed that motor units are always recruited in order of increasing size of the α-motoneuron. This "size principle" of Henneman et al. [34] was based on results from cat motoneurons and is supported by strong evidence that in muscle contraction there is a specific sequence of recruitment in order of increasing motoneuron and motor unit (MU) size [18,24,54,69]. Goldberg and Derfler [28] later showed positive correlations among recruitment order, spike amplitude, and twitch tension of single MUs in human masseter muscle. Because of the great wealth of data supporting this size-based recruitment order in a variety of experimental conditions, it is often referred to as the "normal sequence of recruitment" or "orderly recruitment" [32]. Recent data further confirm the presence of this "size principle," and that transcortical stimulation generates normal orderly recruitment [3].

It is well documented that motor unit recruitment and firing frequency (rate coding) depend primarily on the level of force and the speed of contraction. When low-threshold MUs are recruited, this results in a muscular contraction characterized by low force-generating capabilities and high fatigue resistance. With requirements for greater force and/or faster contraction, high-threshold fatigable MUs are recruited [24,33]. The technical difficulties associated with single motor unit recordings at high forces in humans and the difficulty in generating controlled forces in animal preparations limit the accuracy with which the precise motor unit recruitment and rate coding can be established. However, Kukulka and Clamann [54] and Moritani et al. [74] demonstrated in human adductor pollicis that for a muscle group with mainly type I fibers, rate coding plays a prominent role in force modulation. For a muscle group composed of both types I and II fibers, MU recruitment seems to be the major mechanism for generating extra force above 40% to 50% of maximal voluntary contraction (MVC). Thus, in the intrinsic muscles of human hands, motor unit recruitment appears to be essentially complete at about 50% of maximal force, but recruitment in the biceps, brachialis, and deltoid muscles may continue until more than 80% of maximal force is attained [18,54,72,81].

The number of MUs recruited and their mean discharge frequency of excitation determine the electrical activity in a muscle, that is, there are the same factors that determine muscle force [5,72]. Thus a direct relationship between the electromyogram (EMG) and exerted force might be expected. Under certain experimental conditions this relationship can be demonstrated by recording the smoothed rectified or integrated EMG (iEMG) [20,68,71,73]. The reproducibility of EMG recordings is remarkably high, as the test–retest correlation ranges from 0.97 to 0.99 [52,71,73].

Figure 1.5 represents a typical set of raw surface EMG recording together with the corresponding force curve during force-varying isometric muscle contraction. Surface EMG frequency power spectral data are also shown. It can be readily seen that EMG activity increases progressively as a function of force generated, suggesting a gradual MU recruitment and MU firing rate modulation taking place in order to match the required force demand. Thus the increase in EMG amplitude might represent MU recruitment and/or MU firing frequency modulation whereas the increase of mean frequency (MPF) of the power spectrum might represent, at least in part, the additional recruitment of superficial high threshold MUs that most likely possess large and sharp spikes affecting high frequency bands of the surface EMG power spectrum [72].

However, the change in the surface EMG should not automatically be attributed to changes in either MU recruitment or MU firing frequencies as the EMG signal amplitude is further influenced by the individual muscle fiber potential, degree of MU discharge synchronization, and fatigue [5,6,7,44,75,82]. A direct single motor unit recording with bipolar wire electrodes is shown for comparison (Fig. 1.6) during the same experimental condition previously described. Note that the isolated MU spikes can be observed. Additional motor units as represented by greater spike amplitudes could be identified even at near 80% of maximal force production.

Several previous studies [69,98] have demonstrated that the firing rates of active motor units increase monotonically with increasing force output. This may imply that increased excitation to the active muscle motoneuron pool increases the firing rates of all the active motor units. In addition to this common increase, common fluctuations of firing rates are

POWER SPECTRA

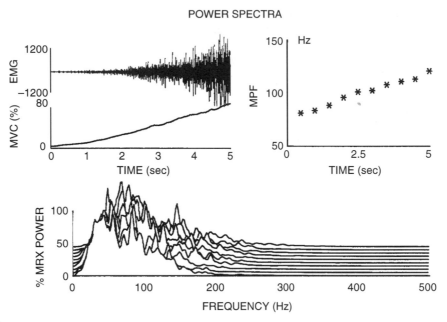

Figure 1.5. Typical set of computer outputs showing the changes in raw EMG signal recorded from the biceps brachii muscle and the corresponding frequency power spectra during linearly force-varying isometric muscle action.

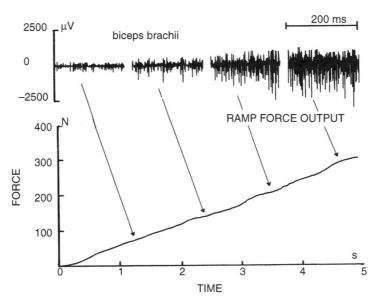

Figure 1.6. Intramuscular spike recordings obtained from the biceps brachii muscle during linearly force-varying isometric muscle contraction.

often present. De Luca et al. [18] investigated this commonality in the fluctuations of the firing rates of up to eight concurrently active motor units during various types of isometric muscle action: attempted constant force, ramp force increase, and force reversals. Their results strongly indicated that there was a unison behavior of the firing rates of motor units, both as a function of time and force. This property has been termed the "common drive." The existence of this common drive implies that the nervous system does not control the firing rates of motor units individually. Instead, it modulates the pool of motoneurons in a uniform fashion; a demand for force modulation can be achieved by modulation of the excitation and/or inhibition on the motoneuron pool as a whole.

1.2.3 Factors Affecting Motor Unit Recruitment and Firing Frequency

Muscle Action. It is well established that eccentric (lengthening) contraction requires less oxygen and lower amount of ATP than concentric contraction [15,42]. Both surface [53] and intramuscular EMG studies [76] have demonstrated that MU recruitment patterns are qualitatively similar in both types of contractions, but for a given MU, the force at which MU recruitment occurs is greater in eccentric contractions than in either isometric or concentric (shortening) contractions.

On the other hand, selective recruitment of type FF motor units has been most clearly demonstrated by comparing different types of muscle actions of the human gastrocnemius and soleus muscles [86,87]. Motor units with low recruitment thresholds observed during concentric (shortening) muscle action were apparently suppressed during eccentric (lengthening) muscle action. Interestingly muscle soreness that has a delayed onset is a common feature among both athletes and untrained individuals. A number of investigators have demonstrated that the eccentric component of dynamic work plays a critical role in determining the occurrence and severity of exercise-induced muscle soreness [25,66,91]. It has been also demonstrated that type II fibers are predominantly affected by this type of muscular contraction [25]. Based on these findings and the results of EMG studies cited earlier, it is most likely that muscle soreness associated with eccentric component of dynamic exercise might be in part due to high mechanical forces produced by relatively few active MUs, which may in turn result in some degree of disturbance in structural proteins in muscle fibers, particularly those of high recruitment threshold MUs.

As movement speed increases, the force supplied by type S motor units decreases much more rapidly than that supplied by type FF units because of differences in their force-velocity relations. As a consequence it has been proposed that rapid movements may be accomplished by selective recruitment of type FF motor units. This selective recruitment of either slow or fast ankle extensor muscles has been documented during a variety of locomotor tasks in cats [37,95]. For example, Smith et al. [95] demonstrated selective recruitment of the fast LG muscle during rapid paw shaking without concomitant recruitment of the slow SOL muscle, possibly due to the time constraints imposed by the rapid movements during which the recruitment of slow muscle would be incompatible with the demands of the movement. Studies in humans have generally not supported this idea. Moritani et al. [78,79], however, reported some evidence of phase-dependent and preferential activation of the relatively "fast" gastrocnemius muscle (as compared to "slow" soleus) with increasing demands of force and speed during different types of hopping in man. In the arm muscles the order of recruitment appears dependent on the specific task requirements, especially movement speed and direction [99]. DeLuca and Erim [17] have recently proposed a model of common drive of motor units that provides a possible scheme for the control of MUs, unifying various seemingly different or isolated past research findings. According to this model the pool of MUs that makes up a muscle is controlled not individually but collectively during a contraction. The unique firing patterns of individual MUs are affected not by separate command signals sent to these units but by one common motor drive to which MUs respond differently. Considering the simplicity of this "common drive" and the previously described size principle, the control of the MUs within a muscle represents a functional elegance that relates the specifics of the hierarchical grading to the local size-related excitation of the MUs. This would obviously free the central nervous system to provide a global input to the motoneuron pool corresponding to the intended output of the muscle (see [17] for more details). A proposal for the special behavior of the motoneuron action potential as a dominant stage in this process was recently made by Kleine et al. [50] on the basis of experimental work of Matthews [65].

Muscle Fatigue. Earlier electromyographic studies [5,75,82] indicated that the amplitude of EMG signals increases progressively as a function of time during sustained fatiguing submaximal contractions. In the wide range of submaximal contractions not all of the available motor units of the pool are recruited. Simultaneous recordings of single motor unit spikes and surface EMG analysis demonstrated that there was a progressive decrease in mean power frequency (MPF) of the surface EMG signal during sustained contractions at 50% of MVC, but this decline was accompanied by a significant increase in the root mean square value of the EMG amplitude and a progressive MU recruitment as evidenced by an increased number of MUs with relatively large intramuscular spike amplitude [75]. It was generally assumed that additional MUs were progressively recruited to compensate for the loss of contractility due to some degree of impairment of fatigued MUs. However, this increased amplitude of the surface EMG could not be demonstrated during sustained maximal voluntary contractions (MVC) [7,8,82]. There was some evidence that a progressive reduction occurs in MU firing rates during sustained MVC in the absence of any measurable neuromuscular transmission failure [9,75]. This finding suggests the existence of different MU recruitment and rate-coding mechanisms during sustained maximal and submaximal voluntary contractions.

On the other hand, an important feature of the neuromuscular system is its plasticity and capability to evolve, adapt, and repair itself. This feature makes it possible to compensate for age-related deterioration, to benefit from training in sport, to adapt to a particular job or physical activity, and so on [88]. It is known that athletes active in resistance sport (e.g., long-distance runners and swimmers) have a higher percentage of type I fiber cross-sectional area than athletes involved in explosive performances (e.g., sprinters). Interestingly aging skeletal muscles exhibit a progressive decrement in cross section and maximal isometric contraction force. The effects of aging on the characteristics of skeletal muscles are fiber type specific. The decline in fiber number and size is more prominent for type II fibers than for type I [13], leading to the paradox of an aged muscle being more fatigue resistant than a young one.

Energy Metabolism and Oxygen Availability. During cycling exercise an increase in plasma lactate concentration occurs already at 50% to 70% of maximal oxygen uptake (V_{O_2max}) and well before the aerobic capacity is fully utilized [103,104]. However, at the exercise intensity with a considerable amount of lactate production, the actual torque is usually less than 20% of maximal voluntary contraction. Despite these relatively low force output and moderate speed of contractions (60 revolution per minute) during cycling, glycogen content of types IIa and IIb fibers is progressively decreased (first in type IIa and finally type IIb) [100], suggesting a decrease in the motor unit recruitment threshold force of these fibers during development of fatigue.

Gollnick et al. [30] showed that in isometric muscle actions, slow-twitch fibers are the only ones to be depleted of glycogen at force developments up to 15% to 20% MVC. Above this level, fast-twitch fibers are also depleted of glycogen. This suggests that above 20% MVC, the availability of oxygen and the developed force influences recruitment of fast motor units, since blood flow is usually restricted during sustained contraction about 20% MVC [39]. These available data suggest that not only the force and speed of contraction but also the availability of oxygen and energy substrates affect the recruitment of high threshold motor units. An early recruitment of slow-twitch MU recruitment in 30% MVC fatiguing tibialis anterior muscle exercise was found in an 31P-NMR spectroscopy study by Houtman et al. [38].

To further shed a light on this issue of energy supply and motor unit recruitment and rate coding pattern, Moritani et al. [83] experimentally determined the interrelationships among oxygen supply, motor unit activity, and blood lactate during intermittent isometric contractions of the hand grip muscles. Subjects performed for 2 seconds 20% of MVC followed by 2 seconds of rest repeated for 4 minutes under free circulation and/or arterial occlusion between the first and second minutes. The constancy of both intramuscular motor unit spikes and surface EMG activity during isometric contraction indicated no electrophysiological signs of muscular fatigue with free circulation condition. However, significant changes in the above-mentioned parameters were evident during contractions with arterial occlusion. Since the availability of oxygen and blood-borne substrate, such as glucose and free fatty acids, are severely reduced during occlusion, a progressive recruitment of additional motor units might have taken place so as to compensate for the deficit in force development [6,75]. This may occur if motor units become depleted of glycogen [30,100] or affected by some degree of intramuscular acidification [67,105]. This in turn interferes with the excitation-contraction (E-C) coupling with subsequent decrease in the developed force. Under these physiological conditions, if the force output were to be maintained, a progressive recruitment of additional MUs with possibly more glycolytic (types IIa and IIb) fibers would take place.

These results suggest that not only the force and speed of contraction but also the availability of oxygen and/or energy substrates may affect the recruitment of high threshold motor units. In good agreement with this hypothesis, the 31P NMR experiments [12] demonstrated that the rate of recovery in PCr/inorganic phosphate (Pi) during the contraction was dependent on oxygen delivery. These results and other evidence [48] suggest a causal relationship between oxygen supply and energy state in the contracting as well as recovering skeletal muscles. It is therefore reasonable to believe that oxygen availability may play an important role in regulating MU recruitment and firing frequency as there exists a close link between state of energy supply and types of muscle fibers being recruited [75,80,83].

1.2.4 Peripheral Motor Control System

There are a variety of specialized receptors located in the muscles, tendons, fascia, and skin that provide information to appropriate parts of the central nervous system (CNS) concerning the length and force characteristics of the muscles. The simplest functional element of the motor activity is the so-called stretch reflex. Reflexes are largely automatic, consistent, and predictable reactions to sensory stimuli. A physician taps a patient's

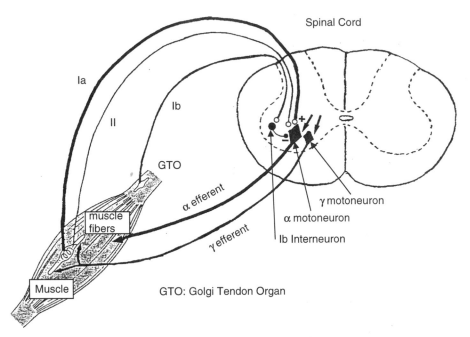

Figure 1.7. Schematic representation of the reflex components.

knee and the leg extends. This is an example of a very simple reflex: the stretch reflex. Figure 1.7 schematically illustrates the basic components involved in this stretch reflex.

Within each muscle are sensory receptors called muscle spindles. Spindles provide information to the nervous system regarding the absolute length of the muscle and the rate of change of the length (velocity) of the muscle. The tap stretches the muscle. This stretch is detected by the muscle spindle and is conveyed directly to a spinal motoneuron via sensory afferent (group Ia afferent). This leads to excitation of motoneuron and efferent impulses that cause contraction of the corresponding muscle. This way the muscle is shortened, the stretching of the muscle spindles is removed and their Ia afferent activity diminishes. In the process only one synapse is involved: a sensory Ia afferent to motoneuron. The term monosynaptic is therefore typically used to describe the stretch reflex. Although the stretch reflex is termed monosynaptic, the sensory afferent from the spindle also contacts interneurons, sensory neurons, and neurons that send ascending projections to higher centers such as thalamus. From there, processed messages return to the motoneurons, closing a longer parallel reflex arc. The stretch reflex therefore also has polysynaptic components (i.e., involving more than one synapse). Note that all reflexes, no matter how simple, can be modified by signals from the brain.

Muscle spindles are composed of intrafusal fibers, sensory endings, and motor axons. Each spindle contains several intrafusal fibers and sensory endings and is innervated by specialized motoneurons (fusimotor neurons or gamma [γ] motoneurons). The neural input to muscle spindles comes from fusimotor neurons located in the spinal cord (see Fig. 1.7). The main function of fusimotor neurons is to control the sensitivity of spindle afferents to dynamic stretches by innervating intrafusal fibers. Some fusimotor neurons (beta [β] motoneurons) innervate both extrafusal (skeletal) and intrafusal (muscle spindle) fibers.

As just described, muscle spindles are only one of many types of receptors that provide information necessary for movement. Control of posture and movement requires

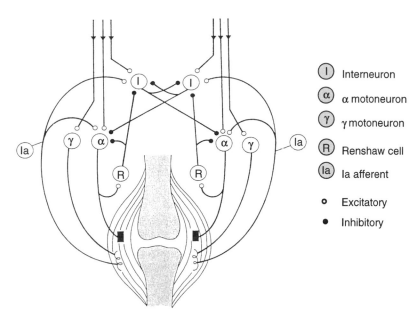

Figure 1.8. Stretch reflex neural circuitry and neural-mechanical coupling between antagonistic pairs of limb muscles. (Modified from [70] with permission)

monitoring not only of muscle length but also of muscle tension. We do possess another specialized receptor called the Golgi tendon organ (GTO). GTOs are specialized sensory receptor organs located primarily in the musculotendinous junction. GTOs provide information regarding the amount of force, or tension, being generated within the muscle. Thus the functioning of these peripheral receptors (muscle spindles and GTOs) is absolutely essential to the control of muscle contraction. GTOs have a low threshold (i.e., they tend to respond to small changes) to contraction-induced changes in muscle tension, and at higher threshold to stretch-induced tension. The sensory information detected by the GTO receptors is conveyed via group Ib sensory afferents (see Fig. 1.7). Group Ib afferents from GTOs mediate nonreciprocal inhibition. Nonreciprocal inhibition, also termed autogenic inhibition, refers to inhibitory input to an agonist (i.e., the prime mover) and its synergists concomitant with an excitatory input to opposing (antagonist) muscles. The inhibition of agonist motoneuron pools and the excitation of antagonist motoneurons are accomplished by Ib interneurons. This type of inhibition assists with the matching of muscle forces to the requirements of a motor task [55]. Ib interneuron can be either facilitatory or inhibitory. GTO activation therefore results in many other responses in addition to nonreciprocal inhibition.

It is now important to realize that smooth coordinated movement relies not only on muscle activation but on muscle deactivation as well. When you are trying to extend the leg, it would be impossible if muscles that opposed that movement were contracting. As previously described, the Ia afferent that convey stretch reflex information branch when they enter the spinal cord (see Fig. 1.8). Some of these branches synapse on interneurons. One type of interneuron that is contacted is the Ia inhibitory interneuron. So when you try to extend the leg, the muscle spindles of the leg extensor will be stimulated and cause stretch reflex together with excitation of this Ia inhibitory interneuron that will have an inhibitory effect on motoneurons innervating muscles that are antagonists to the stretched muscle, in this case the flexor muscles. This process is referred to as reciprocal inhibition or disynaptic inhibition because two synapses are involved in the inhibitory pathway. The Ia inhibitory interneuron receives rich convergent inputs from many other sources and processes in such a way that the appropriate amount of antagonist muscle inhibition is achieved. Obviously different motor tasks require varying degrees of antagonist muscle inhibition and synergist muscle activation.

Inhibition of antagonists and other muscle groups can also be accomplished by Renshaw cell mediated inhibition, Ib-mediated inhibition (presented earlier in GTO description), and presynaptic inhibitory mechanisms. Renshaw cells are interneurons that directly synapse on alpha motoneurons and Ia inhibitory interneurons (see Fig. 1.8). The Renshaw cell will inhibit the alpha motoneuron of a contracting muscle and its synergists. In addition it will inhibit the antagonist muscle's Ia inhibitory interneuron (disinhibition). This aids in grading muscle contractions and assisting task-appropriate agonist/antagonist cocontraction [55].

During various stages of ontogenetic development, neural projections exist that are not normally present in the adults [56]. These neonatal "exuberant" projections appear to retract or become physiologically latent during development. Neural activity definitely plays a critical role in retraction and segregation within the nervous system. The gradual reduction of exuberant neural projections is related to competition for synaptic sites [29] and early redundancy yields to refinement based on a proper matching of afferents and efferents. Competitive interactions between projections are modulated by neural activity and thus removal of competition may result in an abnormal retention of neonatal neural exuberance.

Damaging the motor cortex of cats during the neonatal stage, for example, results in developmental delays and reflex changes similar to those seen in children with cerebral palsy [56,57]. A series of studies by Leonard [59,60] have investigated the effects of neonatal neuronal exuberance and its retention following damage to the central nervous system (CNS) in humans. Figure 1.9 demonstrates a typical set of results obtained from deep tendon reflex testing and surface EMG recordings obtained from a normal infant and a child with cerebral palsy. In normal infants a single tendon tap to either the patellar or Achilles tendon elicited reflex responses in agonist, antagonist, and other leg muscles (see Fig. 1.9). By 5 years of age this reflex overflow was greatly reduced. Children with cerebral palsy did not have a reduction of reflex overflow. Instead, the infantile pattern of reflex distribution to antagonist and other muscles was maintained. Children with cerebral palsy thus seem to have a delayed or absence of reflex development rather than an aberrant development.

The reflex overflow recorded in normal infants and children with cerebral palsy can best be explained by (1) exuberant Ia projections that extend to motoneurons innervating muscles other than the one being stimulated, (2) exuberant motoneuron projections, and/or (3) a lack of supraspinal input that normally would depress reflex activity. The abolition of reflex overflow in normal children and adults may thus reflect either retraction of exuberant neural projections and/or gating of spinal cord reflexes by maturing supraspinal input [59]. H-reflex studies enabling the quantification of spinal motoneuron pool excitability have also indicated that children with cerebral palsy have impairments in reciprocal inhibition, both before and during voluntary movement [60]. These deficits, which involve damage to supraspinal centers, contribute to their inability to perform smooth, coordinated movements [58].

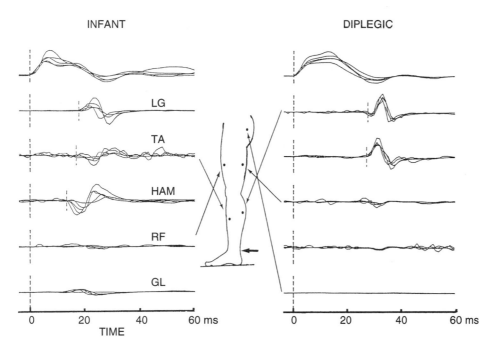

Figure 1.9. Five superimposed reflex responses following an Achilles tendon tap in a normal 8-month-old infant (*left*) and a 13-year-old child with spastic diplegia (*right*). Note that the infant had monosynaptic reflex responses in several muscles following the Achilles tendon tap. The 13-year-old child with diplegia had reflex activation similar to those of the normal infant. *Top plot*: Tendon-tap pressure. LG: Lateral gastrocnemius; TA: Tibialis anterior; HAM: Hamstrings; RF: Rectus femoris; GL: Gluteus maximus. (Modified from [58] with permission)

1.2.5 Muscle Energetics and Neuromuscular Regulation

The transition from aerobic to anaerobic muscle metabolism has been a subject for special focus in human experiments. The level of muscular work just below that at which metabolic acidosis occurs has been called the anaerobic threshold (AT) [103]. The physiological requirements for performing muscular work above AT are considerably more demanding than for lower intensities. Lactic acidosis (anaerobic) threshold occurs at a metabolic rate that is specific to the individual and is usually caused by an inadequate oxygen supply [4,48,104]. Thus the AT can be considered an important assessment of the ability of the cardiovascular system to supply oxygen at a rate adequate to prevent muscle anaerobiosis (lactate acidosis) during muscular work [104].

Earlier EMG and ventilatory studies [11,77] demonstrated a sharp and well-defined rise in the previously stable EMG amplitude and ventilation upon applying an arterial occlusion cuff to the leg while working at a constant level of power output on the bicycle ergometer. Since the EMG activity levels were very constant and showed no electrophysiological sign of fatigue prior to the occlusion, it can be assumed that some shift in the MU recruitment and/or firing frequency takes place due to local muscle hypoxia caused by the occlusion [77,83]. On the other hand, the observed sharp increase in ventilation from its steady state level after the occlusion may be mediated through some neural pathways to the respiratory center [19,46], since the abrupt occlusion of the circulation to and from the exercising limb would isolate the respiratory center and the central chemore-ceptors from the effects of chemical products of muscle metabolism. If one could assume that the sharp increases in EMG activity during the occlusion represent the summation of a progressively increasing MU recruitment and firing frequency due to the compensation of reduced contractility of some fatigued MUs, a progressive increase in the extracellular K^+, for example, could be expected as a result of increased MU activities. By this means the ventilatory response could be stimulated via a neural pathway in the absence of cir-culation. In agreement with these results, Busse et al. [11] showed some evidence that ven-tilatory and thereby $[H^+]$ regulation can be accounted for by the plasma $[K^+]$ concentration and mechano-physiological properties of the peripheral muscles, respectively.

The above-mentioned discussion clearly suggests that there must be a tight link between muscle energy metabolism and neural excitation processes. Prolongation and reduction in the evoked action potential have been reported during high-frequency nerve stimulation [73,74] or during ischemic contractions [21], indicating a possible dependency on energy supply for muscle membrane function. Furthermore the depletion of extracel-lular Na^+ has been shown to accelerate the rate of force fatigue in the isolated curarized preparation [45]. This reduction of extracellular $[Na^+]$ or accumulation of K^+ may reduce the muscle membrane excitability sufficiently during high-frequency tetani to account for the excessive loss of force [45,74]. Thus energy metabolism clearly plays an important role in regulating neural excitation and electrolyte balance within the cell.

There has been some evidence that a decrease in intracellular pH can interfere with muscular contractile function. For example, the increase in $[H^+]$ has been shown to inter-fere with Ca^{++} binding to troponin, suggesting the possible participation of $[H^+]$ in excitation-contraction coupling with subsequent deficit in the developed tension [85]. The findings of Karlsson et al. [47] suggest that at tensions of 30% to 50% MVC, the increase in lactate is responsible for fatigue by direct or indirect changes in pH. However, at higher and lower tensions the possibility that lactate is directly implicated in the development of fatigue is remote, as electrical and metabolic factors can further complicate this phenom-enon [7,8,75,82,89].

For example, the results of NMR study [22] demonstrated that PFK-deficient patients who could not produce lactic acid showed virtually no change in pH during muscle fatigue. Hence other possible mechanisms must be considered. Accumulation of inorganic phos-phate (Pi) and ammonia (NH4+), for example, has also been shown to occur during mus-cular activity as possible inhibitory metabolites contributing to fatigue [35,84]. It has been suggested that Pi may bind to myosin in such a way so as to increase the forward rate of cross-bridge cycling and thereby to reduce force output [14]. Other evidence of Pi-induced force reduction is that patients with McArdle's disease demonstrated greater fatigability than normal individuals and a concomitantly larger increase in Pi accumulation [61].

At present, the specific messenger to which the motor control system responds with varying patterns of motor unit recruitment and firing frequency has not yet been clearly established. There is some evidence that MU recruitment order might be modified by changes in the proprioceptive afferent activity [31,49]. Other, and possibly more likely, explanations might be that the stretch receptors in muscle spindles and Golgi tendon organs could signal the need for adding more motor units [23,26,40,49,90] as a result of a fall in contractility of some motor units affected by the reduced oxygen supply and/or depletion of intramuscular glycogen store [12,30,100]. Or there may be some influences on the motoneuron pool activity from the sensory afferent nerve fibers originating in the "metaboreceptors" [62,94].

1.3 BASIC ELECTROPHYSIOLOGY OF THE MUSCLE CELL MEMBRANE

1.3.1 The Hodgkin-Huxley Model

The skeletal muscle fiber membrane is the seat of the bioelectric phenomena that result in the EMG signal. Most studies on excitable cell electrophysiology have been focused on nerve cells (in particular on the squid giant axon). The membrane of a muscle cell is more complex than the membrane of a nerve cell, it has layers and invaginations and electrical parameters different from those of a nerve cell. The invaginations are referred to as the tubular system, or T-system, which is a network of branching tubules radially oriented into the fibers. These tubules form a pathway for radial current flows that conduct the action potential from the outer membrane (the sarcolemma) into the sarcoplasmatic reticulum and play a role in the excitation contraction coupling. A correct analysis of the electrical properties of the muscle fiber should include the tubular system, which considerably complicates a model of the muscle membrane [64].

For the sake of simplicity, in many studies and in this book, the same model describing nerve cell membrane is used as a first approximation for muscle cell membrane. This model was first proposed in the classical work of Hodgkin and Huxley in 1952 [36] and describes, with an equivalent nonlinear electrical circuit, the behavior of the three main ionic channels, as depicted in Figure 1.10. The key point of the model is the dynamic voltage-dependent behavior of the membrane permeability to the three main ions. The time course of the sodium and potassium conductances is qualitatively depicted in the left portion of Figure 1.10. Because of this voltage dependence and the different dynamic behaviors of the sodium and potassium conductances, a transient membrane voltage phenomenon takes place whenever a membrane voltage threshold value is crossed (see Fig. 1.10). This phenomenon, which can be thought of as the one generated by a "one-shot" electronic circuit, is referred to as the action potential. It can be initiated by a chemically induced change of the sodium conductance (e.g., triggered in the skeletal muscle fiber by the neurotransmitter "acetilcholine") or by an externally applied electrical current (e.g., in direct muscle electrical stimulation). Its waveform, amplitude, and duration are strictly determined by the behavior of the ionic channels of the sarcolemma, the outer layer of the muscle fiber membrane.

Recently many neurological diseases have been identified, also genetically, that originate in the dysfunction of ion channels (channelopathies), whereby the identification of defects in the muscle fiber membrane play an important role [16]. For the EMG signal, the K^+, the Na^+, and the Cl^- channels in the sarcolemma (and in the T-system, as a second approximation) are essential. Force production is dominated by channels through which Ca^{++} ions are released. Electric membrane properties were already studied in detail around the mid-twentieth century with voltage clamp techniques. The understanding of individual membrane ion channels is an important challenge for modern cell biology and electrophysiology [2], for which the so-called patch clamp techniques were a crucial development [96].

Another essential tool for understanding membrane electrophysiology came by way of the development of nonlinear dynamic membrane models, in which the mutual interaction between the different channels are described quantitatively [101]. The classical work of Hodgkin and Huxley (1952 [36]) stood at the basis of these models as well. Their description of the giant axon of the squid gave an amazing replication of real membrane experiments. Later, with modern electrophysiological and histological experiments, it was found that the model dynamics appeared to reflect real life structural elements in the axon's

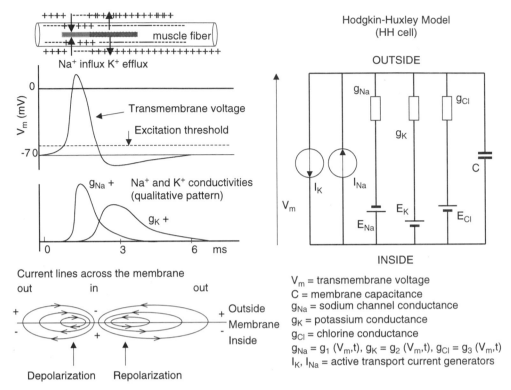

Figure 1.10. Hodgkin-Huxley model (*right*) and the generation of action potential in an excitable cell (*left*).

ion channels, one of the main examples of how even relatively abstract models can robustly relate to the structure and function of elements of living tissue. The importance of these achievements for the understanding of EMG may seem not obvious at first sight. Nevertheless, the source of all extracellular bioelectric activity can be found in the ion channel dynamics. This applies to the initiation, the propagation, and extinction of action potentials, the latter being especially important in surface EMG [51]. For practical purposes, knowledge of the waveform in time and space of the intracellular or transmembrane action potential at the sarcolemma suffices for the understanding of surface EMG. The insight attained is by way of the volume conduction mediated potential field set up by the ionic membrane current densities entering and leaving the sarcolemma [101] (Fig. 1.11). Which aspects of an action potential wave shape are essential for a proper description of the extracellular EMG signal can hardly be predicted without the use of a quantitative model of volume conduction (see Chapters 4 and 8).

Figure 1.11 expands these concepts showing a muscle membrane section subdivided in elementary areas, each described with the simplified HH cell model depicted in Figure 1.10. The figure shows the current lines generated by the local depolarization and caused by the events described in Figure 1.10. These currents produce a number of effects: first, by flowing into the nearby membrane areas, ahead of the depolarization front (see Fig. 1.12), they extend and propagate the action potential; second, by flowing into the volume conductor external to the fiber, they generate voltage drops into such volume, including the skin, and generate the extracellular action potential. These voltage drops constitute the single fiber contribution to the EMG signal, which can be detected either with intramus-

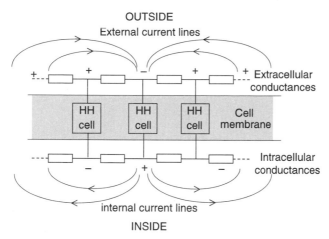

Figure 1.11. Segment of an excitable cell described as a sequence of Hodgkin-Huxley (HH) cells. Current lines resulting from a local depolarization are indicated.

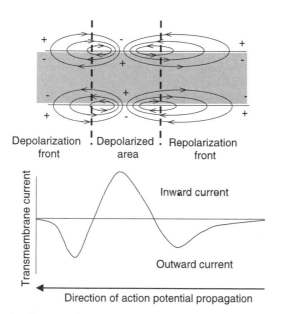

Figure 1.12. Schematic diagram of an action potential propagating along a muscle fiber. The current lines, depicted as ellipses, flow into the volume conductor generating voltages detectable between two electrodes placed into the volume conductor or on the skin. These voltages are contributions of the single fiber to the EMG.

cular needles or thin wires. (Needle and wire detection techniques will be discussed in Chapter 2 while surface detection will be discussed in Chapter 4.)

1.3.2 Propagation of the Action Potential along the Muscle Fiber

As indicated in Figure 1.12, the transmembrane current flow in the depolarization front of the action potential is in the direction that causes the membrane voltage to approach and

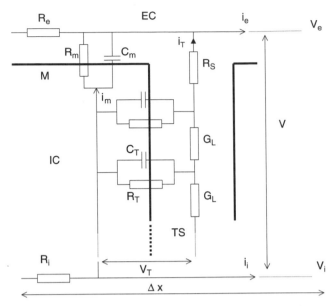

Figure 1.13. Model of a section of length Δx of a muscle fiber, showing the membrane (M), the tubular system (TS), the extracellular space (EC), the intracellular space (IC). The equivalent electrical circuit of a patch of the sarcolemma and T-system is described as a nonlinear resistance and a capacitor. The conductance of a tubule is G_L and I_T is the current flowing in it through the access resistance R_S. The transmembrane voltage V is the difference between the local extracellular and intracellular voltages V_e and V_i while V_T is the tubular potential. R_i and R_e are the intra- and extracellular resistances connecting to the next section. (Redrawn from [92] with permission)

cross the threshold for excitation. As a consequence the action potential moves to the left, with a velocity that depends on the fiber diameter and whose physiological range is between 3 m/s and 5 m/s with an average around 4 m/s. This propagation velocity is referred to as conduction velocity (CV) and is related to membrane properties. Such properties, which are reflected by the muscle fiber action potential, are different in different fiber types. Contradictory findings are reported in the literature [43,63] concerning the effect of fiber diameter (and length) on CV although most reports indicate that larger fibers show higher CV. The role of membrane folds and T-tubule system in determining CV is still unclear.

Figure 1.13 shows a schematic diagram of a T-tubule invagination readrawn from [92]. The role of the T-system is to allow flow of calcium ions into the sarcoplasmatic reticulum, a fact leading to cell contraction through a series of biochemical events. The characteristics of this system may affect (and therefore be reflected by) membrane electrical properties and phenomena. The tubular system is an important component, indicating that the classical membrane models developed for the squid axon may not be suitable for an accurate description of muscle-fiber membrane phenomena.

REFERENCES

1. Alway, S. E., J. Stray-Andersen, W. J. Grumbt, and W. J. Gonyea, "Muscle cross sectional area and torque in resistance-trained subjects," *Eur J Appl Physiol* **60**, 86–90 (1990).

2. Ashcroft, F. M., *Ion channels and disease*, Academic Press, San Diego, 2000.

3. Bawa, P., and R. N. Lemon, "Recruitment of motor units in response to transcranial magnetic stimulation in man," *J Physiol (Lond)* **471**, 445–464 (1993).

4. Beaver, W. L., K. Wasserman, and B. J. Whipp, "Bicarbonate buffering of lactic acid generated during exercise," *J Appl Physiol* **60**, 472–478 (1986).

5. Bigland-Ritchie, B., "EMG/force relations and fatigue of human voluntary contractions," *Exer Sports Sci Rev* **9**, 75–117 (1981).

6. Bigland-Ritchie, B., E. Cafarelli, and N. K. Vollestad, "Fatigue of submaximal static contractions," *Acta Physiol Scand* **28** (suppl), 137–148 (1986).

7. Bigland-Ritchie, B., D. A. Jones, and J. J. Woods, "Excitation frequency and muscle fatigue: electrical responses during human voluntary and stimulated contractions," *Exp Neurol* **64**, 414–427 (1979).

8. Bigland-Ritchie, B., R. Johansson, O. J. C. Lippold, and J. J. Woods, "Contractile speed and EMG changes during fatigue of sustained maximal voluntary contraction," *J Neurophysiol* **50**, 313–324 (1983).

9. Bigland-Ritchie, B., R. Johansson, O. J. C. Lippold, S. Smith, and J. J. Woods, "Changes in motoneuron firing rates during sustained maximal voluntary contractions," *J Physiol* **340**, 335–346 (1983).

10. Burke, R. E., "Motor units: Anatomy, physiology and Functional organization," in V. B. Brooks, ed., *Handbook of physiology: The nervous system*, American Physiological Society, Bethesda, 1981, pp. 345–422.

11. Busse, M. W., N. Maassen, and H. Konrad, "D. Boning, "Interrelationship between pH, plasma potassium concentration and ventilation during intense continuous exercise in man," *Eur J Appl Physiol* **59**, 256–261 (1989).

12. Bylund-Fellenius, A. C., P. M. Walker, A. Elander, J. Holm, and T. Schersten, "Energy metabolism in relation to oxygen partial pressure in human skeletal muscle during exercise," *Biochem J* **200**, 247–255 (1981).

13. Capodaglio, P., and M. V. Narici, "The aging nervous system and its adaptation to training," Advances in Rehabilitation, Maugeri Foundation Books, Pavia, Italy (2000).

14. Cooke, R., and E. Pate, "Inhibition of muscle contraction by the products of ATP hydrolysis: ADP and phosphate," *Biophys J* **47**, 25 (1985).

15. Davies, C. T. M., and C. Barnes, "Negative (Eccentric) work. II. Physiological responses to walking uphill and downhill on a motor-driven treadmill," *Ergonomics* **15**, 121–131 (1972).

16. Davies, N. P., and M. G. Hanna, "Neurological channelopathies: diagnosis and therapy in the new millennium," *An Med* **31**, 406–420 (1999).

17. DeLuca, C. J., and Z. Erim, "Common drive of motor units in regulation of muscle force," *Trend Neurol Sci* **17**, 299–305 (1994).

18. DeLuca, C. J., R. S. LeFever, M. P. McCue, and A. P. Xenakis, "Behavior of human motor units in different muscles during linearly varying contractions," *J Physiol (Lond)* **329**, 113–128 (1982).

19. Dempsey, J. A., H. V. Forster, N. Dledhill, and G. A. dePico, "Effects of moderate hypoxemia and hypocapnia on CSF [H^+] and ventilation in man," *J Appl Physiol* **38**, 665–674 (1975).

20. deVries, H. A., "Efficiency of electrical activity as a measure of the functional state of muscle tissue," *Am J Phys Med* **47**, 10–22 (1968).

21. Duchateau, J., and K. Hainault, "Electrical and mechanical failures during sustained and intermittent contractions," *J Appl Physiol* **58**, 942–947 (1985).

22. Edwards, R. H. T., M. J. Dawson, D. R. Wilkie, R. E. Gordon, and D. Shaw, "Clinical use of nuclear magnetic resonance in the investigation of myopathy," *Lancet* **1**, 725–731 (1982).

23. Enoka, R. G., and D. G. Stuart, "The contribution of neuroscience to exercise studies," *Federation Proc* **44**, 2279–2285 (1985).

24. Freund, H. J., H. J. Budingen, and V. Dietz, "Activity of single motor units from human forearm muscles during voluntary isometric contractions," *J Neurophysiol* **38**, 993–946 (1975).

25. Friden, J., M. Sjostrom, and B. Ekblom, "Myofibrillar damage following intense eccentric exercise in man," *Int J Sports Med* **4**, 170–176 (1983).

26. Garnett, R., and J. A. Stephens, "Changes in the recruitment threshold of motor units produced by cutaneous stimulation in man," *J Physiol (Lond)* **311**, 463–473 (1981).

27. Garnett, R. A. F., M. J. O'Donovan, J. A. Stephens, and A. Taylar, " Motor unit organization of human medial gastrocnemius," *J Physiol (Lond)* **287**, 33–43 (1979).

28. Goldberg, L. J., and B. Derfler, "Relationship among recruitment order, spike amplitude, and twitch tension of single motor units in human masseter muscle," *J Neurophysiol* **40**, 879–890 (1977).

29. Goldberger, M. E., "Spared-root deafferentation of a cat's hindlimb: hierarchical regulation of pathways mediating recovery of motor behavior," *Exp Brain Res* **73**, 329–342 (1988).

30. Gollnick, P. D., J. Karlsson, K. Piehl, and B. Saltin, "Selective glycogen depletion in skeletal muscle fibers of man following sustained contractions," *J Physiol (Lond)* **241**, 59–66 (1974).

31. Grimby, L., and J. Hannerz, "Disturbances in voluntary recruitment order of low and high frequency motor units on blockades of proprioceptive afferent activity," *Acta Physiol Scand* **96**, 207–216 (1976).

32. Heckman, C. J., and M. D. Binder, "Computer simulation of the effects of different synaptic input system on motor unit recruitment," *J Neurophysiol* **70**, 1827–1840 (1993).

33. Henneman, E., and L. M. Mendell, "Functional organization of the motoneuron pool and its inputs," in V. B. Brooks, ed., *Handbook of physiology: The nervous system*, American Physiological Society, Bethesda,1981, pp. 423–507.

34. Henneman, E., G. Somjem, and D. O. Carpenter, "Functional significance of cell size in spinal motoneurons," *J Neurophysiol* **28**, 560–580 (1965).

35. Hibberd, M. G., J. A. Dantzing, D. R. Trentham, and Y. E. Goldman, "Phosphate release and force generation in skeletal muscle fibers," *Science* **228**, 1317–1319 (1985).

36. Hodgkin A. L., and A. F. Huxley, "A quantitative description of membrane current and its application to conduction and excitation in nerve," *J Physiol* **117**, 500–544 (1952).

37. Hodgson, J. A., "The relationship between soleus and gastrocnemius muscle activity in conscious cats—A model for motor unit recruitment?" *J Physiol* **337**, 553–562 (1983).

38. Houtman, C. J., A. Heerschap, M. J. Zwarts, and D. F. Stegeman, "pH heterogeneity in tibial anterior muscle during isometric activity studied by(31)P-NMR spectroscopy," *J Appl Physiol* **91** 191–200 (2001).

39. Humphreys, P. W., and R. A. Lind, "The blood flow through active and inactive muscles of the forearm during sustained hand-grip contractions," *J Physiol (Lond)* **166**, 120–135 (1963).

40. Hutton, R. S., and D. L. Nelson, "Stretch sensitivity of Golgi tendon organs in fatigued gastrocnemius muscle," *Med Sci Sports Exer* **18**, 69–74 (1986).

41. Ikai, M., and T. Fukunaga, "A study on training effect on strength per unit cross-sectional area of muscle by means of ultrasonic measurements," *Int Angew Physiol Arbeitsphysiol* **28**, 173–180 (1970).

42. Infante, A. A., D. Klaupiks, and R. E. Davies, "Adenosine triphosphate changes in muscle doing negative work," *Science* **62**, 595–604 (1964).

43. Jack, J., D. Noble, and R. Tsien, *Electrical current flow in excitable cells*, Clarendon Press, Oxford, 1985.

44. Jessop, J., and O. C. J. Lippold, "Altered synchronization of motor unit firing as a mechanism for long-lasting increases in the tremor of human hand muscles following brief, strong effort," *J Physiol (Lond)* **269**, 29P–30P (1977).

45. Jones, D. A., B. Bigland-Ritchie, and R. H. T. Edwards, "Excitation frequency and muscle fatigue: mechanical responses during voluntary and stimulated contractions," *Exp Neurol* **64**, 401–413 (1979).

46. Kao, F. F., "Experimental study of the pathways involved in exercise hyperpnea employing cross-circulation techniques," in D. J. C. Conningham, and B. B. Lloyd, eds., *Regulation of Human Respiration*, Blackwell Scientific, Oxford, 1963.

47. Karlsson, J., C. F. Funderburk, B. Essen, and A. R. Lind, "Constituents of human muscle in isometric fatigue," *J Appl Physiol* **38**, 208–211 (1975).

48. Katz, A., and K. Sahlin, "Regulation of lactic acid production during exercise," *J Appl Physiol* **65**, 509–518 (1988).

49. Kirsch, R. F., and W. Z. Rymer, "Neural compensation for muscular fatigue: Evidence for significant force regulation in man," *J Neurophysiol* **57**, 1893–1910 (1987).

50. Kleine, B. U., D. F. Stegeman, D. Mund, and C. Anders, "The influence of motoneuron firing synchronization on surface EMG characteristics in dependence of electrode position," *J Appl Physiol* **91**(4), 1588–1599 (2001).

51. Kleinpenning, P. H., T. H. Gootzen, A. Van Oosterom, and D. F. Stegeman, "The equivalent source description representing the extinction of an action potential at a muscle fiber ending," *Math Biosci* **101**, 41–61 (1990).

52. Komi, P. V., and E. R. Buskirk, "Reproducibility of electromyographic measurements with inserted wire electrodes and surface electrodes," *Electromyogr* **4**, 357–367 (1970).

53. Komi, P. V., and J. T. Viitasalo, "Changes in motor unit activity and metabolism in human skeletal muscle during and after repeated eccentric and concentric contractions," *Acta Physiol Scand* **100**, 246–256 (1977).

54. Kukulka, C. G., and H. P. Clamann, "Comparison of the recruitment and discharge properties of motor units in human brachial biceps and adductor pollicis during isometric contractions," *Brain Res* **219**, 45–55 (1981).

55. Leonard, C. T., *The neuroscience of human movement*, Mosby, St. Louis, 1998.

56. Leonard, C. T., and M. E. Goldberger, "Consequences of damage to the sensorimotor cortex in neonatal and adult cats. I. Sparing and recovery of function," *Dev Brain Res* **32**, 1–14 (1987).

57. Leonard, C. T., and M. E. Goldberger, "Consequences of damage to the sensory motor cortex in neonatal and adult cats. II. Maintenance of exuberant projections," *Dev Brain Res* **32**, 15–30 (1987).

58. Leonard, C. T., H. Hirschfeld, and H. Forssberg, "The development of independent walking in children with cerebral palsy," *Dev Med Child Neurol* **33**, 567–577 (1991).

59. Leonard, C. T., H. Hirschfeld, T. Moritani, and H. Forssberg, "Myotatic reflex development in normal children and children with cerebral palsy," *Exp Neurol* **111**, 379–382 (1991).

60. Leonard, C. T., T. Moritani, H. Hirschfeld, and H. Forssberg, "Deficits in reciprocal inhibition of children with cerebral palsy as revealed by H reflex testing," *Dev Med Child Neurol* **32**, 974–984 (1990).

61. Lewis, S. F., R. G. Haller, J. D. Cook, and R. L. Nunnally, "Muscle fatigue in McArdle's disease studied by 31P-NMR: effect of glucose infusion," *J Appl Physiol* **59**, 1991–1994 (1985).

62. Mahler, M., "Neural and humoral signals for pulmonary ventilation arising in exercising muscle," *Med Sci Sports Exer* **11**, 191–197 (1979).

63. Martin, A., "The effect of change in length on conduction velocity in muscle," *J Physiol* **125**, 215–220 (1954).

64. Mathias, R., R. Eisenberg, and R. Valdiosera, "Electrical properties of frog skeletal muscle fibers interpreted with a mesh model of the tubular system," *Biophysical J* **17**, 57–93 (1977).

65. Matthews, P. B., "Relationship of firing intervals of human motor units to the trajectory of post-spike after-hyperpolarization and synaptic noise," *J Physiol* **492**, 597–628 (1996).

66. McCully, K. K., and A. Faulkner, "Characteristics of lengthening contractions associated with injury to skeletal muscle fibers," *J Appl Physiol* **61**, 293–299 (1986).

67. Metzger, J. M., and R. L. Moss "Greater hydrogen ion-induced depression of tension an velocity in skinned single fibers of rat fast than slow muscles," *J Physiol (Lond)* **393**, 724–742 (1987).

68. Milner-Brown, H. S., and R. B. Stein, "The relation between the surface electromyogram and muscular force," *J Physiol (Lond)* **246**, 549–569 (1975).

69. Milner-Brown, H. S., R. B. Stein, and R. Yemm, "Changes in firing rate of human motor units during linearly changing voluntary contractions," *J Physiol (Lond)* **230**, 371–390 (1973).

70. Moore, M. A., and R. S. Hutton, "Electromyographic investigation of muscle stretching tecnique," *Med Sci Sports Exer* **12**, 322–329 (1980).

71. Moritani, T., and H. A. deVries, "Neural factors versus hypertrophy in the time course of muscle strength gain," *Am J Phys Med* **58**, 115–130 (1979).

72. Moritani, T., and M. Muro, "Motor unit activity and surface electromyogram power spectrum during increasing force of contraction," *Eur J Appl Physiol* **56**, 260–265 (1987).

73. Moritani, T., and H. A. deVries, "Reexamination of the relationship between the surface integrated electromyogram (IEMG) and force of isometric contraction," *Am J Phys Med* **57**, 263–277 (1978).

74. Moritani, T., M. Muro, and A. Kijima, "Electromechanical changes during electrically induced and maximal voluntary contractions: Electrophysiologic responses of different muscle fiber types during stimulated contractions," *Exp Neurol* **88**, 471–483 (1985).

75. Moritani, T., M. Muro, and A. Nagata, "Intramuscular and surface electromyogram changes during muscle fatigue," *J Appl Physiol* **60**, 1179–1185 (1986).

76. Moritani, T., S. Muramatsu, and M. Muro "Activity of motor units during concentric and eccentric contractions," *Am J Phys Med* **66**, 338–350 (1988).

77. Moritani, T., A. Nagata, and M. Muro, "Electromyographic manifestations of neuromuscular fatigue of different muscle groups during exercise and arterial occlusion," *J Phys Fitness Jpn* **30**, 183–192 (1981).

78. Moritani, T., L. Oddsson, and A. Thorstensson, "Activation patterns of the soleus and gastrocnemius muscles during different motor tasks," *J Electromyogr Kinesiol* **1**, 81–88 (1991).

79. Moritani, T., L. Oddsson, and A. Thorstensson, "Phase dependent preferential activation of the soleus and gastrocnemius muscles during hopping in humans," *J Electromyogr Kinesiol* **1**, 34–40 (1991).

80. Moritani, T., T. Takaishi, and T. Matsumoto, "Determination of maximal power output at neuromuscular fatigue threshold," *J Appl Physiol* **74**, 1729–1734 (1993).

81. Moritani, T., M. Muro, A. Kijima, and M. J. Berry, "Intramuscular spike analysis during ramp force output and muscle fatigue," *Electromyogr Clin Neurophysiol* **26**, 147–160 (1986).

82. Moritani, T., M. Muro, A. Kijima, F. A. Gaffney, and A. Persons, "Electromechanical changes during electrically induced and maximal voluntary contractions: Surface and intramuscular EMG responses during sustained maximal voluntary contraction," *Exp Neurol* **88**, 484–499 (1985).

83. Moritani, T., W. M. Sherman, M. Shibata, T. Matsumoto, and M. Shinohara, "Oxygen availability and motor unit activity in humans," *Eur J Appl Physiol* **64**, 552–556 (1992).

84. Mutch, B. J. C., and E. W. Banister, "Ammonia metabolism in exercise and fatigue: A review," *Med Sci Sport Exer* **15**, 41–50 (1983).

85. Nakamaru, Y., and A. Schwartz, "The influence of hydrogen ion concentration on calciumbinding and release by skeletal muscle sarcoplasmic reticulum," *J Gen Physiol* **59**, 22–32 (1972).

86. Nardone, A., and M. Schieppati, "Shift of activity from slow to fast muscle during voluntary lengthening contractions of the triceps surae muscles in humans," *J Physiol* **395**, 363–381 (1988).

87. Nardone, A., C. Romano, and M. Schieppati, "Selective recruitment of high-threshold human motor units during voluntary isotonic lengthening of active muscles," *J Physiol* **409**, 451–471 (1989).

88. Narici, M. V., G. S. Roi, L. Landoni, A. E. Minetti, and P. Cerretelli, "Changes in force, cross-sectional area and neural activation during strength training and detraining of the human quadriceps," *Eur J Appl Physiol* **59**, 310–319 (1989).

89. Nassar-Gentina, V., J. V. Passonneau, J. L. Vergara, and S. I. Rapaport, "Metabolic correlates of fatigue and recovery from fatigue in single frog muscle fibers," *J Gen Physiol* **72**, 593–606 (1978).

90. Nelson, D. L., and R. S. Hutton, "Dynamic and static stretch responses in muscle spindle receptors in fatigued muscle," *Med Sci Sports Exer* **17**, 445–450 (1985).

91. Newham, D. J., K. R. Mills, B. M. Quigley, and R. H. T. Edwards, "Pain and fatigue after concentric and eccentric muscle contractions," *Clin Sci* **64**, 55–62 (1983).

92. Reichel, M., W. Mayr, and F. Rattay, "Computer simulation of field skeletal muscle fiber interpreted with a mesh model of tubular system," *Biophisical J* **17**, 57–93 (1977).

93. Sale, D. G., "Neural adaptation to strength training," in P. V. Komi, ed., *Strength and power in sport*, Blackwell Publishing, Oxford, 1991.

94. Sjøgaard, G., "Exercise-induced muscle fatigue: The significance of potassium," *Acta Physiol Scand* **140** (Suppl), 1–63 (1990).

95. Smith, J. L., B. Betts, V. R. Edgerton, and R. F. Zernicke, "Rapid ankle extension during paw shakes: Selective recruitment of fast ankle extensors," *J Neurophysiol* **43**, 612–620 (1980).

96. Sperelakis, N., "Patch clamp and single-cell voltage clamp techniques and selected data. Mol Cell" *Biochem* **80**, 3–7 (1998).

97. Stephens, J. A., and T. P. Usherwood, "The mechanical properties of human motor units with special reference to their fatigability and recruitment threshold," *Brain Res* **125**, 91–97 (1977).

98. Tanji, J., and M. Kato, "Firing rate of individual motor units in voluntary contraction of abductor digiti minimi in man," *Exp Neurol* **40**, 771–783 (1973).

99. Van Bolhuis, B. M., W. P. Medendorp, and C. C. A. M Gielen, "Motor-unit firing behaviour in human arm flexor muscles during sinusoidal isometric contractions and movements," *Brain Res* **117**, 120–130 (1997)

100. Vollestad, N. K., O. Vaage, and L. Hermansen, "Muscle glycogen depletion patterns in type I and subgroups of type II fibers during prolonged severe exercise in man," *Acta Physiol Scand* **122**, 433–441 (1984).

101. Wallinga, W., S. L. Meijer, M. J. Alberink, M. Vliek, E. D. Wienk, and D. L. Ypey, "Modelling action potentials and membrane currents of mammalian skeletal muscle fibres in coherence with potassium concentration changes in the T-tubular system," *Eur Biophys J* **28**, 317–329 (1999).

102. Wallinga-De Jonge, W., F. F. Gielen, P. Wirtz, P. De Jong, and J. Broenink, "The different intracellular action potentials of fast and slow muscle fibers," *Electroenc Clin Neurophysiol* **60**, 539–547 (1985).

103. Wasserman, K., B. J. Whipp, S. N. Koyal, and W. L. Beaver, "Anaerobic threshold and respiratory gas exchange during exercise," *J Appl Physiol* **35**, 236–243 (1973).

104. Wasserman, K., W. L. Beaver, and B. J. Whipp, "Gas exchange theory and the lactic acidosis (anaerobic threshold)," *Circulation* **81** (Suppl), II14–II30 (1990).

105. Wilkie, D. R., "Muscular fatigue: effects of hydrogen ions and inorganic phosphate," *Fed Proc* **45**, 2921–2923 (1986).

2

NEEDLE AND WIRE DETECTION TECHNIQUES

J. V. Trontelj

Department of Neurology
Institute of Clinical Neurophysiology
University Medical Center of Ljubljana, Slovenia

J. Jabre

Department of Neurology, Boston University
Acting Chief, Harvard-BU Neurology Service
Boston VA Medical Center, Boston

M. Mihelin

Department of Biomedical Engineering
Institute of Clinical Neurophysiology
University Medical Center, Ljubljana, Slovenia

2.1 ANATOMICAL AND PHYSIOLOGICAL BACKGROUND OF INTRAMUSCULAR RECORDING

Before considering the recording of EMG signals from inside the muscle, it is appropriate to recall some of its relevant anatomical and physiological features. In anatomical terms, the motor unit (MU) consists of a motor neuron, its axon, and all the muscle fibers innervated by the axonal branches. A motor unit may contain widely different numbers of muscle fibers, ranging from 10 to 15 in the extraocular muscles to over 500 in large limb muscles such as the gastrocnemius. The muscle fibers of each motor unit are intermingled with fibers of other motor units, forming a mosaic pattern. In each fascicle of muscle fibers,

Electromyography: Physiology, Engineering, and Noninvasive Applications, edited by Roberto Merletti and Philip Parker.
ISBN 0-471-67580-6 Copyright © 2004 Institute for Electrical and Electronics Engineers, Inc.

usually several motor units are represented in a scattered fashion. A single motor unit may occupy a relatively large portion of a cross section of a muscle, called the motor unit territory. In human limb muscles the average motor unit territory has an irregular round shape with a diameter of about 10 mm [1,13].

In neurogenic disorders, when some motor neurons degenerate, the surviving motor neurons grow new axonal sprouts establishing synaptic contacts with the denervated muscle fibers (collateral reinnervation). As a result the motor unit may become much denser and larger, often containing several times more than normal number of muscle fibers, although its territory does not increase [14]. It seems that the axonal sprouts cannot cross the boundaries of the muscle-fiber fascicles. On the other hand, the motor unit may lose some of its muscle fibers to various disease processes and become smaller. In some disorders there may be disturbances in conduction of impulses along the axonal branches, impairment of neuromuscular transmission across the motor end plates, and rarely, disturbed depolarization of the muscle-fiber membrane. Needle electromyography is well suited for the detection of changes in motor unit size and its internal structure, and can often reveal abnormal function. Needle EMG also makes it possible to record action potentials of spontaneously contracting single muscle fibers (fibrillation potentials). Such potentials, which are an important sign of denervation, cannot be detected with surface electrodes. Needle EMG can distinguish between other types of normal and abnormal spontaneous activity (fasciculations, complex repetitive discharges, myotonia, and neuro-myotonic discharges).

In physiological terms, all the muscle fibers of a motor unit work in unison; that is, all are discharged nearly synchronously upon the arrival of a nerve impulse along the axon and through its terminal branches to the motor end plates. A motor unit action potential (MUAP) is recorded by a needle electrode inserted in the contracting muscle (Fig. 2.1). Depending on the type of electrodes used, the recorded MUAP can be derived from action potentials (APs) of a small number of muscle fibers (1–3), a moderate number of muscle fibers (15–20), or a great majority of muscle fibers belonging to the MU (several hundreds).

The objective of recording of a muscle's electrical activity by intramuscular recording techniques is to study the physiology and pathology of the MU. At the peripheral level, one can study the effects of lesions such as loss of nerve supply to the muscle (denervation), the ability of the nerve to regenerate (reinnervation), diseases affecting the muscle fiber itself (myopathies, e.g., muscular dystrophies), and diseases of the neuromuscular junction (e.g., myasthenia gravis). At the central level, one can study MU recruitment and firing patterns, both of which give information about central nervous system motor control and its disturbances.

On the other hand, surface electrodes are usually better suited for studies in which information is sought on the various aspects of behavior, temporal pattern of activity, or fatigue of the muscle as a whole or of muscle groups. Modern surface electrode arrays also allow the mapping of the MUAP propagation from the innervation zone to the tendon endings, high-accuracy estimation of muscle-fiber conduction velocity, location of the innervation zone, length and orientation of the fibers (see Chapter 7), and to some degree, recruitment and decomposition. They find important applications in sport, rehabilitation, and occupational medicine where needle techniques are not acceptable or when assessments have to be repeated frequently.

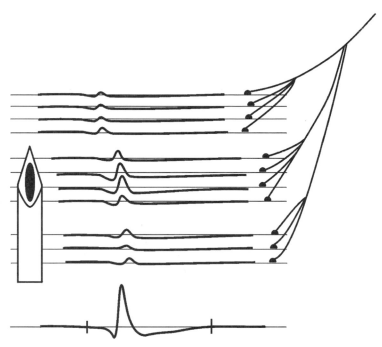

Figure 2.1. Motor unit action potential as detected by a concentric needle electrode. The APs of muscle fibers closest to the active surface are recorded with higher amplitude than those of more distant fibers. The arrival of individual fibers' APs at the electrode is not perfectly synchronous. (Reproduced with permission from [13])

2.2 RECORDING CHARACTERISTICS OF NEEDLE ELECTRODES

Various needle electrodes have been designed for recording muscle fiber(s) APs from the contracting muscle as well as muscle at rest. In nearly all cases intramuscular needle electrodes are used to make extracellular recordings of the AP generated by a MU or a muscle fiber; on the other hand, intracellular recordings made mainly in vitro will not be considered in this chapter.

Typically the shape of the extracellularly recorded AP of a muscle fiber is triphasic, and the signal amplitude decreases exponentially as the distance between the electrode tip and the muscle fibers from which the recording is made increases. Thus the contribution of any individual muscle fiber to the MUAP amplitude crucially depends on its distance from the active electrode surface. The changes of amplitude with distance vary with the different types of electrodes. The electrodes with a small recording surface show a steep decline of the recorded AP amplitude with increasing distance from the source and are therefore more selective, detecting mostly activity of the closest muscle fibers. Conversely, electrodes with large recording surfaces are less selective, picking up activity from a larger area within the muscle (see below).

The size and construction of the needle electrode will affect the uptake area and consequently the MUAP morphology (Fig. 2.2). Electrodes with a larger active surface have a larger uptake area; the action potentials are recorded with smaller amplitude, but the rate of decline of the AP's amplitude with increasing distance between the muscle fiber and

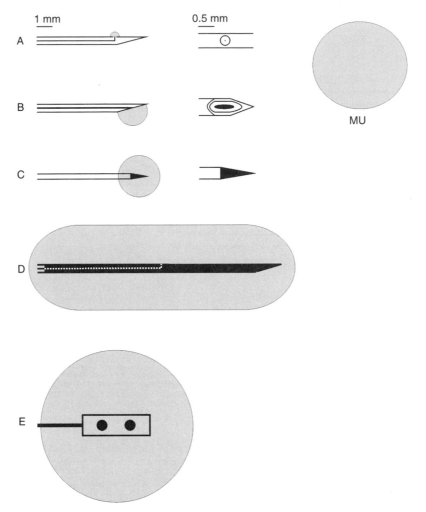

Figure 2.2. Different types of needle electrodes have different recording areas, covering different portions of a typical motor unit territory (inset). (*A*) Single fiber electrode, (*B*) concentric electrode, (*C*) monopolar electrode, (*D*) macro electrode. Also shown is the recording area of a plate-mounted surface electrode (E). (Modified from [13] with permission)

the recording surface is slower. In other words, the recorded amplitude and the rate of its decline are inversely proportional to the size of the active surface and the uptake radius of the recording electrode (Fig. 2.3). In the paragraphs below we will discuss the types and recording characteristics of EMG needle electrodes commonly used in electromyography, both in clinical medicine and in research.

2.3 CONVENTIONAL NEEDLE EMG

In clinical EMG two types of needle electrodes are used to record spontaneous, voluntary, and occasionally stimulus-induced muscle activity: the *concentric* needle electrode and

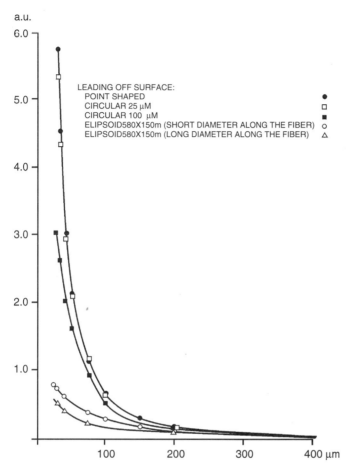

Figure 2.3. Amplitude (ordinate in arbitrary units) versus recording distance for simulated action potentials recorded with different types of leading-off surfaces. Note the low amplitude obtained with large electrodes for short electrode-fiber distance. The curves for point-shaped and 25 µm electrode practically overlap. (Reproduced with permission from [3])

the *monopolar* needle electrode. A needle electrode of either type inserted inside the territory of a discharging motor unit records from all the muscle fibers active within its uptake area. The summating APs generated by different muscle fibers in the motor unit are not perfectly synchronous (see Fig. 2.1). This is due to the fact that conduction times from the end plates to the electrode for the individual muscle fibers of the same MU differ. There are differences in velocity among the nerve and muscle fibers (as a result of different diameters of axonal branches and muscle fibers) and in traveling distance (since terminal axonal branches have different lengths and the end plates are somewhat scattered along the different muscle fibers, resulting in different lengths of the muscle fiber segments to the electrode). When temporal dispersion becomes significant, the individual fibers' APs are less well aligned in time; the resulting partial phase cancellation will reduce the MUAP amplitude and produce a serrated or polyphasic shape.

Consequently the MUAP can be defined as a result of the summation in both space and time of the APs generated by the muscle fibers that lie within the uptake area of the needle electrode. The morphology of the MUAP is therefore greatly influenced by these

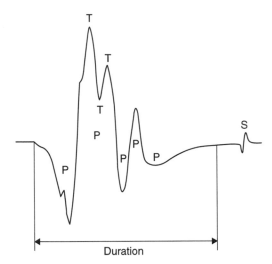

Figure 2.4. Main MUAP parameters. Phases (*P*) and turns (*T*) reflect the degree of asynchronic-ity; the peak to peak amplitude of the main spike mostly depends on the few closest muscle fibers; the duration is correlated to the number of muscle fibers seen by the electrode. A late component, "satellite" (*S*), is generated by a muscle fiber that is either innervated by a slowly conducting terminal axon or, more significantly, itself conducts slowly due to atrophy. (Modified from [13] with permission)

factors that can be described as MU architecture. Disturbances of this architecture following injury or in disease will generate morphological changes in the MUAP that can be studied with needle electrodes.

2.3.1 MUAP Parameters and Their Changes in Disease

The different MUAPs recorded in a muscle show great variation of shape and other parameters, partly depending on the electrode position. Therefore a sufficient number of MUAPs has to be studied in order to obtain a representative sample. A number of parameters can be quantitatively assessed using automatic digital techniques offered by modern EMG equipment.

Parameters of the raw signal useful in diagnostic EMG include the *MUAP amplitude*, *MUAP duration*, *duration of the main spike*, and *number of voltage turns and phases* (Fig. 2.4). In addition *degree of instability* of the MUAP, the so-called jiggle, which reflects uncertainties of impulse travel in the terminal nerve tree across the motor end plates and along the muscle fibers on consecutive discharges (Fig. 2.5), has recently been quantified [17].

The *amplitude*, usually measured from peak to peak, reflects the number of active muscle fibers of a motor unit within the uptake area of the electrode, and the degree of synchronicity of their firing, whereby the closest 2–5 fibers contribute the largest share in the normal muscle when the conventional concentric needle electrode is used. The amplitude is increased in large and dense motor units following reinnervation, and is often decreased in myopathies.

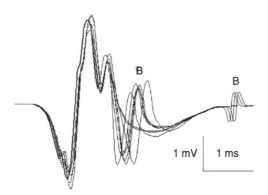

Figure 2.5. MUAP showing an abnormal "jiggle" of 8 superimposed traces. Some of the components show an abnormal jitter, also resulting in vertical shifts. Two components (marked with *B*) display intermittent blocking.

The *duration* measured between the onset of the slow initial phase of the MUAP and the end of the slow terminal phase mainly reflects the number of muscle fibers detected. Contrary to earlier assumptions, MUAP duration does not change much with desynchronization of individual fibers' APs (e.g., due to abnormally large variation in muscle fiber diameters as in myopathies), unless this is unusually excessive.

The *number of voltage turns and phases* shows the degree of temporal dispersion of the Aps from different muscle fibers in the motor unit; it is increased with increased differences in impulse conduction times along the terminal nerve twigs and particularly along the muscle fibers, in case of increased variation in muscle fiber diameters such as in myopathies.

An abnormal jiggle is the result of increased jitter (see below) or intermittent blocking of individual single-fiber Aps or groups of Aps. A disturbed conduction may take place in the immature axonal sprouts, and transmission may be impaired across sick or immature motor end plates. Other parameters have been developed to extract more information from the signal and reduce its dependence on the electrode position [15].

In nerve lesions, where the muscle loses a certain number of the motor neurons innervating it, the muscle fibers become denervated; within hours to days the surviving motor axons start to grow new sprouts and gradually reinnervate the neighboring denervated muscle fibers (collateral reinnervation). Thus the mother unit now contains an additional number of adopted muscle fibers, whose original metabolic, electrophysiological, and mechanical characteristics may have been different from those of its own fibers. In addition the newly reinnervated muscle fibers have an immature motor end plate and are atrophic. Both conditions improve after a few months, resulting in an improved, although often still deficient, synchronization of the contributions to the MUAP.

The MUAP of a reinnervated MU shows increased *amplitude* (because of the additional muscle fibers); a *polyphasic* shape (due to delayed arrival of impulses along the recently acquired muscle fibers); increased *duration* (due to increased number of muscle fibers); and at least initially, an unstable shape, that is, an abnormal jiggle. In addition the *interference pattern* on maximum voluntary effort (see below) is reduced. These changes help the electromyographer assess the nature and age, as well as the localization and extent of the lesion.

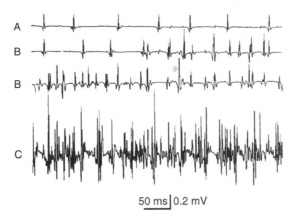

50 ms | 0.2 mV

Figure 2.6. EMG recording at increasing force of contraction. (*A*) At low effort, only a few MUs are active; (*B*) with increasing force, their discharge frequency increases and new MUs are recruited; (*C*) at strongest voluntary contraction, individual MUs can no longer be discerned. There is a full interference pattern. (From [13] with permission)

Another type of change is seen with conventional needle EMG in primary muscle disease. The muscle fiber itself, the generator of the AP, is diseased and may become a weaker electrical generator. Different muscle fibers in the same motor unit may undergo variable changes at different time intervals, depending on the disease process; one of the consequences is that the discharges of individual fibers in the motor unit are less well synchronized. In addition the myopathic motor unit may lose a part of its muscle fibers to degeneration and necrosis and become smaller. These results in a MUAP that typically has a lower amplitude and shorter duration and is serrated or polyphasic in shape with a frequency content shifted toward higher values: the so-called myopathic MUAP. In contrast to neurogenic disorders, the number of motor units recruited at maximum effort often remains normal in myopathy.

2.3.2 Needle EMG at Increasing Voluntary Contraction

With increasing voluntary effort, the active Mus fire at increasing discharge rates and new motor units with progressively higher recruitment thresholds appear in the tracing. Individual MUAPs start to overlap and become less and less discernable in a dense electrical activity, called the summation or interference pattern (Fig. 2.6). Recruitment of new motor units may be delayed when there is partial loss of motor units and the remaining ones increase by collateral reinnervation; furthermore the recruitment pattern on maximum voluntary effort is reduced. This is in contrast to muscle diseases where motor units become small and weak: recruitment is abnormally early, but the interference pattern may remain full (Fig. 2.7). Several quantitative methods have been developed to analyze the interference pattern, for example, based on counts of voltage turns and amplitudes between the turns [8].

2.3.3 The Concentric Needle Electrode

Developed by Adrian and Bronk in 1929, the concentric needle electrode (CNE) is still one of the most commonly used electrodes in clinical recordings of EMG signals (Fig.

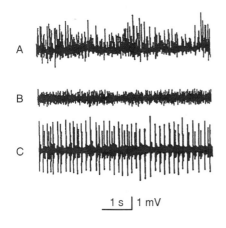

A

B

C

1 s | 1 mV

Figure 2.7. (A) Interference pattern in normal muscle, (B) myopathy-dense pattern of low amplitude, and (C) neuropathy-reduced pattern with high amplitude components. (From [13] with permission)

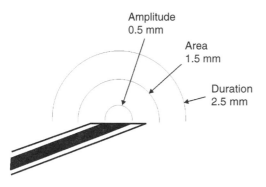

Amplitude
0.5 mm

Area
1.5 mm

Duration
2.5 mm

Figure 2.8. Uptake area of the concentric needle electrode. The MUAP amplitude depends on muscle fibers located in a semiglobe above the active surface with a radius of 0.5 mm; the radii for MUAP area and for MUAP duration are 1.5 mm and 2.5 mm, respectively.

2.2B). The electrode has an elliptical ($150 \times 580 \mu$m) active recording surface located in the beveled tip of a metal cannula. Signals detected by the active recording surface are referenced to the cannula. The cannula is thus not "inactive," but its large recording size ensures that the amplitudes of the signals recorded by it remain generally small in comparison to those recorded by the core (see Fig. 2.3).

Because the concentric electrode's recording surface is located in the beveled tip, the electrode has directional asymmetry, which means that electromyographers may need to rotate the needle to maximize the AP amplitude. (This applies to the SFEMG electrode as well but not to the monopolar electrode.) Simulation studies have shown that a MUAP amplitude recorded with this electrode is directly related to the number and size of muscle fibers located within a semiglobe with a radius of 0.5 mm at the needle tip while the MUAP area is related to muscle fibers located within 1.5 mm; the MUAP duration to those located within 2.5 mm (Fig. 2.8). On average, the concentric MUAP is derived from 15 to 20 muscle fibers of a motor unit that lie within the uptake area of the needle electrode.

2.3.4 The Monopolar Needle Electrode

The monopolar needle electrode (MNE) consists of a solid Teflon™-insulated pin, and the recording area is the denuded tip of the cone. The standard recording surface is between 0.15 and 0.20 mm². A separate surface electrode is needed as a reference.

The MNE recording characteristics differ from those of the CNE. Because it is not a directional electrode, the MNE can "see" more muscle fibers in the neighborhood of its active recording area; as a result the MUAPs tend to be more polyphasic and to have a longer duration and higher amplitude. On average, the MNE's recording surface can see anywhere from 20 to 30 muscle fibers at an individual insertion site.

The MNE is used (most often by US physiatrists) for the same purposes as the CNE, with similar efficiency and results, although somewhat less frequently. Slightly different reference values apply. The MNE can also be used for intramuscular stimulation, when only a single or a few motor units need to be activated at a time (e.g., in jitter studies; see below).

2.4 SPECIAL NEEDLE RECORDING TECHNIQUES

Two techniques based on the use of special needle electrodes will be briefly described: single-fiber EMG and macro EMG. They have both found clinical application, and perhaps more important, have essentially contributed to the present understanding of the motor unit electrophysiology. Decomposition EMG with the quadrifillar electrode will also be described; it has contributed to development of algorithms for automatic MUAP analysis. In addition scanning EMG will be mentioned as a technique that has essentially increased our knowledge of the motor unit territory and the anatomical organization of its constituents.

2.4.1 Single-Fiber EMG

Sometimes, in order to resolve a physiological question, it may be necessary to selectively observe the activity of just a single motor end plate, one or a few muscle fibers, or a single motor unit. Such recording from a restricted area in situ is possible with a fine wire electrode inserted into the muscle by means of an injection needle and then left in place. The wire is insulated except for a small area where the insulation coating is broken, such as with an electrical spark, to provide a point-size recording area. The size of the bare area, however, is difficult to standardize. Furthermore, once in place, the electrode cannot be moved in a controlled manner to optimize the recording position. Whenever prolonged recording from a single site is not needed, a better approach to studies of the fine details within the motor unit is to use a specially designed single-fiber electrode [18].

In this electrode a needle cannula contains a $25\,\mu$m platinum wire exposed in a side port near the tip (Fig. 2.3). With this electrode, which has an uptake radius of about $300\,\mu$m, it is possible to record action potentials of single muscle fibers of relatively large amplitude, 1 to 6 mV. For special studies, multi-electrodes have been constructed containing up to 11 electrode surfaces arranged in two rows in the side port [16]. Single-fiber EMG has been used to study a variety of physiological and morphological parameters of the motor unit and its constituents.

Fiber Density. This is a morphological term describing the degree of grouping of muscle fibers of a single motor unit in the cross section of a muscle. As described above, the muscle fibers of different motor units are normally arranged in a kind of a mosaic pattern, namely interdigitated with each other. This normal arrangement is disrupted in disorders involving denervation followed by reinnervation. Then muscle fibers of the same MU appear adjacent to each other, forming smaller or larger clusters. In fiber density estimation (FD, expressed as the average number of recorded single muscle fibers of motor units observed per recording site), the SFEMG needle is inserted repetitively into the

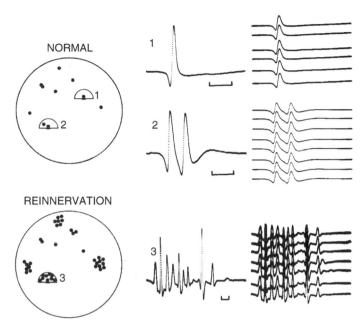

<u>Figure 2.9.</u> Fiber density measurement. The uptake area of the SFEMG electrode is shown as semicircles. In the normal muscle 1 or 2 fibers from the same motor unit are recorded. After denervation followed by reinnervation many more fibers are recorded within the uptake area. (From [18])

slightly contracted muscle and recordings are made from a number of randomly sampled sites (Fig. 2.9). All spike components time-locked to a selected SFAP are counted and an average from 20 recording sites is obtained. In a normal limb muscle only one fiber is recorded in about 70% of insertions, and two or three at the remaining sites. Normal FD values are close to 1.5, with slight differences between muscles. s a sensitive measure of spatial rearrangement of muscle fibers in the motor unit and is increased early in both neurogenic and myopathic disorders. Increased FD is the electrophysiological counterpart of the histochemical fiber type grouping.

Jitter and Neuromuscular Transmission. Single-fiber action potentials can be elicited with intramuscular electrical stimulation of small bundles of motor axons through a pair of monopolar needle electrodes. On consecutive stimulations the latency between stimulus and response varies by a few tens of microseconds (Fig. 2.10A). This is called the neuromuscular jitter. Most of the jitter is due to the variation in time needed for end plate potentials at the neuromuscular junction to reach the depolarization threshold. During voluntary activation of the muscle, the SFEMG electrode can be positioned to record from two (or more) muscle fibers in one active motor unit (Fig. 2.10B). The jitter in this case is seen as the varying time interval between action potentials from two fibers on consecutive discharges. It represents the combined variation of delays at two end plates.

The jitter is expressed as the mean value of consecutive differences (MCD) of successive interpotential intervals (in case of voluntary activation) or latencies (in case of axonal stimulation). It is usually measured and computed by means of jitter meters incorporated in the EMG equipment. Some algorithms use the mathematically detected peak

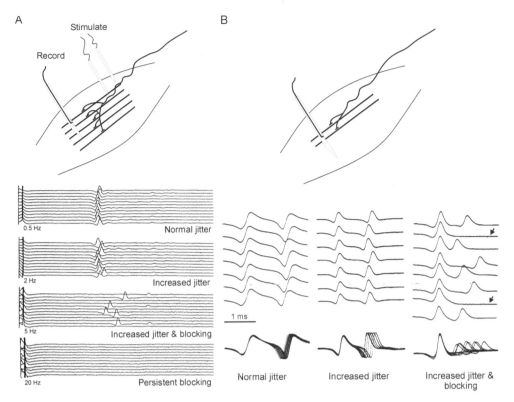

Figure 2.10. Jitter measurement during electrical stimulation of the motor axon (*A*) and during voluntary activation (*B*). In (*A*) the jitter is measured between the stimulus and the SF AP on consecutive discharges. In (*B*) the recording is made from a pair of muscle fibers innervated by the same axon and the jitter is measured on the second of the two SF APs. (Reproduced with permission by Elsevier Science from [19])

of the SF AP as the point for time measurements; others rely on a manually selected point on the AP slope. Usually 20 potential pairs or 30 to 40 SF APs from different muscle fibers are recorded to assess the status of neuromuscular transmission in a muscle.

The jitter value is a measure of the safety factor of neuromuscular transmission. It is increased whenever the ratio between the end plate potential and the depolarization threshold (i.e., the safety margin) is smaller than normal. This may be seen even in muscles with no clinical weakness, where transmission of impulses does occur, although with decreased safety margin. Increased jitter is thus a subclinical sign of impaired neuromuscular transmission. With further impairment, intermittent or persistent impulse blocking occurs and weakness becomes clinically manifest.

SFEMG is considered to be the most sensitive method to detect disordered neuromuscular transmission, and is widely used to diagnose myasthenia gravis and other diseases of neuromuscular junction. The axonal stimulation technique makes it possible to distinguish between the pre- and postsynaptic abnormality. In postsynaptic disorders, such as myasthenia gravis, maximum abnormality is manifest at 5 to 10 Hz, while transmission is more reliable at 0.5 to 1 Hz, and as a result of intratetanic facilitation of transmitter release, often also improves at 15 to 20 Hz. In a presynaptic disorder, exemplified by the Lambert-Eaton myasthenic syndrome, the jitter and blocking are most pronounced at low activation rates (e.g., 1–5 Hz) and may improve dramatically as the rate is increased to 15 to 30 Hz [23]. However, mixed findings may be seen in either of the two disorders, in particular, in Lambert-Eaton myasthenic syndrome, perhaps pointing to a dual pathophysiology [24].

Muscle-Fiber Conduction Velocity. Velocity of propagation of the action potentials can be studied at the level of single muscle fibers using either intramuscular stimulation and recording with an ordinary SFEMG electrode or by recording from two surfaces in a multi-electrode. During activity, velocity changes as a function of preceding interdischarge intervals [9]. In response to a pair of stimuli of increasing lag, subnormal values of conduction velocity are observed, followed, for most muscle fibers, by supernormal values starting at interstimulus intervals of 3 to 12 ms, with a peak at 5 to 15 ms decayed to normal velocity at 1 s [7]. Muscle-fiber propagation velocity was also shown to depend on its length status—the degree to which it is lax or stretched at the moment of depolarization [20]. Muscle-fiber conduction velocity can also be estimated using surface electrode arrays, although not at the level of single fibers. The surface technique is simpler and noninvasive but somewhat affected by the end-of-fiber effect that (if not removed or corrected for) causes overestimates of velocity.

Other Studies. SFEMG is conveniently used whenever recordings are needed from single muscle fibers, such as in the denervated muscle [22], or from single motor units, such as in the studies of various spinal or brain stem reflexes or corticospinal system physiology [21,26]. The interested reader is referred to a more comprehensive text on SFEMG [18].

2.4.2 Macro EMG

This method is based on the use of a special Macro needle electrode [10]. The purpose of this technique was to record from a majority of muscle fibers of the entire MU (the macro MUAP). Recordings are made between a large bare segment of the cannula of a needle electrode and a distant surface reference electrode. The macro electrode uses a modified SFEMG electrode to trigger on an action potential from a nearby muscle fiber on one amplifier channel while the time-locked macro potential is recorded by the needle's cannula on a second channel. An external reference electrode is required. Averaging must be used to eliminate the activity of other motor units (Fig. 2.11). The resulting AP, the macro MUAP, is the compound action potential of a large portion of muscle fibers in the MU, indeed most often the majority of fibers. This potential has been shown to correlate well with the size and number of muscle fibers of the MU. Because of its large size, the macro recording surface has a pronounced shunting effect on the action potentials of muscle fibers that are closest to the recording electrode, and these are recorded with low amplitude (see Fig. 2.1). On the other hand, the distant fibers are relatively well represented because of a slow rate of amplitude decline with increasing distance. In contrast to APs recorded with the CNE and MNE, the macro MUAP remains relatively stable at different positions of the needle electrode within the motor unit territory. This makes it particularly useful for the study of MU size, while it is rather insensitive to the changes affecting the micro architecture of the MU. However, information of this kind (e.g., on FD) is simultaneously obtained on the SFEMG channel.

A modification of the macro electrode, the concentric macro or ConMac electrode was also developed [4]. The ConMac electrode uses a modified CNE (coated with Teflon to within 15 mm of its tip) to trigger an action potential from nearby muscle fibers on one

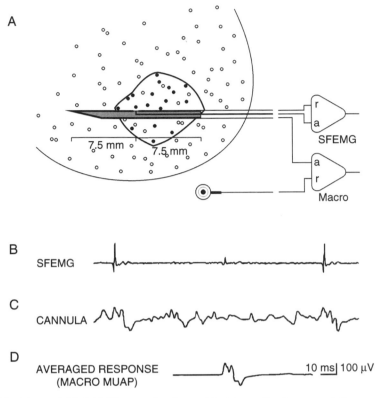

Figure 2.11. Macro EMG. SFEMG recording is used to trigger the traces and the cannula referenced to a distant electrode records activity of most muscle fibers within the motor unit (marked with full circles). The recording is averaged to eliminate non-synchronized activity of other motor units. "a" active surface, "r" reference. (Reproduced with permission from [25])

channel and average the time-locked macro potential recorded by the needle's cannula on the other channel. In concentric macro EMG, however, the concentric MUAP signal performs the task of the single-fiber EMG signal in triggering the oscilloscope sweep. The result is a simultaneous display on the screen of both the concentric and the macro MUAP. Since most nerve and muscle pathology has been studied with the concentric MUAP as a reference, relevant information is gained when this action potential can be studied alongside the macro action potential (Fig. 2.12).

A major difference between the concentric and single-fiber macro electrode is that in the former, the triggering surface is located at the tip, while in the latter, it is in the middle of the macro recording surface. Consequently the chance of having the recording corridor reach all the way through the MU territory is slightly better with the single-fiber Macro electrode.

Macro EMG allows the study of the motor unit at the focal, local, and global levels. It yields useful information on the size of the motor unit in terms of the quantity of its muscle fibers. This may be particularly valuable in conditions where nonuniform focal changes in a motor unit (reinnervation by sprouting, myopathic involvement, or sparing) seen in conventional or single-fiber EMG give a false impression and lead to misinterpretation.

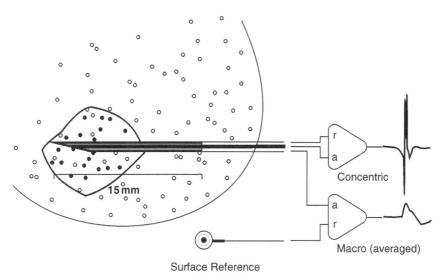

<u>Figure 2.12.</u> Macro EMG with the Con-Mac electrode. Averaging in this case is triggered with a concentric MUAP.

2.4.3 EMG Decomposition Technique with Quadrifilar Needle Electrode

The quadrifillar needle electrode (QNE), developed by De Luca [2], consists of a cannula with four 75 μm leads arranged in a square configuration with 200 μm sides and located in a side port 7.5 mm from the tip of the needle electrode. The electrode also includes a concentric recording wire at the tip and a macro recording surface (by shielding it with Teflon™ leaving bare 15 mm of cannula). This electrode is used for EMG signal decomposition studies (the decomposition of the raw EMG signal into its constituent MUAP components). Decomposition is achieved using an operator interactive algorithm which scans the EMG signal and identifies unique MUAP shapes. Each MUAP is then classified by matching its morphology to that of the template of one or more of the Mus thus identified. The technique, allowing for a high level of decomposition accuracy, is useful in the study of central and peripheral motoneuronal firing patterns, MU recruitment thresholds, and firing rates.

Similar algorithms have been developed to identify different MUAPs recorded with CNE or MNE during slight to moderate degrees of muscle contraction. At the same time automatic analysis of their parameters can be performed [15]. The method, also known as multi-MUP EMG analysis, is available in some EMG equipment and is supported by a large reference material for many muscles. It represents a modern version of the classical manual MUAP analysis of Buchthal but is incomparably faster and more sensitive in detecting mild pathological changes in neurogenic and primary muscle disorders.

2.4.4 Scanning EMG

This method was designed to study the anatomical organization of the motor unit. A CNE is used to record the electrical activity from a few muscle fibers of a motor unit to trigger the display and the step motor to which the scanning electrode is connected. A position is

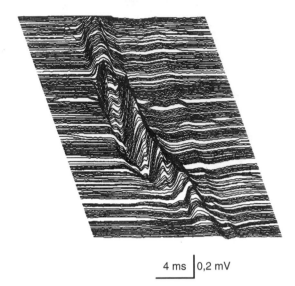

4 ms | 0,2 mV

Figure 2.13. Scanning EMG of a motor unit in a normal biceps brachii. The electrode is pulled through the muscle in 50 μm steps. The MUAP amplitude and shape change considerably. (Reproduced from [13] with permission)

found from which the scanning electrode records activity from the same motor unit as the triggering electrode. The scanning electrode is then pushed through the motor unit territory. Scanning is performed by pulling the electrode back by means of a step motor in 50 μm steps per discharge. The recording gives an electrophysiological cross section through the motor unit [11] (Fig. 2.13). This method has shown "silent areas" and "fractions" of a motor unit, the former of which are seen more commonly in myopathies, apparently due to focal loss of muscle fibers [12]. Scanning EMG has confirmed the asumption that collateral reinnervation does not cross the original fascicular boundaries of the reinnervating motor unit [5].

2.5 PHYSICAL CHARACTERISTICS OF NEEDLE EMG SIGNALS

From the point of view of signal analysis, the electrical activity recorded by needle EMG may be pseudorandom (e.g., so-called interference pattern, which results from a number of independently firing MUAPs, overlapping in time and space) or semiperiodic (e.g., discrete voluntary, or stimulus evoked motor unit activity with recognizable repetitive appearance of individual waveforms). In the former case, statistical methods are applied for signal analysis; in the latter, deterministic parameters describing individual MUAP or other waveforms, namely amplitude, shape and time parameters can be analyzed.

The *frequency content* of the recorded signal depends on several factors. Because of the electrical properties of intramuscular tissues affecting volume conduction, the higher frequencies within the signal will be attenuated significantly more than the lower ones with increasing distance of the AP generator from the active electrode surface. Electrodes with small active surface and high-input impedance have a better high-frequency response characteristic, and vice versa. In contrast to the surface recorded EMG signals, where almost all the frequency content lies below 500 Hz, this limit is at about 2 kHz for the CNE and 10 kHz for the SFEMG electrode. The frequency content of the EMG signal has a physiological significance, and frequency analysis has been used in studies of muscle fatigue, mainly with surface EMG [6], although the same type of changes is seen with needle EMG. In needle EMG, a shift toward high frequencies is a typical feature in myopathies, while a shift toward low frequencies is often prominent in chronic neurogenic conditions.

The amplitude of the recorded signals depends on the same physical factors as the frequency content, the most important being size of the active electrode surface and the distance between the electrode and the AP generator. Thus amplitudes of single-muscle fiber APs recorded with a SFEMG electrode are greatly variable (0.3–10 mV) even when the fibers act as equally strong electrical generators. For this reason SF AP amplitude is not a diagnostically relevant parameter in SFEMG. The MUAP amplitude recorded with CNE, however, is more significant, as the variability related to distance is smaller. With macro EMG, dependence on distance is even much smaller; the macro MUAP amplitude faithfully reflects the strength of the generator.

2.6 RECORDING EQUIPMENT

2.6.1 Principles of Instrumentation (see also Chapter 5)

The last three decades were marked by gradual transition from fully analog instrumentation to digital technology. The earliest EMG equipment consisted of an oscilloscope with high-input impedance amplifier and with an audio amplifier and a loudspeaker as output. Further development was triggered by the introduction of delay lines and averagers, an early step into digital technology, which offered very significant advantages in signal analysis. At that time commercial EMG machines were mainly analog, with a few digital modules. The first fully digitalized equipment contained specially developed digital units, including processors, memory, and dedicated built-in software. The most recent generation, however, is based on general purpose digital technology and software, mostly PC and standard graphics environment. The amplifier thus remains the only analog part. EMG equipment nowadays uses network facilities, including remote data storage and printing, and even remote servicing. The equipment can be tailored to the needs of individual consumers and can easily be upgraded or changed. The same applies to software. The prevailing standard today is based on the PC technology and Windows environment. This relatively open system offers high flexibility but is unfortunately not as reliable as it would be desirable.

2.6.2 Features of EMG Equipment (see also Chapter 5)

Modern digital EMG equipment consists of a mainframe and three peripheral modules: an amplifier, an electrical stimulator, and a dedicated keyboard.

The EMG signals detected require voltage and current amplification. The high-input impedance amplifiers change the incoming weak, low voltage signals into stronger, higher voltage signals suitable for analog to digital (AD) conversion and further processing.

The *amplifier*s are built as differential units, with two electronically symmetrical inputs, so the difference between the two signals is what is actually detected (differential mode). When the signals are detected in the monopolar mode, the other amplifier input is connected to the reference electrode. The components of the amplifiers must be low noise; the two inputs must be as symmetrical as possible. The common mode rejection ratio (CMRR) is a parameter defining the degree of symmetry and thus the ability of the amplifier to reject common signals. An isolation unit based on optical coupling technology is often incorporated for electrical safety; it ensures galvanic separation between the equipment and the patient. Following amplification, the signal undergoes an anti-aliasing filtering and AD conversion; the AD converters are usually placed in the same unit as the amplifiers, and close to the patient in order to minimize the effects of external (mostly power line) electrical disturbances. Most EMG machines are multichannel, and following AD conversion, their signals are multiplexed and serially transmitted to the mainframe, either by copper or optic fiber cable. The signal is displayed on a screen after digital-to-analog (DA) conversion. From the user's point of view, the important parameter is the *system sensitivity*, expressed as input signal amplitude on the display (e.g., mV/screen-division).

The *mainframe* is essentially a PC computer with industry grade motherboards and additional hardware for communication, equipped with the amplifiers and an electrical stimulator. In addition to a standard keyboard, it has a dedicated control panel containing the standard EMG investigation commands, such as audio volume, gain and time base, start-stop of signal acquisition, trigger adjustment, and stimulus controls.

The EMG machine *software* follows the same developmental cycles as the general purpose computer software. A variety of specialized medical diagnostic programs are incorporated, whereby only those actually purchased are activated. Additional programs can be made accessible at a later date by entering activation codes supplied by the manufacturer. One of the developments in the recent decade was that of the knowledge-based expert systems, which should assist and occasionally even guide the user's decisions during clinical investigations. Large databases have been accumulated to support these expert systems.

The *electrical stimulator* is a peripheral unit digitally controlled by the mainframe's software. Like the amplifier, it has an isolation unit to provide safety for the patient. A feedback system measures the actual strength of the delivered stimuli. The stimulus output mode may be either constant voltage or constant current.

Auxiliary connections may be necessary if the EMG machine is used as part of a more complex measurement system, and synchronization of stimulation and acquisition is needed. Higher end equipment usually offers this option. The acquisition control may be performed by the EMG machine or by some other external unit (e.g., magnetic stimulator). In the latter case the EMG equipment can serve as a slave unit acquiring data under the control of an external master unit.

The PC technology itself offers *network connections* that can be used to incorporate the EMG equipment into a wider information system. The EMG data may be stored on a network server, as patient and signal database, and printing may be performed either locally or through a network printer. In addition the network allows remote screen displays, which can be used for teaching purposes or for interpretation by a remote expert (for consultation or a second opinion).

The future development of traditional EMG machines seems to be in the direction of greater hardware independence of the peripheral units from the mainframe, which should allow for easy replacement of the central computer with more advanced types, once they become available on the market.

2.6.3 Features of Digitized Signals

While the digital format of the EMG data offers a number of advantages, the digitization process also involves a modification of the original signal, depending on the resolution of the system. The *time resolution* of EMG equipment is usually based on 512 or 1024 points per sweep, and this is constant in most cases. With 1024 points per sweep, a sweep time of 100 ms means a time resolution of 100/1024 ms (i.e., close to 100 μs). This implies a sampling frequency of 10 kHz, which, according to the Nyquist theorem (which defines sampling rate as at least twice the highest frequency of interest), means an upper frequency limit of the recorded signals at 5 kHz. In most equipment, sampling rate is automatically adjusted to the set sweep time. A better time resolution can only be achieved by shortening the sweep time. Some equipment offers the option of "long sweep mode," in which the sampling rate is increased to improve time resolution.

The *amplitude resolution* depends on the quality of the A/D converter. Older digital equipment contained only 8 bit converters, yielding no more than $2^8 = 256$ amplitude levels on the screen. A 200 mm high screen had a vertical resolution of 200/256 = 0.8 mm/point. This produced steps in traces visible to the naked eye. For this reason 12 or more bit converters are built into more recent equipment, making the traces smooth ($2^{12} = 4096$; 200/4096 = 0.05 mm/point).

The quality of the screen is a further factor limiting the visual quality of the displayed signals. Modern EMG equipment uses standard computer screens with 1600×1280 pixels.

2.6.4 Data Format

Already older digital EMG equipment had made it possible to store the recorded signals on a built-in magnetic memory medium. The format of data storage used to be specially designed, making it difficult to export data for further computer analysis. The most recent generation of EMG equipment makes use of one of the standard data formats or at least allows data export in a standard format.

REFERENCES

1. Buchthal, F., C. Guld, and P. Rosenfalck, "Multielectrode study of the territory of a motor unit," *Acta Physiol Scand* **39**, 83–103 (1957).
2. De Luca, J. C., "Precision decomposition of EMG signals," *Methods Clin Neurophysiol* **4**, 1–28 (1993).
3. Ekstedt, J., and E. Stålberg, "How the size of the needle electrode leading–off surface influences the shape of the single muscle fibre action potential in electromyography," *Comput Progr Biomed* **3**, 204–212 (1973).
4. Jabre, J. F., "Concentric macro electromyography," *Muscle Nerve* **14**, 820–825 (1991).
5. Kugelberg, E., L. Edström, and M. Abbruzzese, "Mapping of motor units in experimentally reinnervated rat muscle," *J Neurol Neurosurg Psychiatry* **33**, 319–329 (1970).
6. Lindström, L., R. Kadefors, and I. Petersen, "An electromyograhic index for localized muscle fatigue," *JAP* **43**, 750–754 (1977).
7. Mihelin, M., J. V. Trontelj, and E. Stålberg, "Muscle fiber recovery function studied with double pulse stimulation," *Muscle Nerve* **14**, 739–747 (1991).
8. Sanders, D. B., E. Stålberg, and S. D. Nandedkar, "Analysis of the electromyographic interference pattern," *J Clin Neurophysiol* **13**, 385–400 (1996).

9. Stålberg, E., "Propagation velocity in single human muscle fibres," *Acta Physiol Scand* **287**(suppl), 1–112 (1966).

10. Stålberg, E., "Macro EMG, a new recording technique," *J Neurol Neurosurg Psychiatry* **43**, 475–482 (1980).

11. Stålberg, E., and L. Antoni, "Electrophysiological cross section of the motor unit," *J Neurol Neurosurg Psychiatry* **43**, 469–474 (1980).

12. Stålberg, E., and P. Dioszeghy, "Scanning EMG in normal muscle and in neuromuscular disorders," *Electroencephalogr Clin Neurophysiol* **81**, 403–416 (1991).

13. Stålberg, E., and B. Falck, "The role of electromyography in neurology," *Electroencephalogr Clin Neurophysiol* **103**, 579–598 (1997).

14. Stålberg, E., and G. Grimby, "Dynamic electromyography and biopsy changes in a 4 year follow up: Study of patients with history of polio," *Muscle Nerve* **18**, 699–707 (1995).

15. Stålberg, E., S. D. Nandedkar, D. B. Sanders, and B. Falck, "Quantitative motor unit potential analysis," *J Clin Neurophysiol* **13**, 401–422 (1996).

16. Stålberg, E., M. S. Schwartz, B. Thiele, and H. Schiller, "The normal motor unit in man," *J Neurol Sci* **27**, 291–301 (1976).

17. Stålberg, E., and M. Sonoo, "Assessment of variability in the shape of the motor unit action potential, the 'jiggle,' at consecutive discharges," *Muscle Nerve* **17**, 1135–1144 (1994).

18. Stålberg, E., and J. V. Trontelj, *Single fiber electromyography: Studies in healthy and diseased muscle*, 2nd ed, Raven Press, New York, 1994.

19. Stålberg, E., and J. V. Trontelj, "The study of normal and abnormal neuromuscular transmission with single fibre electromyography," *J Neurosci Methods* **74**, 145–154 (1997). (Quoted in legend to Fig. 2.8, part of which is reproduced with permission from Elsevier Science.)

20. Trontelj, J. V., "Muscle fiber conduction velocity changes with length," *Muscle Nerve* **16**, 506–512 (1993).

21. Trontelj, J. V., "A study of the H-reflex by single fibre EMG," *J Neurol Neurosurg Psychiat* **36**, 951–995 (1973).

22. Trontelj, J. V., and E. Stålberg, "Responses to electrical stimulation of denervated human muscle fibers recorded with single fiber EMG," *J Neurol Neurosurg Psychiatry* **46**, 305–309 (1983).

23. Trontelj, J. V., and E. Stålberg, "Single motor end plates in myasthenia gravis and LEMS at different firing rates," *Muscle Nerve* **14**, 226–232 (1991).

24. Trontelj, J. V., and E. Stålberg, "The effect of firing rate on neuromuscular jitter in Lambert-Eaton myasthenic syndrome," *Muscle Nerve* **15**, 258 (1992).

25. Trontelj, J. V., and E. Stålberg, "Single fiber and macro electromyography," in T. Bertorini, ed., *Textbook of electromyography*, Butterworth–Heineman, New York, 2002.

26. Zidar, J., J. V. Trontelj, and M. Mihelin, "Percutaneous stimulation of human corticospinal tract: A single fiber EMG study of individual motor unit responses," *Brain Res* **422**, 196–199 (1987).

3

DECOMPOSITION OF INTRAMUSCULAR EMG SIGNALS

D. W. Stashuk

Department of Systems Design Engineering
University of Waterloo, Ontario, Canada

D. Farina

Laboratory for Engineering of the Neuromuscular System
Department of Electronics, Politecnico di Torino, Italy

K. Søgaard

Department of Physiology
National Institute of Occupational Health, Copenhagen, Denmark

3.1 INTRODUCTION

As the smallest functional parts of the muscle, motor units (MUs) are the basic building blocks of the neuromuscular system. Monitoring MU recruitment, derecruitment, and firing rate, by either invasive or surface techniques, leads to the understanding of motor control strategies and of their pathological alterations. "EMG signal decomposition" is the process of identification and classification of individual motor unit action potentials (MUAPs) in the interference pattern detected with either intramuscular or surface electrodes so that the constituent motor unit trains that make up the signal can be determined (Fig. 3.1).

When needle or wire detection techniques are used, this process is complex because the different MUAPs show different degrees of overlapping in time and may change shape because of changing MU properties or MU-electrode relative location. When surface elec-

Electromyography: Physiology, Engineering, and Noninvasive Applications, edited by Roberto Merletti and Philip Parker.
ISBN 0-471-67580-6   Copyright © 2004 Institute for Electrical and Electronics Engineers, Inc.

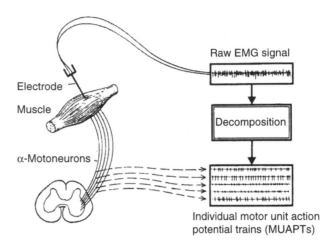

Figure 3.1. Schematic representation of the detection and decomposition of the intramuscular EMG signal. (From [8] with permission)

trode systems are used, this process is considerably more complex because the surface MUAPs, filtered through the equivalent tissue transfer function, strongly reduces the shape differences between MUAPs and increases their duration, thereby increasing the likelihood of superposition and the difficulty of shape classification.

Some techniques, such as spike-triggered averaging, combine the advantages of intramuscular and surface detection by using the first to detect and classify MUAPs and the second to estimate features of single MUs associated to such MUAPs. In many ways invasive and surface approaches complement each other and are not alternative. They provide different views of the muscle and different types of information concerning the central controller and the peripheral actuator.

Since the pioneer work completed by De Luca and his co-workers both in the development of detection needles, processing techniques and in the use of EMG signal decompositions for understanding fundamental physiological issues [12,13,30,31], many techniques have been developed to implement this process for intramuscularly detected signals with various degrees of automation [57]. These methods involve the application of complex signal processing and pattern recognition techniques and in most cases an interaction with an operator.

This chapter is devoted to the description of the techniques for resolving the interference intramuscular EMG signal into its constituent MUAP trains (MUAPTs), to the issue of validation of the performance of these techniques and to their applications.

3.2 BASIC STEPS FOR EMG SIGNAL DECOMPOSITION

The resolution of a composite EMG signal into its significant, constituent MUAPTs requires the ability to detect the discharges (i.e., MUAPs) of the MUs significantly contributing to the composite signal and to correctly associate each detected MUAP with the MU that generated it. EMG signal decomposition therefore involves the two basic steps of detecting MUAPs and recognizing detected MUAPs. To identify the occurrences of

MUAPs within a composite signal or to segment the composite signal into portions that contain significant MUAPs, common characteristics of MUAP shapes, which differentiate them from background signal noise, are used. To recognize detected MUAPs, it is essential that the MUAPs produced by the same MU be more similar in shape than MUAPs produced by different MUs, that differences in MUAP shapes can be quantified, and that the distribution of the shapes of the MUAPs of each active MU can be characterized. The inherent variability of the MUAPs produced by a single MU and the inherent difference between the MUAPs of different MUs will depend on aspects of muscle physiology and on the methods used to detect the EMG signal. The features used to represent MUAPs and the distance measures used to compare MUAPs affect the resolution with which differences in MUAP shapes can be determined. Both clustering and supervised classification techniques are used to characterize the distributions of MUAPs of different MUs. Furthermore MUAPs from several MUs can overlap in time or superimpose, and therefore their superimposed waveforms need to be resolved. Finally, because resulting candidate MUAPTs represent the activity of MUs, information based on their temporal pattern is useful. Successful EMG signal decomposition can therefore involve the following steps: (1) EMG signal acquisition, (2) MUAP detection or signal segmentation, (3) MUAP representation, (4) MUAP clustering, (5) supervised MUAP classification, (6) resolution of superimposed MUAPs, and (7) the discovery of temporal relationships between MUAPTs.

3.2.1 EMG Signal Acquisition

Choice of Electrodes. The electrodes used to detect composite intramuscular EMG signals that are to be decomposed should be configured so that they will have a strong chance of acquiring uniquely shaped MUAPs for each active MU. This can be accomplished by using electrodes with small detection surfaces such that the fibers of different MUs will be at unique distances from the electrode surface. However, as a detection surface becomes smaller, surface movements will cause larger changes to the shapes of MUAPs detected. This nonstationarity can make MUAP recognition difficult if not impossible. On the other hand, as a detection surface increases, more MU fibers of each MU become essentially equidistant from the detection surface and the MUAPs generated by different MUs become more similar. Consequently concentric, monopolar, fine-wire or single-fiber needle electrodes are preferred, with the latter being most sensitive to electrode movement. In order to increase the probability of detecting unique MUAPs for each MU, special multichannel needle electrodes can be used [10,30,31].

Protocol for Signal Detection. The complexity of a detected EMG signal and the ease with which it can be decomposed depend on the type of electrode, electrode positioning, profile of muscle contraction, and muscle selected. The electrode should be positioned so that it is close to active muscle fibers and detects MUAPs of maximum amplitude and sharpness in order to maximize the relative differences in the distances between the fibers of different MUs and the electrode surface. This way MUAPs that are distinct from the background noise can be detected. The recommended procedure is to initially position the electrode in a minimally contracting muscle to detect MUAPs of maximum amplitude and sharpness, and then to increase muscle contraction as isometrically as possible and initiate data acquisition once the contraction is at the desired level. If the decomposition system can process signals acquired during force-changing contractions, data acquisition should start immediately after needle positioning.

The distribution of the MUAP shapes of each active MU must be characterized in order for the detected MUAPs to be recognized. This consequently requires that MUAPs of each MU be individually detected (i.e., not in superposition with the MUAPs of any other MUs) several times. This places constraints on methods and protocols for detecting decomposable EMG signals. Thus the peak level of force is often below 50% of maximal voluntary contraction (MVC). However, signal complexity at a certain %MVC level of contraction is related to a number of physiological factors such as the number of MUs in the muscle, the muscle fiber diameter, the density of MU muscle fibers, the MU recruitment thresholds, and the MU rate coding strategies. Smaller muscles, such as the first dorsal interosseous, compared to larger muscles, such as the biceps brachii, have fewer MUs, tend to have a narrow as opposed to a broad range of recruitment thresholds, and use large amounts of rate coding. Therefore, for comparable %MVC contractions, smaller muscles will on average generate more complex signals than signals detected in larger muscles [2]. In addition the need to minimize needle movement usually requires that the rate of change of force be less than 10% MVC/s.

Interesting EMG Signal Attributes. The ideal EMG signal decomposition system should be able to successfully process signals with the following attributes: (1) five or more MUAPTs, (2) nonstationary MUAP shapes, usually due to electrode movement, (3) variable MUAP shapes due to variability in the operation of neuromuscular junctions and background biological noise due to the activity of other MUs, (4) two or more MUs with similar MUAP shapes, (5) frequent superpositions of MUAPs, (6) nonstationary MU firing pattern statistics, and (7) intermittent recruitment and decruitment of MUs.

3.2.2 Detecting MUAPs or Signal Segmentation

In theory, complete EMG signal decomposition requires the detection of all MUAPs generated by MUs active during signal acquisition. In practice, however, there are many MUAPs produced by MUs with no fibers close to the detection surface. These MUAPs are generally small, primarily of low-frequency content, and similarly shaped. Therefore, it is difficult to consistently assigning such MUAPs to their correct MUAPT and it is easy to miss the small MUAPs when they occur in close temporal proximity to larger MUAPs. Consequently, it is more useful to only detect MUAPs that can be consistently correctly assigned. MUAP detection usually involves calculating, for each sample of the composite signal acquired, a statistic and comparing its value to a pre-set threshold. Some of the signal statistics used include the raw or bandpass-filtered signal amplitude [5,16,30,40,42] or variance [22,38,64,65], or a combination of both raw signal slope and amplitude [56,58]. When the threshold value is exceeded, a candidate MUAP can be defined as a fixed length section of a neighboring signal or a variable-length signal section, assumed to possibly contain several significant MUAP contributions [21,22,29,33,38,64,65]. Any selected signal section may be an isolated MUAP, a superposition of MUAPs from two or more MUs, only a portion of a single MUAP, or a spurious noise spike. Therefore, before further processing, it is required that the composition of a detected section be determined, and properly aligned and represented.

Bandpass filtering or low-pass differentiation, in general, shortens the duration of MUAPs, which in turn reduces their temporal overlap and the number of superimposed waveforms. This type of signal conditioning also reduces the amplitude of the many similarly shaped MUAPs of different MUs that do not have fibers close to the electrode. In addition, by reducing the variability of the MUAP shapes, it removes nondiscriminative,

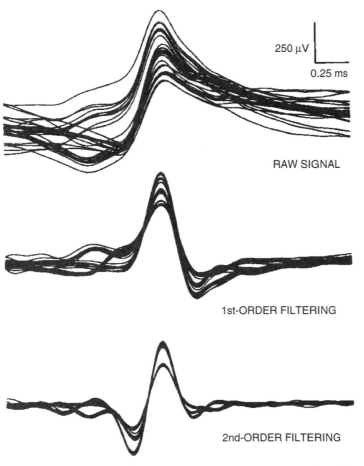

250 µV

0.25 ms

RAW SIGNAL

1st-ORDER FILTERING

2nd-ORDER FILTERING

Figure 3.2. Effects of bandpass filtering on the shapes and distinguishability of MUAPs. Low-pass differentiation reduces MUAP durations and allows the MUAPs created by each of the three contributing MUs to be more easily discriminated. (From [34] with permission)

low-frequency information to make it easier to discriminate between the MUAPs of different MUs. Therefore the slope of the signal [21,29,33] or the signal after it has been passed through a lowpass differentiator [25,26,28,34,35,37,57] can be used to select segments that contain MUAPs that can be consistently correctly assigned. Figure 3.2 shows the effects of first- and second- order difference filtering. After filtering, the MUAPs have shorter durations and some of the nondiscriminative baseline noise is removed. The difference filters suggested by McGill [35] are structured so that high-frequency noise (above ~3 kHz) as well as low-frequency signal components (below ~1 kHz) are suppressed. The second-order filter attenuates low-frequency signal components more than does the first-order filter (see [35] for more details). With regard to facilitating MUAP detection and classification, the phase responses of these filters are equally, or more important, than their amplitude responses because of the relevance of phase in the definition of shape.

Absolute detection-threshold values used to select MUAPs can have a fixed set of characteristics across several analyzed signals [21,42,56,58]. Alternatively, threshold values can be relative to the noise [28,29,33,35,38] or the largest signal components

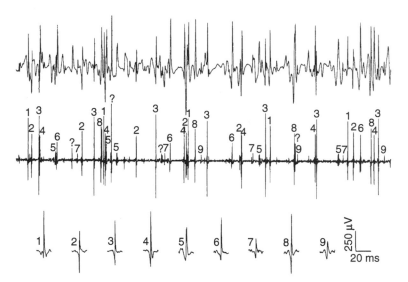

Figure 3.3. Low-pass differentiation can facilitate the detection of MUAPs. A concentric-needle EMG signal segment detected in a biceps-brachii muscle (*top trace*). The same segment after second-order low-pass differentiation and with the detected MUAPs (*middle trace*). The MUAP waveforms of the MUs significantly contributing to the shown EMG segment (*bottom trace*). (From [34] with permission)

[5,22,36,57] of each composite signal. The latter approach reduces the chance of small MUAPs being "lost" among larger ones but also introduces a possible bias against smaller MUAPs. No matter how the MUAPs are detected, they must be somehow properly aligned. Detected waveforms initially aligned using peak values either of the raw or slope signal can sometimes have major alignment errors that cannot be corrected during later alignment attempts and can cause classification problems. Improved methods of alignment during segmentation, when there is little contextual information, need to be developed and evaluated. Figure 3.3 shows a filtered and segmented section of an EMG signal. In this example the detected MUAPs are also classified. The number above a detected MUAP represents the MUAPT or MU to which it was assigned.

3.2.3 Feature Extraction for Pattern Recognition

For detected MUAPs to be automatically recognized using pattern recognition algorithms, they must be represented by a feature vector in a multidimensional feature space. Some of the important factors to consider when selecting features include (1) the storage and computational requirements of the feature vectors, (2) the computational effort required to extract or compute the feature values, (3) the signal-to-noise ratio and variance of each feature, (4) the amount of correlation between features, (5) the discriminative power of the feature representation, (6) the sensitivity of the representation to waveform alignment, (7) the effect of MUAP superpositions on the feature values, and (8) the effect of MUAP shape variability on the feature values.

A number of different feature spaces can represent MUAPs for signal decomposition. Some of these include (1) morphological statistics such as peak-to peak voltage, number of phases, duration, and number of turns, [22,33,42,56], (2) Fourier transformation coefficients [35,58], (3) coefficients obtained from other transformations (e.g., a wavelet basis) [16,64,65], (4) the time samples of the bandpass-filtered signal [5,21,22,28,29,30,38, 40], and (5) the time samples of the low-pass differentiated signal [25,36,57].

3.2.4 Clustering of Detected MUAPs

A clustering algorithm used for EMG signal decomposition has two main objectives. One is to determine the correct number of MUs contributing significant MUAPs (i.e., determining the correct number of MUAPTs). The second is to assign as many MUAPs as possible to their correct MUAPT so that, for each contributing MU, the distribution of the shapes and times of occurrence of its MUAPs can be estimated along with its prototypical or mean MUAP shape. To meet the objectives, the clustering algorithm attempts to partition a set of patterns or objects into the a priori unknown, correct number of groups or clusters by assigning each member to a specific group. The members of any one group should be more similar to each other than they are to the members of any other group, and each group can be represented by a prototypical shape or template calculated based on its membership. However, since each incorrectly assigned MUAP increases the probability of more assignment errors and supervised classification results (see Section 3.2.5 below) will be more successful if based on accurate clustering results, it is more important that the clustering results are accurate and that superimposed MUAPs are ignored than it is for all detected MUAPs to be assigned. A clustering algorithm therefore does not need to consider all of the detected MUAPs nor classify all MUAPs considered. Rather, it should be conservative and try to minimize the number of erroneous classifications while focusing on correctly determining all of the MUAPTs significantly contributing to the signal. Furthermore, because it is more likely to merge two trains into one than it is to split one train into two during later stages of signal decomposition, overestimation of the number of MUAPTs is preferred to underestimation.

Distance Measures. To be able to make classification decisions, pattern recognition algorithms must use some quantitative measure of the similarity or dissimilarity between objects of a set. For objects represented in a feature space, the most often-used dissimilarity or distance measure is the Euclidean distance. Let $\mathbf{X}_i$ denote the ith N-dimensional pattern or object (MUAP)

$$\mathbf{X}_i = (x_{i1}, x_{i2}, \ldots, x_{iN})^T$$

the Euclidean distance between two objects (X_i, X_k) in an N-dimensional feature space is then

$$d^2(\mathbf{X}_i, \mathbf{X}_k) = (\mathbf{X}_i - \mathbf{X}_k)^T(\mathbf{X}_i - \mathbf{X}_k) = \sum_{j=1}^{N}(x_{ij} - x_{kj})^2 \qquad (4)$$

Let $\mathbf{M}_k$ denote the N-dimensional mean (centroid) or prototype pattern (MUAP) of the kth cluster (MUAPT)

$$\mathbf{M}_k = (m_{k1}, m_{k2}, \ldots, m_{kN})^T$$

The Euclidean distance between an object and a cluster mean in an N-dimensional feature space is then $d^2(\mathbf{X}_j, \mathbf{M}_k)$.

Classical Clustering Techniques. Clustering can be achieved by using either a hierarchical or partitioning strategy. Hierarchical methods consider the complete set of objects at the same time and can be either agglomerative or divisive [27]. An agglomerative hierarchical algorithm initially considers each object as a cluster, computes the distances between all pairs of objects, and stores the results in a distance or proximity matrix. The two most similar objects are combined into one cluster and replaced by their mean shape. The distance matrix for the reduced number of objects (clusters) is then recalculated. This process of joining objects and recalculating the distance matrix is repeated until only a single object (cluster) remains. Tracking this process creates a *dendrogram*, which is a binary tree with a distinguished root. Cutting the dendrogram at some level of dissimilarity creates K clusters with all of the objects represented by each of the K remaining means assumed to belong to each respective cluster. Figure 3.4, from Jain [27], presents examples of distance matrices and their associated dendrograms using single and complete linkage distance metrics for a hypothetical data set containing five objects.

A divisive hierarchical algorithm starts with all of the objects in one cluster and creates a dendrogram by continuing to split the clusters, based on a distance metric, until each object forms a single cluster. Because hierarchical methods consider all of the data at once, they can be computationally expensive, especially for large data sets, but the initial order of the objects in the set does not affect results. Distance metrics used for hierarchical clustering are either complete or single linkage metrics. A complete-linkage distance metric considers the similarity between an object to be assigned to a cluster and each or all of the members of a cluster. For example a complete-linkage metric is the farthest-neighbor distance, which is the distance from the candidate object to the least similar object already assigned to a cluster. Another complete-linkage metric would be the distance between the candidate object and the mean of the cluster. A complete-linkage metric is dependent on the current membership of a cluster and therefore changes as members are added to or deleted from a cluster. Using a complete-linkage metric tends to create compact clusters. A single-linkage distance metric only considers the similarity or distance between the candidate object and one member of a cluster, like the nearest-neighbor distance. Single-linkage metrics do not change as the membership of a cluster changes and their use tends to create elongated clusters. Various complete-linkage hierarchical techniques can differ based on how they combine objects and compute cluster means. In addition single-linkage metrics can be extended to consider more than one cluster member such as by using an N-nearest-neighbor metric.

Partitioning methods create the clustering set partition $(C_1, C_2, \ldots, C_K)$, where C_i represents the set of objects that belong to the ith cluster that minimizes a specified objective function. Often the function to be minimized is the squared error defined as

$$E_k^2 = \sum_{k=1}^{K} e_k^2 = \sum_{k=1}^{K} \sum_{X_i \in C_k} (\mathbf{X}_i - \mathbf{M}_k)^T (\mathbf{X}_i - \mathbf{M}_k) \qquad (6)$$

In all partitioning methods, K, the number of partitions or clusters and initial estimates of the cluster means are required. Once these are obtained, each object is assigned to the cluster with the most similar mean.

$$\mathbf{X}_i \in C_c \quad \text{if} \quad d^2(\mathbf{X}_i, \mathbf{M}_c) < d^2(\mathbf{X}_i, \mathbf{M}_k), \qquad k = 1, 2, \ldots, K, k \neq c \qquad (7)$$

The mean value of each cluster is updated after each assignment or at the end of a pass through all of the data. Multiple passes through the data are required until the cluster means

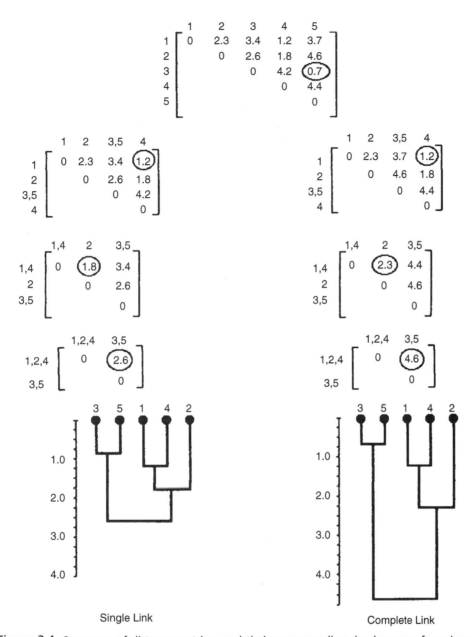

Figure 3.4. Sequence of distance matrices and their corresponding dendrograms for a hypothetical data set of five objects. The matrices include the distances between each pair of objects as measured using a single linkage metric (*left*), such as nearest neighbor, or a complete linkage metric (*right*), such as farthest neighbor. At each successive step, the closest object pair is removed and replaced by their mean value until only one pair of objects remains. The dendrograms represent the sequence of object combinations (i.e., cluster formation). Cutting a dendrogram at a specific level results in a specific clustering. (From [27] with permission)

are suitably stable. With partitioning methods, the order in which the objects are classified can change the results. However, only the cluster means and one object need to be
considered at any one time, which reduces the computational complexity compared to hierarchical methods. Sequential clustering techniques, such as leader-based clustering [61],
can be considered special cases of the partitioning method. Sequential clustering techniques make a single pass through the data using the initial object considered as an initial
cluster template and fixed thresholds of similarity to assign subsequent objects to the
closest template, to update the current template, or to use the current object being classified to initialize a new template.

Clustering Algorithms Used for EMG Signal Decomposition. Most clustering algorithms used for EMG signal decomposition have been based on the single-linkage
or nearest-neighbor concepts [5,9,16,21,24,28,29,33,38,64,65]. Pattichis et al. [42] used a
modified complete-linkage technique that was not based on traditional distance metrics
but rather used successively applied range criteria for each MUAP feature considered.
Others have used partitioning methods such as a modified K-means technique [61] and the
leader-based approach [28,30,35,36,56]. Self-organizing neural networks with learning
vector quantization have been used [5] as well as a recurrent neural network [25,26]. For
many of these applications the standard Euclidean distance is normalized in some way by
the energy of the waveforms being compared [5,24,28,29,30,33,64,65]. Guiheneuc et al.
[22] and McGill et al. [35] instead adjusted acceptance thresholds based on MUAP size.
Distance normalization and threshold adjustment account for the higher biological
variability expected for larger MUAPs. Stashuk and Qu [61] considered a portion of the
detected MUAPs and actively used MUAP shape as well as firing pattern information, if
available, to determine the number of MUAPTs and to make classification decisions. The
algorithms used to calculate the firing pattern information can specifically deal with sparse
MUAPTs that will have missed MU discharges and possibly include erroneous MU discharges [62]. LeFever and Deluca [30], via a hazard function, also actively used firing
pattern information. Many other methods passively use firing pattern information to
detect misclassifications, correct classification errors, and possibly merge MUAPTs
[21,22,24,25,26,28,33,35,58]. Several other distance-metric modifications have been used.
Gerber et al. [21] used an L_1 distance measure that is area instead of energy based.
Nandedkar et al. [36] first compared maximum slope and amplitude of the MUAPs and,
if suitably similar, then calculated their distance measure. Fang et al. [16] used only the
maximal difference between corresponding wavelet transform coefficients, while Stashuk
and Naphan [59] compared probabilistic inference-based classification with classical
distance measures.

One of the most important aspects of any clustering algorithm is the thresholds used
for cutting dendograms to determine the number of clusters or for making decisions to
classify candidate MUAPs or to use them to initiate new templates. This is because of the
variety of EMG signals that may be encountered and the fact that superimposed MUAP
waveforms need to be identified for subsequent resolution and must not be used to update
cluster templates. In this regard many methods use fixed a priori determined thresholds
based on shape similarity [5,30,35,42,56], firing pattern [16], or both [29,36,58]. Alternatively, some methods use thresholds that are dependent on the specific EMG signal being
decomposed [21,25,33,38,61]. In general, data-driven thresholds should lead to performance that is more robust across a variety of EMG signals. Nonetheless, in most cases a
MUAPT must have a minimum number of occurrences, or its members are considered to
be superimposed waveforms. Figure 3.5 presents typical clustering results.

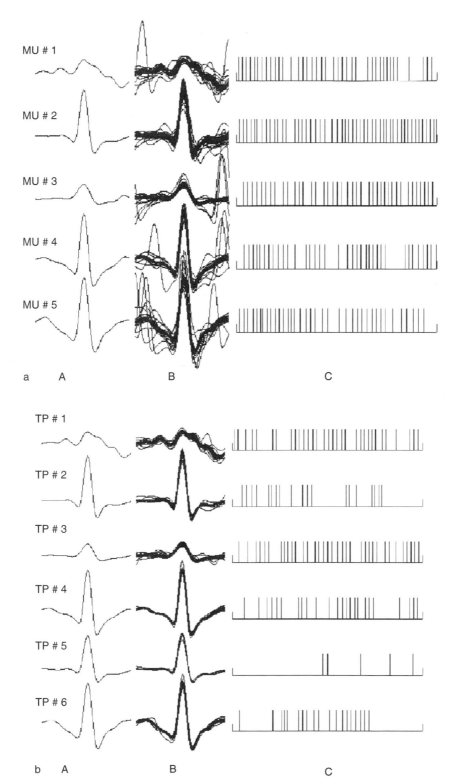

Figure 3.5. Example output of a clustering algorithm that uses both shape and firing pattern information. Manually determined gold standard results (a) for a 5s analysis interval: (A) template plots, (B) shimmer plots of assigned MUAPs, and (C) firing time plots. Gaps in the firing time plots are caused by unresolved MUAP superpositions. (b) The clustering algorithm results (A), (B), and (C) as in panel (a). Clusters 2 and 5 are duplicates and together represent the activity of MU #2. They were created due to signal nonstationarity and could be merged by a subsequent supervised classification algorithm. (From [61] with permission)

3.2.5 Supervised Classification of Detected MUAPs

Most of the reported EMG signal decomposition methods use only clustering techniques to classify all of the individual MUAPs. Only a few use a supervised classification algorithm. Each of these algorithms uses the results of an initial clustering stage as a training set to assist in making supervised classification decisions. Supervised classification involves the use of sets of assigned objects (training sets) in which there is a known number of expected classes. Training sets allow the characterization of the properties of each class and in turn can lead to a more efficient assignment of unassigned objects. Therefore the complete set of detected MUAPs can be classified using supervised classification techniques once estimates of the number of active MUs and the distributions of their MUAP shapes, including their prototypical MUAPs, are available (these estimates are usually based on clustering results). Several decomposition techniques include the application of supervised classification methods following the application of clustering or unsupervised methods [9,21,22,25,26,40,57]. The features used to represent MUAPs for supervised classification are often the same as those used for clustering [21,57]. However, some supervised classification schemes use expanded feature sets [9,22].

Possible Steps for Supervised Classification
1. Check the clustering results for obvious errors.
2. Select the most discriminative feature space and/or features.
3. Derive effective prototypes for each class.
4. Derive thresholds for shape similarity that suit the data.
5. Estimate the firing pattern statistics for each class.
6. Attempt to classify MUAPs unassigned in the clustering phase given shape and firing pattern statistics for each class.
7. Given new classifications, update MUAP shape prototypes and firing pattern statistics for each class.
8. Merge classes that are from presumably the same generating MU (e.g., see Fig. 3.5b).
9. Split classes that contain MUAPs from presumably more than one MU.
10. Repeat steps 6, 7, 8, and 9 until no more classifications are made.

Possible Factors Affecting Supervised Classification.
The variability of the MUAPs within a MUAPT as well as the similarity of MUAPs from different trains greatly affects the performance of supervised classification algorithms. The two factors that cause MUAP shape variability are additive noise and signal nonstationarity. Biological interference, or the energy produced by the MUAPs of other MUs that are active at the same time or nearly the same time as the candidate MUAP, comprise the majority of the signal noise. This includes MUAP superpositions created by the relatively few MUs with fibers close to the detection electrode, as well as the more temporally diffuse background noise contributed by the larger number of MUs located throughout the muscle. Biological variability is the second source of noise. It is due to the variable amounts of time required at a neuromuscular junction to depolarize its muscle fiber membrane, *jitter* [48], which in turn varies the temporal dispersion among the muscle fiber potentials of a MUAP. Therefore shape of the MUAP will vary from discharge to discharge. Biological variation, or *jiggle* [55], arises entirely from within a given MU and is evident even if only a single MU

is active. The third and smallest source of noise is due to the instrumentation required to amplify the detected voltages. Signal nonstationarity, characterized by a trend in the change of the shapes of the MUAPs to either larger or smaller potentials, occurs when the electrode moves relative to the fibers. Despite the several sources of MUAP shape variability, the MUAPs of two different MUs can still be remarkably similar if the MUs have similar muscle fiber geometries. This is more likely if they each have a relatively small number of very close fibers. One of the main challenges for supervised classification algorithms is to accommodate MUAP shape variability within a MUAPT and yet consistently discriminate between similarly shaped MUAPs generated by different MUs.

The use of firing pattern information can also significantly affect the performance of supervised classification algorithms. The successful passive or active use of firing pattern information requires the consideration of factors that affect the validity of any calculated statistics. Several factors that affect the usefulness of firing pattern statistics include (1) the classification error rate, (2) the classification identification rate, (3) the minimum number of interdischarge intervals (IDIs) in the estimation sample, and (4) the inherent variability of the firing pattern in the estimation interval.

Therefore in computing the length of the estimation intervals used to determine firing pattern statistics, one must consider the stationarity of the firing patterns. This is obviously most important when analyzing signals detected during force variable contractions. In addition the degree to which firing pattern information determines a classification decision should match the level of confidence in the information used.

Desired Attributes of a Supervised Classification Algorithm

1. Robust performance across a variety of suitable EMG signals with minimal computational cost.
2. Ability to perform despite having small, imperfectly labeled training sets after clustering.
3. Ability to decompose despite background noise and biological variability.
4. Minimal use of arbitrary thresholds of similarity and minimal sensitivity to any required thresholds.
5. Prudent use of firing pattern information to complement shape-based classification.
6. Robust use of shape and temporal information given nonstationarity.
7. Multi-pass algorithm to allow classification context to grow iteratively.
8. Valid stopping criteria for iterative process.
9. Ability to merge classes that correspond to the same MU.
10. Ability to split classes that correspond to more than one MU.
11. Ability to decouple low-assignment rate versus low-error rate trade-off as much as possible.
12. Accuracy and completeness of classification.
13. Ability to supply some measurement of confidence regarding the classification results.

Supervised Classification Algorithms.
Both Gerber et al. [21] and Guiheneuc et al. [22] used a single-pass supervised classification algorithm to classify individual MUAPs discovered in significant variable-length signal segments. Hassoun et al. [25,26] used a trained neural network to classify individual MUAPs. Gut and Moschytz [23]

modeled the EMG signal as an $[M + 1]$-ary signaling system with intersymbol interference. They use a sparse-sequence constrained Viterbi algorithm and MUAP shape and firing pattern information to obtain a maximum a posteriori probability based estimate of the innervation sequences of the active MUs. Finally, Stashuk and Paoli [60] have developed a multi-pass certainty-based supervised classification algorithm that uses both shape and firing pattern information and which has many of the attributes listed above.

Certainty-Based Classification. The certainty of a particular MUAP assignment roughly approximates the probability of it being correct. Certainty is defined as the product of the results of three decision functions, each with a range of values from 0 to 1, that combines shape and firing-pattern information. Therefore, like probability, certainty has a range of values from 0 to 1. However, certainty does not satisfy other necessary properties of a true probability measure. Nonetheless, using certainty, one evaluates MUAP classifications based on the likelihood that they are correct. Only decisions deemed to be the best possible and to have requisite certainty are executed. Any candidate MUAP assignment with a calculated certainty below a threshold value is not made. Many of the unassigned MUAPs are actually superimposed MUAPs. The threshold of certainty affects the number and accuracy of decisions made, but the performance of the algorithm may not be highly sensitive to the certainty threshold used.

One decision function calculates a relative Euclidean similarity. This is a measure of the distance of a candidate MUAP to its closest MUAPT template relative to the distance of the candidate MUAP to its second closest MUAPT template. A second function measures the absolute normalized Euclidean similarity between a candidate MUAP and a prospective MUAPT template. A third decision function measures the firing time consistency of the candidate MUAP relative to the firing pattern of a prospective MUAPT. The certainties of assigning a candidate MUAP to its closest MUAPT and to its second closest MUAPT are calculated by multiplying the respective decision function values. The MUAP is assigned to the MUAPT which has the greatest certainty value provided the value is greater than the certainty assignment threshold. Figure 3.6a and b, from Paoli [41], displays contour maps of the relative similarity and the product of relative similarity and normalized similarity to the closest template, respectively, for a simulated two-dimensional feature space.

Multiple passes through the set of detected MUAPs are made. With each pass the certainty-based assignment of each MUAP is considered. During initial passes, limited temporal information is available and only MUAPs with shapes similar to the template of a train are assigned. During later passes, as more temporal information is developed, the consistency of classifications with the established firing patterns become more important and more difficult assignment decisions can be made or earlier decisions modified. The iterations continue until a maximum number of iterations have been completed or until the percentage of MUAP assignment changes in total and the maximum percentage of changes within any specific train are both below specified threshold values.

Attempts to track MUAP shape's nonstationarity are made by calculating the MUAPT templates as moving averages of their assigned MUAPs using the certainties with which the respective MUAPs are assigned as weighting factors. At the end of each assignment pass at most two trains are merged if the average certainty of all the classifications within the merged train is greater than the average certainty of all the classifications of each individual train. Figures 3.7 and 3.8 demonstrate the capability of the certainty algorithm to discriminate between very similar MUAPs from different MUAPTs and to allow a large amount of MUAP variability in a single MUAPT.

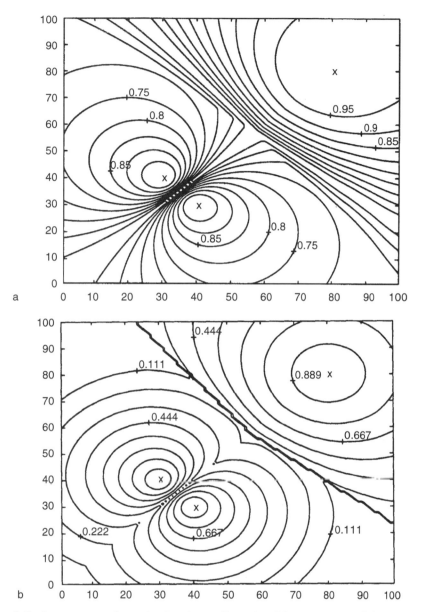

Figure 3.6. Contour maps for a simulated two-dimensional feature space of the relative similarity (*a*) and the product of relative similarity and normalized similarity (*b*) to the closest template. The relative and normalized similarities range from 0 to 1. In this example an object to be classified is represented as a point located at a specified *x* and *y* value. The lines connect points (potential objects) with equal similarity with respect to the closest of the three means (*x*) . (From [41] with permission)

MUAP Raster Plots - Similar MUAPs in Two Different MUAPTs

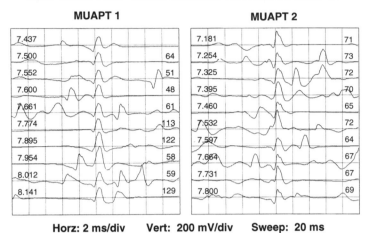

Horz: 2 ms/div   Vert: 200 mV/div   Sweep: 20 ms

Figure 3.7. Raster plots of two different MUAPTs, with very similar MUAPs. Time of MU discharge in seconds is on the left and distance of the firing from the previous, in ms, on the right. Because the MUAPs in each train are similar, both shape and firing pattern information is required to consistently discriminate between the MUAPs created by each MU. (From [57] with permission)

MUAP Raster Plot - Biological Shape Variation

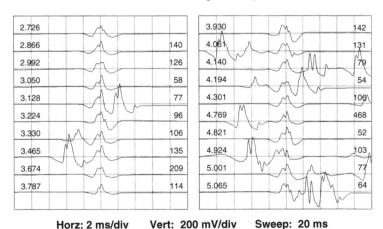

Horz: 2 ms/div   Vert: 200 mV/div   Sweep: 20 ms

Figure 3.8. Raster plots showing the MUAPs of a single MUAPT. The variability in the MUAP shapes is due to biological variability in the times of arrival at the electrode detection surface of the individual muscle fiber action potentials significantly contributing to the detected MUAPs. Relative and absolute shape criteria, firing pattern information, and data-driven assignment thresholds help in discovering this train. (From [57] with permission)

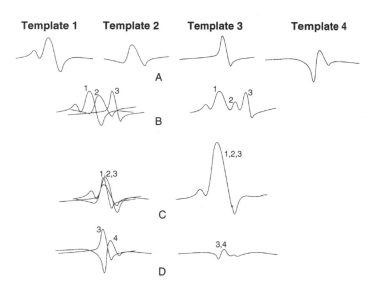

Figure 3.9. MUAP waveforms superimposed in different ways are shown: (A) MUAP waveforms; (B) partially superimposed waveforms; (C) completely supeimposed waveforms; (D) destructively superimposed waveforms. (From [15] with permission)

3.2.6 Resolving Superimposed MUAPs

If more than one MU discharges at the same time or in close temporal succession, a superimposed MUAP, which is the algebraic summation of the respective MUAPs, is detected. Three different types of superimposed MUAPs may be defined. *Partially superimposed* MUAPs overlap without their peaks being obscured. *Completely superimposed* MUAPs have peaks that combine to make one large peak. *Destructively superimposed* MUAPs combine in such a way that their out-of-phase peaks sum together to partially or completely cancel each other. Figure 3.9 provides examples of the various types of superimposed MUAPs. In the clinical use of EMG signal decomposition systems, the resolution of superimposed MUAPs is not a highly critical requirement if only mean firing rate and firing rate variability during constant contractions are to be studied [62]. However, as the use of firing-pattern information becomes more important, the clinical importance of resolving superimposed MUAPs will increase. In addition, as computing power available in the clinic increases, it will become more practical to consider resolving superimposed MUAPs. Nonetheless, for the detailed analysis of MU firing patterns required for the study of the basic mechanisms of MU control, superimposed MUAPs must be resolved.

There are fundamentally two different strategies for resolving superimposed MUAPs. The first, based on matching MUAPs, one at a time, with the superimposed waveform or a residual form of it, is the *peel-off* or *sequential* approach. A suitably matched MUAP is assumed to have contributed to the superposition and subtracted from it. The created residual waveform is then used to search for other contributing MUAPs. The final residual can be used to accept the combination of subtracted MUAPs or potential individual MUAP contributions can be evaluated as correct after each subtraction. A number of algorithms for resolving superimposed MUAPs using the peel-off approach have been reported [5,16,22,29,30]. These algorithms differ in how they align candidate templates with the superimposed waveform, the order in which they align and subtract templates, and the thresholds used to accept suggested resolutions. Some methods align using MUAP peak values [22,30], while others use peak correlation values [5,16]. The methods used for matching and subtracting or *peeling-off* MUAPs vary. The MUAP waveform with the most similar amplitude may be subtracted first [22]. Instead, the MUAP that matches best [5] or the MUAP that produces the smallest residual signal [22] may be first *peeled-off*. Finally, the MUAP with the best combination of smallest residual and most likely firing time may be subtracted first [30]. Some methods accept proposed contributions individually, without resolving the complete superimposed waveform [5,16], while others only accept combinations of contributing MUAPs that can sufficiently account for the energy of the entire superimposed waveform [22,29,30].

The second strategy, or *modeling* approach, synthesizes superimposed waveform models by adding up combinations of MUAPT templates shifted in time by different amounts and searches for an optimal or acceptable match between a model-superimposed waveform and the actual superimposed MUAP. The *modeling* approach is theoretically capable of resolving all types of superimposed waveforms, while the *peel-off* approach is not very powerful when the superimposed waveforms exhibit destructive properties. The *modeling* approach, however, generally requires many more computations than the *peel-off* approach. The *modeling* approach requires $N(N-1)M^2/2$, where N is the number of possibly contributing MUs and M is the number of time shifts considered for matching. By comparison, the *peel-off* approach requires only $2NM$ tests to resolve a superimposed MUAP. Several similar *modeling* approaches have been reported [7,21,24,35]. They all (1) reduce the space required to be searched by first selecting a subset of possibly contributing MUAPs, (2) limit the number of assumed contributing MUAPs to 2 or 3, (3) initially align the MUAPs using either peak values [67] or maximal correlation [16], and (4) use optimization techniques to solve for the model parameters. The superimposed waveform is resolved using the best-fitting, optimized model.

Methods based on limited modeling and parameter optimization that combine the *peel-off* and *modeling* strategies have also been proposed [15,33,38]. In these methods a list of possible MUAP combinations is constructed, and limited MUAP alignment optimizations are performed to search within the list for the best match. Knowledge extracted from the EMG signal being decomposed can be used to construct the list [15,33], or the list can simply be composed of a limited number of combinations that provide the best match [38]. Loudon et al. [33] used net energy, MUAPT firing-pattern information, and a rule-based expert system. Etawil and Stashuk [15] used template energy and firing-pattern information, both within and across MUAPTs, to prioritize the constructed list and direct the search. Both of these techniques alter the order of MUAP template subtraction in an attempt to find the minimal residual. Etawil and Stashuk [15] also used intermediate searches to refine alignment. Figure 3.10 shows the resolution of some example superimposed MUAPs.

3.2.7 Uncovering Temporal Relationships between MUAPTs

Importance of Discovering Temporal Relationships between MUAPTs. During the course of EMG signal decomposition, portions of complex, long-duration MUAPs can be detected as separate MUAP sections. Dividing complex MUAPs into sections, and subsequently treating each section as a separate MUAP, simplifies MUAP clustering and supervised classification. However, for each actual MUAPT that has *multiply-detected* MUAPs, one, or more, extra, temporally linked trains will be created. Therefore algorithms that can discover linked trains are required so that the correct number

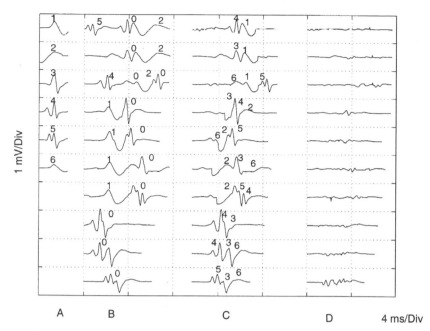

Figure 3.10. Example MUAP superposition resolutions: (A) Candidate template MUAP waveforms; (B) signal segments containing a superimposed MUAP (indicated with 0) and other classified MUAP waveforms; (C) resolved superimposed MUAPs with other classified MUAPs removed; (D) residuals. (From [15] with permission)

of active MUs can be determined. In addition during detection there is no contextual information available to allow efficient MUAP alignment. As such, if a MUAP has two peaks of similar amplitude with less than approximately 1 ms temporal separation, slight biological variation in the shape of the MUAP may cause the peak used to define its firing time and therefore its alignment to vary. This can also occur for single peaked MUAPs if the biological variability is great. Improperly aligned or *disparately detected* MUAPs created by the same MU are often placed in separate trains by clustering and supervised classification algorithms creating one, or more, *exclusive* MUAPT. Exhaustive alignment searches may overcome these detection problems, but they are computationally expensive, and it may be more efficient to have MUAPTs containing such *disparately detected* MUAPs to be processed by dedicated algorithms that can discover *exclusive* MUAPTs. Both of these problem situations, which are due to detection errors, can be corrected if the temporal relationships between MUAPTs can be determined.

Defining and Measuring MUAPT Temporal Relationships or Interdependence. Stochastic point processes can be used to model MUAPTs. The MUAPT of a MU that has discharge times that are independent of previous and future firings and identically distributed can be characterized as a renewal point process. The renewal point process represents the discharge times or the intervals between discharge times of such MUAPTs as random variables [43]. The random variables representing the activity of two MUAPTs can be either independent or dependent [44]. Independent MUAPTs have firing times of one train that have no effect on, or are not related to, the firing times of the other train.

Dependent MUAPTs have firing times of one train that affect, or are related to, the firing times of the other train.

During and after MUAP classification it is important to determine the temporal relationships between MUAPTs. Three kinds of dependent behavior, defined as *linked*, *exclusive*, or *synchronized*, are possible. *Linked* trains fire with a definite, essentially constant interval relative to each other. For example, as suggested above, *linked* trains would be created if a single, long-duration MUAP is consistently *multiply-detected* as two or more distinct MUAPs each representing only a portion of the single MUAP. This may be caused by the physiological phenomenon, termed satellite potential, that occurs when a muscle fiber is reinnervated by an axon normally supplying a dense group of distant fibers providing the main MUAP in the signal. A satellite potential may either follow or proceed the main action potential. This has been described especially as an age-related effect, when high threshold MUs lose their innervating axon and are reinnervated by axons from low threshold MUs. At the other extreme, *exclusive* trains never fire "together." This is to say that the absolute difference between the firing times of any two MUAPs each selected from a different train is never less than some threshold amount (e.g., 25 ms). As suggested above, a set of two or more *exclusive* trains may represent the activity of a single MU whose MUAPs have been *disparately detected* and thus have been erroneously separated into two or more trains by the classification algorithms. *Synchronized* trains have behaviors somewhere in between *linked* and *exclusive* and may represent a true biological tendency of MUs to have dependent firing behaviors.

Recurrence histograms have been used to study the relationships between spike trains [39]. However, this type of analysis requires assumptions regarding the probability distributions of the bins of the resulting histograms. Mutual information, on the other hand, is a nonparametric measure of the interdependency between two random variables [49]. Using concepts of mutual information, one can determine the temporal relationship between two MUAPTs by studying the probabilities of either MUAPT or both MUAPTs having an occurrence within an analysis window randomly positioned in time. For considering MUAP durations and peak MU firing rates, a window of 25 ms duration is sufficient [3,6]. The event of a MUAPT firing within an analysis window can be represented by a discrete random variable that has a value of 1, if a firing of the MUAPT is in the analysis window, and 0 otherwise. By defining A and B as such discrete random variables, representing $MUAPT_i$ and $MUAPT_j$, respectively, of a selected pair of trains, the interdependency redundancy of the trains can be measured. The interdependency redundancy can then be used along with individual and joint discharge probabilities of the trains to determine if the trains are independent or dependent and, if dependent, whether they are *exclusive* or *linked* [57,67,68].

When applied to pairs of MUAPTs created following clustering and supervised classification, the mutual information-based interdependency redundancy measure described above is able to successfully discover relationships that exist between trains detected in EMG signals. Trains associated with *multiply detected* MUAPs are found to be *linked*, and subsequently they are considered a single train. Trains associated with *disparately detected* MUAPs are found to be *exclusive* and become candidates for merging. Therefore the application of a temporal-relationships discovering algorithm often allows the number of MUAPTs discovered following signal decomposition to more closely match the number of MUs consistently contributing significant MUAPs to the composite EMG signal. However, in some research situations it may not be appropriate to excessively use such temporal information, since apparently "strange" relationships between MU firing patterns may indeed reveal real physiological phenomena.

3.3 EVALUATION OF PERFORMANCE OF EMG SIGNAL DECOMPOSITION ALGORITHMS

Verification of the accuracy of an intramuscular EMG signal decomposition requires the availability of signals for which the decomposition result is known and the definition of quantitative indexes that allow comparison of performance. Moreover, for completeness and to assess robustness, the performance of a specific algorithm should be evaluated based on a number of signals of different complexity.

The reference result can be obtained by manual decomposition of a number of experimental signals by expert operators. However, different patterns may result when the same or different operators attempt to decompose the same signal twice, especially if the MU firing rates are irregular, the MUAPs are similar, superpositions of MUAPs are frequent, and some MUs may be intermittently recruited [45]. In addition different algorithms (or operators) may weigh different information, such as waveform similarity or firing regularity, differently and therefore produce different results. Furthermore specific algorithms may be more appropriate in certain cases and others in other cases. To assess accuracy, DeLuca [8] proposed to detect signals (using multiple electrode surfaces) from the same MU at different locations and to compare the results of the decomposition of the two signals obtained. This way the probability of incorrectly decomposing the different signals and yet having the same firing pattern for an investigated MU is low. When the decomposition results agree for all the channels, the decomposition is considered correct.

The reference decomposition result can also be obtained from synthetic signals generated by a model. In this case the crucial issue is to describe all the relevant characteristics of the experimental signals. A model is the only way to test the algorithms with signals having selected characteristics in order to evaluate the sensitivity of the decomposition algorithms to different EMG signal parameters. Whatever the approach for the generation of reference decomposition results, it is necessary to introduce indexes of performance computed from the comparison of the results obtained by the application of the algorithm under test and the reference.

3.3.1 Association between Reference and Detected MUs

The use of a reference decomposition result (either given by a model or by manual decomposition) introduces problems related to the definition of event occurrences and correspondence between reference and detected classes (MUs). In particular, problems arise when the number of detected MUs differs from the number of reference MUs [19]. The establishment of correspondence between classes creates a criterion for assigning a detected occurrence to an effective one. This is because the firing instant is defined in different ways by different algorithms and in the reference result. For example, the firing instant can be defined as the time corresponding to the median of energy, to the peak value, to the "beginning" of or "end" of the MUAP, to a zero crossing. The difference in the firing instant definition can be as large as 3 to 4 ms (e.g., in the difference between the beginning and the end of the waveform) and not negligible. Therefore deterministic (offset in the detection time) and random differences between the reference and the detected times of occurrence are expected. The deterministic difference should not be taken into account (it must be estimated and compensated for in some way) while the random difference should be limited because it determines an error in instantaneous firing frequency estimation. If an error δt_o is present in the detection of firing time, the error δf_r in instantaneous firing frequency estimation is given by

$$\delta f_r = 2 f_r^2 \delta t_o$$

with f_r being the instantaneous firing frequency. If one considers 40 Hz to be the upper limit for firing rate, in the case of $\delta t_o = 0.5$ ms, a relative error in f_r of 4% is obtained, which is acceptable. Thus, after the estimation and compensation for the systematic error, a detection can be considered correct if the estimated time of occurrence is within ±0.5 ms from a real one. The problem is how to associate a detected class with a reference one. Note that the averaged template of each detected class cannot be used to define a correspondence between classes because it is influenced by classification errors and its definition can change with the algorithm. Moreover, if the shapes of the MUAPs within a train change over time (i.e., if there are nonstationarities), the definition of an averaged template over the whole train is meaningless. To determine the correspondence between a reference and a detected class (MUAPT), each reference class should be compared with all the detected classes (determined by the algorithm under test) by computing the differences between the reference times of occurrence and the estimated ones. The time differences are computed in order to assign a detected MUAP to the closest (in time) reference MUAP. The detected versus reference class assignment can then be done on the basis of the maximum number of correspondent firings [19]. The offset in the detection time for a particular class is not known a priori but can be estimated from the mean of the time differences between the occurrences in the associated reference and detected classes, respectively.

3.3.2 Indexes of Performance

The use of indexes to synthetically describe the quality of a decomposition process is partly due to traditional detection theory [14] adapted to the multi-class problem of EMG signal decomposition. As indicated previously, any decomposition algorithm operates on the signal in two basic steps: the segmentation and classification phase. The segmentation phase consists of the detection of all the MUAPs contributing significantly to the signal, and the classification phase consists of their assignment to classes (MUAPTs). It is useful to characterize an algorithm's performance separately for these two steps. Note that the decomposition process may be divided in more than two steps (e.g., superpositions are usually resolved separately), but in view of the result, the whole process can be divided into segmentation—the capability of detecting an event—and classification—the ability to differentiate between MU potentials.

3.3.3 Evaluation of the Segmentation Phase Performance

The segmentation phase can be handled as a classic two-class detection problem [14]. True positives are in this case defined as the detected occurrences corresponding to actual occurrences (independently of the classification step) and consist of the detected occurrences correctly classified plus the occurrences detected but incorrectly classified. False negatives (or missed detections) are defined in the segmentation phase as actual occurrences not detected by the algorithm. The number of false negatives $^{(s)}FN$ is thus the difference between the total number of reference MUAPs and the number of true positives $^{(s)}TP$. False positives (or false detections) are detected occurrences not corresponding to actual ones; their number $^{(s)}FP$ is the difference between the total number of detected occurrences and

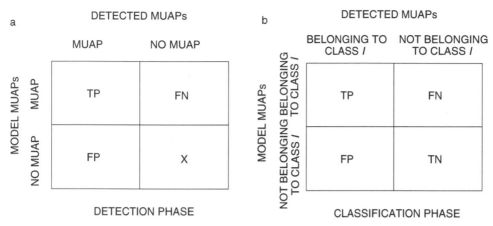

Figure 3.11. Definition of true positives, false positives, true negatives, and false negatives for the segmentation (*a*) and classification (*b*) phase (for class *I*). (From [19] with permission)

$^{(s)}TP$. True negatives cannot be computed for the segmentation phase. The significance of these definitions is reported in Figure 3.11*a*. Sensitivity $^{(s)}Se$ and positive predictivity $^{(s)}P$ [14] for the segmentation phase can therefore be defined as

$$^{(s)}Se = \frac{^{(s)}TP}{^{(s)}TP + {}^{(s)}FN} \qquad \text{for each discovered reference class}$$

$$^{(s)}P = \frac{^{(s)}TP}{^{(s)}TP + {}^{(s)}FP}$$

3.3.4 Evaluation of the Classification Phase Performance

True positives, false positives, true negatives, and false negatives are defined differently for the classification phase with respect to the detection phase. True positives for class *I* are the detected occurrences correctly classified as belonging to class *I* and their number $^{(c)}TP$ is computed directly during the class assignment procedure. False positives for class *I* are detected occurrences corresponding to actual occurrences but incorrectly classified as belonging to class *I*. Their number $^{(c)}FP$ is the difference between the number of correctly detected occurrences classified in class *I* and $^{(c)}TP$ of class *I*. True negatives are, for each discovered reference class *I*, the number of detected occurrences not belonging to class *I* and not assigned by the algorithm to class *I*. Note that the classification of an occurrence can be incorrect and that the occurrence can be a true negative. For example, in the case of four classes, if a MUAP belonging to class 3 is classified in class 4, it is a true negative for class 1 and 2. False negatives for class *I* are correctly detected occurrences of class *I* incorrectly classified to a class different from *I*. Figure 3.11*b* summarizes the meaning of the definitions introduced.

From $^{(c)}TP$, $^{(c)}FP$, $^{(c)}TN$, and $^{(c)}FN$ it is possible to compute sensitivity, specificity, and accuracy of the algorithm for each class. They are defined as follows [14]:

$$^{(c)}Se = \frac{^{(c)}TP}{^{(c)}TP + \,^{(c)}FN} \qquad \text{for each not missed reference class}$$

$$^{(c)}Sp = \frac{^{(c)}TN}{^{(c)}TN + \,^{(c)}FP} \qquad \text{for each not missed reference class}$$

$$^{(c)}Ac = \frac{^{(c)}TP + \,^{(c)}TN}{^{(c)}TP + \,^{(c)}TN + \,^{(c)}FP + \,^{(c)}FN} \qquad \text{for each not missed reference class}$$

3.3.5 Reference Decomposition

The indexes described can be used in any situation where a reference result is available, for example, when a manual decomposition is performed and considered correct. Manual decomposition for evaluating performance has drawbacks (discussed above). On the other hand, when using synthetic signals as reference, it is very important that the generated signals closely resemble real ones. Thus a generation model should be able to simulate the main characteristics of experimental signals. In particular, the features of intramuscular EMG signals that are important to simulate and quantify from the decomposition point of view are as follows:

- Shape similarity of a MUAP detected at the same time from different detection surfaces (if multichannel recordings are simulated).
- Shape similarity of MUAPs of different MUs.
- Shape changes of MUAPs created by the same MU during time.
- MU firing statistics.
- Recruitment and de-recruitment of individual MUs.
- Degree of superposition of the MUAPs in the signal.
- Amount of additive noise.

Farina et al. [17] recently proposed a phenomenological intramuscular EMG signal model that includes the previously listed characteristics. In that model a generic MUAP is represented in a 16-dimensional space using the Associated Hermite expansion with coefficients that vary in time-simulating shape changes in the MUAPs. Figure 3.12 shows examples of simulated MUAP shape changes. The proposed model allows one also to quantitatively evaluate the complexity of the signal with respect to different features.

3.4 APPLICATIONS OF RESULTS OF THE DECOMPOSITION OF AN INTRAMUSCULAR EMG SIGNAL

The decomposition of the EMG signal provides detailed information about the recruitment thresholds and firing rates of single MUs. This information has been used in the past for the investigation of a number of basic physiological issues related to the central control of MUs. The availability of the firing times of active MUs also allows the estimation of single MU properties from other signals in which the separation of the contributions of single MUs is difficult. In these cases the concomitant recording of the intramuscular EMG signal and other signals, such as the joint force or the surface EMG, provides more com-

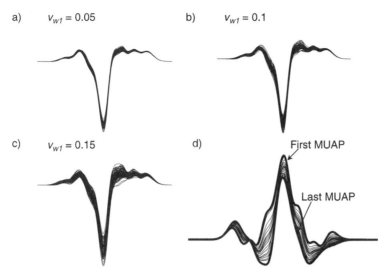

Figure 3.12. Synthetic waveforms with shape variability generated by the model described in [17]. (*a*), (*b*), and (*c*) report waveforms with different degrees of random shape changes while (*d*) shows waveforms with systematic shape variability. The parameter v_{w1} (defined in [17]) quantifies the variability. (From [17] with permission)

plete information about single MU properties. Usually the signals recorded are averaged using the firing instants detected from the intramuscular EMG signal as triggers.

3.4.1 Firing Pattern Analysis

A relation between the activity of a certain MU and the exerted force of the muscle can be described from isometric contractions with different percentages of maximal voluntary contraction (MVC) as the target force or from increasing and decreasing force ramps. For each MU a force threshold for recruitment and derecruitment can be estimated, and the relation between mean firing rate and level of force can be established. Thus it is possible to order the recruited MU hierarchically according to recruitment threshold force (Fig. 3.13). However, the absolute recruitment force levels must be carefully interpreted, since the measured forces normally are not produced only by the agonist but also involve a muscle synergy that is influenced by antagonist muscle activity, the latter being considerable for instance during a force tracking task. Changes in firing rates and concurrent recruitment and derecruitment of MU may be used to estimate the relative importance of rate modulation and recruitment in a given muscle and in a certain condition.

Firing patterns may be qualitatively described as continuous or tonic compared to intermittent, sporadic, or phasic. A fixed criterion for such a classification may be that a continuous MUAPT does not show any inter-firing intervals longer than three times the preceding interval or longer than a fixed period, such as 200 ms.

Due to the technical difficulties outlined in this chapter, the available experience on decomposition has been acquired during relative short duration, isometric contractions at constant low force levels. For this type of contraction the activity of each identified MU may be characterized by the mean firing rate and the standard deviation. Since the standard deviation is dependent on the actual firing rate, the coefficient of variance (i.e., the

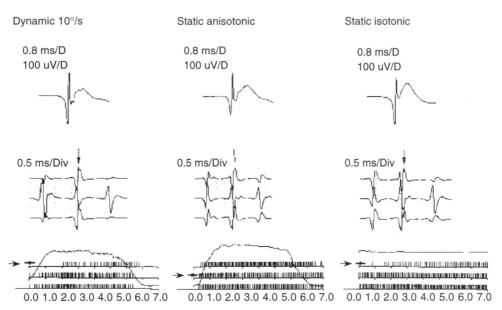

Figure 3.13. Three recordings from various contraction modes of the biceps brachii muscle. In the upper part is shown the shape of the MUAP in signal sampled with a larger pickup area. In the middle a short section of the raw recording of each contraction shows the shape used for identification of the MU. In the bottom is the firing pattern of all identified MUs (time is in seconds) and the arrow indicates the activity patterns of the identical MU during the three different types of contraction. The solid line shows time history of the force. (From [51] with permission)

standard deviation relative to the mean firing rate) often is a better description of the variability in the firing pattern.

The mean firing rate may be calculated as the mean of the inverse of all inter-firing intervals identified in a contraction, the inverse of the main component of the inter-firing interval histogram or the median value of all inter-firing intervals, the two latter methods being least sensitive for missing classifications in a MUAPT. Normally the mean firing rate in an isometric contraction reflects the order of recruitment, so the first recruited MUs show the largest mean firing rates.

If the firing patterns are studied during a long-term contraction or during force- and length-changing conditions, much larger variation and irregularities are present in the discharge patterns [40,52]. Mean firing rate must then be used with caution and it may be convenient to be able to describe trends in firing rate over time.

Long-term recordings at low-force levels have been used to investigate overloading of MUs during the development of muscle fatigue. At low-force levels the activity of the whole muscle (e.g., estimated from the surface EMG) is relatively small compared to maximal. However, the general lack of load sharing among all MUs during a low-force long-term contraction may impose a relatively high load on a few active MUs. The prevailing hypothesis is that prolonged inhomogeneouos activation of the same single muscle fibers may cause degenerative changes in the muscles [50]. However, data are scarce and many questions are still unsolved regarding motor control during low-force contractions typical for many natural movements and occupational activities. Among these questions, for instance, are whether a preventive mechanism exists substituting or rotating activity

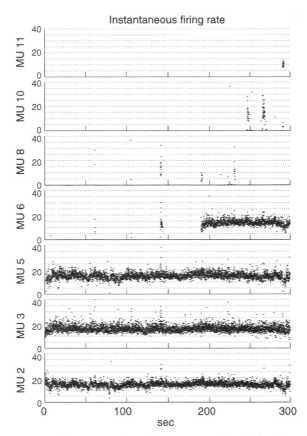

Figure 3.14. Instantaneous firing rate (in pulses per second) calculated as the inverse of each inter-spike interval plotted.

among MUs (i.e., the phenomenon that continuously active MUs are derecruited and higher threshold MUs are recruited simultaneously) and whether just silent periods in an otherwise continuously active MU are present during long-term contractions.

From the decomposition of the intramuscular EMG it is possible to extract the simultaneous behavior of several MUAPTs and their temporal relations with the generated force and movements. However, only recently has it become possible to obtain such reliable recordings of several concurrently active MUs over long-term contractions involving also force and length changes. Figure 3.14 shows an example of continuously active MUs during a long duration contraction.

The most comprehensive description of the firing rates of single MUs is the calculation of the instantaneous firing rate from each interval between adjacent discharges of a MU showing the short-term changes in firing rates. These internal firing rates are useful in a study of the transient in firing patterns of the MU in response to a given change in force or length.

A special phenomenon that becomes clearly visible in the instantaneous firing rate is the *doublet* or double discharge of a MU with inter-firing intervals less than 20 ms (Fig. 3.15). Doublets are typically seen during free ballistic movements, and are thought to represent a way to augment the force output of the muscle fiber. In addition to a large initial force, also a much larger force is elicited during the following discharges, a phenomenon

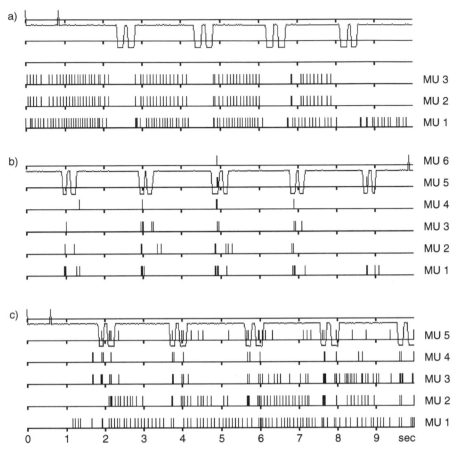

Figure 3.15. Firing patterns of the identified MU in three different subjects during right-hand double clicking shown as an inter-firing interval bar plot. Each horizontal line represents one MU and each vertical bar represents one firing. The solid line shows the signal from the trigger incorporated in the mouse key giving the time of a double click. (From [53] with permission)

termed catchlike properties. This implies that the relation between the force and firing rate is disturbed so that the force produced by the MU is higher than that indicated by the mean firing rate.

Doublets are not only found during new recruitment but also imposed on a continuous firing pattern. Recently it has been found that the fast movement of the index finger consistently induced a large number of doublets in the already activated low-threshold MUs (Fig. 3.15). Therefore doublets may represent a peak tension added on top of a prolonged accumulated load on the MU [53,66].

3.4.2 Investigation of Correlation between MU Firing Patterns

Cross-correlation of the mean firing rates of two MUs provides a measure of the common behavior of the firing rates. This is often referred to as the common drive [11] and is probably reflecting a shared excitatory or inhibitory input to the motor neuron pool supplying this specific muscle or muscle compartment. Thus, although the MU is the smallest functional part of the muscle, the amount of cross correlation between the MUs reveals the extent to which the nervous system relies on the motor neuron pool as one functional unit, where only the MU properties and selective afferent feedback modifies firing behavior. Note that this is not implying a synchronized behavior of the two MUs but only a common behavior in the trend of the firing rates.

As mentioned earlier, a true biological tendency of MUs to have dependent firing behaviors may sometimes be seen as firing patterns somewhere in between linked and exclusive MUAPTs. An analysis of synchronization between two MUAPTs is performed as described above for the quality check of the decomposition result. The degree of synchronization among MUs is determined by the amount of shared synaptic input, which could be, for instance, afferent feedback or recurrent inhibition and thus may reveal important changes in neuromuscular control related to aging or fatigue.

3.4.3 Spike-Triggered Averaging of the Force Signal

Stein et al. [63] proposed a method to estimate single MU contractile properties recording intramuscular EMG signals and joint torque. The method is based on the averaging of the joint torque signal using the detected single MU firing instants as triggers. This technique can be applied only at very low firing rates, since the force twitches of the MUs have durations of the order of hundreds of milliseconds and are fused for frequencies above 8 to 10 Hz. For this reason the spike-triggered averaging technique is usually applied during nonphysiological contractions, with the subject using particular feedback techniques to maintain a constant firing rate of the investigated MU [39]. Even in these ideal conditions the technique of twitch averaging has some drawbacks, and it is not yet clear how much it is affected by MU synchronization, nonlinear force summation, or jitter in the latency between the electrical and mechanical response [4,32]. Despite these limitations twitch averaging has been applied extensively to investigate basic relationships between MU twitch, recruitment threshold and firing rate, and has provided important results on neuromuscular physiology.

3.4.4 Macro EMG

Since the selective needle or wire electrodes only pick up the summed potentials from the few muscle fibers of a specific MU lying within the uptake radius of the electrode, this MUAP does not represent the real MUAP, which will allow interpretation of MU properties. The number of fibers participating in the MUAP from the selective electrode are dependent on the area and geometry of the recording electrode, the fiber density of the MU and the position of the electrode in relation to the center and periphery of the MU territory in the muscle. These parameters will be different for different wires, different muscles, and even for different MUs identified in the same recording. Ideally, to describe the MU properties, a compound signal should be recorded that contains the summed action potentials from the majority of fibers in the MU. To obtain such a standardized signal, Stålberg [54] introduced the averaging of macro intramuscular signals. The technique was based on the concomitant detection of intramuscular EMG signals by a selective electrode and a large electrode 15 mm long (see Chapter 2). The signal detected by the large intramuscular electrode is spike-triggered averaged using the decomposition of the selective fiber recording as trigger. Macro MUAPs are obtained in this way, as described in Chapter 2. The size parameters of the macro MUAPs, such as the peak-to-peak voltage or area, are related to the overall size of the contributing MU. As described in Section 3.4.5 below, the surface EMG could be used as well to obtain these parameters. However, the macro MUAP has the advantage that fewer discharges are needed to obtain a clean averaged potential and that it can be used in muscles that are not easily accessed by surface electrodes.

3.4.5 Spike-Triggered Averaging of the Surface EMG Signal

With a similar approach, the surface EMG signal can also be spike-triggered averaged and single superficial MUAPs can be extracted from the interference surface signal. The surface signal can be detected by linear electrode arrays in order to estimate the conduction velocity (CV) [18], the location in the muscle [47], and the size [46] of single MUs. In particular, the estimation of CV of single MUs from the surface EMG signal can give indications of fatigue and MU contractile properties [1], overcoming many drawbacks of the twitch averaging technique. For example, since a window length of about 30 ms or less for surface EMG averaging is suitable for detecting the potential along its propagation from the neuromuscular junction to the tendon, there is no overlapping of the surface detected MUAPs even at relatively high discharge frequencies. Figure 3.16 shows an example of extraction of multi-channel surface potentials from the spike-triggered average of the interference signal and Figure 3.17 reports spike-triggered averaged surface EMG signals together with the estimation of CV of a single MU.

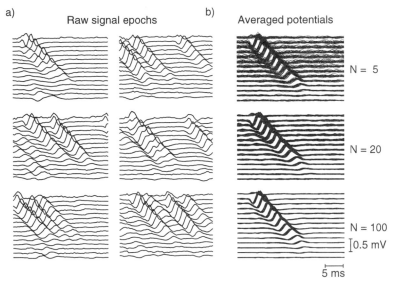

Figure 3.16. Examples of spike-triggered averaged surface multi-channel action potentials detected with a linear array of 16 electrodes in single differential detection modality (refer to Chapter 7). (a) Six of the raw signal epochs from which the average potentials are computed. The detected intramuscular potential can be recognized at the surface, but it is often masked by other potentials. (b) Examples of spike-triggered averaged surface-detected action potentials belonging to the same MU. The effect of different numbers of averages is shown. A rather small number of averages is required in this case for obtaining a clear MUAP at the surface. The distance from the first and last single differential signals of the surface recordings is 70 mm. (From [18] with permission)

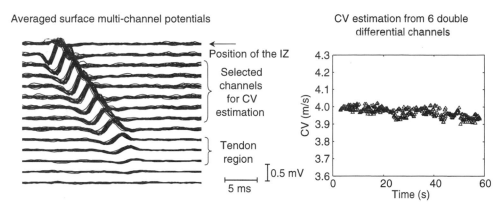

Figure 3.17. Example of estimation of single MU CV from averaged MUAPs during a 60s long contraction of the tibialis anterior muscle. The averaged surface potentials obtained using 20 triggers each are shown. The estimation of CV is performed as proposed by Farina et al. [20] (refer also to Chapter 7). The distance from the first and last single differential signals of the surface recordings is 70 mm. IZ stands for innervation zone. (From [18] with permission)

3.5 CONCLUSIONS

The decomposition of the intramuscular EMG signal is a complex task that involves advanced signal processing and pattern recognition techniques. Its applications cover the fields of basic physiology, neurology, motor control, ergonomics, and many others. Currently available techniques allow a reliable decomposition at low–medium force contraction levels during short and long contractions in static and dynamic conditions. The availability of such methods for automatic intramuscular EMG signal analysis allows the completion of experimental studies that were unthinkable some years ago, such as the investigation of MU activity during very long contractions (up to hours). Intramuscular EMG signal decomposition is, however, still carried out mainly in research environments while it finds limited clinical application. This is mainly due to the limitations that EMG signal decomposition still has, such as the amount of time required to obtain clinically useful information (especially if high reliability on a number of conditions is required), the necessity in most cases of an interaction with an expert operator, the applicability to only low-to-medium contraction levels, and the need of specially trained persons for the proper positioning of the needle electrode to obtain the high-quality signals required for reliable decomposition. These limitations are being addressed by current research efforts.

REFERENCES

1. Andreassen, S., and L. Arendt-Nielsen, "Muscle fiber conduction velocity in motor units of the human anterior tibial muscle: A new size principle parameter," *J Physiol (Lond)* **391**, 561–571 (1987).
2. Basmajian, J. V., and C. J. DeLuca, *Muscles alive: Their functions revealed by electromyography*, 5th ed., William and Wilkins, Baltimore, 1985.

3. Bischoff, C., E. V., B. Falck, and K. E. Eeg-Olofsson, "Reference values of motor unit action potentials obtained with multi-MUAP analysis," *Muscle Nerve* **17**, 842–851 (1994).

4. Calancie, B., and P. Bawa, "Limitations of the spike-triggered averaging technique," *Muscle Nerve* **9**, 78–83 (1986).

5. Christodoulou, I. C., and C. S. Pattichis, "Unsupervised pattern recognition for the classification of EMG signals," *IEEE Trans BME* **46**, 169–178 (1999).

6. Cohen, H., and J. Brumlik, *A manual for electroneuromyography*, Harper and Row, New York, 1968.

7. De Figueiredo, R. J. P., and A. Gerber, "Separation of superimposed signals by a cross-correlation method," *IEEE Trans Acousti Speech Sig Proc* **31**, 1084–1089 (1983).

8. De Luca, C. J., "Reflections on EMG signal decomposition," In: J. E. Desmedt, ed., *Computer-aided electromyography and expert system*, 1989, pp. 33–37.

9. DeLuca, C. J., "Precision decomposition of EMG signals," *Meth Clin Neurophysiol* **4**, 1–28 (1993).

10. DeLuca, C. J., and A. Adam, "Decomposition and analysis of intramuscular electromyographic signals," in U. Windhorst and H. Johannson, eds., *Modern techniques in neuroscience research*, Springer, Heidelberg, 1999.

11. De Luca, C. J., and Z. Erim, "Common drive of motor units in regulation of muscle force," *Trends Neurosci* **17**, 299–305 (1994).

12. De Luca, C. J., R. S. LeFever, M. P. McCue, and A. P. Xenakis, "Behaviour of human motor units in different muscles during linearly varying contractions," *J Physiol* **329**, 113–128 (1982).

13. De Luca, C. J., R.S. LeFever, M. P. McCue, and A. P. Xenakis, "Control scheme governing concurrently active human motor units during voluntary contractions," *J Physiol* **329**, 129–42 (1982).

14. Egan, J. P., *Signal detection theory and ROC analysis*, Academic Press, San Diego, 1975.

15. Etawil, H. A. Y., and D. W. Stashuk, "Resolving superimposed motor unit action potentials," *Med Biol Eng Comput* **34**, 33–40 (1996).

16. Fang, J., G. C. Agarwal, and B. T. Shahani, "Decomposition of multiunit electromyographic signals," *IEEE Trans BME* **46**, 685–697 (1999).

17. Farina, D., A. Crosetti, and R. Merletti, "A model for the generation of synthetic intramuscular EMG signals to test decomposition algorithms," *IEEE Trans BME* **48**, 66–77 (2001).

18. Farina, D., L. Arendt-Nielsen, R. Merletti, and T. Graven-Nielsen, "Assessment of single motor unit conduction velocity during sustained contraction of tibials anterior muscle with advanced spike triggered averaging," *J Neurosci Meth* **115**, 1–12 (2002).

19. Farina, D., R. Colombo, R. Merletti, and H. Baare Olsen, "Evaluation of intra-muscular EMG decomposition algorithms," *J Electromyogr Kinesiol* **11**, 175–187 (2001).

20. Farina, D., W. Muhammad, E. Fortunato, O. Meste, R. Merletti, and H. Rix, "Estimation of single motor unit conduction velocity from the surface EMG signal detected with linear electrode arrays," *Med Biol Eng Comput* **39**, 225–236 (2001).

21. Gerber, A., R. M. Studer, R. J. P. Figueiredo, and G. S. Moschytz, "A new framework and computer program for quantitative EMG signal analysis," *IEEE Trans BME* **31**, 857–863 (1984).

22. Guiheneuc, P., J. Calamel, C. Doncarli, S. Gitton, and C. Michel, "Automatic detection and pattern recognition of single motor unit potentials in needle EMG," in J. E. Desmedt, ed., *Computer-aided electromyography*. Karger Basel, 1983; *Prog Clin Neurophysiol*, **10**, 73–127.

23. Gut, R., and G. S. Moschytz, "High-precision EMG signal decomposition using communication techniques," *IEEE Trans Sig Proc* **48**, 2487–2494 (2000).

24. Haas, W. F., and M. Meyer, An automatic EMG decomposition system for routine clinical examination and clinical research—ARTMUAP, in J. E. Desmedt, ed., *Computer-aided electromyography and expert systems*, Elsevier Science, Amsterdam, 1989, pp. 67–81.

25. Hassoun, M. H., C. Wang, and R. Spitzer, "NNERVE: Neural network extraction of repetitive vectors for electromyography—Part I: Algorithm," *IEEE Trans BME* **41**, 1039–1052 (1994).

26. Hassoun, M. H., C. Wang, and R. Spitzer, "NNERVE: Neural network extraction of repetitive vectors for electromyography—Part II: Performance analysis," *IEEE Trans BME* **41**, 1053–1061 (1994).

27. Jain, A. K., *Algorithms for clustering data*, Prentice Hall, Englewood Cliffs, NJ, 1988.

28. Joynt, R. L., R. F. Erlandson, S. J. Wu, and C. M. Wang, "Electromyography interference pattern decomposition," *Arch Phys Med Rehabil* **72**, 567–572 (1991).

29. Kadefors, R., M. Forsman, B. Zoega, and P. Herberts, "Recruitment of low threshold motor-units in the trapezius muscle during different static arm positions," *Ergonomics* **42**, 359–375 (1999).

30. LeFever, R. S., and C. J. DeLuca, "A procedure for decomposing the myoelectric signal into its constituent action potentials—Part I: Technique, theory and implementation," *IEEE Trans BME* **29**, 149–157 (1982).

31. LeFever, R. S., and C. J. DeLuca, "A procedure for decomposing the myoelectric signal into its constituent action potentials—Part II: Execution and test for accuracy," *IEEE Trans BME* **29**, 158–164 (1982).

32. Lim, K. Y., C. K. Thomas, and W. Z. Rymer, "Computational methods for improving estimates of motor unit twitch contraction properties," *Muscle Nerve* **18**, 165–174 (1995).

33. Loudon, G. H., N. B. Jones, and A. S. Sehmi, "New signal processing techniques for the decomposition of EMG signals," *Med Biol Eng Comput* **30**, 591–599 (1992).

34. McGill, K. C., and L. J. Dorfman, "Automatic decomposition electromyography (ADEMG): Validation normative data in brachial biceps," *Electroenc Clin Neurophysiol* **61**, 453–461 (1985).

35. McGill, K. C., K. L. Cummins, and L. J. Dorfman, "Automatic decomposition of the clinical electromyogram," *IEEE Trans BME* **32**, 470–477 (1985).

36. Nandedkar, S. D., P. E. Barkhaus, and A. Charles, "Multi-motor unit action potential analysis (MMA)," *Muscle Nerve* **18**, 1155–1166 (1995).

37. Nandedkar, S. D., E. V. Stålberg, and D. B. Sanders, "Simulation techniques in electromyography," *IEEE Trans BME* **32**, 775–785 (1985).

38. Nikolic, M., J. A. Sørensen, K. Dahl, and C. Krarup, "Detailed analysis of motor unit activity," *Proc 19th An Int Conf IEEE Eng Med Biol Soc*, Chicago, 1997, pp. 1257–1260.

39. Nordstrom, A., and T. S. Miles, "Fatigue of single motor units in human masseter," *J Appl Physiol* **68**, 26–34 (1990).

40. Olsen, H. B., H. Christensen, and K. Søgaard, "An analysis of motor unit firing pattern during sustained low force contraction in fatigued muscle," *Acta Physiol Pharmacol Bulg* **26**, 73–78 (2001).

41. Paoli, G. M., "Estimating certainty in classification of motor unit action potentials," MS thesis, University of Waterloo, 1993.

42. Pattichis, C. S., C. N. Schizas, and L. T. Middleton, "Neural network models in EMG diagnosis," *IEEE Trans BME* **42**, 486–496 (1995).

43. Perkel, D., and G. Gerstein, "Neural spike trains and stochastic point processes: I. The single spike train," *Biophys J* **7**, 391–418 (1967).

44. Perkel, D., and G. Gerstein, "Neural spike trains and stochastic point processes: II Simultaneous spike trains," *Biophys J* **7**, 391–418 (1967).

45. Pilegaard, M., B. R. Jensen, G. Sjøgaard, and K. Søgaard, "Consistency of motor-unit identification during force-varying static contractions," *Eur J Appl Physiol* **83**, 231–234 (2000).

46. Roeleveld, K., D. F. Stegeman, B. Falck, and E. V. Stålberg, "Motor unit size estimation: confrontation of surface EMG with macro EMG," *Electroenc Clin Neurophysiol* **105**, 181–188 (1997).

47. Roeleveld, K., D. F. Stegeman, H. M. Vingerhoets, and A. Van Oosterom, "The motor unit potential distribution over the skin surface and its use in estimating the motor unit location," *Acta Physiol Scand* **161**, 465–472 (1997).

48. Sanders, D. B., and E. V. Stålberg, "AAEM minimonograph #25 single-fiber electromyography," *Muscle Nerve* **19**, 1069–1083 (1996).

49. Shannon, C. E., "A mathematical theory of communications," *Bell Syst Techni J* **27**, 379–423 (1948).

50. Sjøgaard, G., U. Lundberg, and R. Kadefors, "The role of muscle activity and mental load in the development of pain and degenerative processes at the muscle cell level during computer work," *Eur J Appl Physiol* **83**, 99–105 (2000).

51. Søgaard, K., "Motor unit recruitment pattern during low-level static and dynamic contractions," *Muscle Nerve* **18**, 292–300 (1995).

52. Søgaard, K., H. Christensen, B. R. Jensen, L. Finsen, and G. Sjøgaard, "Motor control and kinetics during low level concentric and eccentric contractions in man," *Electroenceph Clin Neurophysiol* **101**, 453–460 (1996).

53. Søgaard, K., G. Sjøgaard, L. Finsen, H. B. Olsen, and H. Christensen, "Motor unit activity during stereotyped finger tasks and computer mouse work," *J Electromyogr Kinesiol* **11**, 197–206 (2001).

54. Stålberg, E., "Macro EMG, a new recording technique," *J Neurol Neurosurg Psychiatry* **43**, 475–482 (1980).

55. Stålberg, E. V., and M. Sonoo, "Assessment of variability in the shape of the motor unit action potential, the 'jiggle' at consecutive discharges," *Muscle Nerve* **17**, 1135–1144 (1994).

56. Stålberg, E., B. Falck, M. Sonoo, S. Stålberg, and M. Astrom, "Multi-MUAP EMG anlysis—A two year experience in daily clinical work," *Electroenc Clin Neurophysiol* **97**, 145–154 (1995).

57. Stashuk, D. W. "Decomposition and quantitative analysis of clinical electromyographic signals," *Med Eng Phys* **21**, 389–404 (1999).

58. Stashuk, D. W., and H. de Bruin, "Automatic decomposition of selective needle-detected myoelectric signals," *IEEE Trans BME* **35**, 1–10 (1988).

59. Stashuk, D. W., and R. Naphan, "Probabilistic inference-based classification applied to myoelectric signal decomposition," *IEEE Trans BME* **39**, 346–355 (1992).

60. Stashuk, D. W., and G. M. Paoli, "Robust supervised classification of motor unit action potentials," *Med Biol Eng Comput* **35**, 1–8, (1998).

61. Stashuk, D. W., and Y. Qu, "Adaptive motor unit action potential clustering using shape and temporal information," *Med Biol Eng Comput* **34**, 41–49 (1996).

62. Stashuk, D. W., and Y. Qu, "Robust method for estimating motor unit firing-pattern statistics," *Med Biol Eng Comput* **34,** 50–57 (1996).

63. Stein, R. B., A. S. French, and R. Yemm, "New methods for analysing motor function in man and animals," *Brain Res* **40**, 187–192 (1972).

64. Wellig, P., and G. S. Moschytz, "Classification of time varying signals using time-frequency atoms," *Proc 21th An Int Conf IEEE Eng Med Biol Soc*, Atlanta, 1999.

65. Wellig, P., G. S. Moschytz, and T. Laubli, "Decomposition of EMG signals using time-frequency features," *Proc 20th An Int Conf IEEE Eng Med Biol Soc*, Hong Kong, 1998.

66. Westgaard, R. H., and C. J. De Luca, "Motor unit substitution in long-duration contractions of the human trapezius muscle," *J Neurophysiol* **82**, 501–504 (1999).

67. Wong, A. K., and A. Ghahraman, "A statistical analysis of interdependence in character sequences," *Inform Sci* **8,** 173–188 (1975).

68. Wong, A. K., T. S. Liu, and C. C. Wang, "Statistical analysis of residue variability in cytochrome c," *J Mol Biol* **102**, 287–295 (1976).

4

BIOPHYSICS OF THE GENERATION OF EMG SIGNALS

D. Farina and R. Merletti

Laboratory for Engineering of the Neuromuscular System
Department of Electronics
Politecnico di Torino, Italy

D. F. Stegeman

Department of Clinical Neurophysiology
University Medical Center, Nijmegen
Interuniversity Institute for Fundamental, and
Clinical Human Movement Sciences (IFKB), Amsterdam, The Netherlands

4.1 INTRODUCTION

The EMG signal is a representation of the electric potential field generated by the depolarization of the outer muscle-fiber membrane (the sarcolemma). Its detection involves the use of intramuscular or surface electrodes that are placed at a certain distance from the sources. The tissue separating the sources and the recording electrodes acts as a so-called volume conductor. The volume conductor properties largely determine the features of the detected signals, in terms of frequency content and of distance beyond which the signal can no longer be detected.

In this chapter basic concepts of generation and detection of EMG signals will be described. In particular, attention is devoted to the issues of the appearance at the end plate, the propagation along the sarcolemma, and the extinction of the intracellular action potential at the tendons, crosstalk between nearby muscles, selectivity of the detection systems in relation to the signal sources and the volume conductor properties. The last

Electromyography: Physiology, Engineering, and Noninvasive Applications, edited by Roberto Merletti and Philip Parker.
ISBN 0-471-67580-6 Copyright © 2004 Institute for Electrical and Electronics Engineers, Inc.

part of the chapter presents the relationships between the developed force and the characteristics of the detected surface EMG signal.

To fully understand some parts of the Chapter, the reader should consult the paragraphs devoted to spatial filtering and sampling in Chapter 7. The issues discussed in this chapter provide a basis for the EMG modeling approaches presented in Chapter 8.

4.2 EMG SIGNAL GENERATION

The EMG signal is generated by the electrical activity of the muscle fibers active during a contraction. The signal sources are thus located at the depolarized zones of the muscle fibers. The sources of the signal are separated from the recording electrodes by biological tissues, which act as spatial low-pass filters on the (spatial) potential distribution [3]. In case of intramuscular recordings the effect of the tissues between electrodes and muscle fibers is relatively small due to the closeness of the recording electrodes to the sources. On the contrary, for surface recordings the volume conductor constitutes an important low-pass filtering effect on the EMG signal. As indicated, the volume conductor mediated filtering effect is principally spatial. It is a widespread misunderstanding that this filtering acts directly in the temporal domain. This will be elucidated later extensively, especially in relation to the possibilities to reduce crosstalk.

4.2.1 Signal Source

If a micropipette electrode is inserted intracellularly into a muscle fiber, a membrane resting potential of 70/90 mV (which is negative inside the cell with respect to the extracellular environment) is measured [38] (see Fig. 1.10, Chapter 1). The maintenance of this potential mainly depends on the activity of the energy taking sodium-potassium pumps working against the concentration gradients of ions flowing through the membrane [19]. The electric impulse that is propagated along the motoneuron arrives at its terminal and causes the emission of acetylcholine in the gap between the nerve terminal and the muscle fiber membrane, which excites the fiber membrane at this neuromuscular junction. In this case a potential gradient in a part of the fiber is generated. An inward current density (*depolarization zone*) corresponds to this potential change. The depolarization zone propagates along the muscle fibers from the neuromuscular junctions to the tendons' endings. The propagating intracellular action potential (IAP) causes an ionic transmembrane current profile also propagating along the sarcolemma. In case of nerve and muscle fibers the total length of the depolarization zone, along with the zone in which the membrane is repolarizing, is on the order of a few millimeters to several centimeters for fast conducting nerve fibers. This means that in terms of an electric circuit a fiber can be considered as a very thin tube in which current is only flowing axially. If this *line source model* [37] is assumed, the transmembrane current that is generated is proportional to the second spatial derivative of the IAP. The model depicted in Figure 4.1a represents a portion Δz of fiber membrane in the assumption of line source condition. The decrease in potential per unit length is equal to the product of the resistance per unit length and the current flowing through the resistance. For the extracellular and intracellular path we have

$$\frac{\partial \phi_e}{\partial z} = -I_e r_e, \quad \frac{\partial \phi_i}{\partial z} = -I_i r_i \tag{1}$$

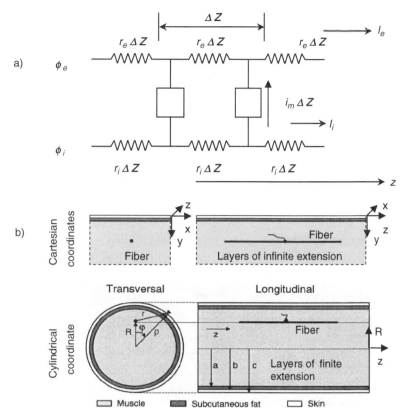

Figure 4.1. (a) Linear core-conductor model representing a portion of the fiber membrane. For graphical representation the structure is shown as a repetitive network of finite length Δz, but in fact Δz → 0; the analysis is based on the continuum. The open box is a symbol representing the equivalent circuit of the membrane, which depends on the membrane state, namely a passive structure during the resting period and a circuit with time-dependent components during the active phase, as described in Chapter 1. (b) Representation of muscle fiber position in Cartesian and cylindrical coordinate systems.

with r_e and r_i the resistance per unit length (Ω/cm) of the extracellular and intracellular path, respectively. Moreover conservation of current requires that the axial rate of decrease in the intracellular longitudinal current be equal to the transmembrane current per unit length:

$$\frac{\partial I_i}{\partial z} = -i_m \tag{2}$$

The extracellular longitudinal current may decrease with increasing z either because of a decrement of current that crosses the membrane (transmembrane current) or a loss that is carried outside by indwelling electrodes:

$$\frac{\partial I_e}{\partial z} = i_m + i_p \tag{3}$$

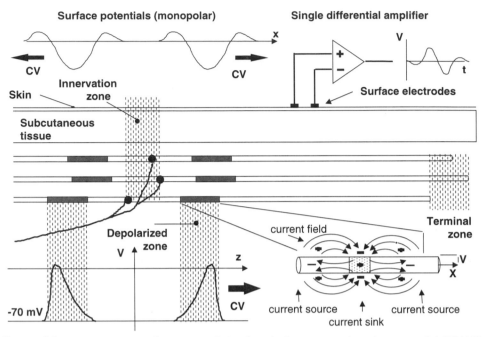

Figure 4.2. Representation of a motor unit (MU) and of a motor unit action potential (MUAP). Zoomed view of the source.

where i_p is the current flowing through the electrodes. The transmembrane voltage is given by

$$V_m = \phi_i - \phi_e \tag{4}$$

thus, its first derivative is obtained as (substituting Eq. 1):

$$\frac{\partial V_m}{\partial z} = \frac{\partial \phi_i}{\partial z} - \frac{\partial \phi_e}{\partial z} = -r_i I_i + r_e I_e \tag{5}$$

and deriving a second time (substituting Eqs. 2 and 3):

$$\frac{\partial^2 V_m}{\partial z^2} = r_i i_m + r_e (i_m + i_p) = (r_e + r_i) i_m + r_e i_p \tag{6}$$

which shows that the second derivative of the transmembrane potential is proportional to the transmembrane current in the hypothesis of the line source model. Figure 4.1b shows two coordinate systems (rectangular and cylindrical) adopted in the literature to study the field generated by a fiber's depolarized area. Figure 4.2 depicts the surface potential generated by a motor unit.

The velocity with which the action potential propagates depends on the fiber diameter and type and is referred to as conduction velocity (CV). It is generally accepted that CV increases with increasing fiber diameter [51]. The observed relation between fiber diameter and CV even seems to be valid for affected muscles in neuromuscular diseases [8].

The IAP shape can be approximated by simple functions, such as the analytical expression provided by Rosenfalck [57] or a triangular approximation. In general, the IAP can be characterized by a depolarization phase, a repolarization phase, and a hyperpolarizing long after potential. The IAP shape may change due to the conditions of the muscle. In particular, during fatigue few stages of IAP alteration can be distinguished [35]. In the beginning of fatigue, the IAP spike width in space increases mainly because of the slowing of the repolarization phase. In this phase the rate of increase of the IAP remains practically unchanged while the amplitude decreases slightly. The amplitude of the after-potential increases. In the following stages, the absolute value of the resting potential, spike amplitude, and rate of the IAP rise decrease together, with further slowing of the IAP falling phase and increasing of the after-potential amplitude.

4.2.2 Generation and Extinction of the Intracellular Action Potential

The generation and extinction of the IAP can be described in different ways [12,15,32,33,34,47]. Generally, it is assumed that the integral of the transmembrane current density over the entire muscle fiber length is zero at all times. On this basis Gootzen et al. [31,32] replaced the current density source at the end plate and at the tendons with an equivalent source proportional to the first derivative of the IAP. The same approach has been used in a model in which the volume conductor is simulated numerically by a finite element approach [45].

Dimitrov and Dimitrova [13] started their description by considering the first derivative of the IAP and assumed its progressive appearance at the end plate and disappearance at the tendons. This second approach is computationally and conceptually attractive and has been also used by a number of researchers who applied different computation techniques [24,47]. One approximates the IAP with a triangular function, whose first derivative is a function that assumes two constant values of opposite sign. The current density source is, in this case, approximated by a current tripole, sometimes divided conceptually in a leading and a trailing dipole part [18]. When the tripole reaches the tendon the first pole stops and the other two poles get closer to the first; when the second pole coincides with the first, the leading dipole disappears, and finally also the trailing dipole disappears at the tendon. A similar mechanism in opposite direction occurs at the end plate. The elementary life cycle of the IAP is schematically presented in Figure 4.3. From the basic understanding of the extinction of an IAP, independent of the way in which it is described, it can be concluded that the end-of-fiber potential wave shape equals the IAP wave shape [64]. On the basis of the concepts outlined by Dimitrov and Dimitrova [13], Farina and Merletti [24] proposed a general description of the current density source traveling at velocity v along the fiber with an origin and an end point:

$$i(z, t) = \frac{d}{dz}\left[\psi(z - z_i - vt)p_{L_1}\left(z - z_i - \frac{L_1}{2}\right) - \psi(-z + z_i - vt)p_{L_2}\left(z - z_i + \frac{L_2}{2}\right)\right] \quad (7)$$

where $i(z, t)$ is the current density source, $\psi(z) = dV_m(z)/dz$ ($V_m(z)$ is the IAP), $p_L(z)$ is a function that takes value 1 for $-L/2 \leq z \leq L/2$ and 0 otherwise, z_i is the position of the end plate, L_1 and L_2 are the length of the fiber from the end plate to the right and to the left tendon, respectively.

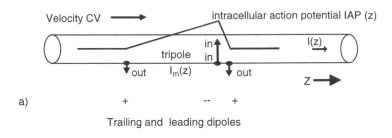

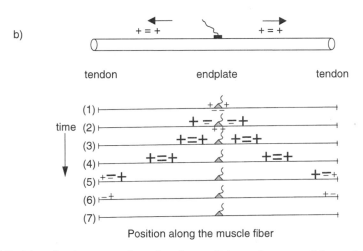

Position along the muscle fiber

Figure 4.3. (*a*) Stylized presentation of an intracellular action potential as a function of position along a muscle fiber during steady action potential propagation. Indications of propagation velocity *CV*, the short depolarization and the (longer tail) repolarization phase are included, as well as the tripole transmembrane ionic current I_m and its ultimate simplification in terms of a leading (− +) and trailing (+ −) dipole pair. (*b*) The spatiotemporal development of the transmembrane current along a muscle fiber. The transmembrane ionic current starts as two dipoles at the endplate after excitation by a motoneuron action potential (*1*). After full development of these first (leading) dipoles, a second (trailing) pair emerges (*2*). The balanced, double pair then propagates as two tripoles in opposite directions (*3, 4*). On arrival at the tendon, the leading dipoles decline in strength and disappears, leaving the trailing dipole (*5*). Subsequently also the trailing dipoles decline (*6*) and disappear (*7*).

Equation (7) is general and does not assume any approximation of the current density source. Special cases, such as the tripole approximation, are included in Eq. (7). From the equation we note that the sources of EMG signals are not plane waves traveling at constant velocity from minus to plus infinity. As a consequence EMG signals at different positions along the fiber's length are not simply delayed versions of each other.

In particular, when the potential stops at the tendon junction, it generates a signal component that is not propagating and will have different shapes at different locations along the muscle. These components are referred to as end-of-fiber signal or end-of-fiber effect. Their properties will be discussed in the next sections.

4.2.3 Volume Conductor

The generation of the intracellular action potential determines an electric field in the surrounding space. The potential generated by a motor unit (MU) can thus be detected also in locations relatively far from the source. The biological tissues separating the sources and the detecting electrodes are referred to as *volume conductor* and their characteristics strongly affect the detected signal.

Under the static hypothesis, in a volume conductor, the current density, the electric field, and the potential satisfy the following relationships [54]:

$$\nabla \cdot J = I, \quad J = \sigma E, \quad E = -\nabla \varphi \tag{8}$$

where J is the current density in the volume conductor (A·m^{-2}), I the current density of the source (A·m^{-3}), E the electric field (V·m^{-1}), and φ the potential (V).

From Eq. (8), Poisson's equation is obtained:

$$-\frac{\partial}{\partial x}\left(\sigma_x \frac{\partial \varphi}{\partial x}\right) - \frac{\partial}{\partial y}\left(\sigma_y \frac{\partial \varphi}{\partial y}\right) - \frac{\partial}{\partial z}\left(\sigma_z \frac{\partial \varphi}{\partial z}\right) = I \tag{9}$$

where σ_x, σ_y, and σ_z are the conductivities of the medium in the three spatial directions. Equation (9) is the general relation (in Cartesian coordinates) between the potential and the current density in a nonhomogenous and anisotropic medium. If the medium is homogenous, the conductivities do not depend on the point and the following equation is obtained:

$$-\sigma_x \frac{\partial^2 \varphi}{\partial x^2} - \sigma_y \frac{\partial^2 \varphi}{\partial y^2} - \sigma_z \frac{\partial^2 \varphi}{\partial z^2} = I \tag{10}$$

Equation (10) can be also written in cylindrical coordinates (ρ, z), resulting in

$$\frac{\partial^2 \varphi}{\partial \rho^2} + \frac{1}{\rho}\frac{\partial \varphi}{\partial \rho} + \frac{1}{\rho^2}\frac{\partial^2 \varphi}{\partial \phi^2} + \frac{\partial^2 \varphi}{\partial z^2} = -\frac{I}{\sigma} \tag{11}$$

The solution of Eq. (10) or (11) provides, theoretically, the potential in any point in space when the characteristics of the source (I) and of the medium (σ) are known. This solution can be obtained only if the boundary conditions can be described in simple coordinate systems. This problem has been solved for different degrees of simplification. The simplest assumption for the solution of Poisson's equation is to deal with a homogeneous, isotropic, infinite volume conductor. In this case, assuming a source distributed along a line in z coordinate direction, the potential distribution in the volume conductor is given by the following relationship:

$$\varphi(r, z) = \frac{1}{2\sigma} \int_{-\infty}^{+\infty} \frac{I(s)}{\sqrt{r^2 + (z-s)^2}} ds \tag{12}$$

where $I(z)$ is the current density source, and σ is the conductivity of the medium. In case of a semi-infinite medium with different conductivities in the longitudinal and radial directions (e.g., muscle tissue), we get

$$\varphi(r, z) = \frac{1}{\sigma_r} \int_{-\infty}^{+\infty} \frac{I(s)}{\sqrt{r^2(\sigma_z/\sigma_r) + (z-s)^2}} ds \qquad (13)$$

where σ_z and σ_r are the longitudinal and radial conductivities, respectively. The method of images may be applied to compute the surface potential distribution in case of a semi-space of conductive medium (tissue) and a semispace of insulation material (air). It is one of the basic observations from electrostatics that in that case the surface potential doubles with respect to the case of an infinite medium [47].

More complex descriptions of the volume conductor have been proposed and include non-homogeneous medium comprised of layers of different conductivities. In case of layered geometries, Eqs. (10) and (11) can be solved independently in the different layers. The final solution is then obtained by imposing the boundary conditions at the surfaces between the layers. Boundary conditions are the continuity of the current in the direction perpendicular to the boundary surface and continuity of the potential itself over the boundary. Additional conditions are obtained by imposing the absence of divergence of the potential in all the points of the volume conductor, except for the locations of the sources.

In case of Cartesian coordinates and layered medium (infinite layers parallel to the xz plane), the boundary conditions are

$$\sigma_{i+1} \frac{\partial \varphi_{i+1}(x, y, z)}{\partial y}\bigg|_{y=h_i} = \sigma_i \frac{\partial \varphi_i(x, y, z)}{\partial y}\bigg|_{y=h_i}$$

$$\frac{\partial \varphi_{i+1}(x, y, z)}{\partial x}\bigg|_{y=h_i} = \frac{\partial \varphi_i(x, y, z)}{\partial x}\bigg|_{y=h_i}$$

$$\frac{\partial \varphi_{i+1}(x, y, z)}{\partial z}\bigg|_{y=h_i} = \frac{\partial \varphi_i(x, y, z)}{\partial z}\bigg|_{y=h_i} \qquad (14)$$

for the interfaces (here $y = h_i$) between adjacent layers. Similar expressions can be derived in case of cylindrical coordinate system [9].

Farina et al. [24,28] computed, in the spatial frequency domain, the surface potential distribution over the skin plane in a three-layer volume conductor model. The layers were separated by planes; thus the solution was provided in the Cartesian coordinate system.

A cylindrical description of the volume conductor is more realistic in the sense that it takes into account the finiteness of the volume conductor. A three-layer model (muscle, fat, and skin) has been developed by Blok et al. [9] as an elaboration of the two-layer model described by Gootzen et al. [31,32]. The general solution of the Poisson's equation for each of the three concentric cylindrical layers of this configuration, for the potential detected over the skin layer reads as follows:

$$\Phi_3(\rho, \phi, k) = \frac{d}{2\sigma_{3r}} K_0\left(r\sqrt{\frac{\sigma_{3z}}{\sigma_{3r}}} |k| \right) G(k)$$

$$+ \sum_{n=-\infty}^{\infty} e^{-in\phi} \left[E_n(k) I_n\left(\rho\sqrt{\frac{\sigma_{3z}}{\sigma_{3r}}} k \right) + F_n(k) K_n\left(\rho\sqrt{\frac{\sigma_{3z}}{\sigma_{3r}}} |k| \right) \right] \qquad (15)$$

In Eq. (15), ρ and ϕ are the cylindrical coordinates, r is the distance of the fiber axis from the detection point, k is the spatial angular frequency in the axial direction, and $G(k)$ is the Fourier transform of the electric current source function to the spatial frequency domain. The conductivity of the skin layer in the radial direction is represented by the parameter σ_{3r}, that in the axial direction by σ_{3z}. All three layers were allowed to be anisotropic in the cylindrical coordinates r and z. The functions K_n and I_n are modified Bessel functions of order n, of the first and second kind, respectively, and $E_n(k)$, $F_n(k)$ are unknowns that have to be determined for each n and k from boundary conditions. Together with expressions similar to Eq. (15), for the two other layers (muscle, subcutaneous fat), five of such unknowns have to be determined by the use of five boundary conditions. Finally, d is the diameter of the fiber. Because the K_n and I_n Bessel functions tend to very large or very small values for increasing values of n and for small or large values of k, the solution system becomes ill-conditioned and its solution inaccurate. As described by Gootzen et al. [31], it is possible to condition the linear system by rewriting it.

It has to be noted that the finiteness of the volume conductor has peculiar consequences especially for the appearance of the end-of-fiber potentials, since they then do belong to the category of so-called far-field potentials [18,32,61], indicating that the potential of a dipolar source can principally be of a nondecaying character [61]. Depicted in Figure 4.4 are examples of monopolarly recorded single muscle fiber action potentials, showing the increasing and differential influence of volume conduction on the propagating and non-propagating components with increasing observation distance. More advanced descriptions of the volume conductor may involve tissue in-homogeneities [59] or the presence of the bones [45] and, in case of needle EMG, the presence of the needle itself [62].

4.2.4 EMG Detection, Electrode Montages and Electrode Size

EMG can be detected by intramuscular electrodes (Chapter 1) or by electrodes attached to the skin surface. The insertion of electrodes directly into the muscle allows the detection of electric potentials very close to the source, so the influence of the volume conductor on the current sources at the fiber membranes is minimal. For this reason the action potentials of the different MUs can be separated relatively easily at medium/low force levels (Chapter 3).

When surface electrodes are applied, the distance between the source and the detection point is significant, and the spatial low-pass filtering effect of the volume conductor becomes relevant. To remove the common mode components caused by technical interference (e.g., a power line) and to partially compensate for the low-pass filtering effect of the tissue separating sources and electrodes (see Chapter 5), the surface signals are usually detected as a linear combination of the signals recorded at different electrodes. This operation can be viewed as a spatial filtering of the monopolar surface EMG signal (Chapters 2 and 7) [17]. The simplest form is the differential detection, the "classical" bipolar montage.

The interpretation of the effect of the electrode configuration as a spatial filtering operation and the relationship between the time and space domains (Chapter 2) led in the past to the observation that by properly selecting the weights of the different electrodes, it is possible to introduce zeros of the transfer function of the filter. If they are within the spatial bandwidth of the EMG signal and assuming pure propagation of the IAP along the muscle fibers, these zeros are reflected in the (time-related) frequency spectrum of the EMG and are referred to as spectral dips. The presence of spectral dips has been theoretically shown by Lindstrom and Magnusson [43] for 1-D differential detection systems. Spectral dips

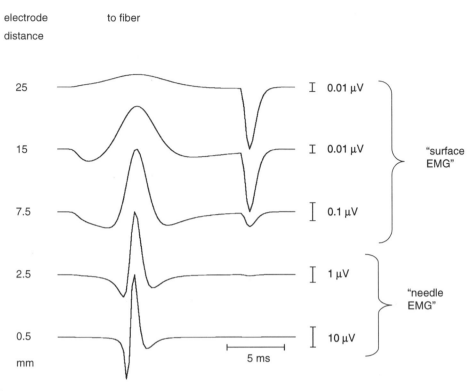

Figure 4.4. Calculated single-fiber action potentials from the same muscle fiber. The observation distance to the muscle fiber decreases from 25 mm (upper trace, monopolar skin recording from a single fiber in a deep MU) to 0.5 mm (lowest trace, monopolar needle electrode recording near to the fiber). Note the large amplitude and waveform differences, and also the differences in vertical scaling. The propagation velocity is 4 m/s, the half fiber length is 60 mm, and the electrode is positioned at 20 mm from the endplate zone. Positive deflection downward. (Adapted from [9] with permission)

were found experimentally for 1-D differentially detected signals by Lindstrom et al. [44] and later by a number of investigators. More in general, considering both one- and two-dimensional systems with point or with physical dimension electrodes, Farina and Merletti [24] demonstrated that a sufficient (but not necessary) condition for having a dip in the EMG spectrum at the spatial frequency f_{z0} is

$$H_{ele}(f_x, f_{z0}) = 0 \qquad \forall f_x \tag{16}$$

where f_x and f_z are the spatial frequencies in the direction transversal and parallel to the muscle fibers, respectively, and $H_{ele}(f_x, f_z)$ is the transfer function of the detection system (including the spatial filter due to the linear combination of signals and to the physical dimensions of electrodes).

Condition (16) cannot hold in the case of detection systems that result in a nonconstant $H_{ele}(f_x, f_z)$ for f_z fixed at the dip frequency. Special cases are those resulting in an isotropic transfer function (for which $H_{ele}(f_x, f_z)$ cannot be constant for any fixed f_z), such as a circular electrode or a concentric ring system (see Chapter 7). In Chapters 2 and 5 the concept of spectral dip and its use in estimating CV for one-dimensional electrode systems is presented.

Also the physical size of an electrode influences the EMG signal. When the electrode-skin impedance between electrode material and the skin surface is equally distributed and when it is (1) low compared to the input impedance of the amplifier but (2) high compared to the impedances within the tissue, it can easily be argued that the potential measured by an electrode equals the average of the potential distribution over the skin under it [14][22][25]. As a consequence the influence of the electrode size can also be described as a spatial low-pass filter whereby the electrode's dimensions define the filter shaping. As in the case of electrode montages, here the influence of electrode size is largely dependent on structural elements of the EMG sources, like the direction of the muscle fibers with respect to the electrode length or width.

4.3 CROSSTALK

4.3.1 Crosstalk Muscle Signals

Crosstalk is the signal detected over a muscle but generated by another muscle close to the first one. The phenomenon is present exclusively in surface recordings, when the distance of the detection points from the sources may be relevant and similar for the different sources. Crosstalk is due to the volume conduction properties in combination with the source properties, and it is one of the most important sources of error in interpreting surface EMG signals. This is because crosstalk signals can be confounded with the signals generated by the muscle, which thus may be considered active when indeed it is not. The problem is particularly relevant in cases where the timing of activation of different muscles is of importance, such as in movement analysis.

If it is clear that crosstalk is a consequence of volume conduction; it is less easy to identify the signal sources that are mostly responsible for it (through volume conduction). The identification of crosstalk sources is crucial for the development of methods for its quantification and reduction. In the past there have been many attempts to investigate crosstalk.

Morrenhof and Abbink [50] used the cross-correlation coefficient between signals as an indicator of crosstalk, assuming minor shape changes of the signals generated by the same source and detected in different locations over the skin. This assumption was based on a simple model of surface EMG signal generation that did not take into account the generation and extinction of the IAP at the end plate and tendon. Their approach for the verification of the presence of crosstalk was also based on the joint recording of intramuscular and surface EMG signals. The same method was followed by Perry et al. [52], who proposed crosstalk indexes based on the ratio between the amplitudes of the surface and intramuscular recordings.

De Luca and Merletti [11] investigated crosstalk by electrical stimulation of a single muscle and detection from nearby muscles. They provided reference results of crosstalk magnitude for the muscles of the leg and proved the theoretically higher selectivity of the double differential with respect to the single differential recording (at least for propagating signal components). A similar technique was more recently applied for crosstalk quantification by other investigators with more complex spatial filtering schemes [67].

4.3.2 Crosstalk and Detection System Selectivity

The selectivity of a detection system for surface EMG recording can be defined as the volume of muscle from which the system records signals that are above noise level. The more selective a system is, the smaller is the number of sources detected. Roeleveld et al. [55] did a comprehensive experimental study of the contribution of the potentials of single MUs of the biceps brachii muscle to the surface EMG. A cross-sectional impression of the influence of bipolar electrode montage versus a monopolar recording for a superficial and a deep MU respectively is presented in Figure 4.5. Intramuscularly detected MU potentials (MUAPs) served as triggers for an averaging process aimed at extracting the surface potentials in different locations over the muscle. The difference between the monopolar and the bipolar recording system is obvious as is the influence of MU depth. Farina et al. [21] recently compared in a similar way the selectivity of a number of spatial filtering systems for surface EMG recording on the basis of joint intramuscular and surface recordings. In particular, these authors detected single MU activities from the intramuscular recordings in different muscle contractions, during which surface systems were located at different locations over the muscle. A representative result of this processing method is shown in Figure 4.6. The rate of decay of the peak potential amplitude of a single MUAP is different for the different filters. It has also to be noted from Figures 4.5 and 4.6 that, in any case, the potentials detected decrease to very small amplitudes within 20 to 25 mm from the source. Similar findings were obtained for a number of MUs and subjects.

The previous findings might suggest that the potential generated by a source decays rather fast in space, and therefore crosstalk should be a limited problem. However, the MUs considered in [21] were all very superficial and with rather long fibers. From Figure 4.5b and d, it can be seen that for deep MUs there hardly is a spatial gradient between different detection locations. Although the contribution of the deep MU in the bipolar montage (Fig. 4.5d) seems negligible, there also is hardly a difference between the electrodes right above the MU and those lying further aside. Since there are many more deep

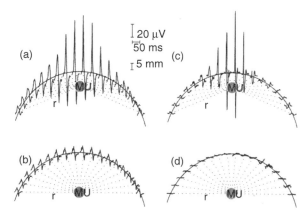

Figure 4.5. Cross-sectional impression, over the surface of the biceps brachii, of the action potentials of a superficial (a, c) and a deep (b, d) MU. Monopolar (far-away reference) (a, b) and bipolar (c, d) detection with a transversal electrode array. These data on individual MUs were obtained in a study in which so-called intramuscular scanning EMG was combined with surface EMG recordings. The interelectrode distance was 6 mm. (From [56] with permission)

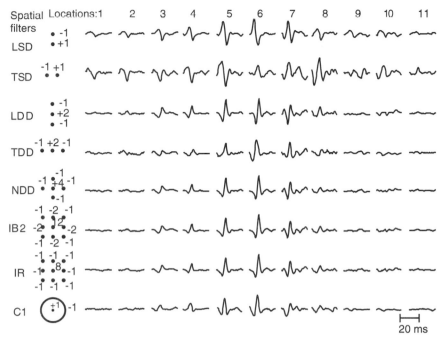

Figure 4.6. Single-surface MU action potentials detected transversally with respect to the muscle fiber direction from the tibialis anterior muscle. The surface potentials were averaged using intramuscularly detected potentials as triggers (contraction level 20% MVC). The results from eight systems for surface EMG detection (schematically represented on the left) are shown. For each spatial filter the waveforms are normalized with respect to the highest peak to peak value among the waveforms detected by that filter. The distance between each recording location is 5 mm, the interelectrode distance of the point electrode spatial filters is 5 mm, and the radius of the ring system is 5 mm. Position 6 is above the selected MU whose intramuscular potential was used as trigger. (Adapted from [21] with permission)

than superficial MUs, this suggests that crosstalk is not a problem that ends 20 to 25 mm away from the source.

Farina et al. [26] applied the technique proposed by De Luca and Merletti [11] (selective muscle stimulation) to the extensor leg muscles and recorded signals by eight contact linear electrode arrays (see also Chapter 7). Their results, in agreement with [20,69], are summarized by the representative signals in Figure 4.7. In this case the signals detected far from the source are generated when the action potentials of the active muscle extinguish at the tendon region and generate nonpropagating components. From Figure 4.7 different considerations can be drawn: (1) crosstalk is mainly due to the so-called far-field signals [61,64] generated by the extinction of the potentials at the tendons, (2) the shape of crosstalk signals is different from that of signals detected over the active muscle, (3) as a consequence of point 2, cross-correlation coefficient is not indicative of the amount of crosstalk, (4) the bandwidth of crosstalk signals may be even larger than that of signals from the active muscle and thus crosstalk reduction cannot be achieved by temporal high-pass filtering of the surface EMG signals [16]. These considerations were proved on a statistical basis in [26], and they indicate that the two methods usually applied to

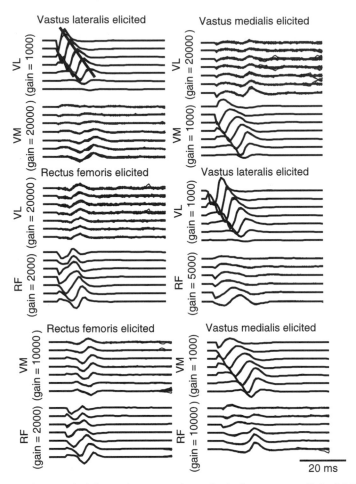

Figure 4.7. Signals recorded from the vastus lateralis (VL), vastus medialis (VM), and rectus femoris (RF) muscles in six conditions of selective electrical stimulation (2 Hz stimulation frequency) of one muscle and recording from a muscle pair (one stimulated and one nonstimulated). In each case the responses to 20 stimuli are shown. For this subject, the distances between the arrays (center to center) are 77 mm for the pair VL-RF, 43 mm for the pair VM-RF, and 39 mm for the pair VL-VM. Note the different gains used for the stimulated and non-stimulated muscles. In this case the ratio between average ARV values (along the array) is, for the six muscle pairs, 2.8% (VL-VM), 2.5% (VM-VL), 6.5% (RF-VL), 6.6% (VL-RF), 15.6% (RF-VM), and 7.0% (VM-RF). The signals detected from a pair of muscles are normalized with respect to a common factor (apart from the different gain) and so are comparable in amplitude, whereas the signals detected from different muscle pairs are normalized with respect to different factors. (From [26] with permission)

identify and reduce crosstalk (cross-correlation, high-pass filtering) are not, in general, appropriate. In particular, cross-correlation between signals, often used to verify the absence or presence of crosstalk [50,71], does not seem justified.

It is important that spatial and temporal frequency characteristics are simply related by the conduction velocity v only for components that are traveling at the velocity v

without relevant shape changes along the fiber. For this reason the temporal frequency f_t is related by a scaling factor to the spatial (in the direction of propagation) frequency f_z (see also Chapter 2). As a consequence low (high) spatial frequencies result in low (high) temporal frequencies. For nonpropagating components this does not occur, as it is clear in Figure 4.7. Indeed, if we consider the crosstalk signals (nontraveling components that arise over the nonstimulated muscle), we will observe that their temporal frequencies are often higher that those of the signals detected from the stimulated muscles. However, considering the rate of variation of the signal in the spatial direction (i.e., along the array), we note that the signal is almost constant (being nontraveling), thus that it is spatially almost dc but changes rapidly in time. High-pass filtering in temporal domain may therefore enhance, rather than reduce, the crosstalk components, which have high temporal and low spatial frequency content. For the same reason it is not possible to predict the amount of crosstalk from models that do not take into consideration the end-of-fiber components; indeed, the rate of decay of the propagating part of the signal is considerably faster than that of the nontraveling potentials [18,61], so that the latter are mostly responsible for crosstalk.

Figure 4.8 shows simulated single-fiber action potentials recorded at different locations along the fiber by single differential recordings with 10-mm interelectrode distance. Three lateral distances are shown for two locations along the fiber: in the middle between the end plate and tendon and at the tendon level. With electrodes directly over the fiber ($x_0 = 0$ mm), fat layer thickness 3 mm, and recording between the end plate and tendon (location 1), it is possible to observe the classic biphasic single differential waveform (compare with Fig. 4.7 in the case of stimulated muscles). The effect of the generation of the action potential can be noted at the beginning of the waveform, and a small end-of-fiber potential of opposite sign with respect to the second phase appears. At the tendon (location 2) the end-of-fiber potential is much larger and is the dominant component of the detected surface signal. If the fat layer thickness increases, the end-of-fiber potentials become the dominant signals for both electrode locations along the fiber. If the transversal distance x_0 increases, the end-of-fiber potentials become increasingly important with respect to the propagating components (in particular, the case $x_0 = 75$ mm). The maximum amplitude signal then corresponds to electrode locations nearest to the tendon region, indicating that the signal source is located at such a region. Moreover the increase of fat layer thickness increases the relative amplitude of end-of-fiber potentials with respect to the propagating potentials. From the case of potentials detected at $x_0 = 0$ mm and at $x_0 = 75$ mm (which simulate crosstalk signals), it can be concluded that the characteristics of the signals are completely different in the two conditions. The attenuation of the propagating component with increasing x_0 is greater than that of the nonpropagating component, so that at $x_0 = 75$ mm the second component is much greater than the first. Finally, observe that the frequency content of the signals close and far from the source is rather different, with the far sources showing higher frequency components, as in the experimental case of Figure 4.7. The influence of the end-of-fiber contribution, which is of the so-called far-field type, becomes even more dominant in case of a finite volume conductor. In that situation a nondecaying contribution over long (theoretically infinite) distances away from the muscle can be observed in the monopolar recordings [9,61].

Since crosstalk signals have only low spatial frequency components, one would think that the use of high-pass spatial filtering (refer to Chapters 2 and 7) may help reduce them, with larger reduction being possible through more selective filters. However, the theoretical transfer functions, in the spatial domain, of the spatial filters are based on the assumptions of signals propagating with the same shape along the direction of the filter. It is clear

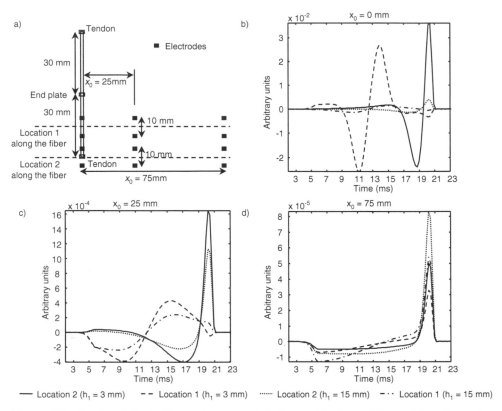

Figure 4.8. Simulated single-fiber action potentials detected in single differential configuration between the end plate and tendon (location 1) and at the tendon (location 2), at three distances from the fiber (0 mm, 25 mm, 75 mm). The interelectrode distance is 10 mm, the depth of the fiber is 1 mm within the muscle. Two values of thickness of the fat layer (h_1) have been simulated. (*a*) Schematic representation of the simulated fiber and of the detection points, (*b*) signals detected over the fiber, (*c*) signals detected at 25 mm laterally from the fiber, and (*d*) signals detected at 75 mm laterally from the fiber. The electrodes are aligned with the muscle fiber; h_1 = thickness of the fat layer. (From [26] with permission)

that the correspondence between time and space is not valid when the signals considered do not travel with the same shape along the fiber direction. Thus the theoretical high-pass transfer function of the spatial filters cannot, in general, be applied to signals detected far from the source. Considering the nontraveling components as signals in the time domain that may be different when detected at different locations along the muscle fibers, the application of a spatial filter with electrodes with weights a_i provide the following signal:

$$V(t) = \sum_{i=0}^{M} a_i R_i(t) \qquad (17)$$

Since the (monopolar) signals $R_i(t)$ depends on a number of anatomical and physical factors, there is no possibility to predict the filter effect on the signal a priori. Equation (17) provides the general expression of a nontraveling signal as detected by a spatial filter

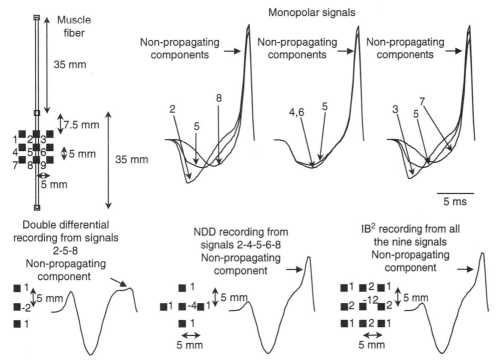

Figure 4.9. Examples of monopolar signals detected with square (1 × 1 mm) electrodes in the longitudinal, transversal, and diagonal direction. The waveform filtered by the double differential (*DD*), Laplacian (*NDD*), and inverse binomial filter of the second-order (*IB*^{2}) filters ([17], Chapter 7) are also shown. The signals are generated by the same fiber with equal semilengths of 35 mm, at a depth of 7 mm within the muscle. Thickness of the fat and skin layer is 4.5 mm and 1 mm, respectively. (From [22] with permission)

comprised of point electrodes. It is not possible to write Eq. (17) as a linear filtering operation and, thus, to derive the transfer function in space domain for components which are not traveling along the fiber direction. In particular, the effect of specific spatial filter transfer function on the signals $R_i(t)$ cannot be predicted, as indicated in Figure 4.9. Figure 4.10 shows the decrease of signal amplitude with increasing distance from the source for different detection systems. Both the propagating and the nonpropagating components of the signal decrease with the distance depending on the spatial filters. Faster decrease of the propagating signal component does not imply also faster decrease of the nonpropagating signal components, however. Selectivity with respect to these components should be addressed separately (see also Chapter 7).

4.4 RELATIONSHIPS BETWEEN SURFACE EMG FEATURES AND DEVELOPED FORCE

4.4.1 EMG Amplitude and Force

Force production in a muscle is regulated by two main mechanisms (see also Chapter 1): the recruitment of additional MUs and the increase of firing rate of the already active MUs.

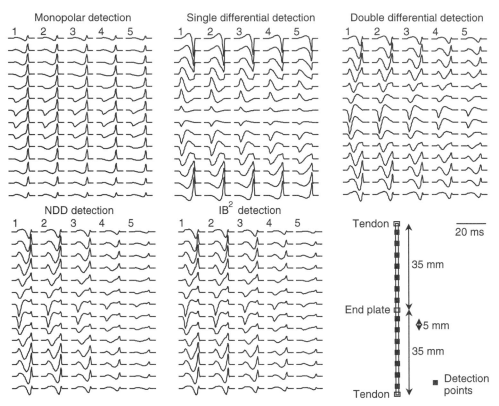

Figure 4.10. Simulated single-fiber potentials detected with linear arrays with 14 longitudinal detection points separated by 5 mm. Monopolar, single differential (*SD*), double differential (*DD*), Laplacian (*NDD*), and inverse binomial filter of the second-order (IB^2) signals ([17], Chapter 7) are shown. Signals detected at a lateral distance from the fiber in the horizontal plane from 0 to 10 mm at steps of 2.5 mm are shown (labeled by numbers from 1 to 5). All signals are generated by the same fiber with equal semilengths of 35 mm, at a depth of 7 mm within the muscle. Thickness of the fat and skin layer is 4.5 mm and 1 mm, respectively. (From [22] with permission)

These two mechanisms are present in different proportions in different muscles. In some muscles the recruitment stops at about 50% of the maximal force, while in others recruitment has been observed almost until the maximal voluntary contraction force [39]. As for the developed force, also the amplitude of the surface EMG signal depends on both the number of active MUs and their firing rates. Since both EMG amplitude (estimated in one of the ways presented in Chapter 6) and force increase as a consequence of the same mechanisms, it is expected that muscle force can be estimated from surface EMG analysis.

The possibility of estimating muscle force from the EMG signal is attractive as it allows the assessment of the contributions of single muscles to the total force exerted by a muscle group. This is the main reason why EMG is and probably always will be the method of choice for force estimation in kinesiological studies. The problem has been addressed experimentally by many researchers in the past (e.g., [48]). In some muscles, such as those controlling the fingers, the relationship between force and EMG amplitude was found to be linear [6,72], while in others the relation is closer to a parabolic shape

[41,72]. Differences in the percentage of recruitment and rate coding have been considered the most likely explanation for these different relations. It should be noted, however, that even a linear relation between EMG and force cannot be explained when starting straightforwardly from the neural drive to a muscle. As a matter of fact, both the force and the EMG amplitude are in most circumstances nonlinearly related to the neural drive. For instance, doubling the number of recruited MUs or doubling the firing rate of an already active population of MUs will, under the simplest signal theoretical assumptions, lead to an EMG increase by a factor of $\sqrt{2}$. A doubling in MU firing rate also does lead to less than a doubling of the force of that MU. Apparently both nonlinearities in the relation between neural drive and EMG, on the one hand, and drive and force, on the other hand, balance each other, leading to an often close-to-linear relation between EMG and force.

Simulation studies are here also relevant in quantifying these relations. Fuglevand et al. [29] performed a simulation study that included a number of recruitment modalities. The relationship between EMG amplitude (average rectified value) and force varied depending on the recruitment strategy and the uniformity among the peak firing rates of the different MUs (Fig. 4.11).

When discussing the issue of the relation between EMG amplitude and force, a number of other factors should be taken into account. First, the surface EMG amplitude depends strongly on the electrode location. For locations in which EMG amplitude is very sensitive to small electrode displacements it is expected that the relation between EMG and force may be poorer than in other locations [27]. Jensen et al. [36] elegantly addressed the issue of electrode location in relation to the EMG/force characteristic for the upper trapezius muscle. These authors showed that while an almost linear relationship between EMG and force could be observed for electrode locations far from the innervation zone, the relationship was far from linear when considering the electrode location resulting in the minimal EMG amplitude (over the muscle belly in the case of the upper trapezius).

Considering an "optimal" electrode placement, the relation between force and EMG may depend on the subcutaneous fat layer thickness, the inclination of the fibers with respect to the detection system, the distribution of conduction velocities of the active MUs, the interelectrode distance, the spatial filter applied for EMG recording, the presence of crosstalk, and the degree of synchronization of the active MUs. Moreover there is a basic factor of variability among different subjects, being the location of the MUs within the muscle. The same control strategy may generate signals with different amplitude trend due to the different locations of the MUs within the muscle [23]. It is clear that all these factors make it impossible to consider any specific EMG-force relation valid, in general. This relation should be identified on a subject by subject and muscle by muscle basis and is likely to be poorly repeatable in different experimental sessions, even on muscles of the same subject. The experimental construction of such a relation may be, however, difficult, if not impossible, when selective activation of a specific muscle of a group is problematic and other muscles (agonists or antagonists) contribute to force but not to EMG.

Importantly, in addition to all these limitations, the relationship (if any) between force and amplitude should be adapted to the muscle condition, including muscle length (joint angle), muscle temperature, fatigue, and so on. In particular, under submaximal contractions, the fatigued muscle generates EMG signals with larger amplitude compared to the unfatigued condition, although maintaining a constant force (see Chapter 9). Given the large variability of behavior of EMG amplitude depending on these factors, an adaptive relation EMG-force is currently not possible, and it will hardly become feasible in the near future. EMG amplitude will likely remain an indicator of muscle activation and a qualitative index of activation change with respect to a reference condition.

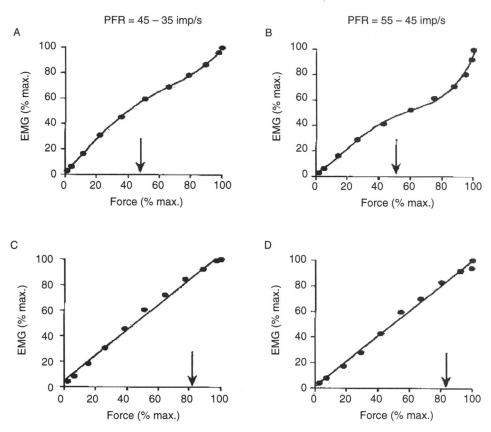

Figure 4.11. EMG amplitude–force relationship in simulated signals for narrow (*A* and *B*) and broad (*C* and *D*) recruitment range conditions. Peak firing rates were inversely related to the motor unit recruitment threshold and ranged from 35 to 45 pps (*A* and *C*) and from 45 to 55 pps (*B* and *D*). Arrows on force axes show the force at which last unit was recruited for each condition. When recruitment operated over a narrow range (*A* and *B*), the EMG-force relation was nonlinear. When recruitment operated over a broad range (*C* and *D*), the relationship between EMG and force was linear. (From [29] with permission)

4.4.2 Estimated Conduction Velocity and Force

Under isometric conditions the recruitment of MUs proceeds from the smallest (slow twitch) to the largest (fast twitch) and from the lowest to the highest CV [1], although in dynamic circumstances this simple scheme does not always hold [66]. The estimation of CV is ideally not affected by the MU depth in case of infinite length fibers. However, for finite fibers, the deeper the MU is, the larger the relative influence of the end-of-fiber components is on CV estimation. This phenomenon might positively bias CV estimation in case of the inhomogeneous recruitment of MUs in the muscle, such as from deep to superficial MUs. Moreover the CV of a muscle fiber is a function of the firing rate of that fiber; therefore the global CV estimates depend on the firing rates of the active MUs, which can easily be visualized with direct muscle fiber stimulation techniques [73]. Thus during an isometric increasing force contraction the global CV estimates usually increase through the entire contraction. At the beginning of a ramp contraction CV estimate reflects the CV of the low threshold MUs, whereas at the end it is an averaged value which depends on the characteristics (depth of the fibers, firing rate, size) of all the active MUs.

4.4.3 EMG Spectral Frequencies and Force

The power spectral density of the interference EMG signal for independent MU firing patterns is the summation of the spectral densities of the MU action potential trains generating the signals. It can be shown (see also Chapter 6) that the power spectrum of a MU action potential train is not influenced significantly by the train of impulses defining the instants of activation of the MUs. Rather, it depends mainly on the shape of the potentials of the active MUs.

Many past studies investigated the relationship between characteristic frequencies of the surface EMG signal and force [4,5,7,30,68]. In particular, in some it was hypothesized that mean and median power spectral frequencies (MNF and MDF; Chapter 6) do reflect the recruitment of new, progressively larger and faster MUs and therefore increase until the end of the recruitment process. They should then reach a constant value (or slightly decrease) when only rate coding is used to track the desired target force level [5]. The latter hypothesis is based on two theoretical considerations: (1) the CV of a single MU action potential scales the power spectrum of that MU action potential [10,43,63,65] (see also Chapter 2), and (2) the MU firing rates do not affect significantly the frequency content of the surface EMG signal [40]. Both observations are simplifications of the real situation. However, observation 1 was experimentally validated within reasonable approximations in case of isometric constant force fatiguing contractions [2,46]. Using intramuscular EMG detection, Solomonow et al. [60] showed experimentally that both observations held when controlled physiological MU recruitment was obtained by a particular stimulation technique. However, as indicated by these authors, extrapolation of these results to surface EMG was not implied since the detection was very selective. The volume conductor has, indeed, a large influence both on the amplitude and on the frequency content of surface detected MUAPs.

The volume conductor theory indicates that any relationship between recruitment and spectral features can be masked by anatomical or geometrical factors. When a deep MU is recruited it will contribute to the low-frequency region of the EMG power spectrum even if its CV is higher than the mean CV of the previously recruited MUs. Thus a recruitment order from the slow to the fast MUs does not necessarily imply an increase of characteristic spectral frequencies. As a consequence the behavior of MNF or MDF with force may change significantly among different subjects and muscles [23].

Along the same line it is not surprising that the works focusing on the investigation of surface EMG power spectrum during ramp contractions reported very different results. Some authors found an increase of spectral variables with increasing force level [30,49], others showed no increase of these variables with force level [53,68], and still others observed a decrease of MNF with increasing force [70]. A dependence of spectral features on electrode location has also to be considered when discussing information extraction from characteristic surface EMG frequencies.

4.5 CONCLUSIONS

The IAP has a shape that depends on the muscle status and that changes with fatigue. The IAPs are the sources of EMG signals, and they propagate without shape changes along the muscle fibers. At the end plate and tendon the IAP originates and extinguishes, so that the total current density over the entire muscle fiber length is zero at all times. The generation and extinction of the IAP produces the so-called end plate and end-of-fiber effects, which are simultaneously detected over the entire muscle length (nonpropagating) and define the distance beyond which the EMG signal can no longer be detected. This distance is related to the selectivity of the detection, although care should be taken when discussing selectivity with respect to different (propagating and nonpropagating) signal sources. The rate of decrease of signal amplitude with observation distance for the propagating and the nonpropagating components (the two main signal components) of the EMG signal can be very different. These considerations are of paramount importance when the issue of crosstalk is discussed.

Considering all the factors related to the volume conductor and the signal sources that affect the characteristics of the EMG signal, it is not surprising that some relationships between EMG signal variables and muscle force are critical and should be carefully considered in practical applications.

REFERENCES

1. Andreassen, S., and L. Arendt-Nielsen, "Muscle fiber conduction velocity in motor units of the human anterior tibial muscle: A new size principle parameter," *J Physiol* **391**, 561–571 (1987).
2. Arendt-Nielsen, L., and K. R. Mills, "The relationship between mean power frequency of the EMG spectrum and muscle fibre conduction velocity," *Electroencephalogr Clin Neurophysiol* **60**, 130–134 (1985).
3. Basmajian, J. V., and C. J. DeLuca, *Muscle alive*, Williams and Wilkins, Baltimore, 1985.
4. Bernardi, M., F. Felici, M. Marchetti, F. Montellanico, M. F. Piacentini, and M. Solomonow, "Force generation performance and motor unit recruitment strategy in muscles of contralateral limbs," *J Electromyogr Kinesiol* **9**, 121–130 (1999).
5. Bernardi, M., M. Solomonow, G. Nguyen, A. Smith, and R. Baratta, "Motor unit recruitment strategies changes with skill acquisition," *Eur J Appl Physiol* **74**, 52–59 (1996).
6. Bigland, B., and O. C. J. Lippold, "The relation between force, velocity and integrated electrical activity in human muscles," *J Physiol* **123**, 214–224 (1954).
7. Bilodeau, M., A. B. Arsenault, D. Gravel, and D. Bourbonnais, "EMG power spectra of elbow extensors during ramp and step isometric contractions," *Eur J Appl Physiol* **63**, 24–28 (1991).
8. Blijham, P. J., B. G. M. Van Engelen, and M. J. Zwarts, "Correlation between muscle fiber conduction velocity and fiber diameter in vivo," *Clin Neurophysiol* **113**, 39 (2002).
9. Blok, J. H., D. F. Stegeman, and A. van Oosterom, "Three-layer volume conductor model and software package for applications in surface electromyography," *An Biomed Eng* **30**, 566–577 (2002).
10. DeLuca, C. J., "Physiology and mathematics of myoelectric signals," *IEEE Trans BME* **26**, 313–325 (1979).
11. DeLuca, C. J., and R. Merletti, "Surface myoelectric signal cross-talk among muscles of the leg," *Electroencephalogr Clin Neurophysiol* **69**, 568–575 (1988).
12. Dimitrov, G. V., "Changes in the extracellular potentials produced by unmyelinated nerve fibre resulting from alterations in the propagation velocity or the duration of the action potential," *Electromyogr Clin Neurophysiol* **27**, 243–249 (1987).
13. Dimitrov, G. V., and N. A. Dimitrova, "Precise and fast calculation of the motor unit potentials detected by a point and rectangular plate electrode," *Med Eng Phys* **20**, 374–381 (1998).

14. Dimitrova, N. A., A. G. Dimitrov, and G. V. Dimitrov, "Calculation of extracellular potentials produced by an inclined muscle fiber at a rectangular plate electrode," *Med Eng Phys* **21**, 583–588 (1999).

15. Dimitrova, N. A., G. V. Dimitrov, and V. N. Chihman, "Effect of electrode dimensions on motor unit potentials," *Med Eng Phys* **21**, 479–486 (1999).

16. Dimitrova, N. A., G. V. Dimitrov, and O. A. Nikitin, "Neither high-pass filtering nor mathematical differentiation of the EMG signals can considerably reduce cross-talk," *J Electromyogr Kinesiol* **12**, 235–246 (2002).

17. Disselhorst-Klug, C., J. Silny, and G. Rau, "Improvement of spatial resolution in surface-EMG: a theoretical and experimental comparison of different spatial filters," *IEEE Trans BME* **44**, 567–574 (1997).

18. Dumitru, D., "Physiologic basis of potentials recorded in electromyography," *Muscle Nerve* **23**, 1667–1685 (2000).

19. Dumitru, D., *Electrodiagnostic medicine*, Hanley and Belfus, 1995.

20. Dumitru, D., and J. C. King, "Far-field potentials in muscle," *Muscle Nerve* **14**, 981–989 (1991).

21. Farina, D., L. Arendt-Nielsen, R. Merletti, B. Indino, and T. Graven-Nielsen, "Selectivity of spatial filters for surface EMG detection from the tibialis anterior muscle," *IEEE Trans BME* **50**, 354–364 (2003).

22. Farina, D., C. Cescon, and R. Merletti, "Influence of anatomical, physical and detection system parameters on surface EMG," *Biol Cybern* **86**, 445–456 (2002).

23. Farina, D., M. Fosci, and R. Merletti, "Motor unit recruitment strategies investigated by surface EMG variables," *J Appl Physiol* **92**, 235–247 (2002).

24. Farina, D., and R. Merletti, "A novel approach for precise simulation of the EMG signal detected by surface electrodes," *IEEE Trans BME* **48**, 637—645 (2001).

25. Farina, D., and R. Merletti, "Effect of electrode shape on spectral features of surface detected motor unit action potentials," *Acta Physiol Pharmacol Bulg* **26**, 63–66 (2001).

26. Farina, D., R. Merletti, B. Indino, M. Nazzaro, and M. Pozzo, "Cross-talk between knee extensor muscles. Experimental and model results," *Muscle Nerve* **26**, 681–695 (2002).

27. Farina, D., R. Merletti, M. Nazzaro, and I. Caruso, "Effect of joint angle on surface EMG variables for the muscles of the leg and thigh," *IEEE Eng Med Biol Mag* **20**, 62–71 (2001).

28. Farina, D., and A. Rainoldi, "Compensation of the effect of sub-cutaneous tissue layers on surface EMG: A simulation study," *Med Eng Phys* **21**, 487–496 (1999).

29. Fuglevand, A. J., D. A. Winter, and A. E. Patla, "Models of recruitment and rate coding organization in motor unit pools," *J Neurophysiol* **70**, 2470–2488 (1993).

30. Gerdle, B., N. E. Eriksson, and L. Brundin, "The behaviour of the mean power frequency of the surface electromyogram in biceps brachii with increasing force and during fatigue. With special regard to the electrode distance," *Electromyogr Clin Neurophysiol* **30**, 483–489 (1990).

31. Gootzen, T. H., "Muscle fibre and motor unit action potentials: A biophysical basis for clinical electromyography," PhD thesis, University of Nijmegen, 1990.

32. Gootzen, T. H., D. F. Stegeman, and A. Van Oosterom, "Finite limb dimensions and finite muscle length in a model for the generation of electromyographic signals," *Electroenc Clin Neurophysiol* **81**, 152–162 (1991).

33. Griep, P., F. Gielen, K. Boon, L. Hoogstraten, C. Pool, and W. Wallinga de Jonge, "Calculation and registration of the same motor unit action potential," *Electroencephalogr Clin Neurophysiol* **53**, 388–404 (1982).

34. Gydikov, A., L. Gerilovski, N. Radicheva, and N. Troyanova, "Influence of the muscle fiber end geometry on the extracellular potentials," *Biol Cybern* **54**, 1–8 (1986).

35. Hanson, J., and A. Persson, "Changes in the action potential and contraction of isolated frog muscle after repetitive stimulation," *Acta Physiol Scand* **81**, 340–348 (1971).

36. Jensen, C., O. Vasseljen, and R. H. Westgaard, "The influence of electrode position on bipolar surface electromyogram recordings of the upper trapezius muscle," *Eur J Appl Physiol* **67**, 266–273 (1993).

37. Johannsen, G., "Line source models for active fibers," *Biol Cybern* **54**, 151–158 (1986).

38. Katz, B., "The electrical properties of the muscle fibre membrane," *Proc R Soc Br (B)*, **135**, 506—534 (1948).

39. Kukulka, C. G., and H. P. Clamann, "Comparison of the recruitment and discharge properties of motor units in human brachial biceps and adductor pollicis during isometric contractions," *Brain Res* **219**, 45–55 (1981).

40. Lago, P., and N. B. Jones, "Effect of motor unit firing time statistics on EMG spectra," *Med Biol Eng Comput* **15**, 648–655 (1977).

41. Lawrence, J. H., and C. J. De Luca, "Myoelectric signal versus force relationship in different human muscles," *J Appl Physiol* **54**, 1653–1659 (1983).

42. Li, W., and K. Sakamoto, "The influence of location of electrode on muscle fiber conduction velocity and EMG power spectrum during voluntary isometric contractions measured with surface array electrodes," *Appl Human Sci* **15**, 25–32 (1996).

43. Lindstrom, L., and R. Magnusson, "Interpretation of myoelectric power spectra: A model and its applications," *Proc IEEE* **65**, 653–662 (1977).

44. Lindstrom, L., R. Magnusson, and I. Petersen, "Muscular fatigue and action potential conduction velocity changes studied with frequency analysis of EMG signals," *Electromyography* **10**, 341–356 (1970).

45. Lowery, M. M., N. S. Stoykov, A. Taflove, and Kuiken, "A multiple-layer finite-element model of the surface EMG signal," *IEEE Trans BME* **49**, 446–454 (2002).

46. Merletti, R., L. Lo Conte, E. Avignone, and P. Guglielminotti, "Modelling of surface EMG signals. Part I: model and implementation," *IEEE Trans BME* **46**, 810–820 (1999).

47. Merletti, R., M. Knaflitz, and C. J. De Luca, "Myoelectric manifestations of fatigue in voluntary and electrically elicited contractions," *J Appl Physiol* **69**, 1810–1820 (1990).

48. Milner-Brown, H. S., and R. B. Stein, "The relation between the surface electromyogram and muscular force," *J Physiol* **246**, 549–569 (1975).

49. Moritani, T., and M. Muro, "Motor unit activity and surface electromyogram power spectrum during increasing force of contraction," *Eur J Appl Physiol* **56**, 260–265 (1987).

50. Morrenhof, J. W., and H. J. Abbink, "Cross-correlation and cross-talk in surface electromyography," *Electromyogr Clin Neurophysiol* **25**, 73–79 (1985).

51. Nandedkar, S. D., D. B. Sanders, and E. V. Stålberg, "Simulation techniques in electromyography," *IEEE Trans BME* **32**, 775–785 (1985).

52. Perry, J., C. Schmidt Easterday, and D. J. Antonelli, "Surface versus intramuscular electrodes for electromyography of superficial and deep muscles," *Phys Ther* **61**, 7–15 (1981).

53. Petrofsky, J. S., and A. R. Lind, "Frequency analysis of the surface EMG during sustained isometric contractions," *Eur J Appl Physiol* **43**, 173–182 (1980).

54. Plonsey, R., "Action potential sources and their volume conductor fields," *IEEE Trans BME* **56**, 601–611 (1977).

55. Roeleveld, K., D. F. Stegeman, H. M. Vingerhoets, and A. Van Oosterom, "Motor unit potential contribution to surface electromyography," *Acta Physiol Scand* **160**, 175–183 (1997).

56. Roeleveld, K., D. F. Stegeman, H. M. Vingerhoets, and A. Van Oosterom, "The motor unit potential distribution over the skin surface and its use in estimating the motor unit location," *Acta Physiol Scand* **161**, 465–472 (1997).

57. Rosenfalck, P., "Intra and extracellular fields of active nerve and muscle fibers: A physico-mathematical analysis of different models," *Acta Physiol Scand* **321**, 1–49 (1969).

58. Roy, S. H., C. J. DeLuca, and J. Schneider, "Effects of electrode location on myoelectric conduction velocity and median frequency estimates," *J Appl Physiol* **61**, 1510–1517 (1986).

59. Schneider, J., J. Silny, and G. Rau, "Influence of tissue inhomogeneities on noninvasive muscle fiber conduction velocity measurements investigated by physical and numerical modelling," *IEEE Trans BME* **38**, 851–860 (1991).

60. Solomonow, M., C. Baten, J. Smith, R. Baratta, H. Hermens, R. D'Ambrosia, and H. Shoji, "Electromyogram power spectra frequencies associated with motor unit recruitment strategies," *J Appl Physiol* **68**, 1177–1185 (1990).

61. Stegeman, D. F., D. Dumitru, J. C. King, and K. Roeleveld, "Near- and far-fields: Source characteristics and the conducting medium in neurophysiology," *J Clin Neurophysiol* **14**, 429–442 (1997).

62. Stegeman, D. F., T. H. Gootzen, M. M. Theeuwen, and H. J. Vingerhoets, "Intramuscular potential changes caused by the presence of the recording EMG needle electrode," *Electroencephalogr Clin Neurophysiol* **93**, 81–90 (1994).

63. Stegeman, D. F., and W. H. J. P. Linssen, "Muscle-fiber action-potential changes and surface emg—a simulation study," *J Electromyogr Kinesiol* **2**, 130–140 (1992).

64. Stegeman, D. F., A. Van Oosterom, and E. J. Colon, "Far-field evoked potential components induced by a propagating generator: Computational evidence," *Electroencephalogr Clin Neurophysiol* **67**, 176–187 (1987).

65. Stulen, F. B., and C. J. DeLuca, "Frequency parameters of the myoelectric signals as a measure of muscle conduction velocity," *IEEE Trans BME* **28**, 515–523 (1981).

66. Van Bolhuis, B. M., W. P. Medendorp, and C. C. Gielen, "Motor unit firing behavior in human arm flexor muscles during sinusoidal isometric contractions and movements," *Exp Brain Res* **117**, 120–130 (1997).

67. Van Vugt, J. P., and J. G. van Dijk, "A convenient method to reduce crosstalk in surface EMG," *Clin Neurophysiol* **112**, 583–592 (2001).

68. Viitasalo, J. T., and P. V. Komi, "Interrelationships of EMG signal characteristics at different levels of muscle tension during fatigue," *Electromyogr Clin Neurophysiol* **18**, 167–178 (1978).

69. Wee, A. S., and R. A. Ashley, "Volume-conducted or 'far-field' compound action potentials originating from the intrinsic-hand muscles," *Electromyogr Clin Neurophysiol* **30**, 325–333 (1990).

70. Westbury, J. R., and T. G. Shaughnessy, "Associations between spectral representation of the surface electromyogram and fiber type distribution and size in human masseter muscle," *Electromyogr Clin Neurophysiol* **27**, 427–435 (1987).

71. Winter, D. A., A. J. Fuglevand, and S. E. Archer, "Cross-talk in surface electromyography: Theoretical and practical estimates," *J Electromyogr Kinesiol* **4**, 15–26 (1994).

72. Woods, J. J., and B. Bigland-Ritchie, "Linear and non-linear surface EMG-force relationships in human muscle," *Am J Phys Med* **62**, 287–299 (1983).

73. Zwarts, M. J., "Evaluation of the estimation of muscle fiber conduction velocity: Surface versus needle method," *Electroencephalogr Clin Neurophysiol* **73**, 544–548 (1989).

DETECTION AND CONDITIONING OF THE SURFACE EMG SIGNAL

R. Merletti

Laboratory for Engineering of the Neuromuscular System
Department of Electronics, Politecnico di Torino, Italy

H. J. Hermens

Roessingh Research and Development
Enschede, the Netherlands

5.1 INTRODUCTION

Despite the large number of clinical reports, the issue of EMG detection (electrode size, distance, and location) is still poorly understood. Generalized confusion about this issue among the users has caused contradictory and nonrepeatable results. This chapter tries to clarify the basic concepts concerning the filtering introduced by finite electrode size, inter-electrode distance, electrode configuration, and location as well as the issues related to the front-end amplifier. These points have been addressed by the European Concerted Action Surface EMG for noninvasive assessment of muscles [32]. The main recommendations produced by this international effort are also reported (the publication is available from Dr. H. Hermens, Roessingh Research and Development, Enschede, the Netherlands, *h.hermens@rrd.nl*).

Electromyography: Physiology, Engineering, and Noninvasive Applications, edited by Roberto Merletti and Philip Parker.
ISBN 0-471-67580-6   Copyright © 2004 Institute for Electrical and Electronics Engineers, Inc.

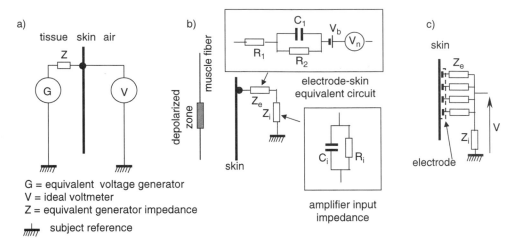

G = equivalent voltage generator
V = ideal voltmeter
Z = equivalent generator impedance
subject reference

Figure 5.1. Basic concepts of skin electrode interface. (*a*) Ideal situation where an infinite input impedance voltmeter monitors the voltage provided by an equivalent generator; (*b*) a model of the real situation with the electrode–skin equivalent circuit and the input impedance of the amplifier, and with the generator as the depolarized zone of a muscle fiber: V_b = dc voltage, V_n = noise voltage; (*c*) a model of electrode with finite area being approximated as a number of point electrodes connected in parallel and to the amplifier.

5.2 ELECTRODES: THEIR TRANSFER FUNCTION

We can consider the skin as the boundary between two media, a conductive layered non-homogeneous and anisotropic semispace (skin, subcutaneous tissue, and muscle) that contains sources of electric field, and an insulating semispace (air). As described in Chapter 4, the sources of electric field generate a two-dimensional potential distribution on the surface of the skin. Such potential is defined with respect to a reference point at a sufficient distance from the sources so that it is not influenced by them. The ideal condition for measuring this potential distribution is to have a point electrode connected to an infinite input impedance voltmeter measuring the voltage with respect to a remote reference where the potential is zero. We can imagine the point electrode moved over the surface to map the "monopolar" potential distribution at a specific instant of time, as described in Figure 5.1*a*.

Although this configuration is often assumed in the literature, it is rarely realistic because the electrode has physical dimensions, the skin-electrode contact has a complex impedance, and the voltmeter (amplifier) has a finite input impedance as described in Figure 5.1*b* and *c*. In addition other sources of potential contribute to the reading. Among these are dc and noise voltages generated at the skin–electrode interface, the capacitively coupled power line voltage, and other electrical phenomena unrelated to EMG [24].

The electrode consists of a metal surface applied to the skin. Four considerations are in order at this point:

1. The skin is a moderately conductive tissue made of cells whose intracellular and extracellular material mostly consists of electrolyte solutions in which current is carried by ions, whereas the metal is a highly conducting material where current is carried by electrons; the resulting interface is intrinsically noisy [22].

2. The electrode–skin interface is very complex, with a capacitive impedance whose R and C components are current and frequency dependent, and incorporate a dc generator accounting for the "battery" potential of the metal-electrolyte interface [28,51].

3. The metal surface in contact with the skin will force the area of contact to be equipotential, therefore modifying the skin potential distribution in the neighborhood.

4. The input impedance of a good EMG amplifier is modeled as a resistor (10^9–$10^{12}\,\Omega$) in parallel with a capacitor (2–10 pF) and is therefore frequency dependent.

It should be noted that in the EMG frequency range, the capacitive component is dominant and cannot be neglected. In other words, the potential under the electrode will be modified by the application of the electrode and the voltage measuring instrument and will be added to an offset and to noise [62]. The output of the amplifier will be a filtered version of the EMG (as modified by the presence of the electrode) added with offset and noise. Figure 5.1 describes these concepts.

The modification of the skin potential distribution introduced by the finite area electrode is too complicated to describe analytically [62] but can be approximated when the electrode size is small with respect to the geometrical extension of the potential distribution [15,16,18]. Toward this end let us assume the contact surface as being made of many small equal and separated contact areas, each offering the same contact impedance as indicated in Figure 5.1c. The detected voltage will then be the average of the individual potentials, that is, the average of the potential over the electrode area. It is intuitive that this phenomenon implies a smoothing effect that is a two-dimensional (2-D) spatial filter transfer function (see also Section 4.2.4 of Chapter 4). Such transfer function depends on the area and the shape of the contact surface and operates on the signal as a convolution in the space domain as well as a product, between the 2-D Fourier transforms of the filter impulse response and of the potential distribution, in the spatial frequency domain.

The following equations describe the simple cases of rectangular electrodes of dimensions a and b and circular electrodes of radius r in the skin plane (x, z) and in the corresponding spatial frequency plane (f_x, f_z):

$h_{size}(x, z) = 1/S$ under the electrode area and $h_{size}(x, z) = 0$ elsewhere

$H(f_x, f_z) = \mathrm{sinc}(af_x)\,\mathrm{sinc}(bf_y)$, for rectangular electrodes

$H(f_x, f_z) = 2J_1(kr)/kr$, for circular electrodes

where S is the electrode area, $\mathrm{sinc}(w) = \sin(\pi w)/\pi w$ for $w \neq 0$ and $\mathrm{sinc}(w) = 1$ for $w = 0$, $J_1(w)$ is the Bessel function of first kind and first order, and $k = 2\pi(f_x^2 + f_z^2)^{1/2}$ [1,17]. The monopolar potential is low-pass filtered by physical electrodes and the resulting smoothing increases with electrode size. The cutoff frequency f_e introduced by a circular electrode is not trivial to compute because the simple correspondence between temporal and spatial frequencies ($f_t = vf_s$) no longer holds in two dimensions. However, as a first approximation and for $v = 4$ m/s, f_e may be estimated as 360 Hz for electrode diameter $\phi = 5$ mm, 220 Hz for $\phi = 10$ mm, and 100 Hz for $\phi = 20$ mm. Small electrodes are therefore preferable to larger, ones and diameters above 5 mm imply loss of information due to the low-pass filtering of the EMG signal. (For further details see [16,18,29].)

5.3 ELECTRODES: THEIR IMPEDANCE, NOISE, AND dc VOLTAGES

The need for stability of the skin-electrode contact and for high-input impedance amplifiers was acknowledged a long time ago [13,25] and addressed in many ways. The most common procedure is to shave, clean, and rub the skin with a slightly abrasive cloth soaked in water or a solvent and apply Ag or AgCl electrodes with some conductive gel to improve and stabilize the contact (see Chapter 11 of [22]). In 1970 Hollis and Harrison tried a paint-on electrode made by mixing a glue with silver powder [33]. Capacitive electrodes, "dry" or "pasteless" electrodes (no gel), and ceramic electrodes have also been tested [7,14,23]. Pre-gelled adhesive electrodes for pediatric ECG applications are easily available and often used for EMG despite their large surface and consequent filtering effect. Miniature electrodes, with diameters below 5 mm, specifically designed for EMG should be preferred.

The skin-Ag or skin-AgCl contact (possibly mediated by conductive gel) has an almost resistive impedance in the EMG frequency range, while other metals present capacitive components that introduce additional filtering [22]. This impedance is highly affected by skin preparation and is inversely related to electrode surface. If gel is used, the important component is the gel-skin portion rather than the metal-gel portion. Measurement of such impedance is not easy, since its value is a nonlinear function of current, and in the detection setup, the electrode current is virtually zero. An estimate can be obtained by extrapolating to zero the impedance values obtained for different currents, as indicated in Figure 5.2b [8]. An increase of dry electrode surface while limiting electrode size may be obtained by making the surface irregular as proposed by Blok et al. [7]. Indications of electrode impedance must be taken with caution because the value strongly depends on the current density and frequency used for the measurement, and these values are often not specified in the literature. Rubbing the skin with medical abrasive paste is without a doubt the best treatment to reduce electrode–skin impedance. The widely used treatment with alcohol is not very effective, as indicated in Figure 5.1a. The electrode area has the effect indicated in Figure 5.1c, which also reflects the change of current density.

The metal-electrolyte contact is intrinsically noisy because of the phenomena taking place at the interface. Because of the averaging effect mentioned in Section 5.2, the noise decreases as the contact surface increases. Since modern high-performance amplifiers have very small intrinsic noise levels (less than $1 \mu V_{RMS}$ in the bandwidth 10–400 Hz), the electrode noise is generally the most important source of noise in EMG recordings and is therefore the limiting factor for detection of very small potentials. Noise is affected by skin treatment (Fig. 5.2d).

Another feature of the metal-electrolyte junction is the generation of a dc voltage due to the "battery" effect [24,28,60]. One might think that since two similar electrodes are generally used in a differential configuration, the two dc voltages indicated in Figure 5.1b would cancel out. This is not the case, since the contact features of the electrode–skin are never the same in the two locations. In addition dc or slowly changing voltages may be present between two points on the skin for a number of physiological reasons [24]. Between two skin electrodes dc voltages as high as a few hundred mV may be observed, and this constrains the design of EMG front-end amplifiers whose dc gain must be limited to prevent saturation. Slight skin abrasion or "peeling" with adhesive tape reduce electrode-skin impedance, noise, dc voltages, and motion artifacts [8,11,60].

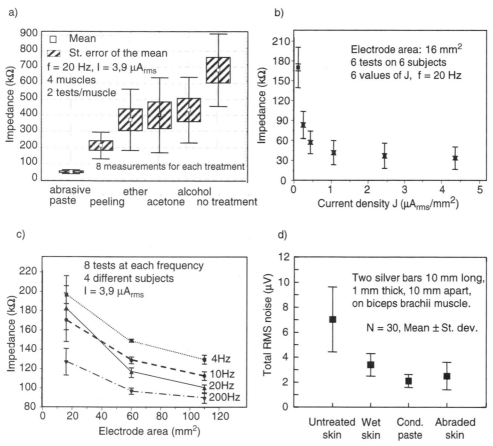

Figure 5.2. Electrode–skin interface properties. (a) Electrode impedance for different skin preparations; (b) electrode impedance versus current density (16 mm² electrode); (c) impedance versus electrode surface area and frequency; (d) electrode noise for different skin preparations. (a) and (c) show impedance values measured between a 1 mm by 10 mm silver electrode and a return metal electrode applied to the skin with a wet cloth of 9 cm² (courtesy of A. Bottin).

5.4 ELECTRODE CONFIGURATION, DISTANCE, LOCATION

The importance of electrode configuration, location, and reproducibility of measurements has been recognized for over 30 years [20,53,63,67] and is still one of the major issues and source of problems and confusion in clinical surface EMG. The monopolar configuration, discussed in the previous sections, contains the entire information available from the detected signal but is used almost exclusively in research applications because of its sensitivity to common mode signals. The differential configuration (Fig. 5.3a), also referred to as bipolar or single differential (SD), is the most widely used configuration, and understanding its features is important for correct EMG detection and interpretation [6]. The double differential configuration (DD) described in Figure 5.3b [9] is used to estimate conduction velocity, limit the detection volume, reduce crosstalk, and increase selectivity. More complex detection configurations are presented and discussed in Chapter 7. The differential configuration (differential amplifier), described in Figure 5.3a, is often

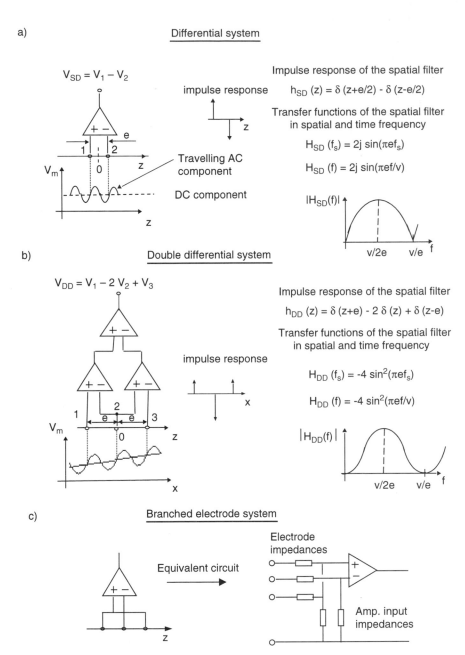

Figure 5.3. Impulse responses and transfer functions of the single (SD) and double differential (DD) configuration (*a* and *b*). The branched electrode [27] configuration is depicted in (*c*) and, for equal electrode impedances, behaves like the DD configuration. Observe that for $e = 10$ mm and $v = 4$ m/s, the first dip is at $v/e = 400$ Hz, that is, at the edge of the EMG spectrum.

considered by engineers and life scientists simply as a way to pick up the voltage between two points. The effect of its transfer function as a spatial filter is often ignored. To clarify this effect, consider an ideal differential amplifier with two-point electrodes separated by a distance *e* and aligned along the fiber direction, with the *z*-axis' zero at the midpoint between the two electrodes. The two electrodes can be considered as two sampling points of the potential distribution, which is changing as a function of both space and time.

For a given time instant the voltage detected between the electrodes and amplified by the system is $V(z)\delta(z + e/2) - V(z)\delta(z - e/2)$, where $\delta(w)$ is the Dirac delta function (Fig. 5.3a). If $V(-e/2) = V(e/2)$, the difference is zero; that is, the system does not theoretically detect common mode voltages in space such as a power line interference or contributions from far-away sources whose field reaches both electrodes with the same intensity and at the same time. The potential distribution $V(z)$ can be seen either in the space domain, as a function of z, or in the spatial frequency domain, as a summation of sinusoidal harmonics representing the Fourier amplitude spectrum (in space) of $V(z)$. Each spectral line, or harmonic, will have a frequency f_s and a corresponding wavelength $\lambda = 1/f_s$.

Let us now consider a potential distribution moving along z at a velocity v and consider one harmonic at a time. If $e = n\lambda$, with $n = 1, 2, \ldots$, and λ is the wavelength of such harmonic, the detected voltage will be zero, regardless of the propagation velocity v, because an integer number of periods in space will fit within e and the two electrodes will detect two identical sine waves. On the other hand, if $e = n\lambda \pm \lambda/2$, the detected voltage will be a sinusoid with the maximal possible peak value, since the two electrodes will detect the difference between two equal and opposite sine waves. Simple trigonometric relations demonstrate that in intermediate situations the output of the system will be a sinusoid whose amplitude is a function of f_s. As a consequence the differential detection system is a spatial filter whose output depends on the (spatial) frequency of the input. Recalling the basic relationships between v, time and spatial frequencies f_t and f_s, period $T = 1/f_t$ and wavelength $\lambda = 1/f_s$ (see Chapter 4), $T = \lambda/v$, or $f_t = v f_s$, the impulse response and the transfer function in time and in space will be given by the equations in Figure 5.3, which hold for point electrodes. The considerations above and the indications of Figure 5.3 hold for propagating potentials (no MUAP generation and extinction effects contribute nonpropagating components).

It is now possible to see that the magnitude of the transfer function of the differential system is a sinusoid; that is, for $\lambda \ll e$ the system behaves as a differentiator, it shows a maximum for $\lambda/2 = e$, it approximates an integrator for $\lambda/2 < e < \lambda$, and it presents its first dip for $\lambda = e$, indicating an overall band-pass filter behavior. The frequency of the dip in space is $f_{s\text{-dip}} = 1/e$ and in time is $f_{t\text{-dip}} = v/e$. Table 5.1 shows the (time) frequency of the first dip as function of v and e [38].

From Table 5.1 it appears that the spatial filter introduces substantial modifications to the monopolar signal spectrum (whose significant harmonics are in the range 10–400 Hz). Further modifications are introduced by the individual electrode transfer function, as discussed in Section 5.2, in particular, for large electrodes. Therefore EMG spectral parameters obtained with different interelectrode distances and electrode sizes cannot be compared because of these geometrical effects [21]. The overall transfer function introduced by an electrode system is the product of the electrode and of the spatial filter transfer functions [19]. For example, at a propagation velocity v of 4 m/s, the transfer function of a pair of electrodes with 15 mm diameter and 30 mm interelectrode distance will show a peak at 66 Hz, a zero at 133 Hz and a low-pass behavior with a pole (3 dB point) at about 160 Hz. If v changes, the first two values will be scaled in proportion to v, while the third will change in a more complex manner because of the two-dimensional nature of the electrode.

In practice, the spectral dips are not easy to observe because they are smoothed out by the distribution of conduction velocity values and by the presence of nonpropagating signals (due to the generation and extinction of the MUAPs) not included in the analysis

TABLE 5.1. Frequency of First Maximum and First Dip of Transfer Function of Single or Double Differential Configuration with Point Electrodes

Conduction Velocity v	Interlectrode Distance e	Frequency of First Max	Frequency of First Dip
3 m/s = 3 mm/ms	0.010 m = 10 mm	150 Hz	300 Hz
3 m/s = 3 mm/ms	0.020 m = 20 mm	75 Hz	150 Hz
3 m/s = 3 mm/ms	0.030 m = 30 mm	50 Hz	100 Hz
4 m/s = 4 mm/ms	0.010 m = 10 mm	200 Hz	400 Hz
4 m/s = 4 mm/ms	0.020 m = 20 mm	100 Hz	200 Hz
4 m/s = 4 mm/ms	0.030 m = 30 mm	66 Hz	133 Hz
5 m/s = 5 mm/ms	0.010 m = 10 mm	250 Hz	500 Hz
5 m/s = 5 mm/ms	0.020 m = 20 mm	125 Hz	250 Hz
5 m/s = 5 mm/ms	0.030 m = 30 mm	83 Hz	166 Hz

Note: Mono-directional source propagation is assumed, with no end effect.

above. In addition misalignment between electrodes and muscle fibers affects dip location [38]. However, when large interelectrode distances are used, spectral shape modifications are evident. For these (as well as other) reasons it is important to reach agreement about a standard. The recommended interelectrode distance is 20 mm [32] (or less for short muscles), with both electrodes on the same side of the innervation zone; this is a compromise between the need to limit spectral modifications and the need to enhance signal amplitude and signal/noise ratio. The magnitude of the differential detection system transfer function has a small value for low spatial frequencies that are (not exclusively) associated to propagating far-away sources. To reduce detection volume and increase spatial selectivity, it may therefore be desirable to achieve a further reduction of low spatial frequency sensitivity. This goal may be obtained either by reducing the interelectrode distance or with the double differential (DD) detection system depicted in Figure 5.3b. The impulse response and transfer function of the DD system may be obtained by computing the difference between the outputs of two single differential systems having a common electrode. The impulse response of the double differential system is a set of three Dirac delta functions (with weights 1, −2, and 1), and its transfer function is a sinusoid raised to power 2. That is, for $\lambda \ll e$ the system behaves as a second-order differentiator, its transfer function shows a maximum for $\lambda/2 = e$, and it presents its first dip for $\lambda = e$, as indicated in Table 5.1. Other more selective systems are described in Chapter 7. A word of caution: Crosstalk is not necessarily reduced by the spatial filtering characteristics of the detection system (see Section 7.2.6 of Chapter 7).

A variation (and a predecessor) of the DD system is the "branched electrode" proposed by Gydikov in his pioneer work [27] and described in Figure 5.3c. The SD, DD, and branched electrode systems are usually placed in the direction of the muscle-fiber (on one side of the innervation zone) but can also be placed in the transversal direction [12]. One configuration of special interest is a set of two DD or branched electrode systems placed at right angle and sharing the central electrode (Laplacian configuration). In this case the transfer function is the 2-D Fourier transform of the impulse response (see Chapter 7).

In the literature it has often been suggested that a decrease of interelectrode distance (IED) can limit the detection volume of the electrode system and consequently limit crosstalk. The effect of the distance between motor unit and recording electrodes has often been described in terms of the following function:

$$V = \frac{V_0}{(r/r_0)^D}$$

where D and V_0 are constants and r_0 some reference distance from the electrical center of the motor unit, where $V = V_0$. This electrical center is the location of the equivalent generator of the MUAP. This equation was found by Buchthal [10] and Gydikov [26] in fitting experimental data obtained with surface electrodes. The exponent D is a function of IED and of the detection system (SD, DD, etc.). This would imply higher values of D for decreasing IED. In the literature no evidence can be found for this. Hermens [30] collected action potentials using a spike-triggered averaging method at IEDs of 20 and 40 mm. They calculated values for D but did not find any significant differences. More recently Roeleveld [56,57] performed a detailed experimental study of the motor unit potentials contributing to the surface EMG using bipolar recordings while varying IED from 6 to 84 mm. Roeleveld found a relation between MUAP area, IED, and the exponent D, with a maximal area showing around IED = 20. The relative contribution of superficial and deep motor units to the recorded SEMG signal was found to be unrelated to IED as long as IED < 40 mm. However, with IED > 40 mm, deeper motor units are represented relatively better in the SEMG signal then superficial ones. This finding also suggests that decreasing IED, at least up to 6 mm, is not a proper technique to limit the view of the electrodes. It should also be considered that EMG amplitude is not only a function of IED and source strength but also of the relative position between electrodes and innervation zone, as indicated in Figure 5.4.

The preceding results apply for electrodes detecting mono-directionally propagating signals that are not changing shape as they propagate. This is the case where motor units with long fibers are superficial and the electrode location is between the innervation zone and the muscle-tendon junctions. A very different situation exists at the motor end plates (innervation zone) and at the muscle-tendon junctions where signals are respectively generated and extinguished and where the scaling factor v between time and space variables, indicated in Figure 5.3, no longer applies.

The SD signal generated by a single fiber and detected with electrodes placed symmetrically with respect to the neuromuscular junction will theoretically be zero (see Chapter 4). Because of the scatter of the neuromuscular junctions this electrode location corresponds to a minimum of signal amplitude and must be avoided in practical applications. When applied over the innervation zone of a muscle SD electrode systems detect signals that are small and noisy, carry little information because of the cancellation effect due to the bidirectional propagation, and are extremely sensitive to small displacements between electrodes and muscle [40]. Figures 5.4 and 5.5 depict this situation and show the importance of a correct electrode placement. Other variables, such as spectral characteristic frequencies and estimates of conduction velocity, are heavily altered when electrode pairs are placed on or near the innervation zone [20,34,37,39,53,58,67]; the criticality of electrode location depends on the muscle and the electrode size (see Fig. 15.1).

5.5 EMG FRONT-END AMPLIFIERS

It has already been mentioned that high input impedance, high CMRR, and low noise are mandatory characteristics of any SEMG front-end amplifier. These parameters are affected

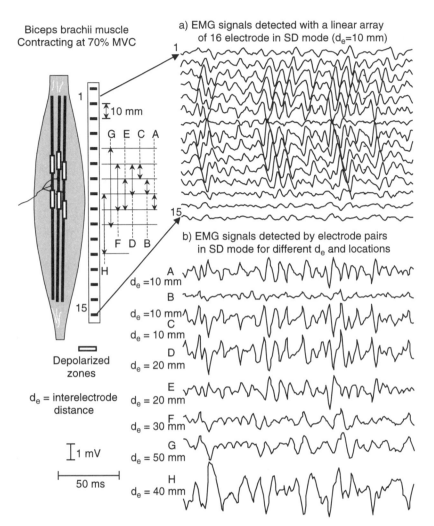

Figure 5.4. Effect of interelectrode distance and electrode location on signal amplitude. (*a*) SD signals detected between adjacent electrodes of a linear array (see Chapter 7) with 10 mm interelectrode distance. Electrodes are 1 mm thick and 10 mm long. (*b*) Single differential signals obtained by summation of adjacent channels and corresponding to interelectrode distances of 10, 20, 30, 40, and 50 mm. (From R. Merletti, A. Rainoldi, D. Farina, Surface electromyography for non-invasive characterisation of muscles, *Exercise and Sport Sciences Reviews* 29:20–25, 2000, with permission)

by the specific circuit configuration adopted. For example, a relatively low-input impedance operational amplifier (OA) connected in the voltage follower configuration may offer an input impedance hundreds of times higher than its own.

The input impedance of a SEMG amplifier must be at least two orders of magnitude greater than the largest expected electrode–skin impedance. Impedances higher than $100\,\text{M}\Omega$ are usually considered acceptable, but $1000\,\text{M}\Omega$ is preferable in case of small dry electrodes whose contact impedance may reach $1\,\text{M}\Omega$ [7,8]. This specification restricts the choice of the front-end amplifier to basically three types: (1) OAs in voltage follower configuration, (2) the classic three OA instrumentation amplifier configuration, and (3) the

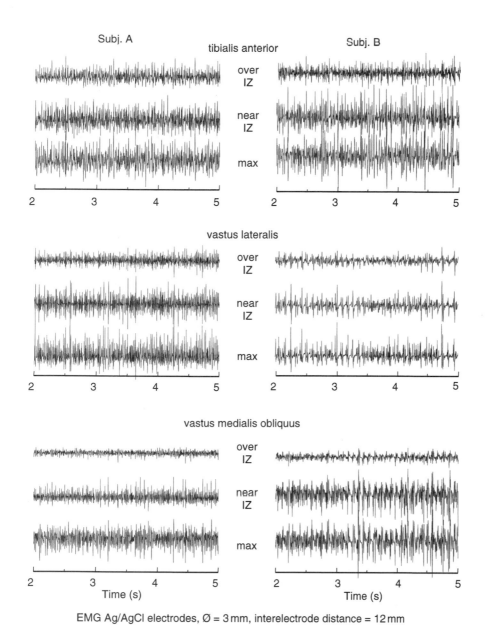

EMG Ag/AgCl electrodes, Ø = 3 mm, interelectrode distance = 12 mm

Figure 5.5. Example of EMG recordings obtained simultaneously with three pairs of electrodes of 3 mm diameter and 12 mm interelectrode distance placed over or near (1 interelectrode distance away) the innervation zone (IZ), and on the location of maximal amplitude (between the IZ and the tendon region) in three muscles of two subjects. Similar results are obtained with larger electrodes.

classic two OA instrumentation amplifier configuration. These circuits, as well as differential FET stages, have been used since the early 1970s [2,3,5,14,35,59,61]. Variations and applications for multichannel detection were developed more recently [42,43,44,48]. Some basic schematics are depicted in Figure 5.6. The input impedance of the circuits depicted in Figure 5.6a and b (see also Fig. 5.1b) comprises a 10^9 to $10^{12}\,\Omega$ resistance in

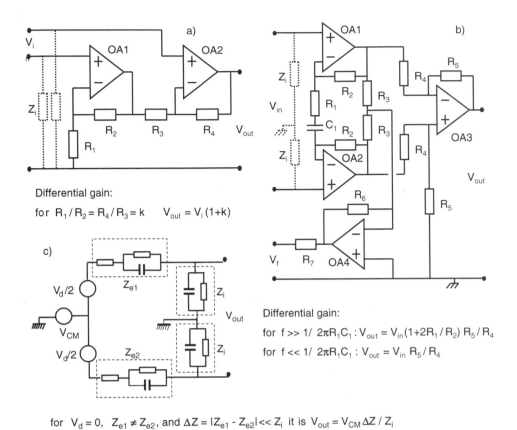

Differential gain:

for $R_1/R_2 = R_4/R_3 = k$ $V_{out} = V_i (1+k)$

Differential gain:

for $f \gg 1/ 2\pi R_1 C_1 : V_{out} = V_{in}(1+2R_1/R_2) R_5/R_4$

for $f \ll 1/ 2\pi R_1 C_1 : V_{out} = V_{in} R_5/R_4$

for $V_d = 0$, $Z_{e1} \neq Z_{e2}$, and $\Delta Z = |Z_{e1} - Z_{e2}| \ll Z_i$ it is $V_{out} = V_{CM} \Delta Z / Z_i$

Figure 5.6. EMG front-end amplifier configurations. (a) Two OA instrumentation amplifier; (b) three OA instrumentation amplifier with common mode feedback; (c) effect of unbalance of the electrode–skin impedances.

parallel with a 1 to 10 pF capacitance. The latter is affecting the impedance at frequencies as low as 50 Hz.

Most of the front-end circuits either present a low gain to avoid saturation by the dc voltage present between the electrodes, or incorporate a high-pass filter to limit or remove it, as indicated Figure 5.6b (R_1 and C_1). The high-pass filter has the disadvantage of introducing a transient artifact in case of electrically elicited EMG.

An important feature of a SEMG amplifier is the common mode rejection ratio (CMRR), which describes the amplifier capability to reject common mode voltages (mainly the power line voltage present between the subject and the mains ground which may be of the order of a few volts, that is over a thousand times the surface EMG signal). The CMRR is defined as $CMRR = 20 \, Log_{10}(A_d/A_c)$, where A_d and A_c are the amplifier's differential and common mode gains respectively. The output due to the common mode input voltage (referred to the input) is given by $V_{out}/A_d \cong V_{cm} (\Delta Z/Z_i + A_c/A_d)$ and is a function of (1) the CMRR of the OAs, (2) the balance of the resistor bridge around the differential to single ended stage, and (3) the unbalance of the two input voltage dividers due to the electrode impedances and the amplifier input impedances (Fig. 5.6c). The $V_{cm} \Delta Z/Z_i$ contribution generates a differential voltage due to the common mode voltage and to the difference between the two electrode–sin impedances. For example, a ratio $\Delta Z/Z_i = 10^{-3}$

would contribute 1 mV to the input referred voltage (V_{out}/A_d) for each volt of common mode input voltage, regardless of the CMRR of the circuit. Common mode input voltages due to parasitic capacitive coupling between the subject and the power line may be of the order of a few volts. As a consequence CMRR in the range of 10^5 to 10^6 (100–120 dB) and $\Delta Z/Z_i$ in the range of 10^{-5} to 10^{-6} are required to limit the equivalent input voltage to a value negligible with respect to EMG. Since these values are not easy to reach, a common mode feedback is often adopted to reduce the common mode voltage on the subject. This technique is frequently applied in ECG recordings and is referred to as "driven right leg" (DRL). It consists in detecting and re-applying the common mode voltage to the subject with opposite phase, as indicated in Figure 5.6b [49,64,65]. Variations of this circuit for array electrodes have been proposed [42,43,44].

Many sources of electronic voltage noise exist within an OA. Their power is uniformly distributed on all frequencies (white noise) with additional contributions below 10 to 20 Hz (flicker noise) at the lower edge of the EMG bandwidth. They are usually referred at the input as equivalent input voltage noise density, V_n, measured as power per unit of bandwidth (V^2/Hz) or RMS voltage per square root of unit of bandwidth ($V_{rms}/\sqrt{\text{Hz}}$). Another important source of noise is the OA input current noise flowing into the equivalent internal impedance of the generator (whose main component is the electrode-skin impedance Z_e). If such impedance is high (Fig. 5.2), as in the case of small dry electrodes, this component, given by $I_n Z_e$ where I_n is the current noise density ($A_{rms}/\sqrt{\text{Hz}}$), may be comparable to the voltage noise density and cannot be neglected. Since the two noise sources are uncorrelated, the total equivalent input noise power is given by $B(V_n^2 + I_n^2 Z_e^2)$, where B is the bandwidth of the amplifier. Limiting the bandwidth of the amplifier will therefore limit the total noise and will also prevent the amplification of disturbances and interferences outside the signal bandwidth.

Amplification of electrically elicited surface EMG (M-waves) requires additional features to reduce the stimulation artifact. Among these are full optical isolation between the stimulation and detection circuits, slew rate limiting, nonlinear filtering, and blanking of the EMG amplifier [36]. Optical isolation serves a second important purpose that is the limitation of leakage currents from either the EMG system or from other (nonmedical) devices connected to it, such as data acquisition and transmission systems, and computers. Figure 5.7 shows a block diagram of such isolated system. The leakage current

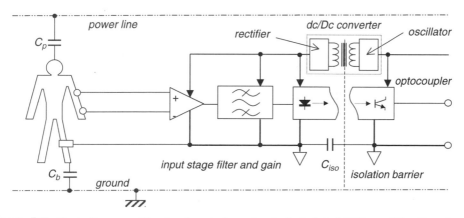

Figure 5.7. Block diagram of the front-end of a galvanically isolated EMG amplifier. Signals are transmitted through an optical isolator and power is provided by a high-isolation dc/dc converter. (From [50] with permission)

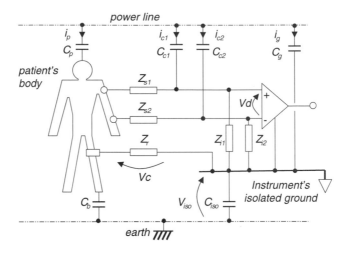

$C_p =$ parasitic capacitance between body and power line
$C_b =$ capacitance between body and earth
$C_{c1}, C_{c2} =$ parasitic capacitances between power line and input cables
$C_{iso} =$ parasitic capacitance of the isolation barrier
$C_g =$ parasitic capacitance of the power supply, between power line and isolated ground
$Z_{s1}, Z_{s2} =$ electrode contact impedances
$Z_r =$ patient reference contact impedance
$Z_{i1}, Z_{i2} =$ amplifier input impedances
$V_d =$ differential input voltage
$V_c =$ common mode input voltage
$V_{iso} =$ isolation mode voltage

Figure 5.8. Detailed analysis of the first stage of a EMG amplifier including parasitic components that are important in determining power line intereference. (From [50] with permission)

flowing from components located to the right of the isolation barrier (power line, computers, etc.), to the left of the barrier (patient), is theoretically zero and, in practice, strongly reduced by the low value of the parasitic capacitance between the isolated and nonisolated part. It is not difficult to keep such current below the prescribed limit of $100\,\mu A$. Power is provided to the isolated part by means of a medical grade dc/dc converter.

Parasitic capacitances play a major role in the design and performance of a EMG front-end amplifier. Figure 5.7 depicts the main capacitances (between patient and power line, patient and ground and across the isolation barrier). Figure 5.8 outlines additional details of the front-end stage, showing impedances and parasitic capacitances often neglected but very important for a good design oriented to the minimization of power line interference [42,43,44].

5.6 EMG FILTERS: SPECIFICATIONS

The SEMG signal conditioning system that follows the front-end includes a high-pass filter (with cutoff frequency near 10–20 Hz) and a low-pass filter (with a cutoff frequency near

400–450 Hz), which usually have a roll-off slope of 40 dB/decade (12 dB/octave). Both filters may be incorporated in the front-end circuitry.

The surface-detected signal often shows slow variations due to movement artifacts and instability of the electrode–skin interface. The harmonics of these unwanted signals are usually in the frequency range 0 to 20 Hz, and the high-pass filter is therefore designed with a cutoff frequency in the 15 to 20 Hz range. The EMG spectrum includes, in this range, information concerning the firing rates of the active motor units, which may be relevant in some applications. In many other cases (e.g., in movement analysis) this information may not be of great interest, and a high-pass filter is used with a cutoff at 25 to 30 Hz. Artifacts due to sliding of the innervation zone below the electrodes, resulting in amplitude modulation of the EMG, are not removed by this filter. Movement artifacts associated with fluctuations of electrode impedances and half-cell potentials may be attenuated but not eliminated. Additional special filtering and/or interference reduction techniques may be applied to remove ECG artifacts from trunk muscles [54,55].

These high-pass and low-pass filters are used to reduce noise and artifacts. In some cases analog notch filters have been used to reduce the 50 or 60 Hz interference. In general, this is not a good practice because (1) it removes power from a frequency band where EMG shows high-power density and (2) introduces phase rotation that extends to frequencies below and above the central frequency thereby dramatically changing the waveform (not so much the power) of the EMG. This latter issue is relevant when waveforms are of interest, but it is not important if only EMG amplitude or power is of interest, such as in EMG biofeedback. Digital adaptive noise cancellation filters may be used to remove power line interference and artifacts either on-line or off-line after signal sampling and A/D conversion. They lock on a predetermined waveform, estimate its amplitude and phase, and remove it by subtraction [45].

5.7 SAMPLING AND A/D CONVERSION

The Nyquist theorem requires that a signal be sampled at a rate at least twice the frequency of its highest harmonic in order to avoid loss of information and the phenomenon called "aliasing." The aliasing effect is described in Figure 5.9, which shows how ambiguity may arise if one harmonic of a signal is sampled at a frequency that is too low. This problem, of course, arises for the signal harmonics of higher frequencies and for wideband noise. For this reason removal of signal or noise components with frequency above those of interest is important. For almost all muscles and most applications, the highest harmonic of interest in surface EMG signal is in the range 400 to 450 Hz, thereby requiring low-pass (anti-aliasing) filters with cut off in this range and sampling at at least 1000 samples per second.

A/D conversion transforms the sampled voltages into "levels" represented in binary code. An A/D converter accepts signals in a specific input range (e.g., ±5 V) which is subdivided into a number of discrete levels given by $2^n - 1$, where n is the number of bits of the A/D converter. Table 5.2 provides an example of the resulting resolution for the input range ±5 V and amplifier gain of 1000.

The amplifier (or filter) gain and the A/D input range cannot be chosen independently. They are tied by the desired resolution, which is in turn related to the noise level. Depending on the maximal peak-to-peak amplitude, which can be expected in a surface EMG signal (e.g., 4–5 mV$_{pp}$), and the noise level (e.g., 1 μV_{RMS}, which can be chosen to corre-

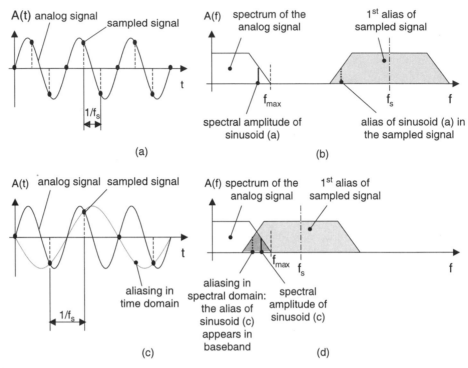

Figure 5.9. Description of the aliasing phenomenon. Consider one harmonic, spectral line indicated in (*a*) of the signal whose spectrum is depicted in (*b*). Proper sampling (above the Nyquist rate) implies the generation of nonoverlapping spectral "aliases" centered around multiples of the sampling frequency (gray areas in *b*). Sampling below the Nyquist rate implies overlapping of the spectral "aliases". An aliased frequency component appears in the signal bandwith with frequency $f_a = f_s - f$, this aliased spectral line is the dotted line depicted in (*c*) and sharing the same samples of the real harmonic. (From [50] with permission)

spond to the least significant bit), it is possible to calculate the number of required levels N (4000 – 5000 in this example), in order to find out how many bits are needed. In the example (1 μV/least bit and 5 mV peak-to-peak amplitude) 13 bits (8195 levels) would be required, and a 14 or 16 bit A/D converter is then sufficient even if the full A/D input range is not exploited. If a 16 bit A/D converter with a ±5V input range is selected, a fixed gain amplifier of 1600 to 2000 will be required to exploit the full A/D range. In such a case the 3 to 4 least significant bits would correspond to noise only. If a lower gain is selected, the most significant bits will never be used. Table 5.2 reports other examples.

It is common practice to select the resolution of the system (4th column of Table 5.2) somewhat below the noise level of the system, and the range somewhat above the maximum expected peak-to-peak amplitude of the signal at the output of the conditioning amplifier-filter. If this requirement cannot be matched for all signals, the amplifier must be designed with adjustable gain in order to obtain the desired resolution by changing the gain on a case-by-case basis.

TABLE 5.2. Relation between Number of Bits of the A/D Converter, Number of Levels, V/Level, and Input Referred Resolution*

Number of Bits n of A/D Converter	Number of Levels N ($N = 2^n - 1$)	V/Level (± 5 V Range) $10/N$	Input Referred Resolution* (Amplifier gain = 1000 and ± 5 V A/D Input range)
8	255	39.06 mV	39.06 μV
10	1 023	9.765 mV	9.765 μV
12	4 095	2.441 mV	2.441 μV
14	16 383	0.610 mV	0.610 μV
16	65 535	0.152 mV	0.152 μV

* Voltage difference corresponding to the least significant bit.

5.8 EUROPEAN RECOMMENDATIONS ON ELECTRODES AND ELECTRODE LOCATIONS

Most SEMG developments in the last decade have taken place locally, resulting in different methodologies among the different clinical research laboratories. This fact hindered the growth of SEMG into a mature well-accepted tool by clinical users as well as large-scale industrial efforts. A standardization effort was needed to make the results comparable and to create a large common body of knowledge on the use of SEMG in its many fields of application.

The European initiative, Surface Electromyography for Noninvasive Assessment of Muscles (SENIAM), was started in 1996 with this objective. The main goal was to create consensus on key items (sensors, sensor placement, signal processing, and modeling) to enable exchange of data and results obtained with SEMG. This section presents the recommendations for the SEMG sensors and sensor placement procedures developed within SENIAM.

"Sensor" is defined as the ensemble of electrodes, electrode construction, and (if applicable) the integrated pre-amplifier. A bipolar device, being a configuration of two electrodes, is the most frequently used sensor to record the SEMG signal. It is defined by the following properties: (1) size and shape of the electrode, (2) material, (3) interelectrode distance (IED), and (4) construction modalities.

The first step was to provide an inventory of the actual use of SEMG sensors in the European laboratories by scanning six volumes (1991–1997) of seven journals, in which studies using SEMG were regularly found [31]. Of the 144 articles were examined, 34 were from the *Journal of Electromyography and Kinesiology* (1991–1996), 24 from the *European Journal of Applied Physiology* (1995–1996), 38 from *EEG and Clinical Neurophysiology* (1992–1996), 20 from *Electromyography and Clinical Neurophysiology* (1993–1996), 13 from the *Journal of Biomechanics* (1992–1997), 6 from *Ergonomics* (1994), and 9 from *Muscle and Nerve* (1992–1996). The complete list of references is available in an appendix of [31]. The following paragraphs summarize the findings:

Electrode Material. The material the electrodes are made of should be indicated in the publications and reports. In 43% of the scanned publications, the electrode material was not mentioned. From the remaining publications, it appeared that Ag/AgCl was by far the most frequently used material. In most cases this was combined with a pre-gelled surface.

number of occurrences in literature

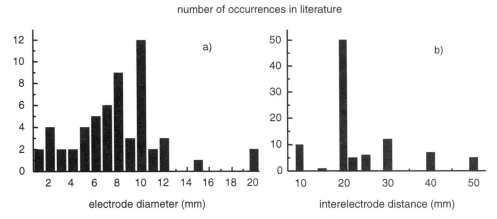

Figure 5.10. Results from an inventory of the actual use of SEMG sensors in the European laboratories obtained by scanning 6 volumes (1991–1997) of 7 journals in which publications about SEMG can be found regularly [31]. 144 articles were examined. (*a*) Histogram of electrode size reported in the examined literature; (*b*) histogram of interelectrode distance reported in the examined literature. (From [31] with permission)

Electrode Shape and Size. "Electrode shape" is defined as the shape of the conductive area of SEMG electrodes. Circular electrodes are used most often. SEMG users should clearly indicate the type, manufacture, and shape of the electrodes used. "Electrode size" is defined as the size of the conductive area of a SEMG electrode. A large variation in electrode size can be found in literature: a total of 57 sizes were found. A slight preference in literature was found for circular electrodes with a diameter ranging from 8 to 10 mm. Results from the scanned articles are reported in Figure 5.10*a*. Theoretical considerations (provided in Section 5.2) demonstrated that smaller electrodes are preferable, whereas larger ones introduce excessive low-pass filtering. It is recommended that the size of the electrodes in the direction of the muscle-fiber does not exceed 10 mm.

Interelectrode Distance. "Interelectrode distance" (IED) is defined as the center-to-center distance between the conductive areas of two electrodes in the bipolar configuration. This is an important property of the SEMG sensor. Widely variable IEDs were found in the literature scan, as indicated in Figure 5.10*b*. Some of muscles and IEDs included were biceps brachii (10–40 mm), biceps femoris (20–50 mm), deltoideus (20–40 mm), gastrocnemius (10–50 mm), and rectus femoris (10–50 mm). The largely preferred distance was 20 mm.

It is recommended that the bipolar SEMG electrodes be applied between the innervation zone and a tendon, with an interelectrode distance of no more than 20 mm. When a bipolar sensor is applied on relatively small muscles, the interelectrode distance should not exceed 1/4 of the muscle-fiber length, and both electrodes should be on one side of the innervation zone. This way tendon and motor end plate effects can be reduced or avoided. In dynamic contractions the conditions above should be met for the full movement range.

Sensor Construction. "Sensor construction" is defined as the (mechanical) construction that is used to integrate the electrodes, the cables, and (if applicable) the pre-amplifier. The construction (and its mass) is not thought to directly affect the SEMG characteristics. However, one has to take into account that when the construction is such that electrodes and cables can move (possibly because of their inertial mass), there is the potential for artifacts to be created from the contact instability. If the IED is allowed to vary during muscle contraction, some feature of the SEMG signal will be affected.

It is recommended that the construction have a fixed interelectrode distance and be built from lightweight material. Cables and connectors can be fixed by double-sided tape or elastic bands so that pulling artifacts can be avoided. If used under water, the sensor assembly should be protected with waterproof tape to prevent the water from creating a low-impedance bridge between the electrodes [52].

Sensor Location and Orientation on the Muscle. In this context *location* is considered to be the position of the center of the sensor on the muscle. The *orientation* is defined as the direction of the bipolar sensor with respect to the direction of the muscle-fibers. In the 144 papers scanned, in total 352 descriptions of the sensor location were counted. These descriptions applied to 53 different muscles. In 58 (16%) reports the sensor location on the muscle was not mentioned. The remaining 294 descriptions mentioned both the name of the muscle and described the sensor location or referred to previous literature. The main literature references are [4,47,66] which contain detailed descriptions of sensor location for a large number of muscles. Table 5.3 shows examples of sensor location descriptions for the biceps brachii muscles. In SENIAM5 [31] additional tables can be found for the soleus and trapezius muscles, as well as the complete references to the papers in these tables. Table 5.3 shows that in total 21 sensor placement descriptions for the biceps brachii muscle were found.

Three placement strategies can be identified for the biceps brachii:

- On the center or on the most prominent bulge of the muscle belly (10 out of 21)
- Somewhere between the innervation zone and the distal tendon (6 out of 21)
- On the motor point (1 out of 21)

In the remaining four publications the sensor location was not mentioned or was unclear. Altogether this shows that half of the authors used a "muscle belly-montage." Certain authors differentiated between the long head and the short head. The orientation of the electrodes with respect to the direction of muscle-fiber was rarely mentioned. In some other muscles (gastrocnemius, trapezius) the sensor placement was reported more clearly. In general, the description of the placement of the SEMG sensor is not good enough to enable replication of the experiment and the theoretical considerations presented in Section 5.4 are too often disregarded.

Sensor Placement Procedure. A general sensor placement procedure has been developed within SENIAM and worked out in detail for 27 different muscles. It consists of the following sequential steps: (1) selection of the SEMG sensor, (2) preparation of the skin, (3) positioning the patient in a starting posture (4) determination of the sensor location, (5) placement and fixation of the sensor, and (6) testing of the connection. For each step recommendations have been developed and are described in detail in the SENIAM reports (see the note at the end of this chapter).

Traditionally sensors have been placed over the belly or over the innervation zone (motor end plate zone), since this was the location to record "large" monopolar SEMG

TABLE 5.3. Overview of Electrode Location Descriptions on Biceps brachii Muscle in Recent Literature

Electrode Location	Author
Middle of muscle belly	van Woensel, W.
To the muscle belly	Martin, A.
Over the belly	Martin, A.
Midpoint of contracted muscle belly	Clarijs, J. P.
Not indicated	Kluth, K.
One of two recording electrodes placed above motor point	Maton, B.
Must bulky part of long head	Christensen, H.
Over muscle in line with main fiber direction of short head of muscle, distal to motor point; location accepted if cross-correlation coefficient between two bipolarly recorded EMG signals was >0.7	Stegeman, D.
Parallel to fiber orientation, halfway between innervation zone and distal tendon	Van der Hoeven, H.
Not indicated	Fellows, S. J.
Between end plate region and tendon insertion	Rau, G.
On short head parallel to muscle-fiber	Vogiatzis, I.
Belly-tendon montage	Logullo, F.
Belly of muscle	Esposito, F.
Parallel to fiber orientation, halfway between innervation zone and distal tendon	Van der Hoeven
In midst of muscle belly	Happee, R.
Over belly	Orizio, C.
Halfway between motor end plate zone and distal tendon aligned in the direction of muscle-fiber	Hermens, H. J.
On muscle after having determined the motor end plates with electrostimulation apparatus	Kahn, J. F.
Over medial belly of each head (long head and short head) parallel to muscle-fiber	Perot, C.
Not indicated	Hummelsheim, H.

Source: H. Hermens and B. Freriks [31].

signals. It is now well known that this location is not suitable for differential recordings, it is not stable nor reproducible because relative small displacements of the sensors with respect to the innervation zone cause large effects on the amplitude of the SEMG signal (see Figs. 5.4 and 5.5).

The EMG amplitude varies while displacing electrodes in longitudinal direction over the muscle. EMG studies on the trapezius, vastus medialis, and lateralis muscles show dips and plateau regions related to the presence of innervation zones [19,53].

With respect to the transversal direction, the vicinity of other active muscles increases the risk of crosstalk. The relevance of this effect will strongly depend on the characteristics of these active nearby muscles, on the location and nature of the muscle-tendon junctions. Theoretical and experimental evidence strongly suggest that crosstalk is due mostly to the fiber-end effect [41]. The availability of an atlas of innervation zones of the most important superficial muscles greatly helps in identifying optimal electrode locations [46].

TABLE 5.4. Summary Table of SENIAM Recommendations

Parameter	Recommended Value or Condition
Electrodes (bipolar montage)	
Electrode size	Diameter <10 mm
Interelectrode distance (IED)	<20 mm, or $<\frac{1}{4}$ the muscle length, whichever is smaller
Electrode location	Between the most distal innervation zone and the distal tendon. Between the most proximal innervation zone and the proximal tendon; not over an innervation zone
Reference electrode location	Wrist, ankle, processus spinosus of C7, or other electrically inactive area
Amplifier	
High-pass filter (low-frequency cutoff)	
For EMG spectral analysis	<10 Hz
For movement analysis only	~20 Hz
Low-pass filter (high-frequency cutoff)	
For general applications	~500 Hz (sampling frequency >1000 samples/s)
For special wideband applications	~1000 Hz (sampling frequency >2000 samples/s)
Input referred voltage noise level	$<1\,\mu V_{RMS}$ (in the 10–500 Hz bandwidth)
Input referred current noise level	$<10\,pA_{RMS}$ (in the 10–500 Hz bandwidth)
Input impedance	>100 MΩ (for conventional electrodes)
	>1000 MΩ (for pasteless "dry" pin electrodes)
Gain	Suitable to bring the signal into the input range of the A/D converter with desired input resolution
Sampler and A/D converter	
Sampling frequency	>1000 samples/s (general applications)
	>2000 samples/s (wideband applications)
n bits of A/D	12 (requires amplifier with variable gain)
	16 (fixed gain amplifiers may be used)

Note: The seven SENIAM volumes, the SENIAM CD ROM, and the book *European Recommendations for Surface Electromyography* are available from the SEIAM Project Coordinator Dr. Hermie Hermens, Roessingh Research and Development, Enschede, NL; e-mail: h.hermens@rrd.nl.

The reference electrode should be placed at a location in which the muscle activity is minimal, preferably on electrically inactive tissue. Depending on the muscle and application, it is recommended to use the wrist, the processus spinosus of C7, or the ankle as the standard location of the reference electrode.

So far SENIAM recommendations (Table 5.4) concern only bipolar sensors. Electrode arrays have not been considered, as they still have limited clinical use. Yet they offer great promise, both for the identification of innervation zones and in the field of neurology, in which diagnosis and monitoring by characterization of the MUAPs is of great importance (see Chapter 12). Standardization efforts always lag behind research and are never completed. Nevertheless, it is important that the clinical community takes them into serious consideration to allow comparison between clinical evaluations performed in different laboratories.

REFERENCES

1. Abramowitz, M., and I. A. Stegun, *Handbook of mathematical functions*, Dover, New York, 1965.

2. Accornero, N., "Active surface EMG probe and contour follower," *EEG Clin Neuroph* **51**, 331–332 (1981).

3. Basmajian, J., and J. Hudson, "Miniature source attached differential amplifier for electromyography," *Am J Phys Med* **53**, 234–235 (1974).

4. Basmajian, J., *Electrodes in EMG biofeedback*, Williams and Wilkins, Baltimore, 1980.

5. Bergey, G., and R. Squires, "Improved buffer amplifier for incorporation within a biopotential electrode," *IEEE Trans BME* **18**, 430–431 (1971).

6. Blok, J., and D. Stegeman, "Simulated bipolar SEMG characteristics," in H. J. Hermens, and B. Freriks, eds., SENIAM 5: *The state of the art on sensors and sensor placement procedures for surface electromyography: A proposal for sensor placement procedures*, Roessingh Research and Development, Enschede, Netherlands, 1997, pp. 60–70.

7. Blok, J., S. van Asselt, J. van Dijk, and D. Stegeman, "On an optimal pastless electrode to skin interface in surface EMG," in H. J. Hermens, and B. Freriks, eds., SENIAM 5: *The state of the art on sensors and sensor placement procedures for surface electromyography: A proposal for sensor placement procedures*, Roessingh Research and Development, Enschede, Netherlands, 1997, pp. 71–76.

8. Bottin, A., and P. Rebecchi, "Impedance of the skin electrode interface in surface EMG recordings," *Proc XIV Congr Int Soc Electrophysiol Kinesiol*, Vienna, 246–247 (2002).

9. Broman, H., G. Bilotto, and C. De Luca, "A note on the non-invasive estimation of muscle fiber conduction velocity," *IEEE Trans BME* **32**, 341–343 (1985).

10. Buchthal, F., C. Guld, and P. Rosenfalck, "Volume conduction of the motor unit action potential investigated with a new type of multielectrode," *Acta Physiol Scand* **38**, 331–354 (1957).

11. Burbank, D., and J. Webster, "Reducing skin potential motion artefact by skin abrasion," *Med Biol Eng Comp* **16**, 31–38 (1978).

12. Christova, P., A. Kossev, I. Kristev, and V. Chichov, "Surface EMG recorded by branched electrodes during sustained muscle activity," *J Electrom Kinesiol* **9**, 2663–2276 (1999).

13. Davis, J., "Manual of surface electromyography," *Lab for psychological studies*, Allan Memorial Inst. of Psychiatry, Montreal, 1952.

14. De Luca, C., R. Le Fever, and F. Stulen, "Pastless electrode for clinical use," *Med Biol Eng Comp* **17**, 387–390 (1979).

15. Dimitrov, C. V., and N. A. Dimitrova, "Precise and fast calculation of the motor unit potentials detected by a point and rectangular plate electrode," *Med Eng Phys* **20**, 374–381 (1998).

16. Dimitrova, N. A., C. V. Dimitrov, and V. N. Chihman, "Effect of electrode dimension on motor unit potentials," *Med Eng Phys* **21**, 479–485 (1999).

17. Farina, D., and C. Cescon, "Concentric ring electrode systems for noninvasive detection of single motor unit activity," *IEEE Trans BME* **48**, 1326–1334 (2001).

18. Farina, D., and R. Merletti, "Effect of electrode shape on spectral features of surface detected motor unit action potentials," *Acta Physiol Pharmacol Bulg* **26**, 63–66 (2001).

19. Farina, D., C. Cescon, and R. Merletti, "Influence of anatomical, physical and detection system parameters on surface EMG," *Biol Cybern* **86**, 445–456 (2002).

20. Farina, D., R. Merletti, M. Nazzaro, and I. Caruso, "Effect of joint angle on surface EMG variables for the muscle of the leg and thigh," *IEEE Eng Med Biol Mag* **20**, 62–71 (2001).

21. Fuglevand, A., D. Winter, A. Patla, and D. Stashuk, "Detection of motor unit action potentials with surface electrodes: influence of electrode size and spacing," *Biol cybern* **67**, 143–153 (1992).

22. Geddes, L. A., and L. E. Baker, *Principles of applied biomedical instrumentation*, Wiley, New York, 1968.

23. Gondran, Ch., E. Siebert, P. Fabry, E. Novakov, and P. Gumery, "Non-polarisable dry electrode based on NASICON ceramic," *Med Biol Eng Comp* **33**, 452–457 (1995).

24. Grimnes, S., "Psychogalvanic reflex and changes in electrical parameters of the skin," *Med Biol Eng Comp* **20**, 734–740 (1982).

25. Grossman, W., and H. Wiener, "Some factors affecting the reliability of surface electromyography," *Psychosomat Med* **28**, 79–83 (1966).

26. Gydikov, A., D. Kosarov, and N. Tankov, "Studying the alpha motoneurone activity by investigating motor units of various sizes," *Electromyography* **12**, 99–117 (1972).

27. Gydikov, A., D. Kosarov, A. Kossev, O. Kostov, N. Trayanova, and N. Radicheva, "Motor unit potentials at high muscle avtivity recorded by selective electrodes," *Biomed Biochim Acta* **45**, S63–S68 (1968).

28. Hary, D., G. Bekey, and J. Antonelli, "Circuit models and simulation analysis of electromyographic signal sources: The impedance of EMG electrodes," *IEEE Trans BME* **34**, 91–96 (1987).

29. Helal, J. N., and P. Buissou, "The spatial integration effect of surface electrode detecting myoelectric signal," *IEEE Trans BME* **39**, 1161–1167 (1992).

30. Hermens, H. J., "Surface EMG," PhD thesis, University of Twente, 1991.

31. Hermens, H. J., and B. Freriks, eds., *SENIAM 5: The state of the art on sensors and sensor placement procedures for surface electromyography: A proposal for sensor placement procedures*, Roessingh Research and Development, Enschede, Netherlands, 1997.

32. Hermens, H., B. Freriks, R. Merletti, D. Stegeman, J. Blok, G. Rau, C. Disselhorst-Klug, and G. Hägg, *European recommendations for surface electromyography*, Roessingh Research and Development, Enschede, Netherlands, 1999.

33. Hollis, L., and E. Harrison, "An improved surface electrode for monitoring myopotentials," *Am J Occup Therapy* **24**, 28–30 (1970).

34. Jensen, C., O. Vasseljen, and R. Westgaard, "The influence of electrode position on bipolar surface electromyogram recordings of the upper trapezius muscle," *Eur J Appl Physiol* **67**, 266–273 (1993).

35. Johnson, S., P. Lynn, J. Miller, and G. Reed, "Miniature skin-mounted preamplifier for measurement of surface electromyographic potentials," *Med Biol Eng Comp* **15**, 710–711 (1977).

36. Knaflitz, M., and R. Merletti, "Suppression of stimulation artifacts from myoelectric evoked potential recordings," *IEEE Trans BME* **35**, 758–763 (1988).

37. Lateva, Z., N. Dimitrova, and G. Dimitrov, "Effect of recording position along a muscle fiber on surface potential power spectrum," *J Electrom Kines* **3**, 195–204 (1993).

38. Lindstrom, L., and R. Magnusson, "Interpretation of myoelectric power spectra: A model and its application," *Proc IEEE* **65**, 653–659 (1977).

39. Lynn, P., N. Bettles, A. Hughes, and S. Johnson, "Influence of electrode geometry on bipolar recordings of the surface electromyogram," *Med Biol Eng Comp* **16**, 651–660 (1978).

40. Masuda, T., H. Myano, and T. Sadoyama," The position of innervatoin zones in the biceps brachii investigated by surface electromyography," *IEEE Trans BME* **32**, 36–42 (1985).

41. Merletti, R., D. Farina, and M. Gazzoni, "The linear electrode array: a tool with many applications," XIV Congr Int Soc Electrophysiol Kinesiol, Vienna, p. 5, 2002.

42. Metting van Rjin, A., C. Peper, and C. Grimbergen, "High quality recordings of bioelectric events. Part 1: Interference reduction theory and practice," *Med Biol Eng Comp* **28**, 389–397 (1990).

43. Metting van Rijn, A., C. Peper, and C. Grimbergen, "The isolation mode rejection ratio in bioelectric amplifiers," *IEEE Trans BME* **38**, 1154–1157 (1991).

44. Metting van Rjin, A., C. Peper, and C. Grimbergen, "High quality recordings of bioelectric events. Part II: Low noise, low power multichannel amplifier design," *Med Biol Eng Comp* **29**, 433–440 (1991).

45. Mortara, D., *Digital filters for ECG signals: Computers in cardiology*, IEEE Press, New York, 1977.

46. Nannucci, L., A. Merlo, R. Merletti, A. Rainoldi, R. Berrgamo, G. Melchiorri, G. Lucchetti, I. Caruso, D. Falla, and G. Jull, "Atlas of the innervation zones of upper and lower extremity muscles," *Proc XIV Congr ISEK*, Vienna, 2002, pp. 353–354.

47. Nieminen, H., "Normalization of electromyogram in the neck-shoulder region," *Eur J Appl Physiol* **76**, 199–207 (1993).

48. Nishimura, S., Y. Tomita, and T. Horiuchi, "Clinical application of an active electrode using an operational amplifier," *IEEE Trans BME* **39**, 1096–1099 (1992).

49. Pallas-Areny, R., "Interference-rejection characteristics of biopotential amplifiers: A comparative analysis," *IEEE Trans BME* **35**, 953–959 (1988).

50. Pozzo, M., D. Farina, and R. Merletti, "Detection, processing and application", in J. Moore, and G. Zourida, eds., *Biomedical technology and devices handbooks*, CRC Press, New York, 2003.

51. Ragheb, T., and L. A. Geddes, "Electrical properties of metallic electrodes," *Med Biol Eng Comp* **28**, 182–186 (1990).

52. Rainoldi, A., C. Cescon, A. Bottin, and R. Merletti, "Surface EMG alterations induced by under water recordings: A case study," *Proc XIV Congr ISEK*, Vienna, 2002, pp. 126–127.

53. Rainoldi, A., M. Nazzaro, R. Merletti, D. Farina, I. Caruso, and S. Gaudenti, "Geometrical factors in surface EMG of the vasus medialis and lateralis muscles," *J Electrom Kinesiol* **10**, 327–336 (2000).

54. Redfern, M. S., "Elimination of EKG contamination of torso electromyographic signals," in S. S. Asfour, ed., *Trends in ergonomics/human factors IV*, Elsevier Science, North-Holland, Amsterdam, 1987.

55. Redfern, M. S., R. E. Hughes, and D. B. Chaffin, "High-pass filtering to remove electrocardiographic interference from torso EMG recordings," *J Clin Biomech* **18**, 44–48 (1993).

56. Roeleveld, K., D. F. Stegeman, H. M. Vingerhoets, and A. van Oostrom, "Motor unit potential contribution to surface electromyography," *Acta Physiol Scand* **160**, 175–183 (1997).

57. Roeleveld, K., D. Stegeman, H. Vingerhoets, and M. Zwarts, "How inter-electrode distance and motor unit depth influence surface potentials," in SENIAM 5: *The state of the art on sensors and sensor placement procedure in surface EMG*, Roessingh Research and Development, Enschede, Netherlands, 1997, p. 55.

58. Roy, S., C. De Luca, and J. Schneider, "Effects of electrode location on myoelectric conduction velocity and median frequency estimates," *J Appl Physiol* **61**, 1510–1517 (1986).

59. Silverman, R., and D. Jenden, "A novel high performance preamplifier for biological applications," *IEEE Trans BME* **18**, 430 (1971).

60. Tam, H., and J. Webster, "Minimizing electrode motion artifact by skin abrasion," *IEEE Trans BME* **24**, 134–139 (1977).

61. Van der Locht, H. M., and J. H. van der Straaten, "Hybrid amplifier-electrode module for measuring surface electromyographic potentials," *Med Biol Eng Comp* **18**, 119–122 (1980).

62. Van Oosterom, A., and J. Strackee, "Computing the lead field of electrodes with axial symmetry," *Med Biol Eng Comp* **21**, 473–481 (1983).

63. Vitasalo, J., and P. Komi, "Signal characteristics of EMG with special reference to reproducibility of measurements," *Acta Physiol Scand* **93**, 531–539 (1975).

64. Webster, J. G., "Reducing motion artifacts and interference in biopotential recordings," *IEEE Trans Biomed Eng* **31**, 823–826 (1984).

65. Winter, B. B., and J. G. Webster, "Reduction of interference due to common mode voltage in biopotential amplifiers," *IEEE Trans BME* **30**, 58–65 (1983).

66. Winter, D., et al., "EMG-profiles during normal human walking: Stride to stride and inter-subject variability," *EEG Clin Neurophysiol* **67**, 402–411 (1987).

67. Zuniga, E., X. Truong, and D. Simons, "Effects of skin electrode position on averaged electromyographic potentials," *Arch Phys Med Rehab* **51**, 264–272 (1970).

6

SINGLE-CHANNEL TECHNIQUES FOR INFORMATION EXTRACTION FROM THE SURFACE EMG SIGNAL

E. A. Clancy

Electrical and Computer Engineering Department
Biomedical Engineering Department
Worcester Polytechnic Institute, Worcester, MA

D. Farina

Laboratory for Engineering of the Neuromuscular System
Department of Electronics
Politecnico di Torino, Italy

G. Filligoi

Department INFOCOM, School of Engineering
CISB, Centro Interdipartimentale Sistemi Biomedici
Università degli Studi "La Sapienza," Roma

6.1 INTRODUCTION

Signal processing techniques are mathematical procedures that can be usefully applied to extract information from biomedical signals. This chapter describes some of the most commonly used techniques for processing single-channel surface EMG signals. Some basic knowledge of signal theory (the concepts of complex numbers, convolution, Fourier transforms, autocorrelation, and stochastic processes) is assumed [94]. The single-channel techniques described in this chapter are used to study the interference pattern that results from the simultaneous activation of many motor units (MUs). These techniques do not

Electromyography: Physiology, Engineering, and Noninvasive Applications, edited by Roberto Merletti and Philip Parker.
ISBN 0-471-67580-6   Copyright © 2004 Institute for Electrical and Electronics Engineers, Inc.

resolve, or decompose, the signal into the individual MUs, rather they provide a global description of the electric potential observed at the recording site. Such information is usually not directly related to physiological phenomena or events. For example, representation of a signal as a sum of sine waves (harmonics) does not imply that the physiology formed the signal by generating and then adding sine waves; modeling of a random signal as the output of a filter with random noise as input does not imply that the signal was generated in such a way. Nevertheless, these mathematical representations are powerful tools used to detect and quantitatively describe the recorded signal resulting from physiological events.

Different approaches—some traditional, others emerging—are discussed in this chapter. In the time domain, the dominant change in the single-channel EMG is a modulation of the signal amplitude due to muscular effort and/or fatigue. As muscle effort increases, the signal strength (or amplitude) grows. Estimates of the EMG amplitude are used as the control input to myoelectric prostheses and as indicators of muscular activity or fatigue. In the frequency domain, the dominant change in the single-channel EMG during sustained contractions is a compression of the signal spectrum toward lower frequencies. Measures of this compression are associated with metabolic fatigue in the underlying muscle. Spectral changes can also be evaluated in the time domain with techniques based on zero crossings of the signal or on spike analysis. Moreover emerging methods based on nonlinear signal analysis are being applied. These techniques, known as recurrence quantification analysis (RQA), are based on detecting deterministic structures in the signals that repeat throughout a contraction.

The current state of the art of these methods is described in the subsequent sections. Two sections will precede these descriptions. The first provides a review of spectral estimation, and the second describes "traditional" stochastic models for the observed EMG signal. These models are used to develop and interpret most of the signal processing techniques described in the chapter.

6.2 SPECTRAL ESTIMATION OF DETERMINISTIC SIGNALS AND STOCHASTIC PROCESSES

6.2.1 Fourier-Based Spectral Estimators

The energy spectral density of a finite energy deterministic discrete time signal $x(k)$ is, by definition, the magnitude squared of the discrete time Fourier transform of the signal $|X(e^{j\omega})|^2$. It represents, as a consequence of Parseval's relation [94], the distribution of signal energy as a function of frequency of the signal's harmonics. The power spectral density (PSD), $S_{mm}(e^{j\omega})$, of a wide sense stationary (WSS) discrete time stochastic process[1] with zero mean is by definition the discrete time Fourier transform of its autocorrelation sequence:

[1] When both the mean and autocorrelation of a discrete time stochastic process are invariant to time shifts (and the second moment is finite), the process is said to be wide-sense stationary (WSS). In this case the mean is a single number μ and the autocorrelation can be represented by a sequence of numbers $r_{mm}(l) = E[m(k + l)m(k)]$, where E[•] is the expectation operator and k and l are discrete time indexes. The EMG signal recorded during isometric constant force contractions can be considered a WWS process, at least for time intervals short enough to exclude fatigue (see below in the text). In the following we will consider only zero mean WSS stochastic processes.

$$S_{mm}(e^{j\omega}) = \sum_{k=-\infty}^{k=+\infty} r_{mm}(k)e^{jkm} \tag{1}$$

where $e^{-jk\omega}$ represents the kth sinusoidal harmonic and $r_{mm}(k)$ is the autocorrelation function defined as: $r_{mm}(l) = E[m(k + l)m(k)]$.

In Eq. (1) the computation of the autocorrelation sequence implies an expectation whose calculation would require the availability of all the realizations of the process. It is clear that in practical applications, collection of these data is not possible. However, in the case of ergodic processes the autocorrelation sequence can be estimated from a single realization by substituting the expectation operation with a temporal average [94]. Given a limited number of samples, the autocorrelation sequence can therefore be estimated as

$$\hat{r}_{mm}(k) = \frac{1}{L}\sum_{l=0}^{L-1-k} m(k+l)m(l), \qquad 0 \le k < L \tag{2}$$

where $m(k)$ is a single process realization and L the number of acquired signal samples. It can be shown that estimator (2) is a biased estimator of the autocorrelation sequence. Replacing $r_{mm}(k)$ with $\hat{r}_{mm}(k)$ in Eq. (1) provides an estimate of the power spectrum of the process. The estimated autocorrelation sequence is generally windowed in order to reduce estimation bias, leading to a class of correlogram-based estimators.

It can be shown that the power spectral estimate based on the discrete Fourier transform of the correlation sequence estimated by (2) is equivalent to the following estimation:

$$\hat{S}_{mm}(e^{j\omega}) = \frac{1}{L}|M(e^{j\omega})|^2 \tag{3}$$

where $|M(e^{j\omega})|^2$ is the energy spectral density of the finite energy signal obtained by windowing one realization of the stochastic process. The estimator defined in (3) is called periodogram. The periodogram is an asymptotically unbiased estimator of the power spectrum (i.e., as $L \to \infty$, the expected value of the periodogram is equal to the true spectrum) but not consistent in the mean square sense, since its variance tends to the square of the spectrum value as $L \to \infty$. To reduce estimation variance, different approaches have been proposed, such as the average of the estimates obtained by different consecutive or partially overlapped signal epochs [109]. Moreover different window shapes have been introduced to enhance frequency resolution (in practical cases, indeed, the expected value of the periodogram estimator is the true spectrum convolved by the spectrum of the observation window).

6.2.2 Parametric Based Spectral Estimators

An alternative approach to spectral estimation is based on the methods referred to as parametric or model based. The theoretical basis for this class of spectral estimation techniques is the representation of the stochastic process under study as the output of a linear time-invariant (LTI) filter with white noise as its input. If the LTI filter, called the generator model, is identified, the spectrum of the process is also known. The parametric approach is based on estimation of the generator model from the available data. While the Fourier approach implicitly considers the signal to be periodic outside the observation window, the parametric methods propose an estimate that is based on the global process whose characteristics are estimated from the available data. Thus, in theory, there are no limitations to the frequency resolution; nevertheless, some assumptions on the generator model are required. In practical applications the generator model is considered physically realizable; thus the system has to be causal, implying that the LTI filter has a rational transfer function and a finite number of poles:

$$H(\mathrm{e}^{j\omega}) = \frac{\sum_{k=0}^{q} b_k \mathrm{e}^{jk\omega}}{\sum_{k=0}^{p} a_k \mathrm{e}^{jk\omega}} \tag{4}$$

It can be shown that the power spectrum of a process generated by filtering white noise with a LTI filter is the multiplication of the power spectrum of the input (a constant in the case of white noise) with the squared magnitude of the filter transfer function. Equation (4) provides the general transfer function that defines a so-called ARMA (autoregressive moving average) model. If $a_i = 0$ ($i = 1, \ldots, p$) and $a_0 = 1$, a MA (moving average) model results, and if $b_i = 0$ ($i = 1, \ldots, q$) and $b_0 = 1$, an AR (autoregressive) model results.

The problem of spectral estimation is thus converted into the problem of estimating a finite number of parameters (from which the term "parametric methods"), which are sufficient to completely describe the entire process from the spectral content point of view. For details about the methods for estimating the model parameters, the interested reader can refer to [51,63].

There are many limitations to the parametric approach. The first issues encountered in dealing with parametric methods are selecting the type (AR, MA, or ARMA) and the order of the model (number of parameters). In the ideal case the model type should always be the most general one (ARMA) and the order must be chosen larger or equal to the real order: the extra parameters will theoretically be estimated to be zero. In practical applications, AR parameters are much easier to compute than MA parameters; moreover it is always possible to represent an ARMA or MA model by an infinite AR model [63] that, in practice, will be truncated at a specific order. Thus AR models are widely used for spectral estimation. In the following we will always refer to AR models. In practice, the choice of an arbitrarily large number of parameters is not appropriate, since the unnecessary parameters are never estimated as zero. As indicated by Parzen [83], the variance of the AR spectrum estimate for large sample sizes is directly related to the number of parameters and inversely related to the number of samples (available data). Thus, in order to make the variance small, the model order p should be small. However, a small p may lead to a poor AR approximation of the true spectrum, resulting in increased bias of the estimated spectrum and lower resolution of closely spaced spectral peaks. The trade-off between estimation variance and model order is the counterpart of the trade-off between variance and frequency resolution of the Fourier-based methods. Many criteria have been proposed to estimate the appropriate model order from the available data. These criteria include Akaike's final prediction error (FPE) [1], Akaike's information criterion (AIC) [1], Parzen's autoregressive transfer function criterion (CAT) [83], and Rissanen's minimum description length criterion (MDL) [91,92]. However, often, as in the case of the surface EMG signal, the selection of the model order is based on signal simulations and is adapted to the specific application.

6.2.3 Estimation of the Time-Varying PSD of Nonstationary Stochastic Processes

If the statistical properties of a process change with time, the process is said to be nonstationary, and spectral analysis with the estimators introduced above may not be appropriate. In particular, the preceding spectral analysis techniques give frequency information without any time localization. For example, in the case of surface EMG, the signal changes its characteristics as a consequence of muscle fatigue (see Chapter 9) or as a consequence of changes in the MU pool. If we consider an EMG signal detected during a prolonged, high-effort muscle contraction of one minute and compute one power spectrum from the entire contraction, we will not obtain any information about the changes that occurred throughout the contraction. We will get some "averaged" information about the frequency content of the signal during the entire contraction. The simplest approach to obtain both time and frequency information is to divide the signal into many segments (epochs) and estimate a power spectrum for each. If the changes we want to monitor are slower than one second, for example, it would be sufficient to divide a one minute duration signal into 60 contiguous epochs (each of one second duration). In each epoch the signal can be considered as a realization of a WSS stochastic process; thus an estimate of its spectrum is feasible. This partitioning is the basic idea of the short-time Fourier transform (STFT), from which the spectrogram is defined; it is currently the most widely used method for studying non-stationary signals. In the same manner a time-varying autoregressive (TVAR) approach results if the spectra of the epochs are estimated with an AR model applied to contiguous signal epochs. The STFT and TVAR approaches can be refined by including epoch overlapping and/or windowing of the data. More advanced time-frequency approaches may be more appropriate in cases of strong nonstationarity of the signals under study [17], and these will be discussed in Chapter 10.

6.3 BASIC SURFACE EMG SIGNAL MODELS

The surface EMG signal detected during voluntary contractions is the summation of the contributions of the recruited MUs that are observed at the recording site. A simple analytical model of the generated signal, $m(k)$ (k being the discrete time index) is the following:

$$m(k) = \sum_{i=1}^{R} \sum_{l=-\infty}^{+\infty} x_{il}(k - \Phi_{i,l}) + v(k) \tag{5}$$

where R is the number of active MUs, $x_{il}(k)$ the lth MU action potential (MUAP) belonging to MU i, $\Phi_{i,l}$ the occurrence time of $x_{il}(t)$, and $v(k)$ an additive noise/interference term [24]. The additive noise/interference represents electrode-electrolyte noise, the noise of the electronic amplifiers, line interference, biological noise and the interference activity of MUs far from the detection point. Equation (5) is an example of a stochastic process represented by an analytical expression containing parameters which are random variables (the occurrence times of the MU firings). For the case of uncorrelated discharges, it can be shown that the resultant spectrum is the summation of the spectra of the MUAP trains. The spectrum of a MUAP train is the product of the spectrum of the MUAP (deterministic finite energy signal) with that of the point process describing the firing pattern (random process). For the case of a Gaussian distributed interpulse interval, the spectrum of the point process is given by [56]

$$S_\phi(\omega) = \frac{1}{\mu} \frac{1 - e^{-\sigma^2\omega^2}}{1 + e^{-\sigma^2\omega^2} - 2e^{-\sigma^2\omega^2/2} \cos(\mu\omega)} \tag{6}$$

where μ and σ are the mean and standard deviation of the interpulse interval, respectively. As $\omega \to \infty$ the spectrum of the point process tends to a constant value. Substituting into (6) values of the mean and standard deviation of the interpulse interval for normal physiological conditions (e.g., a mean firing rate of 8–35 Hz and coefficient of variation of the mean interpulse interval of approximately 15%), one finds that the spectrum of the point process is nonconstant in a rather small frequency region, mainly below 30 Hz. Thus, because high-pass analog filters are used for surface EMG conditioning (see Chapter 5), the influence of the firing patterns of the MUs on the surface EMG power spectrum can be neglected in many applications. The high-frequency range (above 30 Hz) of the EMG power spectrum represents the morphology of the recorded MUAPs, influenced by the relative positions of the MUs with respect to the recording system, the electrode configuration and electrode shape and size, and the conduction velocity (CV) at which the action potentials propagate [30,57,101] (see Chapters 4 and 8). Note that in the case of correlated firing patterns, the global EMG spectrum would also contain cross-terms [111]. Dependent firing patterns are due, for example, to short-term synchronization (MUAPs of different MUs firing at approximately the same time more frequently than would be expected by chance alone) or to common drive (the common modulation of firing rates).

For certain applications, the surface EMG signal can be modeled by functional models in a coarser fashion than that provided by Eq. (5). Functional models of the EMG seek to capture the observed stochastic behavior of the EMG signal without including the complexity that would be involved in modeling the activity of each individual MU (see also Chapter 8). A complete model of this type for a single channel of EMG is shown in Figure 6.1a. This model produces a measured surface EMG (m_k) with statistical properties similar to real EMG, during both fatiguing and non fatiguing contractions. In the model, a zero-mean, WSS, correlation-ergodic (CE), white process of unit variance w_k passes through the stable, inversely stable, linear, time-variant shaping filter $H_{time}(e^{j\omega})$.

The white random Gaussian process and the shaping filter account for the first-order probability density of the EMG and the spectral shape of the EMG, respectively. The signal is then multiplied by the EMG amplitude s_k, which modulates the EMG standard deviation based on the level of muscular activation. The filter $H_{time}(e^{j\omega})$ preserves signal variance so that all modulation in the standard deviation of the noise-free EMG signal (r_k in Figure 6.1a) is attributed to changes in EMG amplitude. Finally, a zero-mean, WSS, CE noise process v_k is added to the signal to form the measured surface EMG m_k. This noise process represents measurement noise (e.g., due to the electrode-amplifier circuitry and due to noise at the electrode–skin interface) and cannot be completely eliminated. The processes w_k and v_k are assumed to be uncorrelated with each other. In Figure 6.1b examples of realizations of the stochastic process defined in Figure 6.1a are reported together with the expected spectra. The shaping filter has been selected as suggested by Shwedyk et al. [97]. The expression of this suggested spectrum has two parameters f_h and f_l that allow to change the shape of the spectrum. Nonstationarity may be generated by changing these parameters during time. The first moment (mean frequency; see also below) of this spectrum can be computed analytically [26]:

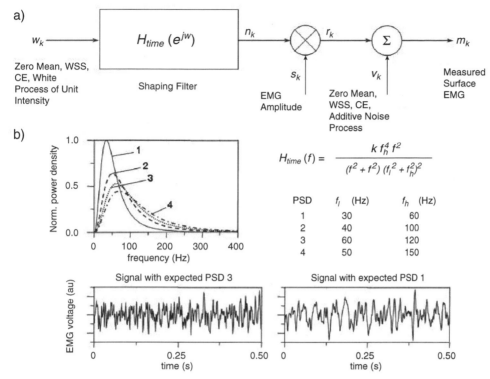

Figure 6.1. Simulation of EMG signals by filtering white Gaussian noise. (a) The output signal is obtained by filtering white Gaussian noise with a shaping filter (from [14]). (b) Examples of expected spectra obtained from Shwedyk's expression [97] for different values of the two parameters f_h and f_l. Two simulated signals obtained by filtering white Gaussian noise with the inverse Fourier transform of the square root of $H_{time}(f)$ are also shown. (From [26] with permission)

$$f_{\text{mean}} = \frac{2 f_h}{\pi} \frac{\alpha+1}{\alpha-1} \left[\frac{2\alpha^2}{\alpha^2-1} \ln \alpha - 1 \right]$$

where α is the ratio f_l/f_h. This model is fundamentally phenomenological and, of course, cannot be used for understanding how physiological events are reflected in the surface EMG signal features. Nevertheless, it assumes importance for the analysis of the statistical properties of estimators of signal features, such as amplitude and frequency content.

6.4 SURFACE EMG AMPLITUDE ESTIMATION

EMG amplitude estimation can be described mathematically as the task of best estimating the standard deviation of a colored random process in additive noise (refer to the model of Fig. 6.1). This estimation problem has been studied for several years. Inman et al. [49] are credited with the first continuous EMG amplitude estimator. They implemented a full-wave rectifier followed by a resistor-capacitor low-pass filter.[2] Subsequent early investi-

---

[2] They termed their processor an "integrator"—a misnomer they acknowledged in their original work. This incorrect term is still frequently used.

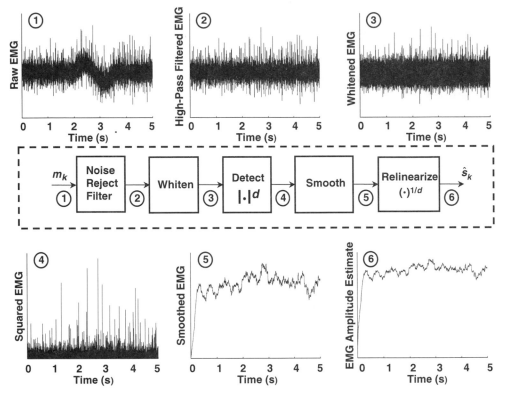

<u>Figure 6.2.</u> Cascade of processing stages used to form an EMG amplitude estimate. The acquired EMG signals are assumed to be from bipolar electrodes. The EMG amplitude estimate is $\hat{s}_k$. In the "detect" and "relinearize" stages, $d = 1$ for MAV processing and $d = 2$ for RMS processing. (Adapted from [16] with permission)

gators studied the type of nonlinear detector that should be applied to the waveform. This work was primarily empirical, and led to the routine use of analog rectification and low-pass filtering to estimate amplitude. Most modern systems are digital, and use mean absolute value (MAV), also called average rectified value (ARV), and root-mean-square (RMS) indicators. Other amplitude estimators, not based on simple rectification or squaring, were later investigated.

From these later works, a standard cascade of sequential processing stages emerged to form a single channel processor for EMG amplitude estimation. The stages are (1) noise and interference attenuation, (2) whitening, (3) demodulation, (4) smoothing, and (5) relinearization (Fig. 6.2). Noise and interference attenuation seek to limit the adverse effects of motion artifacts, electronic noise, power line interference, for example, as described in Chapter 5. The correlation between neighboring EMG samples is a consequence of the limited signal bandwidth, which reflects the actual biological generation of EMG and the low-pass filtering effects of the tissues (see Chapter 8). Decorrelation, that is whitening, makes the samples statistically uncorrelated, increases the "statistical bandwidth" (defined in [8]) and reduces the variance of amplitude estimation. Demodulation rectifies the whitened EMG and then raises the result to a power (either 1 for MAV processing or 2 for RMS processing). Smoothing filters the signal, whereas relinearization inverts the power law applied during the demodulation stage, returning the signal to units of EMG amplitude. Powers higher than 1 or 2 could be used. The quality of EMG amplitude estimates and techniques for implementing the processing stages will be discussed in the following sections.

6.4.1 Measures of Amplitude Estimator Performance

Before describing the processing stages for amplitude estimation, it is important to describe objective assessment measures of amplitude estimator performance. When the contraction is isometric, of constant force, and nonfatiguing, it is generally assumed that the EMG amplitude should be constant. To quantify the "quality" of the amplitude esti- mate, it is common to define a dimensionless signal-to-noise ratio (SNR) as the amplitude estimate mean, computed over a number of signal segments, divided by the standard devi- ation of the estimates. This measure does not vary with the gain of the EMG channel and makes no assumption as to any relationship between EMG and muscle force. Because SNR is a measure of the random fluctuations of the EMG amplitude estimate, better estimators yield higher SNR's. Some authors have used the square of this measure as a performance index (see Chapter 18).

When force or posture is changing, SNR is no longer meaningful and alternative measures of performance must be used. One approach has been to display a real-time amplitude estimate to the subject as a form of biofeedback. The experimenter generates a target display for the subject to track. The target is moved over the range of desired EMG amplitudes, usually via computer control. The tracking error (e.g., RMS error between the target amplitude and the estimate) serves as a performance measure, with better EMG amplitude estimators presumably providing lower error. This technique also makes no assumption of an EMG-force relationship.

Finally, a common application of surface EMG is to estimate joint torques. Again, better amplitude estimation is assumed to provide better EMG-torque estimation. Note that the amplitude of EMG is affected by many confounding factors other than joint torque, such as the subcutaneous layer thickness, the inclination of the fibers with respect to the detection system, and the interelectrode distance selected [30] (Chapter 4). As a conse- quence EMG-based joint torque estimates must account for these confounding factors, for example by normalization with respect to a subject and muscle specific reference value, and results must be appropriately interpreted.

6.4.2 EMG Amplitude Processing—Overview

For the functional model for sampled EMG presented in Figure 6.1, the goal is to estimate $s(k)$ based on samples of $m(k)$. If the myoelectric samples were comprised of independent, identically distributed (IID), noise-free random samples, then estimation of $s(k)$ would be very simple. If the noise-free IID samples were Gaussian distributed, then classic estimation results (e.g., see [45]) show that the maximum likelihood (ML) estimate is the RMS. If the noise-free IID samples were Laplacian distributed (a more centrally peaked distribution than Gaussian), then the ML estimate is the MAV [15]. Unfortunately, EMG samples are neither IID nor noise free. Formal optimal solutions to the complete model do not exist. However, a simplifying approach can be taken to explain existing solutions (Fig. 6.2).

Stage 1: Noise and Interference Attenuation. The goal of the first stage is to eliminate additive noise, artifacts, and power line interference that are acquired along with the "true" EMG. Methods to do so are described in Chapter 5 and will not be repeated here. Note that these methods can incorporate proper skin preparation and electrode setup, analog filtering in the amplifier apparatus, adaptive digital filtering during postprocessing, and so on.

Stage 2: Whitening. Because successive samples of the EMG signal are correlated, direct information extraction from the signal is confounded (in a probabilistic sense). That is, the signal correlation temporally "weights" the information. Whitening resolves this problem by transforming the signal so that successive samples have equal "weight." A frequency domain measure of the degree of correlation in the data is the statistical bandwidth [8]. Hogan and Mann [45] showed that as the statistical bandwidth of a signal is increased (and correlation is decreased, e.g., via whitening), the SNR of the EMG amplitude estimate (for the constant force case) increases as the square root of the statistical bandwidth. Additional details of the relationship between SNR and signal bandwidth are derived in Chapter 18. For contractions above 10% MVC, whitening has led to a 63% improvement in the SNR [14] (Fig. 6.2).

A whitening filter outputs a theoretically constant, or "whitened" power spectrum in response to an input. This filter is formed by first estimating the PSD of the EMG signal. Then the inverse of the square root of the PSD is the magnitude of the whitening filter. As long as the EMG PSD estimate is nonzero at all frequencies below the Nyquist frequency, this inverse will exist (Fig. 6.3). The phase of the whitening filter is arbitrary, but it is treated as causal for the causal EMG processor. For isometric, constant-force, nonfatiguing contractions, it is common to model the EMG as a WSS, amplitude modulated, AR process (software for doing so is readily available [87]). With this model, the PSD of EMG, denoted $S_{mm}(e^{j\omega})$, can be written as

$$S_{mm}(e^{j\omega}) = \frac{b_0}{\left|1 - \sum_{i=1}^{p} a_i e^{-ij\omega}\right|^2}$$

where the a_i are the AR coefficients, p is the model order, and ω is the angular frequency in rad/s. These coefficients are estimated from a calibration contraction that is typically a few seconds in duration. Once these coefficients are determined, whitening can be performed on subsequent recordings with a discrete-time MA filter, which operates as follows (Fig. 6.3):

$$y(k) = \frac{1}{\sqrt{b_0}} x(k) + \frac{-a_1}{\sqrt{b_0}} x(k-1) + \ldots + \frac{-a_p}{\sqrt{b_0}} x(k-p)$$

where $x(k)$ are the generic data input to the whitening filter and $y(k)$ are the whitened output data. Model orders of 4–6 have been found sufficient to model the PSD for whitening purposes [14,43,104].

D'Alessio et al. [11,18,19] used a generalization of this whitening approach based on the assumption that the PSD of the EMG can vary in a general manner (i.e., not just restricted to an amplitude modulated PSD), and thus the MA whitening filter must do so as well. In this case the EMG signal is considered to be nonstationary, and thus the PSD model and the whitening filter must be continuously updated.

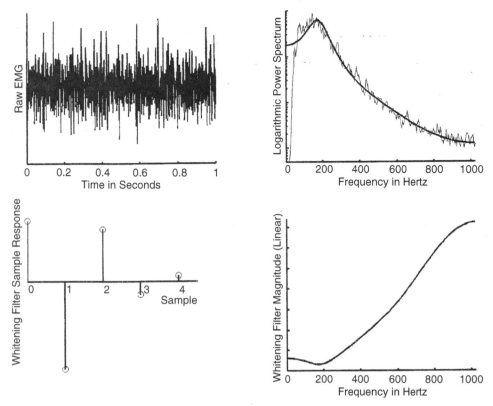

Figure 6.3. Fixed whitening filter design. (*Upper left*) A 1s portion of the EMG signal (total recording is 5s). (*Upper right*) Jagged plot is the discrete Fourier transform estimate of the EMG PSD; smooth plot is the fourth-order AR estimate of the EMG PSD. (*Lower right*) The magnitude response of the whitening filter formed from the AR PSD estimate and (*lower left*) its sample response. (From [14])

The whitening techniques can fail at low contraction levels because of the presence of additive background noise. Clancy and Farry [13] implemented an alternative adaptive MA whitening technique (again based on an amplitude modulated AR PSD model), by incorporating the fact that EMG is invariably acquired in the presence of an additive, broadband background noise. Thus the whitening filter should be adapted, but the adaptation is not as general as that proposed by D'Alessio et al. [19]. The adaptation scheme for the time-varying MA whitening coefficients is determined in a calibration phase based on the estimated PSD of the additive noise (measured from a rest contraction) and a reference contraction (e.g., at 50% MVC).

Stages 3 and 5: Demodulation and Relinearization. After the whitening stage, the signal is assumed to be noise free and uncorrelated. The standard deviation of the "true" EMG is now well approximated by the standard deviation of the resulting signal. In order to estimate this standard deviation of the EMG samples, some form of nonlinearity must be applied to the signal. In general, the nonlinearity consists of raising the absolute value of each sample to a power (demodulation). The two most common powers

(1 and 2) will be discussed. After raising the signal to a power, the signal is smoothed (as discussed in the next section) and then relinearized. Relinearization consists of raising the signal to the inverse of the demodulation power. Hence demodulation and relinearization are considered here together. To approach this problem theoretically, it will also be assumed that the signal is WSS over the short interval being considered (this point will be addressed in more detail in the smoothing section). For testing purposes, WSS EMG samples from isometric constant-force, nonfatiguing contractions will be used.

Usually EMG samples are modeled as conforming to a zero-mean Gaussian probability density function (PDF) [35,45,46]. With a ML estimation, a second-power (or RMS) demodulator is used to give the highest theoretic SNR performance of $SNR_{\text{RMS}} \cong \sqrt{2N}$, where N is the number of statistical degrees of freedom in the EMG [8,45]. For perfectly whitened data, N equals the number of data samples in the smoothing window. More recently Clancy and Hogan [15] modeled an EMG sample with the Laplacian PDF (which is more peaked near zero) and showed that a first-power (or MAV) demodulator can give the best performance in this case, yielding a theoretical SNR of $SNR_{\text{MAV}} = \sqrt{N}$, which is smaller, by a factor of $\sqrt{2}$, than that of the Gaussian distribution. Their experimental comparison of MAV and RMS indicated that MAV processing had a higher SNR than RMS processing, but only by 2.0% to 6.5% [15], suggesting that, in practical conditions, RMS or MAV processing are nearly indistinguishable.

Stage 4: Smoothing. In the smoothing stage, several demodulated samples are time-averaged to form one amplitude estimate. A sliding window selects the demodulated samples for each successive amplitude estimate, thereby forming an averaging filter. Because EMG amplitude is, in general, changing during contraction, an appropriate smoothing window length over which the signal is "quasi-stationary" must be selected. It is found that random fluctuations in the EMG amplitude estimate are diminished with a long smoothing window; however, bias (deterministic) errors in tracking the signal of interest are diminished with a short smoothing window. This trade-off can be expressed by considering the mean square error (MSE) of the amplitude estimate: MSE = $\sigma^2 + b^2$, where σ^2 is the variance and b is the bias of the estimate. An appropriate balance needs to be established. Clancy [12] derived a method for optimal selection of a fixed window length. Both the variance and bias were written as a function of the window length, and then optimization was used to determine the best length. Different results were derived for causal and noncausal (midpoint moving average) processing. For noncausal processing, the optimal window length was found to be

$$\frac{N_{\text{Noncausal}}}{f} = \left[\frac{72}{8}\right]^{1/5} \cdot \left[\frac{s_{\text{Ave}}^2}{(\ddot{s}^2)_{\text{Ave}}}\right]^{1/5}$$

where N is the window length (samples), f is the sampling frequency (Hz), s_{Ave}^2 is the average value of the square of EMG amplitude, and $(\ddot{s}^2)_{\text{Ave}}$ is the average value of the square of the second derivative of EMG amplitude. The quantities s_{Ave}^2 and $(\ddot{s}^2)_{\text{Ave}}$ assume different values for different tasks and must be estimated by the user. The constant g is related to the number of statistical degrees of freedom in the data, determined by the statistical bandwidth of the EMG, the number of EMG channels and the detector type (see [16,19] for details). For causal processing, the optimal window length was found to be

$$\frac{N_{\text{Causal}}}{f} = \frac{1}{g^{1/3}} \cdot \left[\frac{s_{\text{Ave}}^2}{(\dot{s}^2)_{\text{Ave}}} \right]$$

where $(\dot{s}^2)_{\text{Ave}}$ is the average value of the square of the first derivative of EMG amplitude.

Several studies have attempted to improve the amplitude estimate by dynamically adapting the window length to the local characteristics of the EMG (see [12] for a review). In direct comparison to the best fixed-length smoother, these adaptive smoothers have found little or no advantage for generic applications, with a few exceptions. Mathematically other smoothing methods are possible (e.g., Evans et al. [25] used a Kalman filter) and other techniques may prove better than the simple approaches presently discussed. Thus future work in nonstationary EMG processing is needed to develop and compare such methods to existing methods.

6.4.3 Applications of EMG Amplitude Estimation

For many years myoelectrically controlled upper-limb prosthetics have been a driving force in the development of EMG amplitude estimation algorithms. EMG from remnant muscles has been used to control the operation of a prosthetic elbow, wrist, and/or hand. Most frequently the electrical activity of two muscle sites is monitored. If biceps and triceps muscles remain, these sites are usually selected. A common scheme is to estimate the EMG amplitude from the two sites, and determine the difference amplitude. If the biceps amplitude is larger, flexion occurs (i.e., elbow flexion, hand closure). If the triceps amplitude is larger, extension occurs (i.e., elbow extension, hand opening). For these applications small variance estimations are required (for a more detailed description of myoelectrically controlled prosthetics, see Chapter 18).

Clinically EMG amplitude is used to study muscle coordination and activation intervals. For example, EMG amplitude is used in gait analysis to determine when various muscles are active throughout the gait cycle (see Chapter 15). Finally, surface EMG amplitude is computed as an indicator of muscle fatigue, often together with spectral analysis (see Chapter 9), since it is an indicator of CV decrease, MU pool changes, and other mechanisms occurring with fatigue. In occupational field studies, joint amplitude and spectral surface EMG analysis have also been proposed to get better insight into the development of muscle fatigue in non isometric contractions (see below).

6.5 EXTRACTION OF INFORMATION IN FREQUENCY DOMAIN FROM SURFACE EMG SIGNALS

Changes in the power spectrum of the surface EMG signal during muscle contraction were observed for the first time by Piper [86] who detected a decrease in the dominant oscillation of the recorded surface EMG signal during maximal voluntary contractions, as a consequence of muscle fatigue. Subsequent investigators quantified the changes in the frequency content of the EMG signal using various spectral descriptors, such as the centroid frequency [96], the median frequency [100], and the high/low frequency ratio [41]. Moreover parameters extracted from the signal in the time domain, such as the rate of zero crossings [42,48] or the spike properties [36], were proposed as alternative indicators of changes in the surface EMG spectral content.

The theoretical basis for the interpretation of the evolution of the power spectrum of the surface EMG signal during fatigue and for understanding the factors influencing it follow from the works by Lindstrom, DeLuca, and Lago [20,56,57]. These works developed the theory needed to interpret frequency changes in the EMG signal with respect to the underlining physiological events. The conclusions by Lindstrom and Magnusson [57], especially those related to the effect of the spatial filter on the PSD of the detected signal, have been extensively validated experimentally. Lindstrom and Magnusson [57] also provided the basis for the interpretation of the changes of the characteristic spectral frequencies as a consequence of changes in MU CV. They proposed the following expression for the PSD of the surface signal generated by an intracellular action potential traveling along the muscle-fiber:

$$P(f) = \frac{1}{v^2} G\left(\frac{fd}{v}\right) \tag{7}$$

where v is the propagation velocity of the action potential and $G(fd/v)$ includes the spectrum of the action potential and various geometrical factors. This expression implies that the spectrum is scaled by CV. Thus its shape does not change but only the frequency axis is scaled when CV changes.

These concepts were further developed by DeLuca [20,101], who clarified the fundamental role of spectral analysis in the study of muscular fatigue. The work by Lago [56] outlined the effect of the firing pattern of the active MUs on the surface EMG PSD and spectral variables (refer to Eq. (6), to be used together with Eq. (7) to derive the PSD of a MUAP train).

Since these pioneering works, a large number of studies reported the use of spectral analysis of the surface EMG signal for the investigation of muscle fatigue or MU recruitment strategies. From the experimental evidence provided by Merletti, DeLuca, and Arendt-Nielsen et al. [4,5,21,77,73], it is clear that the use of spectral shift of the EMG signal as a measure of muscle fatigue offers a more objective assessment technique compared to the more subjective clinical techniques based on mechanical fatigue. This observation led, in the early 1980s, to the development of analog and digital instruments for monitoring spectral parameters of the signal [39,67,84,102].

This section describes the basic body of knowledge concerning EMG spectral analysis (by Fourier and parametric techniques) and how physiological parameters are reflected by surface EMG power spectra. The section is mostly focused on the assessment of muscle fatigue during isometric constant force contractions, since this assessment is by far the most prevalent application of surface EMG spectral analysis.

6.5.1 Estimation of PSD of the Surface EMG Signal Detected during Voluntary Contractions

Spectral analysis of EMG signals detected during voluntary constant force isometric contractions is usually performed using the STFT with nonoverlapping epochs of 0.25 to 1 s as described in Figure 6.4*a*. A few studies in the literature have reported EMG results from parametric analysis [22,26,64,74,82]. The estimation model chosen in these cases was the AR model with an order between 4 and 11.

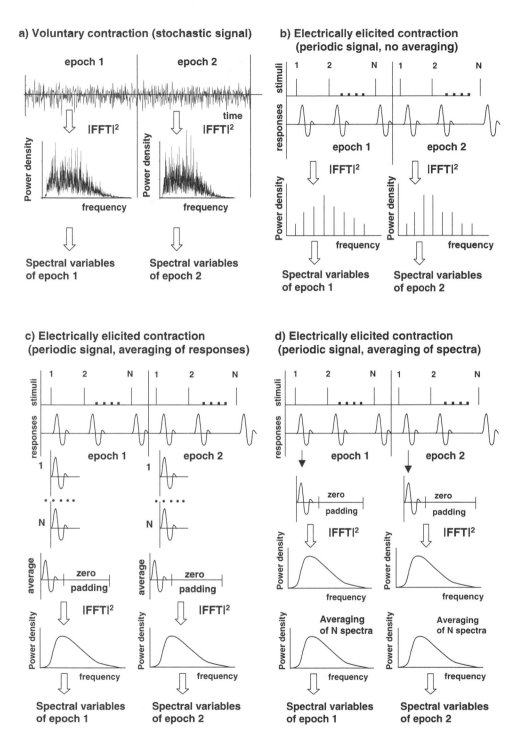

Figure 6.4. Estimation of EMG signal spectrum. (*a*) Spectral estimation during voluntary contractions. The signal is divided into a sequence of epochs during which the signal is assumed stationary. (*b*) Spectral estimation of a signal detected during electrically elicited contractions by dividing the quasi-periodic signal into epochs containing many M-waves (a line spectrum estimate is obtained). (*c*) Spectral estimation of an electrically elicited signal by averaging a number of M-waves within each epoch. (*d*) Spectra obtained by averaging the estimations obtained from each M-wave. (From [72] with permission)

6.5.2 Energy Spectral Density of the Surface EMG Signal Detected during Electrically Elicited Contractions

Electrically evoked signals may be considered deterministic and quasi-periodic signals, with the period determined by the stimulation frequency imposed by the stimulator. Each M-wave is a finite energy, finite duration signal whose frequency content can be described by the energy spectral density. The energy spectral density reflects the properties of the detected MUAPs, in particular, their CV.

Different strategies can be applied to estimate the spectrum of an electrically elicited EMG signal during a prolonged contraction [72] (Fig. 6.4*b*, *c*, and *d*):

1. The signal can be divided into epochs, and the frequency attributes are computed over each epoch. The signal of each epoch includes many M-waves and is periodic; the spectral lines are separated by an interval equal to the stimulation frequency.

2. The spectral features can be computed for each M-wave, which is thus seen as a finite length, nonperiodic signal. In this case the spacing between spectral lines is determined by the epoch length and can be reduced by zero padding, which provides interpolation of an otherwise coarse spectral estimate. The resulting spectra can then be averaged to improve the quality of the estimate.

3. The signal can be divided into epochs, and the M-waves in each epoch are averaged. The spectral features are then computed for each averaged M-wave. Similar considerations as in the previous case can be drawn, but the signal to noise ratio is increased by the averaging process and the computational cost decreased.

The method usually applied is the third one.

Spectral analysis of electrically evoked EMG signals is mainly used for detecting changes of scale of the M-waves. An alternative approach for estimating scale factors in deterministic signals is to process the signals directly in the time domain. Merletti et al. [69] proposed a maximum likelihood approach to solve this problem, while more recently, Muhammad et al. [79] developed a pseudojoint estimator of the time scale factor and time delay between signals for applications of M-wave analysis during fatigue. These approaches were shown to be, in general, more robust than spectral analysis for estimating the scaling of the M-wave due to fatigue for the case of significant truncation of the wave (when the stimulation frequency is above 30–35 Hz, the stimulus interval may be shorter than the total M-wave length and thus the M-wave is truncated). Alternative approaches have been proposed by Lo Conte et al. [58], who decomposed the M-wave into a particular series of functions, and by Olmo et al. [81], who proposed the matched continuous wavelet transform to estimate the M-wave scale factor. The distribution function technique proposed by Rix and Malengé [93] can also be directly applied to pairs of M-waves for estimating the scale factor between them.

6.5.3 Descriptors of Spectral Compression

During both voluntary and electrically elicited fatiguing contractions the PSD of the EMG signal progressively moves toward lower frequencies. This phenomenon can be described to a large extent as a compression of the spectrum, meaning the shape of the spectrum does not change but only the scale factor of the frequency axis changes (refer to Eq. (7)

and to Chapter 9). In the case of a pure scaling, a single parameter would give all the information about the phenomenon of spectral compression; thus any reference spectral frequency can be used as an estimator of spectral compression. One of the possible spectral descriptors is the mean or centroid frequency (MNF) which is defined as

$$ f_{\text{mean}} = \frac{\int_0^{f_s/2} f S(f)\mathrm{d}f}{\int_0^{f_s/2} S(f)\mathrm{d}f} \tag{8} $$

where $S(f)$ is the PSD of the signal and f_s is the sampling frequency.

MNF is the moment of order one of the power spectrum. In general, the central moments of order k are defined as [95]

$$ M_{\text{Ck}} = \frac{\int_0^{f_s/2} (f - f_{\text{mean}})^k S(f)\mathrm{d}f}{\int_0^{f_s/2} S(f)\mathrm{d}f} \tag{9} $$

Merletti et al. [71] proposed a time domain technique to estimate the central moments of any order that avoids direct estimation of the PSD and is particularly suitable for real-time implementation. Other characteristic frequencies f_p are the pth fractile frequencies, indirectly defined as

$$ \int_0^{f_p} S(f)\mathrm{d}f = p \int_0^{f_s/2} S(f)\mathrm{d}f, \qquad 0 < p < 1 \tag{10} $$

Equation (10), with $p = 0.5$, defines the median frequency (MDF), whereas $p = 0.25$ and $p = 0.75$ define the other interquartile frequencies. Although any fractile frequency can be used to estimate spectral compression, it has been recently suggested that a better estimate of spectral compression due to CV decrease may be obtained by computing the average change of a number of properly selected percentile frequencies. Each percentile frequency indeed may provide a different indication of fatigue due to the other factors affecting the PSD of the signal apart from CV. The analysis of a number of percentile frequencies enables a distinction between the spectral changes due to CV and those due to other factors that affect spectral shape. It has been shown [59] that some percentile frequencies are better correlated with CV changes.

In addition to spectral moments and fractile frequencies, all of the parameters from an AR model spectral estimate can be used to characterize spectral changes in the EMG signal [53]. Finally the frequency at which the first dip introduced by the detection system (see Chapter 5) can be used as a descriptor of spectral changes. In this case the spectral descriptor reflects only the CV of the active MUs [57], but its variance of estimation is much higher with respect to spectral moments [65].

MNF and MDF are the most commonly used spectral descriptors. Only a few studies have used higher order moments to describe the EMG spectrum [68]. In the case of voluntary contractions, spectral descriptors are random variables with particular statistical properties that depend on the nature of the descriptor, epoch signal length, amount of epoch overlapping, type of window, and the spectrum estimator adopted. The influence of these parameters can be evaluated by the use of simulation models such as that shown in Figure 6.1 [7,26,66].

Properties of MNF and MDF. MNF and MDF provide some basic information about the spectrum of the signal and its changes versus time. They coincide if the spectrum is symmetric with respect to its center line, while their difference reflects spectral skewness. A tail in the high-frequency region implies MNF higher than MDF. A constant ratio f_{mean}/f_{med} versus time implies spectral scaling without shape change, while a change in this ratio implies a change of spectral skewness or shape. It can be shown that the standard deviation of the estimate of MDF is theoretically higher than that of MNF [102], as confirmed in experimental studies [101]. However, it can also be shown that MDF estimates are less affected by additive noise (particularly if the noise is in the high-frequency band of the EMG spectrum) [101] and more affected by fatigue (since the spectrum becomes more skewed with fatigue). Because of these pros and cons and because of the additional information that is carried jointly by the two variables, researchers often use both in their reports. However, in cases where the signal-to-noise ratio may be very low, at least during particular intervals of time (e.g., at the beginning of a ramp contraction), MDF is often preferred [9,10].

Fourier versus Parametric Approach. The Fourier and parametric approaches have been compared using simulated EMG signals (model of Fig. 6.1) for different window lengths and degree of nonstationarity [26]. It was found that in both stationary and nonstationary conditions the two approaches lead to similar results, in terms of variance and bias of estimation, for MNF and MDF over a large range of epoch lengths. For very short epochs (below 0.25 s) the parametric approach performs better than the Fourier approach, but the difference can be negligible in practical applications. Figure 6.5 shows the comparison between Fourier and AR estimation of the PSD of experimental surface EMG signals for different epoch lengths. MNF and MDF estimates are also reported.

Window Shape. The type of window shape determines the bias in the power spectrum estimation, but it is difficult to predict analytically how the bias in spectral lines is reflected in the bias of spectral descriptors. Again, simulation studies provide an evaluation of this effect [66]. In the case of the generation model in Figure 6.1, it was shown that the choice of the window is not critical for MNF nor MDF estimation. The rectangular window has been used in the majority of the experimental studies.

Epoch Length and Epoch Overlapping in Stationary and Nonstationary Conditions. In stationary conditions (model of Fig. 6.1 with fixed parameters), the larger the duration of the signal epoch, the lower the variance and bias of estimation of the spectral descriptors. In the case of nonstationary conditions, two sources of bias are present, the bias related to the spectral estimator applied to a finite observation window and that due to the nonstationarity. Bias due to both effects increases with epoch length. On the contrary, variance of MNF and MDF estimates decreases with increased window duration. In the case of isometric, constant force, fatiguing contractions, the signal can be considered stationary for epoch durations of the order of 1 to 2 seconds. The spectral descriptors are computed from several sequential (possibly overlapping) epochs. Usually the parameters of the linear regression that best fits the time-changing values of descriptors are used as a fatigue index (see Chapter 9). A large number of short epochs (i.e., of

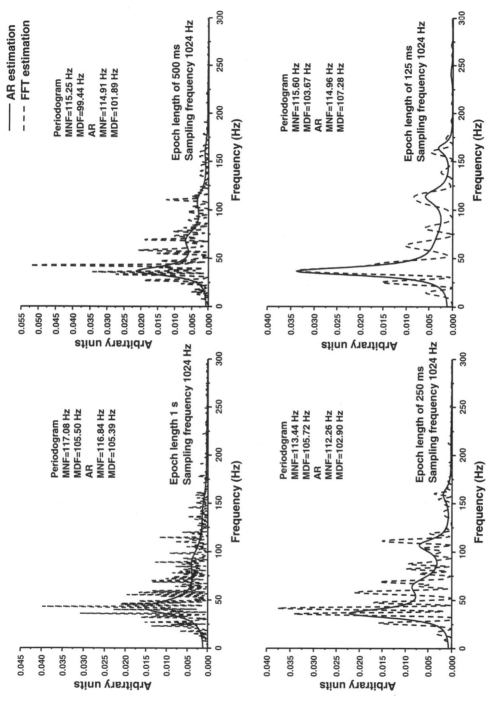

Figure 6.5. Comparison between periodogram (*dashed lines*) and AR (*solid lines*) spectrum estimates of an experimental surface EMG signal (muscle biceps brachii) for different lengths of the observation window. MNF and MDF are also reported. (From [26] with permission)

experimental points) in the regression interval will reduce the variance of the polynomial parameter estimates but will increase the spread of the experimental points. An additional factor of interest is the degree of epoch overlapping, which allows an increase of the number of experimental points without increasing their scatter but increasing their statistical dependence.

The standard deviation of the slope and intercept of the regression line fitting MNF and MDF in simulated fatiguing contractions (with low to medium nonstationarity) has been found to be minimal for epoch durations between 250 and 500 ms, which therefore seems to be the most suitable for regression line parameter estimation [26]. Epochs shorter than 250 ms lead to high variance and bias of estimation. Overlapping does not provide significant benefits as it increases the computational load [26].

6.5.4 Other Approaches for Detecting Changes in Surface EMG Frequency Content during Voluntary Contractions

Another possible indicator of changes in the frequency content of the signal is the rate of zero crossings, meaning the number of sign changes of the signal in the unit time. When the signal is noise free and comprised of only one sinusoidal function, this rate reflects the frequency of oscillation of the function. According to Rice [89], if the signal has a Gaussian stationary amplitude distribution, the expected number Z of zero crossings per second can be expressed by the following relationship:

$$Z = 2 \left[\frac{\int_0^{f_s/2} f^2 S(f) df}{\int_0^{f_s/2} S(f) df} \right]^{1/2} \tag{11}$$

As indicated above, the distribution of amplitudes of the surface EMG signal is in between Gaussian and Laplacian; thus the Gaussian hypothesis is almost verified in practical cases. From (11), it is easy to show that, as MNF and MDF, Z is scaled by CV and can thus be used for evaluating spectral compression. Although the standard deviation of estimation of the zero crossing rate has been shown to be higher than that of MNF (but similar to that of the percentile frequencies) [44], the technique does not require spectral estimation, and it is particularly easy to implement in hardware. Based on this method, Hägg [42] and Inbar et al. [48] developed simple real-time fatigue monitors in the early 1980s.

Another approach to evaluate spectral changes in the surface EMG is based on the automatic detection of spikes in the signal. A spike is defined as a segment of signal shaped by an upward and downward deflection [36]. Both deflections of a spike cross the zero isoelectric baseline and should be at least 100 μV in amplitude. The analysis of spike activity has a long history in both clinical neurophysiology [62] and kinesiology [105] but has received less attention for surface EMG analysis in favor of more sophisticated techniques. Mean spike frequency, that is, the average number of spikes per unit time, has been shown to be highly correlated to MNF [37], and thus, as the characteristic spectral frequencies, it can be used to monitor spectral changes in the signal. This technique has been indicated as potentially useful in EMG analysis, since it does not directly require stationarity of the signal. However, the mean spike frequency, as well as the other spike parameters (see [36] for their definition), imply the use of a signal segment for their computation; thus a trade-off between bias of estimation and variance is necessary also with this method.

6.5.5 Applications of Spectral Analysis of the Surface EMG Signal

Spectral analysis of the surface EMG signal has been extensively applied for the study of muscle fatigue in both voluntary and electrically elicited contractions [73] (see Chapter 9). The preferred application of these techniques has been the analysis of isometric constant force, short duration, and medium–high level contractions. It is now accepted that EMG spectral variables reflect fatigue with the possibility of detecting some differences in muscle fiber composition [55,76,70]. Applications of spectral analysis of the EMG signal for fatigue assessment during nonconstant force and dynamic contractions have been proposed in more recent years together with advances in spectral estimation techniques based on time-frequency representations [54]. Nevertheless, many artifacts, mainly related to geometrical and anatomical factors of the EMG generation system, may be associated with these approaches [31]. The relevance of these artifacts is still not fully clear; thus caution should be taken in extending the considerations drawn for isometric conditions to dynamic exercises.

The analysis of the surface EMG PSD has also been applied to the investigation of MU recruitment strategies, in an attempt to extract information about central nervous system's (CNS) motor control strategies from a global analysis of the surface EMG signal (i.e., without decomposing the individual MU activities). It was speculated that MNF and MDF should reflect the recruitment of new, progressively larger and faster MUs and increase until the end of the recruitment process. They should then reach a constant value (or decrease) when only rate coding is used to track the desired target force level [9,98]. However, Farina et al. [29] indicated that in general, the establishment of a relationship between force and characteristic spectral frequencies is confounded by anatomical factors (see Chapter 4). Figure 6.6 shows CV and MNF of simulated EMG signals detected during ramp contractions.

6.6 JOINT ANALYSIS OF EMG SPECTRUM AND AMPLITUDE (JASA)

In nonisometric contractions that imply changes of muscle activity with recovery periods—a situation typical, for example, in occupational field studies—it is very difficult to interpret amplitude and spectral changes of the surface EMG signal independently. Recently the joint analysis of EMG spectrum and amplitude has been proposed to overcome some of the problems that arise in these situations. As both amplitude and spectral variables depend on muscle force and fatigue, amplitude increases with force and fatigue (see Chapter 9). Spectral variables decrease with fatigue, while their dependency on muscle force is not completely clear and contradictory results are reported in the literature [38,85,110]. Farina et al. [29] recently showed that it is not possible to establish a general relationship between muscle force and surface EMG spectral variables. However, these authors also indicated that, in agreement with past results from the literature, the most probable behavior is an increase with force (see also Fig. 6.6). The joint analysis of spectrum and amplitude (JASA) method [61] is thus based on the assumption of an increase of amplitude with fatigue, and force, a decrease of MDF with fatigue, and an

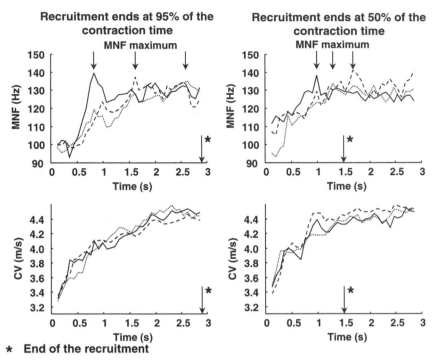

Figure 6.6. MNF and CV computed from three signals obtained by simulating ramp contractions with two different recruitment strategies. CV distribution standard deviation is 0.7 m/s. The set of the three synthetic single differential signals in the two cases have been generated with the same CV distribution, MU sizes and firing pattern; only MU location is different in the three simulations. The end of recruitment is shown in the two cases as well as the MNF maximum point. The two cases represent very different recruitment strategies; in the first case (*left column*) the recruitment of new MUs is present from the beginning until almost the end of the contraction while in the second case (*right column*) recruitment ends at 50% of the contraction time. Note that MNF shows a pattern in time very similar in the two conditions and that the spread of the location of the maximum MNF points is very large in both cases. (From [29] with permission.)

increase of MDF with force. By comparison of MDF and amplitude it is thus possible to identify four regions of muscle activity during a dynamic task: (1) force increase (both amplitude and MDF increase), (2) fatigue (amplitude increases and MDF decreases), (3) force decrease (both amplitude and MDF decrease), and (4) recovery (amplitude decreases and MDF increases). Although based on many simplifying assumptions, the method is easy to apply and does not require complex algorithms [60].

6.7 RECURRENCE QUANTIFICATION ANALYSIS OF SURFACE EMG SIGNALS

Recently nonlinear methods, such as recurrence quantification analysis (RQA), were introduced for the study of single-channel surface EMG signals, mainly in fatigue assessment. In general, the range of applications of nonlinear techniques to problems in biomedicine is rapidly expanding and spans from studies of the heart beat [40,50,88,90,114] to brain rhythms [6,112], from the neuromuscular system [34,52,78,80] to blood pressure regulation [3,106], from the breathing system [2] to cardiorespiratory coordination [47]. Nonlinear analysis was introduced in the study of surface EMG first by Webber et al. [107] and Nieminen and Takala [80]. RQA, described by Eckmann et al. [23], is based on a graphical method originally designed to locate recurring patterns (hidden rhythms) and nonstationarities (drifts) in experimental data sets. By mapping the signal in a bidimensional space (as described below), it is possible to identify time recurrences that are not readily apparent in the original recordings, either by qualitative visual inspection or by evaluating some specific variables, derived from the bi-dimensional maps, which quantify the deterministic structure and complexity of the plot itself. This method was recently used in some experimental surface EMG studies [32,33] that showed its potential in detecting changes of muscle properties due to fatigue.

6.7.1 Mathematical Bases of RQA

If we consider the neuromuscular system to be investigated as a dissipative dynamical system governed by a set of D first-order differential equations, the states of the system can be represented by points in a D-dimensional space where the coordinates are the values of the state variables. In most biological experiments, where it is not possible to measure all components of the vector giving the state of the system, it is feasible to reconstruct the system dynamics from a one-dimensional output of the system by mapping it in a D-dimensional space using delay coordinates. This procedure follows Taken's embedding theorem [103].

The main steps of the algorithm for projecting the original time series of the surface EMG samples into the phase space by means of the time-delay embedding procedure are given in Figure 6.7. The procedure is applied to a signal segment of K samples:

$$s(k) = [s(0)\ s(1)\ldots s(K-1)] \tag{12}$$

Recent applications of RQA to surface EMG [32,34,108,107] have selected a K such that the interval of observation of the signal is equal to that used for spectral or amplitude analysis in order to compare the results obtained by different techniques.

The surface EMG samples in the epoch selected are time shifted by an integer number λ of samples usually estimated as the first zero of the autocorrelation function [80]. This choice uncorrelates the elements of the D-dimensional vectors $\mathbf{v(n)}$ that are extracted from the original myoelectric time series as

$$
\begin{aligned}
\mathbf{v(0)} &= [s(0) & s(\lambda) & \quad \ldots & s((D-1)\lambda) &] \\
\mathbf{v(1)} &= [s(1) & s(\lambda+1) & \quad \ldots & s((D-1)\lambda+1)] \\
\ldots & \ldots & \ldots & \quad \ldots\ \ldots \\
\mathbf{v(N-1)} &= [s(N-1) & s(\lambda+N-1) & \quad \ldots & s(K-1) &]
\end{aligned}
\tag{13}
$$

The last component of the last vector corresponds to the last sample of the original timeseries and the number of vectors is

$$N = K - (D-1)\lambda \tag{14}$$

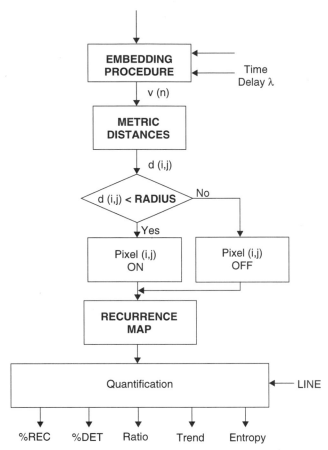

Figure 6.7. Flowchart of the procedure used to project the surface EMG time series into the phase space and get five main variables (%DET, %REC, trend, entropy, ratio, the first two of them described in this chapter; for the others, refer to [107]) of the RQA. The four critical parameters (embedding dimension D, time delay λ, radius, line) are tuned during the analysis.

Choice of the Embedding Dimension D. Nieminen et al. [80] showed that based on a correct evaluation of the correlation dimension, it is possible only to define a minimum necessary embedding dimension $D = d_{min}$, which is related to the muscular task under consideration. More recently Filligoi and Felici [34] have analyzed the problem of a correct and unique evaluation of D remaining valid throughout the course of a complete experiment. The criterion that has been used is based on the evaluation of saturation effects of the nonlinear variables under consideration (see below for their definition), when varying D over a range of values with a technique analogous to that applied to studies of classic chaotic systems, such as the Lorentz–Mackey-Glass differential delay equation or the Henon map (e.g., see [113]). The results showed that the best value to be given to the embedding dimension, as a trade-off between sufficient resolution in the phase space and the need to limit computational time, is $D = 15$ for both constant and nonconstant force contractions.

Distances Matrix DM. The states of the dynamic system under consideration are represented by the vectors $\mathbf{v(n)}$ defined in (13). The closeness of the vectors is then evaluated by the definition of a distance between vectors. The distance commonly used is Euclidean:

$$d(i, j) = \left[< (\mathbf{v(i)} - \mathbf{v(j)})^2 > \right]^{1/2} \tag{15}$$

In order to make the results of the analysis independent of the energy of the observed signal, the usual effective values adopted in RQA are either expressed as a percentage of the maximal distance (considered as 100) or normalized with respect to the average distance between vectors:

$$d_{av} = \frac{\sum_{i=1}^{N} \sum_{i \neq j}^{N} d(i, j)}{N(N-1)/2} \tag{16}$$

where the denominator represents the number of distances $d(i, j)$.

The collection of these normalized distances provides a symmetric matrix called the distance matrix, $\mathbf{DM}_{[N \times N]}$:

$$\mathbf{DM} = \begin{vmatrix} d(1,1) & d(1,2) & ... & d(1,N) \\ d(2,1) & d(2,2) & ... & d(2,N) \\ ... & ... & ... & ... \\ d(N,1) & d(N,2) & ... & d(N,N) \end{vmatrix} \tag{17}$$

Recurrence Map RM. A recurrence plot is finally obtained as a map of pixels, which assume the values 0 or 1 on the basis of a threshold (RADIUS) set on the distance matrix, $\mathbf{DM}$. The following comparison

$$d(i, j) \leq \text{threshold} \begin{cases} \text{YES} \rightarrow \text{pixel } b(i, j) = \text{ON} \\ \\ \text{ON} \rightarrow \text{pixel } b(i, j) = \text{OFF} \end{cases} \tag{18}$$

provides the recurrence map ($\mathbf{RM}$) as the collection of all $b(i, j)$, $\forall i, j$. By the mathematical development above, recurrence maps (1) are symmetric, since $d(i, j) = d(j, i)$ and (2) are characterized by a main diagonal of pixels ON, since obviously $b(i, i) = 1$.

The threshold operation is conceptually equivalent to considering two states of the dynamical system as close to each other when the embedded vectors $\mathbf{v(i)}$ and $\mathbf{v(j)}$ are enclosed in a D-dimensional hypersphere with radius equal to the selected threshold. The choice of high threshold values leads to considering too often the system states as near neighbors (pixel ON). The choice of low values for the threshold leads to opposite results. Fine-tuning of the threshold value will be described below, since it is strictly related to the variables extracted from the procedure of quantifying the $\mathbf{RM}$. Figure 6.8 shows examples of recurrence maps of synthetic signals.

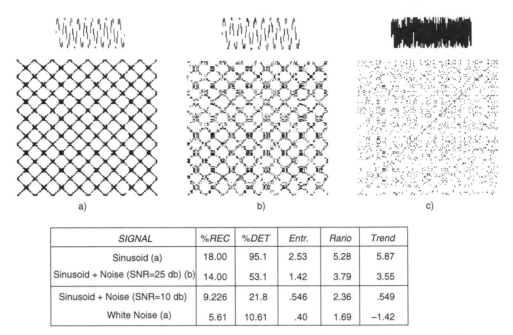

SIGNAL	%REC	%DET	Entr.	Rario	Trend
Sinusoid (a)	18.00	95.1	2.53	5.28	5.87
Sinusoid + Noise (SNR=25 db) (b)	14.00	53.1	1.42	3.79	3.55
Sinusoid + Noise (SNR=10 db)	9.226	21.8	.546	2.36	.549
White Noise (a)	5.61	10.61	.40	1.69	−1.42

Figure 6.8. RQA applied to a sinusoid (*a*), a sinusoid plus additive noise, (*b*) with two SNRs, and white noise (*c*). The values of the five variables extracted in these cases are also indicated.

Quantification of the Recurrence Map. Often recurrence maps contain subtle patterns that are difficult to detect by visual inspection. Hence quantitative descriptors that emphasize different features of the map have been proposed [108]. Among them the percentage of recurrence structures (%REC) and the percentage of determinism (%DET) have been the most used and describe the following quantities:

1. *%REC* is the percentage of pixels ON with respect to the total number of pixels in **RM** and measures the number of embedded vectors close to each other. This parameter indicates how far the time series is from a purely random dynamic system (typically represented by %REC in the range 5–15%) and is highly influenced by the threshold value (see below).

2. *%DET* is the percentage of points that form upward diagonal lines (with length greater than a prefixed cutoff value LINE) with respect to the number of pixels ON in **RM**. Points organized into diagonal patterns represent strings of vectors reoccurring at different times and are indicative of systems progressing through similar states in the phase-space.

Selection of RADIUS and LINE. RADIUS should assume the minimum value compatible with the baseline level of noise. A way to accomplish this task is to record surface EMG signals while the muscle is relaxed and then select RADIUS such that %REC is very low (e.g., between 5% and 15%). LINE value is not critical. Nevertheless, the choice of an excessively low value (2 or 3) could produce an erroneous detection of deterministic structures which are randomly present in the system. Again, when the results obtained on white noise series were compared, a value of 20 of the parameter LINE appeared to be appropriate [32,34].

6.7.2 Main Features of RQA

The aim of the recurrence analysis is to enhance the presence of repetitive patterns within the surface EMG time series. Subtle time correlations are more easily revealed in a bi-dimensional perspective than from direct observation of the original mono-dimensional time series. From this perspective the main features of the recurrence representation **RM** are the following:

- Single isolated points are due to chance recurrences and are characteristic of stochastic behavior.
- Upward diagonal lines result from strings of vector patterns repeating themselves in time. Therefore the presence of diagonal lines indicates that deterministic patterns are present in the phenomenon under observation.
- Downward diagonal lines occur whenever the vector sequences at different locations are mirror images of each other.
- A horizontal or a vertical line results when a specific vector is closely matched with other vectors separated in time.
- Bands of white space are due to the presence of transients (i.e., events that lie far outside the normal distribution of values).
- Non-uniform texture or paling away from the central diagonal line is an indicator of nonstationarities in the time series.

Examples of recurrence maps and their RQA variables applied to computer generated signals (a sinusoid, a sinusoid plus additive noise with signal to noise ratio equal to 10 and 25 dB, and a pure random noise) are given in Figure 6.8. The squared diagonal structure presented by the **RM** in the left side is obviously characteristic of the repetitive structure of the sinusoid. The high value of %DET is due to the fact that most points in the **RM** belong to diagonal lines, as it would be expected for a purely rhythmic signal. When some noise is superimposed on the sinusoid, the regularity of the structure is partially broken and recurrent points are more spread. This phenomenon is detected by %DET with a consistent reduction of its value. In the case of pure noise, points are randomly distributed in the map and %DET is drastically reduced.

6.7.3 Application of RQA to Analysis of Surface EMG Signals

As explained above, RQA primarily extracts information on recurrence structures that are repeated along the signal. For surface EMG, MUAPs are repeating during time regularly, forming an interference pattern when the number of active MUs is high. Thus we have a situation of deterministic structures repeating in a background activity. If the contraction level increases, it is expected that the number of deterministic structures also increases, while with fatigue RQA may detect short-term synchronization among MUAP trains (which should increase the determinism of the signal) or changes in muscle fiber CV. However, due to the nonlinear nature of the RQA approach, it is difficult to predict ana-

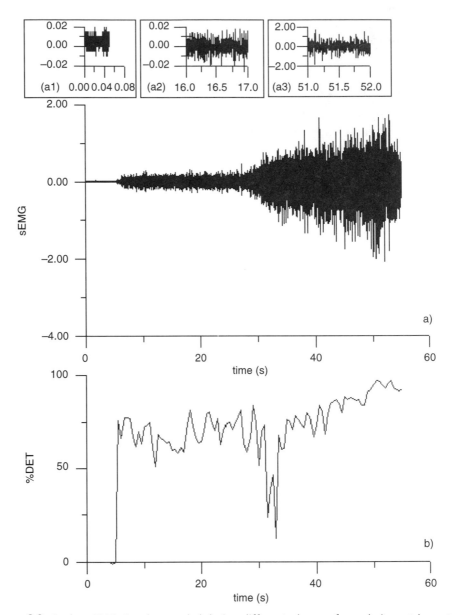

Figure 6.9. Surface EMG signals recorded during different phases of muscle isometric contraction: (*Upper panel*) At rest, on the left side of the figure (0% of the maximal voluntary contraction, MVC), low effort in the middle (20% MVC), high muscular effort on the right side. Within the small boxes a1, a2, and a3 ordinates are in mV and abscissas in seconds. (*Lower panel*) Corresponding values of the nonlinear variable %DET.

lytically the sensitivity of recurrence variables in response to the different changes in muscle properties.

An example application of the RQA technique to an experimental surface EMG signal is provided in Figure 6.9. Signals recorded at very low contraction levels (near 0% of the maximal voluntary contraction, MVC) are often corrupted by several sources of environmental noise (Fig. 6.9, box a1). When the muscle effort increases (Fig. 6.9, box a2, where the muscle load is 20% MVC), several physiological and experimental variables interact in a complicated manner [32,75,99]. When the number of active MUs is large enough for the action potentials to overlap (tract of surface EMG data corresponding to 80% MVC, between second 35 and second 53 in Fig. 6.9), the surface EMG is well described as a Gaussian or Laplacian distributed stochastic process with zero mean and variance related to the muscular effort. Deterministic structures are clearly recognized by RQA in this case (%DET increases with the muscular effort). When this strong muscular effort is exerted for a sufficiently long interval, fatigue phenomena appear (in Fig. 6.9, box a3) and are detected by RQA with a further increase of %DET.

Experimental fatigue studies on sedentary subjects and athletes indicated that RQA is a suitable method for assessing muscle fatigue with some advantages with respect to classic spectral and amplitude analysis [32]. However, the main issue is to verify which properties of the neuromuscular system RQA mostly reflects, and this can be done only by modeling the generation of the surface EMG signal taking into account the effects of anatomical, physiological, physical, and detection system parameters (i.e., with a structure based model rather than a phenomenological model). As mentioned in the introduction to this chapter, most of the techniques and the variables described here are not directly related to physiological phenomena (i.e., they do not provide a direct measurement of any physiological variable). This limitation is particularly true for RQA. The most commonly used variable extracted from the recurrence maps, %DET, does not, of course, directly indicate any physiological events. It is (as are MNF, MDF, zero crossing rate, and almost all the variables previously introduced) a mathematical indicator that may be sensitive to physiological events of potential interest.

The most difficult task is to define the relationships between these mathematical variables and the properties of the system under study. For RQA, this task has yet to be done, except for a recent modeling study by Farina et al. [28]. These authors have generated surface EMG signals with a model varying the mean MU CV and the degree of MU short-term synchronization. Then they applied classic spectral analysis and RQA to the synthetic signals. It was shown that %DET is sensitive to both CV and degree of synchronization changes during fatigue, as it happens with traditional spectral techniques.

In addition, %DET and results of traditional spectral techniques were highly correlated, thus being similarly sensitive to the muscle parameters. However, %DET was shown, both experimentally [28,107] and in simulation [28], to be more sensitive to fatigue-induced changes than spectral analysis. Thus RQA may be a potential alternative technique for muscle assessment with respect to spectral analysis to characterize different muscles and/or subjects in fatiguing conditions.

Figure 6.10 shows the power spectral densities and the recurrence maps of two simulated surface EMG signals (the model used is described in [27]), generated from MUs with different mean CV and degree of short-term synchronization. The recurrence maps clearly show the differences between the two signals, and %DET quantifies these differences. In particular, a higher number of rule-obeying structures in the signal with decreasing CV and increasing the degree of short-term synchronization are clearly shown by a higher regularity in the recurrence map. Other details on the applications of this technique will be described in Chapter 14.

The application of RQA to surface EMG still needs, however, additional research efforts to determine its sensitivity to other factors of variability between subjects and muscles, such as the thickness of the subcutaneous layers, the orientation of the detection system, the interelectrode distance, the electrode location, the recruitment of MUs, and

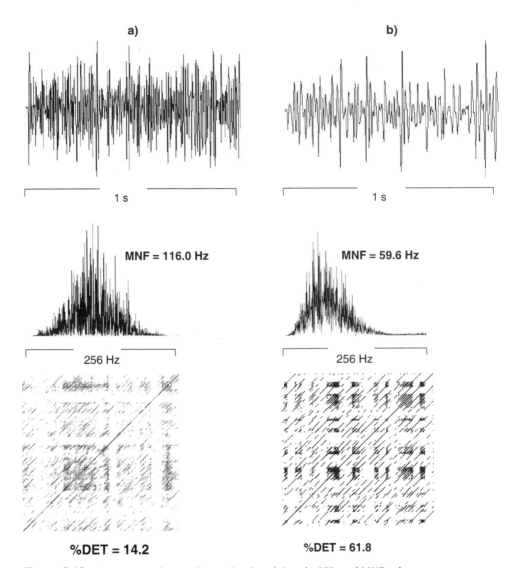

Figure 6.10. (*From top to bottom*) Raw simulated signals, PSD and MNF values, recurrence maps and their %DET. The EMG signals are generated by a model [27] with 5 m/s mean CV and 0% MU synchronization level (*a*) and 3 m/s mean CV and 25% MU synchronization level (*b*). The signals in the top row as well as the power spectra in the middle row are in arbitrary units. (From [28] with permission)

additive noise. While the influence of all these parameters on spectral variables has been analyzed in past works, there is a complete lack of this information for RQA.

6.8 CONCLUSIONS

Single-channel processing methods for surface EMG based on the interference pattern of the signal (without aiming at separating the contributions of individual MUs) have been discussed. This research focuses on the establishment of the relationships between the global variables obtained from the signal and the underlying physiological processes. Amplitude and spectral analysis have been used for many years for assessing muscle activity and fatigue. Although they have some limitations (primarily related to the impossibility of extracting information of single MU properties), these techniques are useful in a number of basic and clinical research studies. Other global processing techniques, less commonly applied, have been also proved to be potentially useful for muscle assessment, and they are currently being investigated in experimental and modeling studies. Recently large research efforts have been devoted to the development of advanced signal processing and detection methods (see Chapters 7 and 10) for more detailed analysis of the surface EMG signal. These methods are mostly focused on the extraction of more localized information on muscle activity (for example related to a small number of MUs). However, frequency and amplitude analysis are still by far the most widely used methods both in basic and applied studies. Their great advantage with respect to other techniques is indeed the simple applicability to a number of experimental conditions and recording systems.

REFERENCES

1. Akaike, H., "A new look at the statistical model identification," *IEEE Trans Autom Control* **19**, 716–723 (1974).

2. Akay, M., and E. J. H. Mulder, "Effects of maternal alcohol intake on fractal properties in human fetal breathing dynamics," *IEEE Trans BME* **435**, 1097–1103 (1998).

3. Almong, Y., O. Oz, and S. Akselrod, "Correlation dimension estimation: can this non linear description contribute to the characterization of blood pressure control in rats?" *IEEE Trans BME* **46**, 535–547 (1999).

4. Arendt-Nielsen, L., and K. R. Mills, "The relationship between mean power frequency of the EMG spectrum and muscle fibre conduction velocity," *Electroencephalogr Clin Neurophysiol* **60**, 130–134 (1985).

5. Arendt-Nielsen, L., and K. R. Mills, "Muscle fiber conduction velocity, mean power frequency, mean EMG voltage and force during submaximal fatiguing contractions of human quadriceps," *Eur J Appl Physiol* **58**, 20–25 (1988).

6. Babloyantz, A., and J. M. Slazar, "Evidence of chaos dynamic of brain activity during the sleep cycle," *Phys Lett* **111**, 152–156 (1985).

7. Balestra, G., M. Knaflitz, and R. Merletti, "Comparison between myoelectric signal mean and median frequency estimates," *Proc 10th An Conf IEEE Eng Med Biol Soc* 1708–1709, (1988).

8. Bendat, J. S., and A. G. Piersol, *Random data: Analysis and measurement procedures*, J Wiley, New York, 1971.

9. Bernardi, M., M. Solomonow, G. Nguyen, A. Smith, and R. Baratta, "Motor unit recruitment strategies changes with skill acquisition," *Eur J Appl Physiol* **74**, 52–59 (1996).

10. Bernardi, M., F. Felici, M. Marchetti, F. Montellanico, M. F. Piacentini, and M. Solomonow, "Force generation performance and motor unit recruitment strategy in muscles of contralateral limbs," *J Electromyogr Kinesiol* **9**, 121–130 (1999).

11. Bonato, P., T. D'Alessio, and M. Knaflitz, "A statistical method for the measurement of muscle activation intervals from surface myoelectric signal during gait," *IEEE Trans BME* **45**, 287–299 (1998).

12. Clancy, E. A., "Electromyogram amplitude estimation with adaptive smoothing window length," *IEEE Trans BME* **46**, 717–729 (1999).

13. Clancy, E. A., and K. A. Farry, "Adaptive whitening of the electromyogram to improve amplitude estimation," *IEEE Trans BME* **47**, 709–719 (2000).

14. Clancy, E. A., and N. Hogan, "Single site electromyograph amplitude estimation," *IEEE Trans BME* **41**, 159–167 (1994).

15. Clancy, E. A., and N. Hogan, "Probability density of the surface electromyogram and its relation to amplitude detectors," *IEEE Trans BME* **46**, 730–739 (1999).

16. Clancy, E. A., E. L. Morin, and R. Merletti, "Sampling, noise reduction and amplitude estimation issues in surface electromyography," *J Electromyogr Kinesiol* **12**, 1–16 (2002).

17. Cohen, L., "Time-frequency distributions: a review," *Proc IEEE* **77**, 941–981 (1989).

18. D'Alessio, T., N. Accornero, and A. Berardelli, "Toward a real time adaptive processor for surface EMG signals," *An Int Conf IEEE Eng Med Biol Soc* **9**, 323–324 (1987).

19. D'Alessio, T., M. Laurenti, and B. Turco, "On some algorithms for the tracking of spectral structure in non-stationary EMG signals," *Proc MIE '87*, Rome, Italy, 1987.

20. De Luca, C. J., "Physiology and mathematics of myoelectric signals," *IEEE Trans BME* **26**, 313–325 (1979).

21. De Luca, C. J., "Myoelectric manifestations of localized muscular fatigue in humans," *Crit Rev BME* **11**, 251–279 (1984).

22. Diemont, B., and M. Maranzana, "Spectral analysis of EMG data: Identification methods versus FFT," Politecnico di Milano, Italy, Internal report 87-022, 1987.

23. Eckmann, J. P., S. O. Kamphorst, and D. Ruelle "Recurrence plots of dynamical systems," *Europhys Lett* **4**, 973–977 (1987).

24. Englehart, K. B., and P. A. Parker, "Single motor unit myoelectric signal analysis with nonstationary data," *IEEE Trans BME* **41**, 168–180 (1994).

25. Evans, H. B., Z. Pan, P. A. Parker, and R. N. Scott, "Signal processing for proportional myoelectric control," *IEEE Trans BME* **31**, 207–211 (1984).

26. Farina, D., and R. Merletti, "Comparison of algorithms for estimation of EMG variables during voluntary isometric contractions," *J Electromyogr Kinesiol* **10**, 337–350 (2000).

27. Farina, D., and R. Merletti, "A novel approach for precise simulation of the EMG signal detected by surface electrodes," *IEEE Trans BME* **48**, 637–645 (2001).

28. Farina, D., L. Fattorini, F. Felici, and G. C. Filligoi, "Nonlinear surface EMG analysis to detect changes of motor unit conduction velocity and synchronization," *J Appl Physiol* **93**, 1753–1763 (2002).

29. Farina, D., M. Fosci, and R. Merletti, "Motor unit recruitment strategies investigated by surface EMG variables: An experimental and model based feasibility study," *J Appl Physiol* **92**, 235–247 (2002).

30. Farina, D., C. Cescon, and R. Merletti, "Influence of anatomical, physical and detection system parameters on surface EMG," *Biol Cybern* **86**, 445–456 (2002).

31. Farina, D., R. Merletti, M. Nazzaro, and I. Caruso, "Effect of joint angle on EMG variables in muscles of the leg and thigh," *IEEE Eng Med Biol Mag* **20**, 62–71 (2001).

32. Felici, F., A. Rosponi, P. Sbriccoli, G. C. Filligoi, L. Fattorini, and M. Marchetti, "Linear and non-linear analysis of surface electromyograms in weightlifters," *Eur J Appl Physiol* **84**, 337–342 (2001).

33. Felici, F., A. Rosponi, P. Sbriccoli, M. Scarcia, I. Bazzucchi, and M. Iannattone, "Effect of human exposure to altitude on muscle endurance during isometric contractions," *Eur J Appl Physiol* **85**, 507–512 (2001).

34. Filligoi, G. C., and F. Felici, "Detection of hidden rhythms in surface EMG signals with a nonlinear time-series tool," *Med Eng Phys* **21**, 438–448 (1999).

35. Filligoi, G. C., and P. Mandarini, "Some theoretical results on a digital EMG signal processor," *IEEE Trans BME* **31**, 333–341 (1984).

36. Gabriel, D. A., "Reliability of SEMG spike parameters during concentric contractions," *Electromyogr Clin Neurophysiol* **40**, 423–430 (2000).

37. Gabriel, D. A., J. R. Basford, and K. N. An, "Training-related changes in the maximal rate of torque development and EMG activity," *J Electromyogr Kinesiol* **11**, 123–129 (2001).

38. Gerdle, B., N. E. Eriksson, and L. Brundin, "The behaviour of the mean power frequency of the surface electromyogram in biceps brachii with increasing force and during fatigue: With special regard to the electrode distance," *Electromyogr Clin Neurophysiol* **30**, 483–489 (1990).

39. Gilmore, L. D., and C. J. DeLuca, "Muscle fatigue monitor (MFM): second generation," *IEEE Trans BME* **32**, 75–78 (1985).

40. Goldberger, A. R., and J. W. Bruce, "Application of nonlinear dynamics to clinical cardiology," *An Acad Sci* 195–212, 1985.

41. Gross, D., A. Grassino, W. R. D. Ross, and P. T. Macklem, "Electromyogram pattern of diaphragmatic fatigue," *J Appl Physiol* **46**, 1–7 (1979).

42. Hägg, G., "Electromyographic fatigue analysis based on the number of zero crossings," *Eur J Physiol* **391**, 78–80 (1981).

43. Harba, M. I. A., and P. A. Lynn, "Optimizing the acquisition and processing of surface electromyographic signals," *J BME* **3**, 100–106 (1981).

44. Hof, A. L., "Errors in frequency parameters of EMG power spectra," *IEEE Trans BME* **38**, 1077–1088 (1991).

45. Hogan, N., and R. W. Mann, "Myoelectric signal processing: optimal estimation applied to electromyography—Part I: Derivation of the optimal myoprocessor," *IEEE Trans BME* **27**, 382–395 (1980).

46. Hogan, N., and R. W. Mann, "Myoelectric signal processing: Optimal estimation applied to electromyography—Part II: Experimental demonstration of optimal myoprocessor performance," *IEEE Trans BME* **27**, 396–410 (1980).

47. Hoyer, D., R. Bauer, B. Walter, and U. Zwiener "Estimation of nonlinear coupling on the basis of complexity and predictability—A new method applied to cardio-respiratory coordination," *IEEE Trans BME* **45**, 545–552 (1998).

48. Inbar, G. F., J. Allin, O. Paiss, and H. Kranz, "Monitoring surface EMG spectral changes by the zero crossing rate," *Med Biol Eng Comput* **24**, 10–18 (1986).

49. Inman, V. T., H. J. Ralston, J. B. Saunders, B. Feinstein, and E. W. Wright, "Relation of human electromyogram to muscular tension," *Electroenceph Clin Neurophysiol* **4**, 187–194 (1952).

50. Kaplan, D. T., and R. J. Kohen, "Is fibrillation chaos?" *Circ Res* **67**, 886–900 (1990).

51. Kay, S. M., and L. M. Marple, "Spectrum analysis—A modern perspective," *Proc IEEE* **69**, 1380–1413 (1981).

52. Kearney, R. E., and J. W. Hunter, "Nonlinear identification of stretch reflex dynamics," *An BME* **16**, 79–94 (1988).

53. Kiryu, T., C. J. De Luca, and Y. Saitoh, "AR modeling of myoelectric interference signals during a ramp contraction," *IEEE Trans BME* **41**, 1031–1038 (1994).

54. Knaflitz, M., and P. Bonato, "Time-frequency methods applied to muscle fatigue assessment during dynamic contractions," *J Electromyogr Kinesiol* **9**, 337–350 (1999).

55. Kupa, E., S. Roy, S. Kandarian, and C. J. De Luca, "Effects of muscle fiber type and size on EMG median frequency and conduction velocity," *J Appl Physiol* **78**, 23–32 (1995).

56. Lago, P. J., and N. B. Jones, "Effect of motor unit firing time statistics on e.m.g. spectra," *Med Biol Eng Comput* **5**, 648–655 (1977).

57. Lindstrom, L., and R. Magnusson, "Interpretation of myoelectric power spectra: A model and its applications," *Proc IEEE* **65**, 653–662 (1977).

58. Lo Conte, L., R. Merletti, and G. V. Sandri, "Hermite expansions of compact support waveforms: Applications to myoelectric signals," *IEEE Trans BME* **41**, 1147–1159 (1994).

59. Lowery, M. M., C. L. Vaughan, P. J. Nolan, and M. J. O'Malley, "Spectral compression of the electromyographic signal due to decreasing muscle fiber conduction velocity," *IEEE Trans Rehabil Eng* **8**, 353–361 (2000).

60. Luttmann, A., M. Jäger, and W. Laurig, "Electromyographic indication of muscular fatigue in occupational field studies," *Int J Indus Ergonom* **25**, 645–660 (2000).

61. Luttmann, A., M. Jäger, J. Sökeland, and W. Laurig, "Joint analysis of spectrum and amplitude (JASA) of electromyograms applied for the indication of muscular fatigue among surgeons in urology," in A. Mital, H. Krueger, S. Kumar, M. Menozzi, and J. E. Fernandez, eds., *Advances in occupational ergonomics and safety*, International Society for Occupational Ergonomics and Safety, Cincinnati, 1996, pp. 523–528.

62. Magora, A., and B. Gonen, "Computer analysis of the shape of spikes from the electromyograpic interference pattern," *Electromyography* **10**, 261–271 (1970).

63. Makhoul, J., "Linear prediction: A tutorial review," *Proc IEEE* **63**, 561–581 (1995).

64. Maranzana, M., and M. Fabbro, "Autoregressive description of EMG signals," ISEK Far East Regional Meeting, 1981.

65. McVicar, G. N., and P. A. Parker, "Spectrum dip estimator of nerve conduction velocity," *IEEE Trans BME* **35**, 1069–1076 (1988).

66. Merletti, R., G. Balestra, and M. Knaflitz, "Effect of FFT based algorithms on estimation of myoelectric signal spectral parameters," *11th An Int Conf IEEE Eng Med Biol Soc*, (1989), pp. 1022–1023.

67. Merletti, R., D. Biey, M. Biey, G. Prato, and A. Orusa, "On-line monitoring of the median frequency of the surface EMG power spectrum," *IEEE Trans BME* **32**, 1–7 (1985).

68. Merletti, R., F. Castagno, C. Saracco, G. Prato, and R. Pisani, "Properties and repeatability of spectral parameters of surface EMG in normal subjects," *Rassegna Bioing* **10**, 83–96 (1985).

69. Merletti, R., Y. Fan, and L. Lo Conte, "Estimation of scaling factors in electrically evoked myoelectric signals," *Proc 14th An Conf IEEE Eng Med Biol Soc*, Paris, (1992), pp. 1362–1363.

70. Merletti, R., D. Farina, M. Gazzoni, and M. P. Schieroni, "Effect of age on muscle functions investigated with surface electromyography," *Muscle Nerve* **25**, 65–76 (2002).

71. Merletti, R., A. Gulisashvili, and L. R. Lo Conte, "Estimation of shape characteristics of surface muscle signal spectra from time domain data," *IEEE Trans BME* **42**, 769–776 (1995).

72. Merletti, R., M. Knaflitz, and C. J. DeLuca, "Electrically evoked myoelectric signals," *Crit Rev BME* **19**, 293–340 (1992).

73. Merletti, R., M. Knaflitz, and C. J. DeLuca, "Myoelectric manifestations of fatigue in volunatry and electrically elicited contractions," *J Appl Physiol* **68**, 1657–1667 (1990).

74. Merletti, R., and L. Lo Conte, "Advances in processing of surface myoelectric signals. Part 1," *Med Biol Eng Comp* **33**, 362–372 (1995).

75. Merletti, R., L. LoConte, E. Avignone, and P. Guglielminotti, "Modelling of surface myoelectric signals. Part I: Model implementation," *IEEE Trans BME* **46**, 810–820 (1999).

76. Merletti, R., L. R. Lo Conte, C. Cisari, and M. V. Actis, "Age related changes in surface myoelectric signals," *Scand J Rehab Med* **25**, 25–36 (1992).

77. Merletti, R., M. A. Sabbahi, and C. J. DeLuca, "Median frequency of the myoelectric signal: Effect of muscle ischemia and cooling," *Eur J Appl Physiol* **52**, 258–265 (1984).

78. Moser, A. T., and D. Graupe, "Identification of non-stationarity models with application to myoelectric signals for controlling electrical stimulation of paraplegics," *IEEE Trans Acous Speech Sig Proc* **37**, 713–719 (1989).

79. Muhammad, W., O. Meste, H. Rix, and D. Farina, "A pseudo joint estimation of time delay and scale factor for M-wave analysis," *IEEE Trans BME* **50**, 459–468 (2003).

80. Nieminen, H., and E. P. Takala, "Evidence of deterministic chaos in the myoelectric signal," *Electromyogr Clin Neurophysiol* **36**, 49–58 (1996).

81. Olmo, G., F. Laterza, and L. Lo Presti, "Matched wavelet approach in stretching analysis of electrically evoked surface EMG," *Sig Proc* **80**, 671–684 (2000).

82. Paiss, O., and G. Inbar, "Autoregressive modelling of surface EMG and its spectrum with application to fatigue," *IEEE Trans BME* **34**, 761–770 (1987).

83. Parzen, E., "Some recent advances in time series modeling," *IEEE Trans Automat Control* **19**, 723–730 (1974).

84. Petrofsky, S. J., "Filter bank analyzer for automatic analysis of EMG," *Med Biol Eng Comput* **18**, 585–590 (1980).

85. Petrofsky, S. J., and A. R. Lind, "Frequency analysis of the surface EMG during sustained isometric contractions," *Eur J Appl Physiol* **43**, 173–182 (1980).

86. Piper, H., *Electrophysiologie Menschlicher Muskeln*, Springer-Verlag, Berlin, 1912.

87. Press, W. H., W. T. Vetterling, S. A. Teukolsky, and B. P. Flannery, *Numerical recipes in C: The art of scientific computing*, 2nd ed., Cambridge University Press, Cambridge, UK, 1994, pp. 572–576.

88. Ravelli, F., and R. Antolini, "Complex dynamics underlying the human electrocardiogram," *Biol Cybern* **67**, 57–65 (1992).

89. Rice, R. O., "Mathematical analysis of random noise." in N. Wax, ed., *Selected papers on noise and stochastic processes*, Dover, New York, 1945.

90. Richter, M., T. Schreiber, and D. T. Kaplan, "Foetal ECG extraction with non-linear state space projections," *IEEE Trans BME* **45**, 133–137 (1998).

91. Rissanen, J., "Modeling by shortest data description," *Automatica* **14**, 465–471 (1978).

92. Rissanen, J., "A universal prior for integers and estimation by minimum description length," *An Stat* **11**, 416–431 (1983).

93. Rix, H., and J. P. Malengé, "Detecting small variation in shape," *IEEE Trans Syst Man Cybernet* **10**, 90–96 (1980).

94. Roberts, R. A., and C. T. Mullis, *Digital signal processing*, Addison Wesley, Reading, MA, 1987.

95. Sachs, L., *Applied statistics*, Springer-Verlag, Berlin, 1982.

96. Schweitzer, T. W., J. W. Fitzgerald, J. A. Bowden, and P. Lynne-Davies, "Spectral analysis of human inspiratory diaphragmatic electromyograms," *J Appl Physiol* **46**, 152–165 (1979).

97. Shwedyk, E., R. Balasubramanian, and R. Scott, "A non-stationary model for the electromyogram," *IEEE Trans BME* **24**, 417–424 (1977).

98. Solomonow, M., C. Baten, J. Smith, R. Baratta, H. Hermens, R. D'Ambrosia, and H. Shoji, "Electromyogram power spectra frequencies associated with motor unit recruitment strategies," *J Appl Physiol* **68**, 1177–1185 (1990).

99. Stegeman, D. J., J. H. Blok, H. J. Hermens, and K. Roeleveld, "Surface EMG models: Properties and applications," *J Electromyogr Kinesiol* **10**, 313–326 (2000).

100. Stulen, F. B., "A technique to monitor localized muscular fatigue using frequency domain analysis of the myoelectric signal," PhD thesis, Massachusetts Institute of Technology, Cambridge, 1980.

101. Stulen, F. B., and C. J. DeLuca, "Frequency parameters of the myoelectric signal as a measure of muscle conduction velocity," *IEEE Trans BME* **28**, 512–522 (1981).

102. Stulen, F. B., and C. J. DeLuca, "Muscle fatigue monitor: a non invasive device for observing localized muscular fatigue," *IEEE Trans BME* **29**, 760–769 (1982).

103. Takens, F., "Detecting strange attractors in turbulence in dynamical systems" *Lecture notes in Mathematics*, Springer, Berlin, 366–381, 1981.

104. Triolo, R. J., D. H. Nash, and G. D. Moskowitz, "The identification of time series models of lower extremity EMG for the control of prostheses using Box-Jenkins criteria," *IEEE Trans BME* **36**, 584–594 (1988).

105. Viitasalo, J. H. T., and P. V. Komi, "Signal characteristics of EMG fatigue," *Eur J Appl Physiol* **37**, 111–127 (1977).

106. Wagner, C. D., and P. B. Persson, "Nonlinear chaotic dynamics of arterial blood pressure and renal flow," *Am J Physiol (Heart Circ Physiol)* **268**, 621–627 (1996).

107. Webber, C. L., M. A. Schmidt, and S. M. Walsh, "Influence of isometric loading on biceps EMG dynamics as assessed by linear and nonlinear tools," *J Appl Physiol* **78**, 814–822 (1995).

108. Webber, C. L., and J. B. Zbilut, "Dynamical assessment of physiological systems and states using recurrence plot strategies," *J Appl Physiol* **76**, 965–973 (1994).

109. Welch, P. D., "The use of fast Fourier transform for the estimation of power spectra: A method based on time averaging over short, modified periodograms," *IEEE Trans Electroacoust* **15**, 70–73 (1967).

110. Westbury, J. R., and T. G. Shaughnessy, "Associations between spectral representation of the surface electromyogram and fiber type distribution and size in human masseter muscle," *Electromyogr Clin Neurophysiol* **27**, 427–435 (1987).

111. Weytjens, J. L., and D. van Steenberghe, "The effects of motor unit synchronization on the power spectrum of the electromyogram," *Biol Cybern* **51**, 71–77 (1984).

112. Yaylali, I., H. Koçak, and P. Yayakar, "Detection of seizures from small samples using nonlinear dynamic system theory," *IEEE Trans BME* **43**, 743–751 (1996).

113. Zbilut, J. P., and C. L. Webber, "Embedding and delays as derived from quantification of recurrence plots," *Phys Let* A **171**, 199–203 (1992).

114. Zhang, X. S., Y. S. Zhu, Z. Z. Wang, and N. V. Thakor, "Detecting ventricular tachycardia and fibrillation by complexity measure," *IEEE Trans BME* **46**, 548–555 (1999).

7

MULTI-CHANNEL TECHNIQUES FOR INFORMATION EXTRACTION FROM THE SURFACE EMG

D. Farina and R. Merletti

Laboratory for Engineering of the Neuromuscular System
Department of Electronics
Politecnico di Torino, Torino, Italy

C. Disselhorst-Klug

Helmholtz-Institute for Biomedical Engineering
Aachen, Germany

7.1 INTRODUCTION

The surface EMG signal presents a smaller bandwidth with respect to the intra-muscular EMG, since the tissues separating the muscle fibers and the recording electrodes act as low-pass filters [4,15,51]. This determines low spatial selectivity, which hinders the separation of the contributions of different motor units (MUs). For this reason past research efforts in the surface EMG field were mainly devoted to the development of processing techniques in time and frequency domains, which gave indications about the global EMG activity without aiming at an analysis at the single MU level (see Chapter 6). In recent years the possibility of extracting single MU contributions from the interference EMG signal was shown to be feasible by the use of advanced detection systems and algorithms specifically designed to process information from multi-channel recordings.

The main strategies adopted to separate single MU activities and estimate single MU properties from surface EMG are based on spatial filtering and spatial sampling of the surface potentials. In both cases the detection is based on more electrodes (from four up to hundreds) with respect to the classic two electrode differential recording technique, and

Electromyography: Physiology, Engineering, and Noninvasive Applications, edited by Roberto Merletti and Philip Parker.
ISBN 0-471-67580-6   Copyright © 2004 Institute for Electrical and Electronics Engineers, Inc.

thus it requires more complex recording systems and more sophisticated methods for information extraction.

In this chapter we describe the basic concepts of multichannel surface EMG recording and processing. We focus on spatial filtering, the method to increase spatial selectivity in surface EMG recordings, on spatial sampling, the method to increase the amount of information available from the surface recordings, and on the estimation of MU conduction velocity (CV), which is an important parameter to assess muscle functions and whose estimation requires at least two recorded signals.

In contrast to the techniques presented in Chapter 6, in which the EMG signal was modeled as the realization of a stochastic process, this chapter will mainly focus on the separation of the MU action potential trains from surface EMG signals. Although the techniques presented in this chapter are currently being refined, they have shown many promising applications and can be considered as the most advanced available methods for surface EMG detection and processing.

7.2 SPATIAL FILTERING

One of the most crucial limitations of the conventional surface EMG techniques is their poor spatial resolution. This means that it is very difficult to distinguish, from surface signals, sources that are closely placed in the muscle. As a consequence surface EMG methods allow at medium/high contraction levels only statements about the compound activity of a large number of MUs. However, the analysis of the single MU activities is of great interest, not only in the field of diagnosis and therapy evaluation of neuromuscular disorders but also in research related to muscle control strategies. To get the required information in a noninvasive way, new recording techniques have been recently developed that are based on spatial filtering methods to extract single MU activity from the compound signal [11,18,22,72,73].

7.2.1 Idea Underlying Spatial Filtering

Spatial filters in surface EMG detection are based on the linear combination of signals detected by a number of electrodes placed over the skin in a particular geometrical configuration. It can be shown (see below) that these recording modalities can be mathematically described by linear filters in the two dimensional spatial domain (the skin plane). The effect of the spatial filter depends on the geometry of the electrode configuration, on the electrode shape and size, and on the weights assigned to each electrode in the linear combination. The idea of introducing spatial filters in surface EMG detection is based on the fact that the volume conductor between the excited MU and the recording electrodes acts on the propagating part of the MU potential as a spatial low-pass filter [51]. Due to the low-pass characteristic of the volume conductor, propagating action potentials of MUs located close to the recording electrodes generate a spatially steeper potential distribution than MUs located more distant. This means that the surface potential contribution of the MUs close to the recording site contains higher spatial frequencies than that of deep MUs. Figure 7.1 describes this concept.

In voluntary contractions many MUs distributed over the cross section of the muscle are active. The potential distributions generated by each active MU superimpose on the

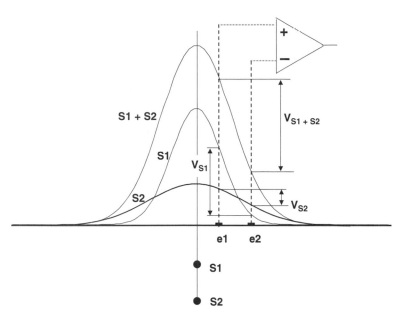

Figure 7.1. Schematic representation of the potential distribution generated on the skin surface by two point-like sources located close to the skin surface, S1 and S2. A bipolar electrode system is also shown. The contribution of the most superficially located source is dominant. If the inter-electrode distance is small with respect to the spatial spread of the potential, the output voltage of the amplifier approximates the first spatial derivative of the voltage distribution.

skin surface to produce the resulting potential distribution. Because of the spatial filter effect of the volume conductor, the spatial high-frequency components of the resulting potential distribution can be assigned to the few MUs located closer than the others to the detection point.

The spatial low-pass filter characteristics of the volume conductor, which depend on the distance between the source and the recording point, allow discrimination in the frequency domain among the contributions of different MUs. This feature can be used to design filters that enhance the propagating potentials of MUs located close to the skin surface while contributions of more distant MUs are reduced. This effect can be achieved by a filter that inverses the spatial low-pass transfer function of the volume conductor and that are consequently spatial high-pass filters. A bipolar lead with interelectrode distance d, which generates a differential signal, is a very simple high-pass spatial filter for wavelengths longer than $d/2$ [43]. Electrode arrangements with smaller interelectrode distances suppress signals containing higher spatial frequencies than electrode arrangements with larger electrode distances. Thus, to shift the cutoff frequency of the spatial filter toward higher frequencies, the interelectrode distance has to be reduced (Chapter 5).

The principle of spatial filtering can be applied to systems with electrodes placed in the transversal or longitudinal direction with respect to the muscle fibers [10,11,39,59] or with two-dimensional electrode arrangements [72,73]. If the surface potential distributions are interpreted as images in the spatial domain, high-pass spatial filters correspond to filters for edge detection. The increase of spatial resolution can indeed be viewed as a de-blurring process. Figure 7.2 shows the potential distributions generated by two muscle

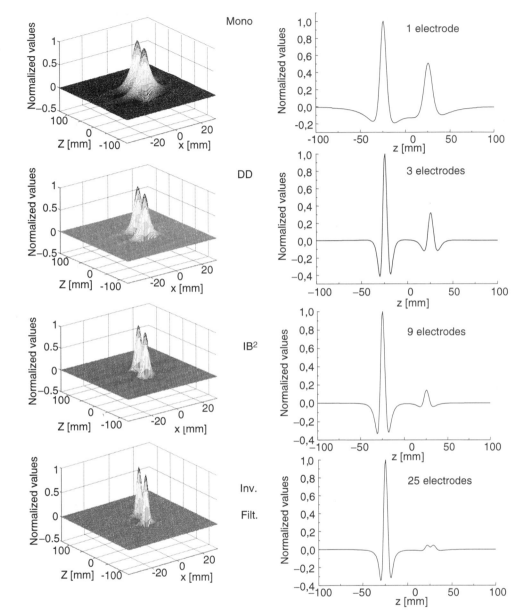

Figure 7.2. Potential distributions (*left*) and sections along the *z* axis (*right*) generated with a simulation model by two current tripoles at a depth of 2 mm in the muscle and at a distance of 5 mm along the *x* axis and 50 mm along the *z* axis, as obtained with different types of detection systems: monopolar, longitudinal double differential, IB^2 (see below in the text for the definition of this two-dimensional spatial filter), and the filter proposed by Farina and Rainoldi (1999). The plots in the right column report the output signal detected by the four systems placed at $x = 0$. (From [26] with permission)

fibers (by a simulation model) at different locations in the muscle. The surface potential values are reported as would be detected by different spatial filters (i.e., the potential detected over the skin by different spatial filters is reported). The signals detected by the different systems, assuming placement over one of the two sources, are also shown. The de-blurring effect of the spatial filters makes the observed potential distributions of the two fibers narrower, thus increasing the selectivity of the detection when compared to the monopolar recording.

7.2.2 Mathematical Basis for the Description of Spatial Filters Comprised of Point Electrodes

Consider a potential distribution $\phi(x, z)$ generated over the skin by a muscle fiber, and suppose that the potential is moving along the fiber direction. If the fiber is aligned along the z axis and is of infinite length, the potential detected over the skin, in case of a single point detection electrode at a distance $x = x_0$ from the fiber and in the location $z = z_0$ along the z axis, at the time instant $t = t_0$ is

$$V_0 = \phi(x_0, z_0 - vt_0) = [\phi(x, z) * (\delta(x)\delta(z - vt_0))]_{\substack{x=x_0 \\ z=z_0}} \tag{1}$$

where v is the velocity of propagation of the potential along the fiber and $\delta(.)$ is the Dirac distribution. Assuming, for simplicity, $t_0 = 0$, $x_0 = 0$, and $z_0 = 0$, we have

$$V_0 = [\phi(x, z) * (\delta(x)\delta(z))]_{\substack{x=0 \\ z=0}} \tag{2}$$

From Eq. (2) we can express the signal detected at time zero in the generic point (x_i, z_i) as

$$V_i = [\phi(x, z) * (\delta(x \mid x_i)\delta(z + z_i))]_{\substack{x=0 \\ z=0}} \tag{3}$$

Thus, if we combine by linear summation the signals detected in a number of different points, we get

$$\sum_{i\_=0}^{M} a_i V_i = \sum_{i=0}^{M} [\phi(x, z) * (a_i \delta(x + x_i)\delta(z + z_i))]_{\substack{x=0 \\ z=0}} = \left[\phi(x, z) * \sum_{i=0}^{M} (a_i \delta(x + x_i)\delta(z + z_i)) \right]_{\substack{x=0 \\ z=0}} \tag{4}$$

where $M + 1$ is the number of electrodes and a_i ($i = 0, \ldots, M$) are the weights given to the linear combination. Including the temporal variable yields

$$\sum_{i\_=0}^{M} a_i V_i(t) = \left[\phi(x, z) * \sum_{i=0}^{M} a_i \delta(x + x_i)\delta(z + z_i) \right]_{\substack{x=0 \\ z=-vt}} = [\phi(x, z) * h(x, z)]_{\substack{x=0 \\ z=-vt}} \tag{5}$$

From Eq. (5) it appears that the operation of linear combination of the potentials detected over the skin by point electrodes can be described by an equivalent two-dimensional spatial impulse response:

$$h(x, z) = \sum_{i=0}^{M} a_i \delta(x + x_i) \delta(z + z_i) \tag{6}$$

whose corresponding transfer function is given by the two-dimensional Fourier transform of Eq. (6):

$$H(f_x, f_z) = \sum_{i=0}^{M} a_i e^{j2\pi f_x x_i} e^{j2\pi f_x z_i} \tag{7}$$

By selecting the weights a_i, it is possible to design transfer functions with particular characteristics. Usually the spatial filters are designed in order to meet the basic condition

$$H(0, 0) = \sum_{i=0}^{M} a_i = 0 \tag{8}$$

which implies the rejection of the dc components in the two spatial directions. This assures the absence of common mode signals. Given condition (8), and given the geometry of the electrode configuration, M degrees of freedom remain for the design of the spatial filter transfer function.

7.2.3 Two-Dimensional Spatial Filters Comprised of Point Electrodes

Indications about filter design in one or two dimensions can be partly derived from the field of image processing, since the enhancement of spatial selectivity of EMG detection can be viewed as an image de-blurring process (Fig. 7.2). It is known that a class of spatial high-pass filters—called Laplace filters—that perform the second spatial derivative are well suited for the detection of spatially steep edges in the direction of differentiation. Such double differentiating filters have been successfully applied to surface EMG [10,11,59]. A one-dimensional Laplace filter can be realized with three electrodes in a row, the central electrode weighted with the factor −2 and the others with +1. If the electrodes are arranged parallel to the muscle fibers, a LDD (longitudinal double differential) filter is obtained. This system has been introduced earlier in Chapter 5, and it is usually referred to as double differential filter, without specifying the direction of differentiation.

Although the excitation moving along the infinite muscle fiber can be envisioned as a current tripole moving along a line (Chapter 8), it is obvious that the propagating potential resulting on the skin surface is a propagating two-dimensional distribution. Indeed, it has been spatially low-pass filtered by the volume conductor in both spatial directions. Since the optimal spatial filter is that which comes closest to the inverse of the filter due to the volume conductor, two-dimensional spatial high-pass filters seem to be more appropriate than one-dimensional filters [18,22,72,73]. Reucher et al. [72,73] have shown that a two-dimensional Laplace filter, called NDD-filter (normal double differentiating filter), allows the separation of single MU activities even at maximum voluntary contractions. The weighting factors of each lead of the NDD-filter can be determined from an approximation of the Laplace operator by a Taylor series [43]. The NDD-filter can be represented by the so-called filter mask:

$$M_{NDD} = \begin{bmatrix} 0 & 1 & 0 \\ 1 & -4 & 1 \\ 0 & 1 & 0 \end{bmatrix} \tag{9}$$

Rows and columns of the filter mask represent the position of the recording electrodes and the entries the respective weighting factors. Thus a NDD-filter can be realized by five crosswise arranged electrodes, whereby the central electrode is weighted with the factor −4 and the surrounding electrodes with the factor +1. In contrast to the LDD-filter, the NDD-filter performs a spatial double differentiation in two orthogonal directions and builds in this way a two-dimensional spatial high-pass filter.

The impulse response of the NDD filter is obtained applying (Eq. 6):

$$h(x, z)_{NDD} = \delta(x, z + d_z) + \delta(x + d_x, z) - 4\delta(x, z) + \delta(x, z - d_z) + \delta(x - d_x, z) \tag{10}$$

where d_x and d_z are the interelectrode distances in the two spatial directions. The extent to which the spatial filter represents an inversion of the spatial low-pass filter of the volume conductor is represented by its transfer function. The transfer function of the filter is computed as the two-dimensional Fourier transform of Eq. (10) (refer also to Eq. (7)). We obtain

$$H(f_x, f_z)_{NDD} = -4 + 2\cos(2\pi f_x d_x) + 2\cos(2\pi f_z d_z) \tag{11}$$

The transfer function (11) is periodic in both spatial directions with periods equal to the inverse of the interelectrode distance in the two directions, respectively. As shown by Eq. (7), the periodicity is a general property of spatial filters comprised of grids of point electrodes. Thus the interelectrode distances d_x and d_z have to be chosen with respect to the smallest wavelength contributing to the potential distribution in the two spatial directions. By selecting the interelectrode distances smaller than twice the reciprocal of the largest spatial frequencies contributing to the potential distribution, we can contain the spectrum of the EMG signal all in one period of the transfer function.

Figure 7.3a shows the two-dimensional transfer function of the NDD-filter. The NDD is a spatial high-pass filter that suppresses spatial frequencies substantially smaller than the cutoff frequency. This is in contrast to spatial frequencies close to the cutoff frequency, which pass through the filter almost unchanged. As the cutoff frequencies are given by the reciprocal of the interelectrode distances, electrode arrangements with smaller interelectrode distances suppress higher spatial frequencies than electrode arrangements with larger interelectrode distances. Figure 7.4 shows examples of experimental signals detected by the single differential, LDD, and NDD filters. The improvement in spatial selectivity with NDD is evident (especially with respect to single differential) and allows the identification of single MU action potentials from the interference signal.

The NDD-filter is anisotropic [43]; thus the filter is not invariant to rotations. This implies that the capacity of the filter to distinguish between signals generated by close and far sources is sensitive to the orientation of the filter with respect to the fiber's orientation. This is a general characteristic of spatial filters based on grids of electrodes. Indeed, the invariance to rotations of the transfer function is a direct consequence of the invariance to rotations of the electrode configuration. It is not possible to design a matrix of electrodes, independently of the weights, which is invariant to rotations. However, whether the anisotropy of the NDD-filter is in practice really a disadvantage or not, depends mainly

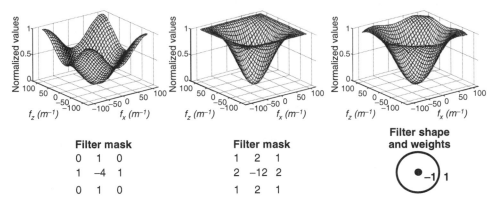

Figure 7.3. Magnitude of the two-dimensional transfer function of the NDD-filter (*a*), the IB^2-filter (*b*), and the concentric 1-ring system (*c*). The filter masks are also shown. The interelectrode distance is 5 mm in (*a*) and (*b*), the radius is 5 mm in (*c*). (Adapted from [18] and [22])

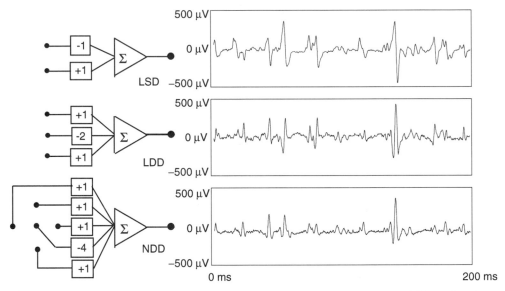

Figure 7.4. Signals detected from the biceps brachii muscle with three spatial filters (longitudinal single differential, longitudinal double differential and NDD). The interelectrode distance is in the three cases 5 mm. (With modifications, from [70])

on the interelectrode distance chosen and on the depth of the sources. If the interelectrode distance is very small, the transfer function of the NDD-filter is almost isotropic in the spatial frequency domain. The spatial bandwidth of the EMG potential decreases with increasing the source depth. There are mainly technical reasons to limit the reduction of the interelectrode distance, and 2 to 2.5 mm is the smallest interelectrode distance realised up to now.

Other spatial filter arrangements, consisting of more than five electrodes, have been developed [18,26]. One of them is the inverse binominal filter (IB^2-filter) proposed by

Disselhorst-Klug et al. [18], which consists of nine electrodes and whose filter mask is given by Eq. (12):

$$M(x, z)_{IB2} = \begin{bmatrix} 1 & 2 & 1 \\ 2 & -12 & 2 \\ 1 & 2 & 1 \end{bmatrix}$$ (12)

The transfer function of the IB^2-filter, which is shown in Figure 7.3b, is given by Eq. (13):

$$H(f_x, f_z)_{IB2} = 1 - \left(\frac{1}{2} + \frac{1}{2}\cos(2\pi f_x d_x)\right)\left(\frac{1}{2} + \frac{1}{2}\cos(2\pi f_z d_z)\right)$$ (13)

The IB^2-filter is more isotropic than the NDD, as can be shown by the two-dimensional Taylor expansion of transfer function (13).[1] More complex filter arrangements [18,26] are more selective but have the drawback of consisting of a large number of electrodes, which limits the practical applicability.

7.2.4 Spatial Filters Comprised of Nonpoint Electrodes

Kossev et al. [50] investigated the selectivity of electrodes in the *branched* configuration (see also Chapter 5), which consists of the connection, by short circuit, of a number of electrodes to obtain the average of the signals they detect. With three electrodes placed equidistantly on a line, making the difference between the signals detected by the central electrode and the two short-circuited extreme electrodes, it is possible to obtain a spatial filter with transfer function equivalent (apart from an amplification factor) to the double differential filter[2] [50]. Kossev et al. [50] also noted that a system comprised of a central point electrode and a ring electrode, with the center coincident with the point electrode, can be interpreted as the limit of a branched configuration in which N point electrodes along a circumference are short-circuited and the signal they collect is subtracted from that recorded by the central electrode. If $N \to \infty$, the point electrodes along the circumference tend to a real circumference. With this method Kossev et al. [50] simulated the concentric ring system and compared it with other spatial filters in the branched configuration (including the Laplacian filters realized by short-circuiting electrodes with the same weight).

Bhullar et al. [6] investigated the same configuration with a ring electrode and a central point electrode, deriving a theoretical basis for its description in the spatial frequency domain. However, their approach was based on a simplification of the problem to the one-dimensional case. The system was studied only in the direction of propagation of the potential, and this way it was concluded that its transfer function is equivalent to that of the longitudinal double differential filter. The theoretical approach followed by Bhullar et al. [6] was in fact based on approximations that reduced the system to a branched configuration with three electrodes, theoretically equivalent to the double differential system.

---

[1] The "degree of isotropy" can be evaluated from the two-dimensional Taylor expansion of the filter transfer functions. A filter is more isotropic than another if, in its series expansion, compared to the second filter, the maximum degree of the terms $(f_x^2 + f_z^2)^k$, with k positive integer, is larger [43].

[2] This holds only if the electrode–skin impedances are identical.

The importance of the study by Bhullar et al. [6] lies in the fact that, first, they developed a theoretical basis for the description of a spatial filter comprised of electrodes with physical dimensions in the spatial frequency domain. Thus they recognized that although comprised of nonpoint electrodes, this system could have been described with the same mathematical tools used to describe the spatial filters based on grids of point electrodes.

The work by Helal and Bouissou [40] is another attempt to describe the effect of non-point electrodes on the detected signal in the spatial frequency domain. These authors focused on circular electrodes but their analytical derivation can be easily extended, for example, to ring electrodes. However, also in this case the analysis was done in only one spatial dimension, since it was derived from the concepts developed by Lindstrom and Magnusson [51] for one-dimensional filters.

Farina and Merletti [24] developed the theory for the description of the effect of the size and shape of the recording electrodes in the two spatial dimensions. The analytical techniques applied are the same used to derive the transfer functions of matrixes of point electrodes, and the assumptions on which the theory is based are the same. Each nonpoint electrode can be characterized by a two-dimensional spatial transfer function. Thus it is possible to develop spatial filters in which the transfer function is designed on the basis of electrode shape rather than from a fixed configuration of point electrodes [22].

Let's consider again the potential distribution $\phi(x, z)$ traveling along the fiber direction. In case of electrodes with physical dimensions, at each spatial detection point, the potential is integrated under the electrode area[3] (Chapter 8 and [40]). This integration can be mathematically described as the two-dimensional convolution of the potential distribution over the skin with a function that depends on the electrode shape [23,24]. This function is particularly simple and assumes a constant value, equal to the inverse of the electrode area, in the region of space corresponding to the electrode area and zero outside:

$$h_{\text{size}}(x, z) = \begin{cases} \dfrac{1}{S} & f(-x, -z) \leq 0 \\ 0 & f(-x, -z) > 0 \end{cases} \tag{14}$$

where S is the electrode area and $f(x, z) < 0$ the mathematical expression which defines the electrode shape. From Eq. (14) the transfer function equivalent to the electrode shape is obtained (analytically or numerically) by computing the two-dimensional Fourier transform $H_{\text{size}}(f_x, f_z)$ of $h_{\text{size}}(x, z)$. In the case of a circular electrode with radius r we get

$$H_{\text{size}}(f_x, f_z) = \begin{cases} \dfrac{J_1(2\pi r \sqrt{f_x^2 + f_z^2})}{\pi r \sqrt{f_x^2 + f_z^2}} & (f_x, f_z) \neq (0, 0) \\ 1 & (f_x, f_z) = (0, 0) \end{cases} \tag{15}$$

where $J_1(k)$ is the first-order Bessel function of the first kind.

From Eq. (14) the physical dimension of an electrode can be described as a two-dimensional spatial filter, so the transfer function of combinations of nonpoint electrodes can be predicted. Applying Eq. (15) and basic integral properties, we obtain the equivalent transfer function of a ring electrode with internal and external radii r_1 and r_2, respectively:

---

[3] This holds under the assumption that the equipotential electrode surface does not significantly alter the field distribution in the nearby volume.

$$H_{\mathrm{ring}}(k_y) = \frac{1}{r_2^2 - r_1^2}\left[r_2 \frac{J_1(r_2 k_y)}{k_y} - r_1 \frac{J_1(r_1 k_y)}{k_y}\right] \tag{16}$$

where $k_y = 2\pi\sqrt{f_x + f_z}$.

It can be demonstrated that, by assuming rings with negligible thickness $[(r_2 - r_1) \to 0]$ and a point central electrode, an approximation of the equivalent transfer function of the system comprised by a point electrode and a concentric ring is given by

$$H_{Mappr}(k_y) = 1 - J_0(rk_y) \tag{17}$$

with $J_0(z)$ the zero-order Bessel function of the first kind. Farina and Cescon [22] extended these concepts to configurations of more than one ring. Systems comprised of concentric rings are isotropic; this can be observed from the transfer function (17) and from the geometrical configuration, which does not depend on the orientation of the system. The ring class is the only configuration of spatial filters, apart from the monopolar detection, that is ideally isotropic. Also note that transfer function (17) is not periodic as is the case for the grid electrode systems. Grid electrode systems represent digital two-dimensional spatial filters, while systems comprised of ring electrodes represent analog two-dimensional spatial filters. Figure 7.3*c* shows the transfer function (17). It has to be underlined that the above derivations hold with the approximation that the presence of an electrode with physical dimension does not alter the electric field outside the electrode.

7.2.5 Applications of Spatial Filtering Techniques

Two-dimensional spatial filters have been successfully used in many applications [17,19, 77,78,81]. Most of them were focused on the noninvasive investigation of the single MU activity. Noninvasive diagnosis of neuromuscular disorders as well as investigations of the MU properties, such as firing rate, recruitment, or CV, have been the main fields of applications.

In clinical applications the NDD-filter has been used in the diagnosis of neuromuscular disorders [19,42,69,70,71]. It has been shown that in 97% of the cases the NDD-filter allows a correct distinction between patients with muscular disorders, patients with neuronal disorders, and healthy volunteers, in a noninvasive way [42]. Figure 7.5 shows

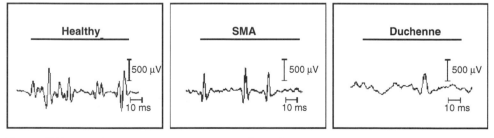

Figure 7.5. NDD-filtered surface EMG signals of healthy volunteers and patients suffering from spinal muscle atrophy (SMA) and Duchenne muscle dystrophy, recorded during maximum voluntary contraction of the abductor pollicis brevis muscle. The pathological changes in the motor unit action potentials can be clearly observed in the NDD-filtered signals. Spatial filtering contributes to a noninvasive detection of pathological changes in the muscles. (From [42] with permission)

representative signals recorded from patients and healthy subjects by high spatial resolution systems. Furthermore recent investigations of patients recovering from nerve lesions have shown that the follow-up of the patients and the evaluation of the treatment outcome is possible using NDD-filters in combination with electrode arrays (see below) [19].

7.2.6 A Note on Crosstalk

Spatial high-pass filters enhance the signals of MUs located close to the recording electrodes and suppress the contributions of more distantly located sources. For this reason spatial filters have been considered for reducing crosstalk in applications of surface EMG in which the detection of the muscular coordination pattern is of interest [49,57,86]. Different kinds and different orders of spatial filters have been applied for crosstalk reduction. However, the experimental results have shown that crosstalk is not reduced by spatial high-pass filtering [84]. This apparent discrepancy between theory and practice becomes clear when the model used for the development of the spatial filter theory is considered in greater detail. In this model the transfer function of the spatial filter has been adapted to the potential distribution generated on the skin surface by a propagating MU action potential. Generation and extinction phenomena of the MU excitation at the motor end plate and fiber ends (Chapter 4) are neglected in this model.

From simulations of the generation and extinction phenomena it is known that they generate a nonpropagating potential distribution on the skin surface [15,38]. Recent simulations of the filter responses to such nonpropagating potentials have shown that most of the spatial high-pass filters do not reduce the contributions of the nonpropagating potentials, in some cases they even enhance them [16,27] (see Figure 4.9 in Chapter 4). Far-field potentials are mainly constituted by end of fiber potentials, as shown in simulation and experimental studies [21,33,85]. Consequently, for the reduction of crosstalk signals, filter arrangements that take also nonpropagating potentials into account have to be developed.

7.3 SPATIAL SAMPLING

Sampling the surface EMG signal over the skin consists in placing a number of detection systems (which may perform a spatial filtering operation) in different locations over the skin. In this case the potential distribution is sampled at particular points and spatially filtered at each point by the detection systems. The availability of many channels may be used to estimate CV and to obtain more information on the processes of generation and extinction of the action potentials. The first systems of this type were proposed by DeLuca, Merletti, and Masuda [10,53,54,55,56,59], who applied linear electrode arrays along the fiber direction to estimate the velocity of propagation of action potentials or to identify innervation zones.

Recently more complex systems, with electrodes located both longitudinally and in the transversal direction with respect to the muscle fibers, have been applied for the estimation of MU anatomical properties [46,47,67,75,76,77,78]. Currently some research groups are using systems of hundreds of electrodes for sampling the surface potential in the two spatial directions. The information extracted from this type of signals was shown to be extremely important both for research and clinical applications [20,47].

7.3.1 Linear Electrode Arrays

Masuda et al. [54] proposed the sampling of the potentials generated by muscle fibers over the skin in many different points along the muscle. The linear electrode arrays they proposed were comprised of point electrodes from which single or double differential signals could be extracted. These systems were shown to provide information about single MU anatomical properties, such as the location of the innervation zones and the length of the muscle fibers. Later linear electrode arrays were applied and further developed by other research groups [5,29,44,45,61,79]. Figure 7.6 shows the linear electrode array detection modality, and Figure 7.7 reports examples of signals detected with this technique from muscles of the thigh and leg. The generation of the action potentials at the motor end plate can be observed from the surface signals as the point from which the potentials start the propagation in the two directions (toward the tendon regions).

It is possible to follow the action potentials from the generation at the neuromuscular junction to the extinction at the tendons. The pattern corresponding to the propagation of the signals in both directions is typical of single MU action potentials and can be used to identify the contributions of the single MUs in the interference signals. In particular conditions, potentials of single MUs can be isolated, extracted, and classified from noninvasive recordings by multichannel pattern recognition techniques. Decomposition methods applied to these signals are being developed and are based on the classic segmentation/classification scheme. Figure 7.8 shows an example of application of one of these techniques. Single MU action potentials are identified and classified during a long contraction at low force level. Recruitment of MUs during the contraction is evident and reflects the decrease of MU recruitment thresholds during fatigue.

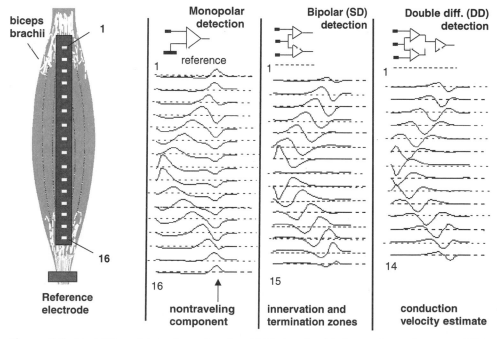

Figure 7.6. Modalities of detection of surface EMG signals with linear electrode arrays. (With modifications from [61])

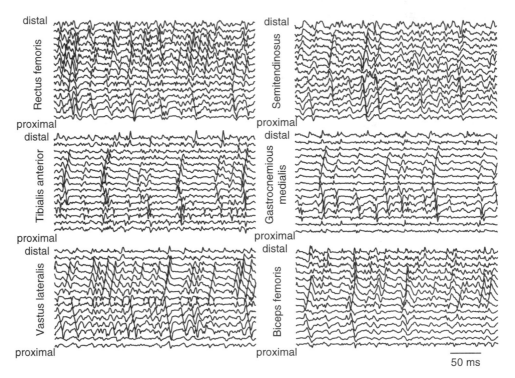

Figure 7.7. Signals detected with linear electrode arrays from muscles of the leg and thigh during voluntary contractions at 50% MVC. Single motor unit action potentials and their innervation zones can be identified. The interelectrode distance is 10 mm for all the recordings, except for those from the rectus femoris muscle (for which the interelectrode distance is 5 mm). (From [31] with permission)

The possibility of following action potentials during their propagation along the muscle fibers is particularly important for investigating MU conduction properties. The issue of estimating muscle fiber CV will be addressed later, and it will be shown that the availability of more than two signals for CV estimation may considerably reduce the variance of estimation. Apart from the estimation of CV, the use of such recording systems was shown to be useful in the detection of abnormalities in the potential propagation. Figure 7.9 shows signals detected by arrays of electrodes from the biceps brachii muscle of a subject and a subject with generalized myotonia. The conduction blocking of MU action potentials can be clearly detected by the multichannel surface recordings [20]. A single pair of electrodes cannot extract such information.

Linear electrode arrays have been shown to be very promising in association with processing techniques specifically designed to automatically extract single MU properties from multichannel surface EMG recordings [28,34,35,36,46]. Farina et al. [28] recently presented a method for the automatic estimation of MU CV distribution from signals detected with linear arrays. This technique has been applied to basic physiological studies [63] and has showed potential use for noninvasive muscle assessment.

Figure 7.10 gives an example of three NDD-filtered channels arranged in a row. The signals were recorded during maximal isometric contraction of the biceps brachii muscle.

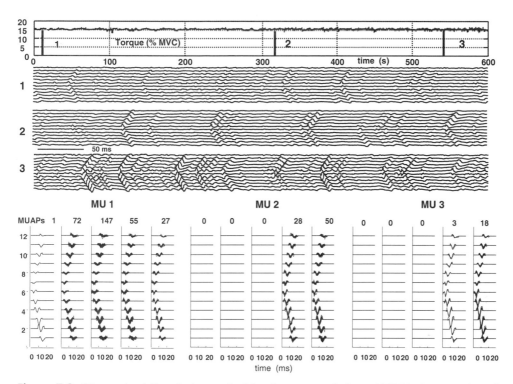

Figure 7.8. (Upper plots) Signals detected with a linear array during a 15% MVC contraction of the biceps brachii muscle. The torque signal is also shown. (Lower plots) Examples of MUAPs automatically detected and classified during the 10 minute long contraction. Only 3 seconds every 30 have been decomposed. In each section, the first column represents the MU action potentials detected and classified in the first 2 minutes, the second column those detected in the next 2 minutes, and so on. The three MUs reported begin firing at different instants of time. (With modifications, from [35] with permission)

The activity of single MUs can be identified as isolated peaks in the signal, due the high selectivity of the recording systems. From the time delay between two or more NDD-filtered channels, the CV of single MUs can be estimated (see below for the techniques used to estimate muscle fiber CV).

Linear electrode arrays have been applied for methodological studies on the influence of electrode location along the muscle fibers on the global variables extracted from the surface EMG signal. This was shown to be of particular importance for the standardization of the recording modalities when a global analysis of the surface EMG signal is performed [5,31,32,44,68,79].

7.3.2 Two-Dimensional Spatial Sampling

The systems for signal detection can be placed in either direction over the muscle. The placement of electrodes in the transversal direction can be used to analyze the rate of decrease of the potential with increasing transversal distance from the source. This

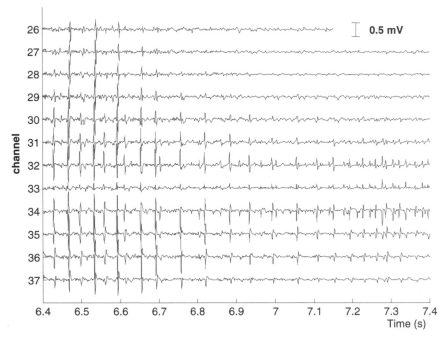

Figure 7.9. Multi-channel surface EMG signals detected with an electrode array from the biceps brachii muscle of a patient with generalized myotonia. The patient is decreasing the force due to a transient paresis typical of this pathology. Note the conduction blocking in the patient. (From [20] with permission)

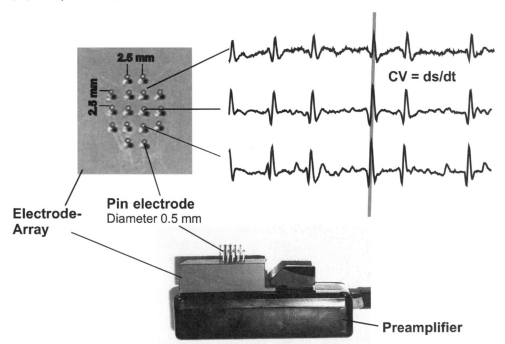

Figure 7.10. Determination of the conduction velocity of single motor units by means of the NDD-filter. Three spatially filtered channels are shown, recorded during maximal voluntary contraction of the biceps brachii muscle. CV can be calculated from the time delay between two or more channels. (From [19] with permission)

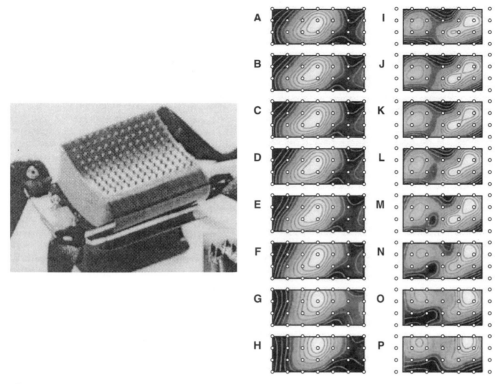

Figure 7.11. High-density surface matrix of electrodes (*left*) and topographical maps of the root mean square value for monopolar (A–H) and single differential signals (I–P) (*right*), at increasing force levels (5% MVC, 10% MVC, 15% MVC, 20% MVC, 35% MVC, 50% MVC, 100% MVC for the different rows). The maps are computed from recordings with a system similar to that shown on the left from the upper trapezius muscle. (With modifications, from [48] with permission)

decrease depends on the depth of the MU, which thus can be estimated from these types of recordings. Roeleveld et al. [78] first proposed such recording techniques (see Fig. 4.5 in Chapter 4).

Systems of more than 100 electrodes have been developed by the group led by Stegeman [88] and are being used in advanced studies on muscle functions. In particular, from these systems it is possible to compute maps of potential amplitude and to investigate the distribution of muscle activity in different tasks. Figure 7.11 shows examples of these amplitude maps obtained from the upper trapezius muscle.

7.4 ESTIMATION OF MUSCLE-FIBER CONDUCTION VELOCITY

Muscle-fiber CV is a basic physiological parameter, and it is related to the type and diameter of muscle fibers, ion concentration, pH, muscle temperature, and MU firing rate [1,3,9]. The velocity of propagation of action potentials is indicative of the MU contractile properties [1] and thus can be used to assess muscle fatigue [2,14,62]. Muscle-fiber CV can be estimated from intramuscular or surface EMG recordings [3].

Buchthal et al. [12] first proposed a method for single-MU CV estimation. Their technique was based on the detection of intramuscular EMG signals at two locations at a known distance and on the estimation of the delay between the two detected potentials. This way CV could be estimated during either voluntary or electrically elicited contractions. The technique is time-consuming and suffers from some methodological limitations, the most important being the difficulty of obtaining potentials of similar shape from two extremely selective detection systems. High selectivity, indeed, implies large changes of the detected waveform shapes as a consequence of small electrode movements. Troni et al. [83] proposed a simpler method based on the estimation of the delay in the detection of action potentials elicited by intramuscular electrical stimulation of the muscle fibers. However, the invasive methods to estimate CV remain time-consuming and of difficult applicability.

Muscle-fiber CV can be also estimated from surface EMG signals. When epochs of signal are considered (global EMG analysis), the estimation is affected by the contributions of all active superficial MUs firing during the selected epoch. The methods based on surface EMG signal analysis are affected by an estimation bias that depends on MU action potential generation and extinction (Chapter 4), and that explains the estimates generally higher with respect to those provided by invasive recordings. Zwarts [87] found, for instance, that CV estimates from surface EMG signals were on average 1 m/s higher with respect to the values obtained from invasive recordings. However, the estimations obtained with the two recording systems were highly correlated and both potentially useful for the diagnosis of particular myopathies [69,87].

We will focus in the following on the noninvasive CV estimation from two or more surface EMG signals detected along the muscle fibers. These techniques can be applied to either signal epochs, to estimate mean muscle-fiber CV, or to single-MU action potentials, extracted from the interference EMG signals.

7.4.1 Two Channel-Based Methods for CV Estimation

Practical Problems in Estimating Delay between Two Sampled and Not Identical Signals. With the availability of two signals detected in two locations over the skin, the CV of the potentials contributing to the signal can be estimated from the delay between the two signals [11]. In ideal conditions of identical shapes, continuous acquisition (analog signals), and absence of noise, the delay can be estimated from any reference point, such as a zero crossing or peak. However, these conditions are never verified in practice. The signals are recorded as sequences of samples (sampling of the analog signals); thus the temporal resolution is limited in the time domain by the sampling period. A simple analytical model of two delayed signals in discrete time domain, without shape difference, is the following:

$$x_1(n) = s(n) + w_1(n)$$

$$x_2(n) = s(n - \theta) + w_2(n) \tag{18}$$

where θ is the delay between the two signals and $w_k(n)$ ($k = 1, 2$) are independent, white, zero mean, additive Gaussian noises, which can be assumed of equal variance σ^2.

The sampling of the signals leads to limitations in the CV estimate resolution if only the available time samples are used for signal alignment. For example, considering a mean

CV of 4 m/s and acquiring two signals with two recording systems at a distance $d =$ 10 mm, the delay between the two signals is $\theta = d/v = 2.5$ ms. Sampling the surface EMG signals at 1000 Hz (within the Nyquist limit), the temporal distance between two samples is 1 ms; thus it is not possible to detect temporal differences smaller than 1 ms. The relative errors in estimating CV and time delay are approximately the same:

$$\frac{\Delta v}{v} \cong \frac{\Delta \theta}{\theta} \tag{19}$$

Thus the absolute error of CV estimation is

$$\Delta v = v \frac{\Delta \theta}{\theta} = v^2 \frac{\Delta \theta}{d} \tag{20}$$

which leads, in the specific case, to $\Delta v = 1.6$ m/s (relative error 40%). Clearly, this is not acceptable in any application. However, it is certainly possible to improve temporal resolution, theoretically limited only by noise, by interpolating the signals (or their cross-correlation function), if the Nyquist limit is met. The problem will be addressed in the following paragraphs.

The two detected signals are never identical and just delayed. Even in ideal recording conditions the two signals cannot be identical and delayed in time due to the distribution of conduction velocities of the active MUs. Other factors determining shape changes of the signals during propagation are the generation and extinction of the action potentials, the inclination of the fibers with respect to the detection system, and the nonhomogeneity of the subcutaneous layers in the direction of the muscle fibers [82]. From the mathematical point of view, the delay between functions that are not the same is not trivially defined. Different definitions of delay lead to different results even in case of noise free signals. In addition each definition of delay, and thus each delay estimator, present a variance and bias of estimation in the presence of additive noise.

The definition of delay between signals most commonly used in EMG field is the time-shift which, applied to one of the two signals, minimizes the mean square error between the two. The cost function to be minimized is

$$e_t(\hat{\theta}) = \sum_{n=1}^{N} \left[x_2(n+\hat{\theta}) - x_1(n) \right]^2 \tag{21}$$

This definition corresponds to the maximum likelihood delay estimation between two signals with additive white Gaussian noise [34]. From this definition it is possible to derive the estimators based on the maximum of the cross-correlation function or on the *spectral matching* [7,52,58,60].

Algorithms for Estimating CV from Two Signals

MAXIMUM LIKELIHOOD ESTIMATORS: SPECTRAL MATCHING APPROACH. In the frequency domain the mean square error (21) can be written as

$$e_f(\hat{\theta}) = \frac{2}{N} \sum_{k=1}^{N/2} \left| X_2(k) e^{j2\pi k \hat{\theta}/N} - X_1(k) \right|^2 \tag{22}$$

where N is the number of samples in the signal epoch, and $X_1(k)$, $X_2(k)$ are the Fourier transforms of the two signals.

As a consequence of the Parseval's theorem, it follows that $e_t(\hat{\theta}) = e_f(\hat{\theta})$; thus Eq. (21) and Eq. (22) are equivalent. However, by Eq. (22), $\hat{\theta}$ is not limited to be a multiple of the sampling interval, but rather, it can be selected as any real value. The multiplication for the exponential function is indeed equivalent to the translation and interpolation in the time domain. The temporal resolution is, thus, in principle not limited if Eq. (22) is used instead of Eq. (21). The problem is solved by finding the minimum of $e_f(\hat{\theta})$ with an iterative algorithm that converges to the optimal point in a few iterations by finding the zero of its first derivative, using a gradient method that is starting from a coarse estimation of the delay and updating this estimate moving along the gradient direction. The iteration becomes [58]

$$\theta_{i+1} = \theta_i - \frac{\mathrm{d}e_f(\theta)/\mathrm{d}\theta \big|_{\theta=\theta_i}}{\mathrm{d}^2 e_f(\theta)/\mathrm{d}\theta^2 \big|_{\theta=\theta_i}} \tag{23}$$

starting from an initial estimate θ_0. It can be demonstrated, from Eq. (22), that

$$\frac{\mathrm{d}e_f(\theta)}{\mathrm{d}\theta} = \frac{4}{N} \sum_{k=1}^{N/2} \left(\frac{2\pi k}{N}\right) \mathrm{Im}\left[X_2(k) e^{j2\pi k \hat{\theta}/N} X_1^*(k) \right] \tag{24}$$

$$\frac{\mathrm{d}^2 e_f(\theta)}{\mathrm{d}\theta^2} = \frac{4}{N} \sum_{k=1}^{N/2} \left(\frac{2\pi k}{N}\right)^2 \mathrm{Re}\left[X_2(k) e^{j2\pi k \hat{\theta}/N} X_1^*(k) \right] \tag{25}$$

where * stands for complex conjugate, and Re[], Im[] for the real and imaginary part of a complex number.

The starting estimate of the delay, necessary for the iteration (23), can be obtained as the maximum of the cross-correlation function of the two signals (see below), or it can be selected randomly in a range of values corresponding to physiological conduction velocities. The algorithm can be improved using the technique developed by Brent [8] and applied by Bonato et al. [7] that avoids the occasional convergence on a local minimum of the error function (22) at the cost of additional computational load. The iteration is stopped when the difference between subsequent estimates of $\hat{\theta}$ is smaller than a given threshold. The resolution of the estimate is thus only limited by this threshold (which is usually very small) and by the additive noise. The quality of the estimation can be indicated by the residual error.

MAXIMUM LIKELIHOOD ESTIMATORS: CROSS-CORRELATION FUNCTION APPROACH. The normalized cross-correlation function of the two signals of model (18) is given by

$$\rho_{1,2}(\tau) = \frac{\sum_{n=1}^{N} x_2(n+\tau)x_1(n)}{\sqrt{\sum_{n=1}^{N} x_1^2(n) \sum_{n=1}^{N} x_2^2(n)}} = \frac{R_{1,2}(\tau)}{\sigma_1 \sigma_2} = \frac{1}{2\sigma_1 \sigma_2}[\sigma_1^2 + \sigma_2^2 - e_t(\tau)] \tag{26}$$

where e_t is given by Eq. (21). Thus the minimum of the mean square error (in time or frequency domain) corresponds to the maximum of the cross-correlation function. Computing the maximum of the cross-correlation function can be an alternative approach for delay estimation, theoretically equivalent to the spectral matching method. The quality of the estimation can be indicated by the value of the correlation coefficient at the optimal point. However, the cross-correlation function can be computed only at time instants that are multiples of the sampling period, and this limits the time resolution. To avoid this problem, it is possible to follow two approaches: (1) interpolating the cross-correlation function with ideal interpolation, using all the lags, or (2) interpolating the cross-correlation function around its peak, for example, with polynomial functions, using only the samples of the cross-correlation function necessary for estimating the coefficients of the interpolating function. The second approach leads to theoretically not limited resolution, since the interpolating function is computed for any time lag and the interpolation is low cost from the computational point of view. However, this approach does not assure to provide the maximum likelihood value.

OTHER APPROACHES: PHASE METHOD. From the model (18) and by simple Fourier transformation of the two delayed signals, it is possible to derive the relationship between the phases of the Fourier transforms of the two signals. In particular, without additive noise, the difference between the phases of the Fourier transforms of the two signals is a linear function in the frequency domain:

$$X_1(k) = S(k), \qquad X_2(k) = S(k)e^{-j2\pi k\theta/N} \tag{27}$$

$$phase[X_1(k)] - phase[X_2(k)] = \frac{j2\pi k\theta}{N} \tag{28}$$

Equation (28) can be used to estimate the delay from the angular coefficient of the linear regression function interpolating the phase difference between the two signals. The method can be improved by selecting a limited range of frequencies for the estimation of the regression line, for example, on the basis of the coherence function [41].

OTHER APPROACHES: DISTRIBUTION FUNCTION METHOD. The distribution function method (DFM) can be used for estimating the time delay and time scale factor between two signals. The distribution function (DF) of a signal (also referred to as cumulative function) is the cumulative sum of the samples of the function. The DFM is based on the invariance of the DF of a signal to any transformation of the type

$$p_1(t) \to p_2(t) = Cp_1[\phi^{-1}(t)]\frac{d}{dt}\phi^{-1}(t) \tag{29}$$

where C is a constant and $\phi(\cdot)$ is an increasing continuous function. It has been shown [74] that the DFs of two signals are equal in abscissa values x and y related by $y = \phi(x)$. On the other hand, two signals $s_1(t)$ and $s_2(t)$ are equal in shape if we can write a relation of the form

$$s_1(t) = Cs_2\left(\frac{t-\theta}{a}\right), \qquad a > 0 \tag{30}$$

which is the same as choosing $\phi(t) = at + \theta$ in Eq. (29). In case of no scale and shape change, we obtain the line $\phi(t) = t + \theta$ ($a = 1$). The distance of the computed curve $\phi(t)$ from this ideal line can be considered a measure of distance between signals of different shape. The delay can be estimated by linear interpolation of the experimental values defining the curve $\phi(t)$.

The comparison of some of the methods described above for CV estimation from two channels is reported in [25].

7.4.2 Methods for CV Estimation Based on More Than Two Channels

If more than two signals detected along the fiber direction are available, new delay estimators can be defined on the basis of the alignment of all the available signals. Assuming that the systems used to detect the EMG signals are placed between the innervation zone and tendon region, then, ideally, all the detected signals are equal in shape but delayed. If the electrodes are equally spaced, in ideal conditions of pure propagation without shape changes, the delay between adjacent signals is constant, and its maximum likelihood estimation is the minimization of the sum of the mean square errors between each signal assumed as the reference and the other signals aligned [34,66]. In this case the signal model is the following:

$$x_k(n) = s(n - (k-1)\theta) + w_k(n), \qquad k = 1, \ldots, K \tag{31}$$

which is the extension of model (18) to the case of more than two propagating signals. In Eq. (31), K is the number of available signals and $w_k(n)$ ($k = 1, \ldots, K$) are, as in (18), independent, white, zero-mean, additive Gaussian noises, which can be assumed of equal variance σ^2.

Algorithms for Estimating CV from More Than Two Signals

MAXIMUM LIKELIHOOD ESTIMATOR. It can be demonstrated [34] that for the model (31) the maximum likelihood estimator of the delay θ is obtained by the minimization of the following mean square error:

$$e_{MLE} = \sum_{k=1}^{K} \sum_{n=1}^{N} \left[x_k(n) - \frac{1}{K} \sum_{m=1}^{K} x_m(n + (m - k)\theta) \right]^2 \tag{32}$$

where N is the number of samples in the signal epoch considered. Equation (32) is the generalization of Eq. (21) to the case of any number of channels. With simple passages, from Eq. (32) we get the equivalent expression of the mean square error for the maximum likelihood estimation [34]):

$$e_{MLE} = \sum_{k=1}^{K} e_k \tag{33}$$

with

$$e_k = \sum_{n=1}^{N} \left[x_k(n) - \frac{1}{K-1} \sum_{\substack{m=1 \\ m \neq k}} x_m(n + (m-k)\theta) \right]^2 \tag{34}$$

Minimization of e_k results in the delay that minimizes the sum of the mean square errors between the signal $x_k(n)$ (addressed later as *reference signal*) and the average of the other resynchronized signals (*modified beamforming*, see also below). From Eq. (34) it is clear that the maximum likelihood estimator is the minimization of the sum of the mean square errors obtained by performing this modified beamforming with all the signals of the array taken as reference one after another.

As already mentioned in the two-channel case, the minimization of the mean square error in discrete time domain will lead to a delay resolution limited by the sampling period. Hence an interpolation technique is required. The frequency domain approach provides a solution to this problem [58] also in the multi-channel case. Equation (34) can be rewritten in the frequency domain where the delay is a continuous variable and no resolution limit is imposed. Recently Farina et al. [34] proposed an algorithm based on the iterative minimization in the frequency domain of Eq. (33) by extending the McGill and Dorfman method [58] to the multi-channel case.

It has been shown [34] that in case of many propagating signals the problem of local minima in the iterative gradient procedure is more serious than in the case of two channels. Lack of convergence to a global minimum may result from an incorrect choice of starting point. The starting point should thus be selected close to the real value (e.g., as the maximum point of the discrete cross-correlation function between two of the available signals).

BEAMFORMING-BASED ESTIMATORS. From Eq. (34) the mean square errors e_k can also be used to define and estimate the delay between signals. In this case, depending on the reference signal used for the estimation, a different definition of delay is obtained. The variance of these estimates strongly depends on the selection of the reference signal [34]. A beamforming-based estimator, with the selection of the first signal as the reference, was used by Farina et al. [25] for estimating single MU conduction velocities from signals detected by linear arrays of electrodes.

Figure 7.12 shows simulation results with the comparison of two-channel and multi-channel based algorithms for the estimation of CV. The performance index reported is the mean square error of the estimation, defined as the sum of the variance and the square of the bias, thus providing a global indication of performance. The comparison is based on simulated signals with additive white Gaussian noise. The improvement of performance with the multichannel methods with respect to the two-channel techniques is significant.

Figure 7.13 shows the comparison between two-channel and multichannel based CV estimates from experimental signals acquired during a voluntary low force level isometric contraction of the biceps brachii muscle. The reduction of the estimation variance is evident also in the experimental case.

7.4.3 Single MU CV Estimation

The same algorithms described above can be used for the estimation of CV of single MU action potentials, if they can be extracted from the interference EMG signal. A method for

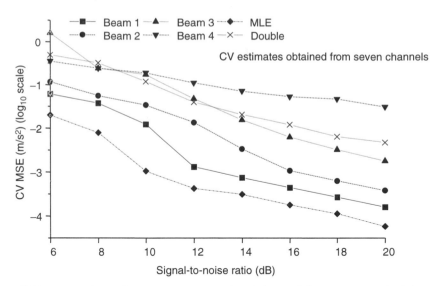

Figure 7.12. Comparison of mean square error (MSE) (sum of the variance and of the square of the bias) of CV estimations (simulated signals) obtained from seven signals in case of iterative procedure with starting point selected with cross-correlation for different multi-channel and two-channel methods. Results of the two-channel delay estimation are indicated as "double," while "Beam1" indicates beamforming method with the first signal as reference, "Beam2" the beam-forming with the second signal as reference, and so on. Results obtained with beamforming are reported for the four choices of the reference signal corresponding to the first four signals used for the estimation. MLE indicates the multi-channel maximum likelihood approach. For each signal-to-noise ratio 200 simulations have been performed. (From [34] with permission)

extracting single MU action potentials from the surface signals is to average the surface signals using intramuscularly detected potentials as triggers. This technique is described in Chapter 3 and in reference [30]. In case of spike-triggered averaged potentials, the signal-to-noise ratio can be enhanced by increasing the number of averages. However, the time resolution of the technique also depends on the number of averages, since it is not possible to detect CV changes taking place within the interval of time used for the averaging process. Thus, in this case, the availability of low-variance methods can be of paramount importance for detecting rapid changes in membrane properties of the muscle fibers. Figure 7.14 reports the comparison of CV estimates obtained from different number of channels in case of averaged multi-channel signals during a fatiguing contraction of the tibialis anterior muscle. Note the reduction of the variance of estimation with increasing number of channels. In particular, using the maximum number of available channels, we find that the CV estimates reveal a fatigue pattern that is hardly detectable when using fewer channels.

Single MU action potentials can be extracted from the interference signal also without an averaging process, in particular conditions (low/medium contraction levels, low level of additive noise, small number of active MUs). This can be obtained by applying spatial filtering techniques together with spatial sampling methods, as described in Section 7.3. In this case the CV of single MU action potentials can be estimated by applying the described algorithms to the extracted multi-channel potentials. Figure 7.10 shows an

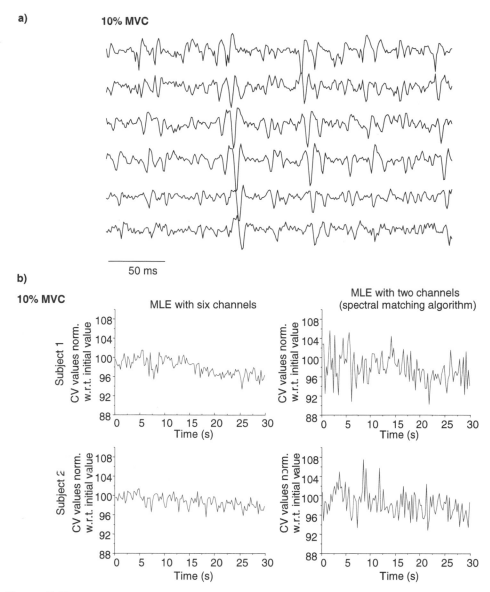

Figure 7.13. EMG signals detected by a linear electrode array from the biceps brachii muscle during a voluntary isometric contraction at 10% MVC (*a*) and mean CV estimations (epoch length 250 ms) computed during two contractions lasting 30 s at 10% MVC of two subjects. Maximum likelihood estimation (MLE) with iterative procedure is shown on the left (using six channels), traditional two-channel estimations are shown on the right (using the central channels of the six used for MLE). CV values are normalized (and expressed in percentage) with respect to the intercept of the regression line (initial value). (From [34] with permission)

example of identification of single MU action potentials detected by high spatial resolution noninvasive techniques. From these signals CV can be estimated with the techniques described in Section 7.4.1. Figure 7.15 shows an example of extraction and classification of single MU action potentials from surface multichannel signals with longitudinal double

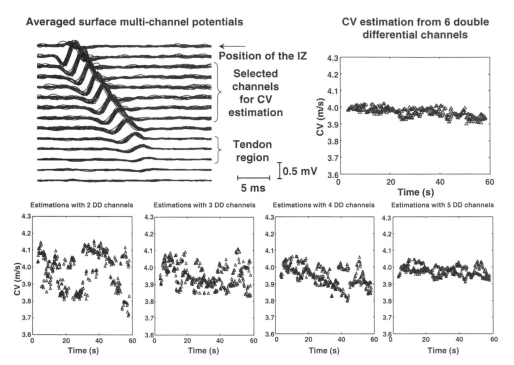

Figure 7.14. Estimation of single MU CV from averaged MU action potentials detected from the tibialis anterior muscle. The surface averaged potentials, obtained by 20 averages each, are also shown. The optimal estimation of CV requires six channels. In this case a clear fatigue pattern is evident (correlation coefficient between CV and time statistically different from zero). The CV estimation obtained by two to five channels is shown. The estimation improves with the number of channels. Note the large difference in estimation quality between the optimal estimation and the classic two-channel estimation. The two-channel estimation is performed in the ideal condition with recording electrodes located in the middle between the innervation zone and tendon region. With standard recording techniques the two-channel estimation would be highly influenced also by electrode positioning, and the difference with respect to the multi-channel method would be even larger. The distance from the first and last single differential signals of the surface recordings is 70 mm. DD stands for double differential recording, IZ for innervation zone (From [30] with permission)

Figure 7.15. (a) Single MU action potentials detected from the biceps brachii muscle at 10% of the maximal voluntary contraction (MVC) with a linear array of 16 electrodes in longitudinal double differential modality. The MU action potentials are classified as belonging to three MUs. (b) Results of the application of a two-channel estimator (spectral matching), the maximum likelihood estimator (MLE) and beamforming to the three MUs. DFM indicates a specific method for choosing the reference signal (see [34] for details). CV has been computed from all the detected firings of the three MUs with two or five double differential signals (in case of MLE and beamforming). In the case of two-channel estimation, the four estimations obtained using the four pairs of adjacent signals are reported for each firing. In the lower row, mean and standard deviation of the 10 estimations for each MU are reported for the three techniques. The p value, which results from the Student t-test for independent samples, is shown for the multi-channel methods and is related to the comparison between CV values of the three MUs. It appears that the three MUs are statistically different in CV value if multi-channel techniques are used. In the case of two-channel method, no significative difference has been obtained. (From [34] with permission)

a

Detection and classification of MUAPs generated by 3 MUs
(voluntary contraction of the biceps brachii muscle)

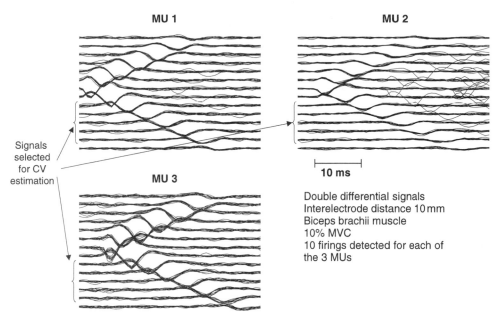

b

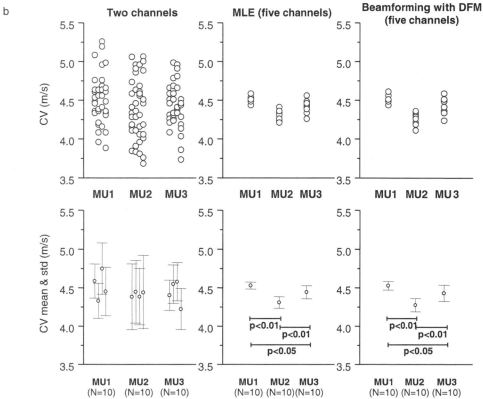

differential detection and the estimation of their CV with three different methods, described above. The three MUs can be in this case differentiated by their CVs if multichannel methods are used for delay estimation.

Methods to estimate single MU CV by using the displacement of the peaks detected by different channels have also been proposed. Moreover techniques have also been investigated to estimate MU CV distribution [13,37].

7.4.4 Influence of Anatomical, Physical, and Detection System Parameters on CV Estimates

The methods for muscle-fiber CV estimation described in Section 7.4 have been developed on the basis of very simple signal generation models, Eqs. (18) and (31). Indeed, the models used are phenomenological and make no attempt at the description of the physical processes underlying the generation of two (or more) delayed signals. The models (18) and (31) of course do not fully describe the characteristics of the experimental signals. Differences between experimental recordings and those models are, for example, the possible inclination of the fibers with respect to the detection system or the end of fiber effects. These non-ideal conditions determine a bias of the estimates, which depends on the estimator used. The estimation bias can be analyzed by the use of more complex models. Different algorithms for CV estimation, applied to signals generated with models that include possible shape changes of the propagating potentials, may provide significantly different results regardless of the additive noise. In some applications the sensitivity of different CV estimators to non-ideal recording conditions may be more important than their statistical properties in terms of variance of estimation. Figure 7.16 shows the dependence of CV estimates on the fiber inclination angle for different methods (two- channel- and multichannel-based techniques). The method with the best theoretical performance (i.e., in case of absence of shape changes and with only additive white Gaussian noise) is the most sensitive to fiber inclination angle in this example. Thus its use should be limited to cases where the fiber orientation can be easily identified. Moreover other important aspects should be noted: (1) the estimate of delay (and therefore CV) for fibers not aligned with the detection systems is very different using different methods, (2) the CV estimates can be either higher or lower than the correct value, and (3) for some methods even a small inclination angle can result in large bias of estimation. Point 2 was investigated in past research works [64,65], and it is particularly important for defining procedures for electrode placements. The selection of the electrode orientation that minimizes the estimated CV, suggested in the past [80], is not a good choice if one-dimensional spatial filters are used, since underestimations of CV may occur for not aligned fibers. Farina et al. [27] showed that the dependence of CV estimates on the fiber inclination angle is strongly affected by the spatial filter used (in particular, it depends on the isotropy property of the spatial filter).

Figure 7.17 reports the sensitivity of multi- and two-channel methods to the generation and extinction of the action potentials at the end plate and tendon. The overestimation of CV approaching the innervation zone and tendon region is due to the nontraveling components (end of fiber effect, Chapter 4).

7.5 CONCLUSIONS

The contributions of different MUs to surface EMG signals can be separated and classified by the use of advanced recording systems and processing techniques. Spatial filtering

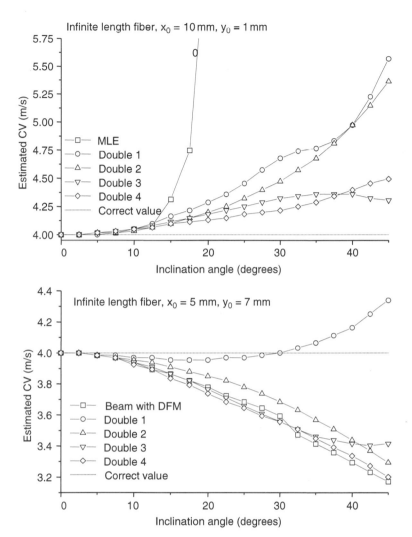

Figure 7.16. Single MU CV estimations as a function of the angle of inclination of the fibers with respect to the detection array (simulated signals) in the case of two-channel estimation, maximum likelihood estimation (MLE) and beamforming (using 5 channels) for particular positions of the fiber in the muscle (*a*) and (*b*). The recording system was the longitudinal double differential. y_0 is the depth of the MU in the muscle and x_0 the distance, in the direction perpendicular to the fiber, between the fiber and the central electrode of the array. In the case of two-channel method, the four estimations corresponding to the four choices of the pair of adjacent double differential signals (from double 1 to double 4) are shown. (From [34] with permission)

of the surface signals reduces the information content of the signal but extracts the contributions of the sources located close to the recording points. Reducing the number of sources makes identification of the remaining ones feasible. Unlike spatial filtering, spatial sampling expands the information available, with the possibility of following the action potential generation, propagation, and extinction along the muscle fibers. The joint use of spatial filtering and spatial sampling is promising for the issue of decomposition of the

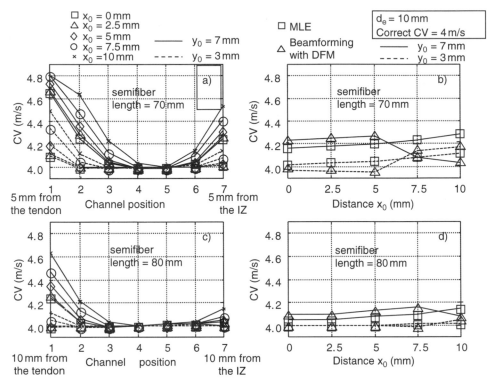

Figure 7.17. CV estimations (simulated signals) in the case of finite length fibers with two-channel spectral matching method (*a, c*), beamforming and maximum likelihood estimation (MLE) using five channels (*b, d*) for two semilengths of the fibers (70 and 80 mm, respectively). DFM indicates a specific method for choosing the reference signal (see [34] for details). The neuromuscular junction and the tendon are at a distance of 5 mm (*a, b*) and 10 mm (*c, d*) from the first and the last electrode of the array, respectively. The MU is at the depths y_0 of 7 and 3 mm in the muscle and is aligned with the detection array. x_0 is the distance between the fiber and the array in the direction perpendicular to the fiber, IZ stands for innervation zone. Interelectrode distance = 10 mm. (From [34] with permission)

surface EMG signal into the constituent MU action potential trains. From the identified MU activities, the multichannel recordings allow the extraction of single MU anatomical, control and conduction properties. This can be useful in a number of applications, including clinical and basic neurophysiology, ergonomics, and rehabilitation medicine. The techniques described in this chapter have been developed and refined in recent years and are still being tested in experimental conditions. However, it is clear from the few applications shown in this chapter, and from those that can be found in literature, that the type and quality of information that can be obtained from these recording systems and processing methods is significantly greater than what is obtained by classic global EMG analysis. One of the most important conclusions that can be drawn from this review of multichannel based methods for surface EMG signal analysis is that the techniques proposed are not intended only as a partial replacement of invasive recordings but rather provide both common and complementary information with respect to intramuscular EMG signal analysis. In particular, precise anatomical information on single MUs and accurate CV estimation are characteristics peculiar to the multichannel surface EMG recordings.

Currently the extraction of single MU action potential trains from surface EMG signals is still a difficult task that can be performed successfully only in particular conditions. Thus joint recording techniques [30] are still used to get reliable indications about MU control properties. However, it is likely that in the near future it will be possible to obtain such information exclusively from noninvasive recordings.

REFERENCES

1. Andreassen, S., and L. Arendt-Nielsen, "Muscle fiber conduction velocity in motor units of the human anterior tibial muscle: A new size principle parameter," *J Physiol* **391**, 561–571 (1987).

2. Arendt-Nielsen, L., and K. R. Mills, "The relationship between mean power frequency of the EMG spectrum and muscle fibre conduction velocity," *Electroencephalogr Clin Neurophysiol* **60**, 130–134 (1985).

3. Arendt-Nielsen, L., and M. Zwarts, "Measurement of muscle fiber conduction velocity in humans: techniques and applications," *J Clin Neurophysiol* **6**, 173–190 (1989).

4. Basmajian, J. V., and C. J. DeLuca, *Muscle alive*, Williams and Wilkins, Philadelphia, 1985.

5. Bergamo, R., M. Gazzoni, D. Farina, A. Lantermo, R. Merletti, C. Sanfilippo, and M. Titolo, "Multichannel surface EMG of muscles of the hand and forearm," *Proc PROCID Symp Muscular Disorders in Computer Users*, Copenhagen, Netherlands, 1999, pp. 198–202.

6. Bhullar, H. K., G. H. Loudon, J. C. Fothergill, and N. B. Jones, "Selective noninvasive electrode to study myoelectric signals," *Med Biol Eng Comp* **28**, 581–590 (1990).

7. Bonato, P., G. Balestra, M. Knaflitz, and R. Merletti, "Comparison between muscle fiber conduction velocity estimation techniques: spectral matching versus crosscorrelation," *Proc 8th Int ISEK Congr*, 1990, pp. 19–22.

8. Brent, R., *Algorithms for minimization without derivatives*, Prentice Hall, New York, 1973.

9. Brody, L., M. Pollock, S. Roy, C. J. De Luca, and B. Celli, "pH induced effects on median frequency and conduction velocity of the myoelectric signal," *J Appl Physiol* **71**, 1878–1885 (1991).

10. Broman, H., G. Bilotto, and C. J. De Luca, "Myoelectric signal conduction velocity and spectral parameters: influence of force and time," *J Appl Physiol* **58**, 1428–1437 (1985).

11. Broman, H., G. Bilotto, and C. J. De Luca, "A note on noninvasive estimation of muscle fiber conduction velocity," *IEEE Trans BME* **32**, 311–319 (1985).

12. Buchthal, F., C. Guld, and P. Rosenfalck, "Propagation velocity in electrically activated muscle fibers in man," *Acta Physiol Scand* **34**, 75–89 (1955).

13. Davies, S., and P. Parker, "Estimation of myoelectric conduction velocity distribution," *IEEE Trans BME* **34**, 98–105 (1987).

14. DeLuca, C. J., "Physiology and mathematics of myoelectric signals," *IEEE Trans BME* **26**, 313–325 (1979).

15. Dimitrov, G. V., and N. A. Dimitrova, "Extracellular potential field of an excitable fibre immersed in anisotropic volume conductor," *Electromyogr Clin Neurophysiol* **14**, 437–450 (1974).

16. Dimitrova, N. A., G. V. Dimitrov, and O. A. Nikitin, "Neither high-pass filtering nor mathematical differentiation of the EMG signals can considerably reduce cross-talk," *J Electromyogr Kinesiol* **12**, 235–246 (2002).

17. Disselhorst-Klug, C., J. Silny, and G. Rau, "Estimation of the relationship between noninvasively detected activity of single motor units and their characteristic pathological changes by modelling," *J Electromyogr Kinesiol* **8**, 323–335 (1998).

18. Disselhorst-Klug, C., J. Silny, and G. Rau, "Improvement of spatial resolution in surface-EMG: A theoretical and experimental comparison of different spatial filters," *IEEE Trans BME* **44**, 567–574 (1997).

19. Disselhorst-Klug, C., J. Bahm, V. Ramaekers, A. Trachterna, and G. Rau, "Non-invasive approach of motor unit recording during muscle contractions in humans," *Eur J Appl Physiol* **83**, 144–150 (2000).

20. Drost, G., J. H. Blok, D. F. Stegeman, J. P. van Dijk, B. G. van Engelen, and M. J. Zwarts, "Propagation disturbance of motor unit action potentials during transient paresis in generalized myotonia: A high-density surface EMG study," *Brain* **124**, 352–360 (2001).

21. Dumitru, D., and J. C. King, "Far-field potentials in muscle," *Muscle Nerve* **14**, 981–989 (1991).

22. Farina, D., and C. Cescon, "Concentric ring electrode systems for non-invasive detection of single motor unit activity," *IEEE Trans BME* **48**, 1326–1334 (2001).

23. Farina, D., and R. Merletti, "Effect of electrode shape on spectral features of surface detected motor unit action potentials," *Acta Physiol Pharmacol Bulg* **26**, 63–66 (2001).

24. Farina, D., and R. Merletti, "A novel approach for precise simulation of the EMG signal detected by surface electrodes," *IEEE Trans BME* **48**, 637–645 (2001).

25. Farina, D., and R. Merletti, "Comparison of algorithms for estimation of EMG variables during voluntary isometric contractions," *J Electromyogr Kinesiol* **10**, 337–350 (2000).

26. Farina, D., and A. Rainoldi, "Compensation of the effect of sub-cutaneous tissue layers on surface EMG: A simulation study," *Med Eng Phys* **21**, 487–496 (1999).

27. Farina, D., C. Cescon, and R. Merletti, "Influence of anatomical, physical and detection system parameters on surface EMG," *Biol Cybern* **86**, 445–456 (2002).

28. Farina, D., E. Fortunato, and R. Merletti, "Non-invasive estimation of motor unit conduction velocity distribution using linear electrode arrays," *IEEE Trans BME* **41**, 380–388 (2000).

29. Farina, D., R. Merletti, and U. Dimanico, "Non-invasive identification of motor units with linear electrode arrays," *Proc XII ISEK Congr*, Montreal, 1998, pp. 50–51.

30. Farina, D., L. Arendt-Nielsen, R. Merletti, and T. Graven-Nielsen, "Assessment of single motor unit conduction velocity during sustained contractions of the tibialis anterior muscle with advanced spike triggered averaging," *J Neurosci Meth* **115**, 1–12 (2002).

31. Farina, D., R. Merletti, M. Nazzaro, and I. Caruso, "Effect of joint angle on EMG variables in leg and thigh muscles," *IEEE Eng Med Biol Mag* **20**, 62–71 (2001).

32. Farina, D., P. Madeleine, T. Graven-Nielsen, R. Merletti, and L. Arendt-Nielsen, "Standardising surface electromyogram recordings for assessment of activity and fatigue in the human upper trapezius muscle," *Eur J Appl Physiol* **86**, 469–478 (2002).

33. Farina, D. R. Merletti, B. Indino, M. Nazzaro, and M. Pozzo, "Cross-talk between knee extensor muscles: Experimental and modelling results," *Muscle Nerve* **26**, 681–695 (2002).

34. Farina, D., W. Muhammad, E. Fortunato, O. Meste, R. Merletti, and H. Rix, "Estimation of single motor unit conduction velocity from surface electromyogram signals detected with linear electrode arrays," *Med Biol Eng Comput* **39**, 225–236 (2001).

35. Gazzoni, M., D. Farina, and R. Merletti, "Motor unit recruitment during constant low force and long duration muscle contractions investigated by surface electromyography," *Acta Physiol Pharmacol Bulg* **26**, 67–71 (2001).

36. Gazzoni, M., D. Farina, A. Rainoldi, and R. Merletti, "Surface EMG signal decomposition," *Proc XIII ISEK Congr*, Sapporo, Jpn, 2000, pp. 403–404.

37. Gonzalez-Cueto, J., and P. A. Parker, "Deconvolution estimation of motor unit conduction velocity distribution," *IEEE Trans BME* **49**, 955–962 (2002).

38. Gydikov, A., and D. Kosarov, "Volume conductor of the potentials from separate motor units in human muscles," *Electromyogr Clin Neurophysiol* **12**, 127–147 (1972).

39. Gydikov, A., L. Gerilovski, and N. Radicheva, "Influence of the muscle fibre end geometry on the extracellular potential," *Biol Cybern* **54**, 1–8 (1986).

40. Helal, J. N., and P. Bouissou, "The spatial integration effect of surface electrode detecting myoelectric signal," *IEEE Trans BME* **39**, 1161–1167 (1992).

41. Hunter, I. W., R. E. Kearney, and L. A. Jones, "Estimation of the conduction velocity of muscle action potentials using phase and impulse response function techniques," *Med Biol Eng Comp* **25**, 121–126 (1987).

42. Huppertz, H. J., C. Disselhorst-Klug, J. Silny, G. Rau, and G. Heimann, "Diagnostic yield of noninvasive high-spatial-resolution-EMG in neuromuscular disease," *Muscle Nerve* **20**, 1360–1370 (1997).

43. Jähne, B., *Digitale Bildverarbeitung*, Springer-Verlag, Berlin 1989, pp. 88–128.

44. Jensen, C., O. Vasseljen, and R. H. Westgaard, "The influence of electrode position on bipolar surface electromyogram recordings of the upper trapezius muscle," *Eur J Appl Physiol* **67**, 266–273 (1993).

45. Kiryu, T., H. Kaneko, and Y. Saitoh, "Influence of an innervation zone on surface EMG signals and its compensation," *Proc 14th An Int Conf IEEE Eng Med Biol Soc*, 1992, pp. 1444–1445.

46. Kleine, B. U., P. Praamsta, J. H. Blok, and D. F. Stegeman, "Single motor unit discharge patterns recorded with multi-channel sEMG," *Proc Symp Muscular Disorders in Computer Users*, Copenhagen, 1999, pp. 175–179.

47. Kleine, B. U., P. Praamstra, D. F. Stegeman, and M. J. Zwarts, "Impaired motor cortical inhibition in Parkinson's disease: Motor unit responses to transcranial magnetic stimulation," *Exp Brain Res* **138**, 477–483 (2001).

48. Kleine, B. U., N. P. Schumann, D. F. Stegeman, and H. C. Scholle, "Surface EMG mapping of the human trapezius muscle: The topography of monopolar and bipolar surface EMG amplitude and spectrum parameters at varied forces and in fatigue," *Clin Neurophysiol* **111**, 686–693 (2000).

49. Koh, T. J., and M. D. Grabiner, "Evaluation of methods to minimize cross-talk in surface electromyography," *J Biomech* **26**, 151–157 (1993).

50. Kossev, A , A. Gydikov, N. Trayanova, and D. Kosarov, "Configuration and selectivity of the branched EMG electrodes," *Electromyogr Clin Neurophysiol* **28**, 397–403 (1988).

51. Lindstrom, L., and R. Magnusson, "Interpretation of myoelectric power spectra: a model and its applications," *Proc IEEE* **65**, 653–662 (1977).

52. Lo Conte, L., and R. Merletti, "Advances in processing of surface myoelectric signals. Part 2," *Med Biol Eng Comput* **33**, 362–372 (1995).

53. Masuda, T., and T. Sadoyama, "The propagation of single motor unit action potentials detected by a surface electrode array," *Electroenc Clin Neurophysiol* **63**, 590–598 (1986).

54. Masuda, T., H. Miyano, and T. Sadoyama, "The distribution of myoneural junctions in the biceps brachii investigated by surface electromyography," *Electroenceph Clin Neurophysiol* **56**, 597–603 (1983).

55. Masuda, T., H. Miyano, and T. Sadoyama, "The position of innervation zones in the biceps brachii investigated by surface electromyography," *IEEE Trans BME* **32**, 36–42 (1985).

56. Masuda, T., H. Miyano, and T. Sadoyama, "A surface electrode array for detecting action potential trains of single motor units," *Electroenc Clin Neurophysiol* **60**, 435–443 (1985).

57. Matthews, B. H. C., "A special purpose amplifier," *J Physiol* **81**, 28–33 (1934).

58. McGill, K. C., and L. J. Dorfman, "High resolution alignment of sampled waveforms," *IEEE Trans BME* **31**, 462–470 (1984).

59. Merletti, R., and C. J. De Luca, "New techniques in surface electromyography," in J. E. Desmedt, ed., *Computer-aided electromyography and expert systems*, Elsevier Science, Amsterdam, 1989.

60. Merletti, R., and L. Lo Conte, "Advances in processing of surface myoelectric signals. Part 1," *Med Biol Eng Comput* **33**, 373–384 (1995).

61. Merletti, R., D. Farina, and A. Granata. "Non-invasive assessment of motor unit properties with linear electrode arrays," in *Clinical neurophysiology: From receptors to perception*, Elsevier, Amsterdam, 1999, pp. 293–300.

62. Merletti, R., M. Knaflitz, and C. J. De Luca, "Myoelectric manifestations of fatigue in voluntary and electrically elicited contractions," *J Appl Physiol* **69**, 1810–1820 (1990).

63. Merletti, R., D. Farina, M. Gazzoni, and M. P. Schieroni, "Effect of age on muscle functions investigated with surface electromyography," *Muscle Nerve* **25**, 65–76 (2002).

64. Merletti, R., L. Lo Conte, E. Avignone, and P. Guglielminotti, "Modelling of surface EMG signals. Part I: Model implementation," *IEEE Trans BME* **46**, 810–820 (1999).

65. Merletti, R., S. Roy, E. Kupa, S. Roatta, and A. Granata, "Modelling of surface EMG signals. Part II: Model based interpretation of surface EMG signals," *IEEE Trans BME* **46**, 821–829 (1999).

66. Meste, O., W. Muhammad, H. Rix, and D. Farina, "On the estimation of muscle fiber conduction velocity using a co-linear electrode array," *Proc 23rd An Int Conf IEEE Eng Med Biol Soc*, Istanbul, 2001 [Full paper available on CD-ROM (ISBN 0-7803-7213-1)].

67. Prutchi, D., "A high resolution large array (HRLA) surface EMG system," *Med Eng Phys* **17**, 442–454 (1995).

68. Rainoldi, A., M. Nazzaro, R. Merletti, D. Farina, I. Caruso, and S. Gaudenti, "Geometrical factors in surface EMG of the vastus medialis and lateralis," *J Electromyogr Kinesiol* **10**, 327–336 (2000).

69. Ramaekers, V., C. Disselhorst-Klug, J. Schmeider, J. Silny, J. Forst, R. Forst, F. Kotlarek, and G. Rau, "Clinical application of a noninvasive multi-electrode array EMG for the recording of single motor unit activity," *Neuropaediatrics* **24**, 134–138 (1993).

70. Rau, G., and C. Disselhorst-Klug, "Principles of high-spatial-resolution surface EMG (HSR-EMG): Single motor unit detection and application in the diagnosis of neuromuscular disorders," *J Electromyogr Kinesiol* **7**, 233–239 (1997).

71. Rau, G., C. Disselhorst-Klug, and J. Silny, "Noninvasive approach to motor unit characterization: Muscle structure, membrane dynamics and neuronal control," *J Biomech* **30**, 441–446 (1997).

72. Reucher, H., G. Rau, and J. Silny, "Spatial filtering of noninvasive multielectrode EMG. Part I: Introduction to measuring technique and applications," *IEEE Trans BME* **34**, 98–105 (1987).

73. Reucher, H., G. Rau, and J. Silny, "Spatial filtering of noninvasive multielectrode EMG. Part II: Filter performance in theory and modelling," *IEEE Trans BME* **34**, 106–113 (1987).

74. Rix, H., and J. P. Malengé, "Detecting small variation in shape," *IEEE Trans Syst Man Cybern* **10**, 90–96 (1980).

75. Roeleveld, K., A. Sandberg, E. V. Stalberg, and D. F. Stegeman, "Motor unit size estimation of enlarged motor units with surface electromyography," *Muscle Nerve* **21**, 878–886 (1998).

76. Roeleveld, K., J. H. Blok, D. F. Stegeman, and A. Van Oosterom, "Volume conduction models for surface EMG: Confrontation with measurements," *J Electromyog Kinesiol* **7**, 221–232 (1997).

77. Roeleveld, K., D. F. Stegeman, B. Falck, and E. V. Stalberg, "Motor unit size estimation: Confrontation of surface EMG with macro EMG," *Electroenc Clin Neurophysiol* **105**, 181–188 (1997).

78. Roeleveld, K., D. F. Stegeman, H. M. Vingerhoets, and A. Van Oosterom, "The motor unit potential distribution over the skin surface and its use in estimating the motor unit location," *Acta Physiol Scand* **161**, 465–472 (1997).

79. Roy, S. H., C. J. De Luca, and J. Schneider, "Effects of electrode location on myoelectric conduction velocity and median frequency estimates," *J Appl Physiol* **61**, 1510–1517 (1986).

80. Sadoyama, T., T. Masuda, and T. Miyano, "Optimal conditions for the measurement of muscle fiber conduction velocity using surface electrode arrays," *Med Biol Eng Comput* **23**, 339–342 (1985).

81. Schneider, J., G. Rau, and J. Silny, "A noninvasive EMG technique for investigating the excitation propagation in single motor units," *Electromyogr Clin Neurophysiol* **29**, 273–280 (1989).

82. Schneider, J., J. Silny, and G. Rau, "Influence of tissue inhomogeneities on noninvasive muscle fiber conduction velocity measurements investigated by physical and numerical modelling," *IEEE Trans BME* **38**, 851–860 (1991).

83. Troni, W., R. Cantello, and I. Rainero, "Conduction velocity along human muscle fibers in situ," *Neurology* **33**, 1453–1459 (1983).

84. van Vugt, J. P., and J. G. van Dijk, "A convenient method to reduce crosstalk in surface EMG," *Clin Neurophysiol* **112**, 583–592 (2001).

85. Wee, A. S., and R. A. Ashley, "Volume-conducted or "far-field" compound action potentials originating from the intrinsic-hand muscles," *Electromyogr Clin Neurophysiol* **30**, 325–333 (1990).

86. Winter, D. A., A. J. Fuglevand, and S. E. Archer, "Cross-talk in surface electromyography: Theoretical and practical estimates," *J Electromyogr Kinesiol* **4**, 15–26 (1994).

87. Zwarts, M. J., "Evaluation of the estimation of muscle fiber conduction velocity: Surface versus needle method," *J Neurophysiol* **83**, 441–452 (1989).

88. Zwarts, M. J., G. Drost, and D. F. Stegeman, "Recent progress in the diagnostic use of surface EMG for neurological diseases," *J Electromyogr Kinesiol* **10**, 287–291 (2000).

8

EMG MODELING AND SIMULATION

D. F. Stegeman

Department of Clinical Neurophysiology
University Medical Center, Nijmegen
Interuniversity Institute for Fundamental and
Clinical Human Movement Sciences (IFKB)
Amsterdam, Nijmegen, The Netherlands

R. Merletti

Laboratory for Engineering of the Neuromuscular System
Department of Electronics
Politecnico di Torino, Italy

H. J. Hermens

Roessingh Research and Development
Enschede, The Netherlands

8.1 INTRODUCTION

A mathematical model of a system describes the relationships between a relevant set of physical variables involved in the system. These relationships involve coefficients and parameters that are specific for the model. A mathematical model is a set of equations that can be implemented on a computer to study the effects of changing parameters and coefficients and to simulate, in order to understand and predict, with some approximations, the behavior of the system in specific conditions.

The use of computer models is inevitable in almost all areas of the natural sciences. Proper model design is often the major step in problem solving. Models are used in two

Electromyography: Physiology, Engineering, and Noninvasive Applications, edited by Roberto Merletti and Philip Parker.
ISBN 0-471-67580-6   Copyright © 2004 Institute for Electrical and Electronics Engineers, Inc.

ways. First, they attempt to extract the essentials of reality for the problem at hand, and second, they permit the estimation of parameters and variables not directly measurable as well as the investigation of the effects of their variations on the measurable variables of the system.

The goal in using models is manifold. They not only provide a better understanding of real life processes by the reductional approach, but they also allow the estimation of internal process characteristics by their so-called inverse use (i.e., the reproduction of experimental results by adapting the model's parameters) [58,75]. In the context of EMG, models are also well suited and often indispensable for testing methods for the analysis of experimental data and comparing their performances. Since model-generated data have known properties, the model's output can serve as a "gold standard" for testing algorithms developed to estimate such properties; indeed, a physiological gold standard can rarely be found [22]. A final important goal of modeling and simulation is a general development in biological and medical education, namely the use of simulations for educational purposes and to test the knowledge and insight acquired by the learner [73].

A model can approach reality at different levels. First, a model may have a *descriptive* character with limited validity. The presumed proportionality between EMG amplitude and muscle force can be considered an example of a descriptive model. Second, a model can operate at a *phenomenological* level (i.e., the model output mimics real phenomena under a wide range of conditions, but the model is not necessarily coupled to any underlying "real world" elements evoking the observed outcomes. Autoregressive models and models that are using adaptation of the parameters of a nonstationary stochastic process to generate EMG like signals [68] are of this kind. Models of this type are not primarily meant to give better understanding of the phenomena behind the EMG signal, they are rather used to reduce the information content in the signal on the basis of a limited number of model variables or parameters not necessarily associated to physiological events (see Section 8.2). Third, a model can be *structure based* or *structural*, which means that it describes elements of the real system's structure and takes them into account in a reductional way in order to represent the system's elements, features or mechanisms that are important for a specific purpose (e.g., investigation of a particular effect or phenomenon). Although descriptive and phenomenological models are widely used in EMG research, in this chapter we will mainly focus on structure-based type of models.

An EMG model may be based on an "integrative" analytical approach or on an approach dividing the system physically in small elements (e.g., finite elements). In the first case, the relations between variables are given by equations. In the finite element case, the volume conductor can be divided into a fine grid of elements whereby boundary conditions are defined at the edges of these elements [53]. The first case implies simplifications and assumptions (e.g., homogeneous tissue layers), whereas in the second different elements may have different properties. These inhomogeneities may be simulated at the price of a much higher computational cost. Most available EMG models are of the first type and only these will be considered in this chapter.

Both intramuscular and surface EMG models have been described in the literature. Some concepts of EMG signal modeling apply to both invasive and noninvasive conditions while some others may be specific for the first or the second. Modeling of the source, in case of needle or wire EMG, is usually based on experimental or mathematically described MUAPs. In principle, both intramuscular and surface EMG signal modeling imply the description of the source, of the effect of the volume conductor (see Chapter 4), and the generation of the motor unit firing patterns. However, the description of the detection system is very different in the two conditions. Some aspects of the simulation may be more or less important in the two conditions; for example, a very accurate description of the volume conductor, including tissues of different conductivities and complex muscle architectures may be relevant for surface EMG simulation but is less important for generating synthetic intramuscular recordings. Moreover detailed modeling of signal features is usually more important for surface EMG because, in this case, many physical quantities are not accessible and only modeling can assess their effect on signal characteristics. Modeling of intramuscular recordings has been used for testing algorithms for the decomposition of EMG signals into the constituent motor unit action potential trains (see Chapter 3).

In this chapter we describe general concepts of simulation of EMG signals that apply to both intramuscular and surface recordings; however, most of the applications we present will refer to surface EMG.

8.2 PHENOMENOLOGICAL MODELS OF EMG

At relatively high level of muscle contractions the EMG detected with one electrode pair resembles a noise that has a Gaussian or Laplacian [9] probability density function and is bandlimited to 10 to 400 Hz. A colored noise of this type can be easily generated with a computer either in stationary or nonstationary modalities. Many of its features (e.g., its power spectral density and the statistical moments) may be calculated theoretically and used as a gold standard to test the accuracy of algorithms designed to identify activation intervals, mean and median frequency, average rectified value, and root mean square value (see Chapter 6). Examples of this approach were very common in the 1990s where many authors (Merletti and Lo Conte [52,56], among many others) used the power spectral density shape proposed by Shwedick [68] or autoregressive models to generate or represent EMG signals with known features. These approaches have been (and still are) very useful to assess the performances of algorithms designed to estimate amplitude or spectral variables but do not provide a faithful representation of the EMG generating mechanism. They do not allow to simulate changes of physiological parameters, such as the geometry of MUs, the propagation velocity of the depolarized zones, the MU's depth, the angle between fiber and electrode directions, and MU synchronization. In particular, these models are no longer valid at low contraction levels when the EMG is random, due to the random nature of the innervation processes, but is not Gaussian and is the summation of a sparse pattern of MUAP firings.

8.3 ELEMENTS OF STRUCTURE-BASED SEMG MODELS

Surface EMG recorded during voluntary muscle activity can be considered a signal where the contributions of all active motor units (MUs) are intermingled in a so-called interference pattern. The term "interference" suggests that the contributions of the individual MUs can barely be recognized in the signal because of the extensive overlapping taking place at high contraction levels. Nevertheless, the surface EMG signal global characteristics are largely dependent on the properties of the contributing MUs, their firing patterns, and their interdependence. A complete model for the EMG regards the interference pattern as a summation of the motor unit action potential (MUAP) trains. Each MUAP train can be described as the mathematical convolution of the firing instants (a sequence of Dirac functions) with the MUAP wave shape. An EMG model that describes the interference pattern should therefore consider both the firing behavior of the activated MUs and the MUAP wave shapes. MUAP wave shapes themselves are, in turn, part of the modeling effort. The MUAP characteristics (i.e., shapes and distribution of amplitude and duration) are determined by morphological and functional properties of the activated MUs, together with passive and active properties of the depolarized fiber membranes (sources) and surrounding tissue (volume conductor).

The firing patterns reflect the motor control of the central nervous system and can also be incorporated in an EMG model. One of the intriguing and complicating factors from a modeling point of view in the firing sequence behavior is the so-called short time synchronization (STS) between mutually active MUs, revealed by cross-correlation methods [29]. The subpopulation of recruited MUs and their firing behavior reflect the motor control strategy of the central nervous system.

It should be realized that synchronization makes the contributions of different MUs interdependent. An "extreme" form of synchronized activity between MUs occurs when the MUs are simultaneously electrically activated via their motor axons or their terminal branches. This is a frequently applied technique in clinical neurophysiology to test motor nerve conduction velocity, but it is also suitable to test fatigue behavior and other muscle properties. This form of activation, leading to the so-called compound muscle action potentials (CMAPs) or M-wave, can be implemented in a surface EMG model [57,58].

The major model elements concerning the sources, that is, the MUAP wave shapes, the voluntary MU firing, the activation patterns, and the recording configuration (described in Fig. 8.1), will be discussed in the next sections. We will call a "complete"(structural) EMG model a model in which both the aspects of MUAP shape and

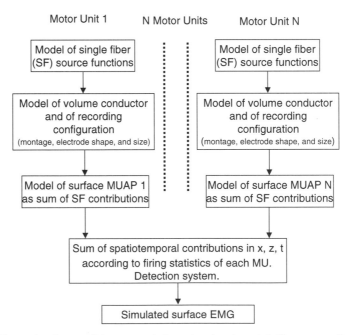

Figure 8.1. Example of a possible approach to a structural model. The source, the volume conductor, and the electrode transfer functions are regarded separately. Results from the surface contributions of the individual sources are computed and added to form the MUAP's contributions and to obtain the EMG signal (in time) from which finally the relevant variables may be computed.

those concerning the activation patterns are incorporated. As elucidated below, for a number of questions, "partial" models with limited complexity will be useful to deal with limited aspects of the muscle structure and function and/or part of the EMG signal generation mechanisms.

8.4 BASIC ASSUMPTIONS

To reduce the inherent complexity of the theory of muscle potential fields in the model implementation, three important assumptions are usually made (see also Chapter 4).

First, it is assumed that the measured potential distribution is the linear summation of the potential distributions of the contributing muscle fibers. The governing laws of electrostatics support this assumption under the condition that the volume conductor properties are not modified by the currents flowing trough it.

Second, it is assumed that muscle-fiber sources do not interact (ephaptic transmission) at the level of the sarcolemma or at the level of the terminal motor nerve branches. It should be noted that this assumption can be questioned for some conditions, for instance, muscle cramps [46].

A third general assumption is that the problem can be described quasi-statically; that is, the potential field at any moment is determined by the sources at that same instant. Only the movement and the intensity changes of the bioelectric sources cause the changes of the potential field in time, not the electrodynamic phenomena nor the possible non-resistive elements in the tissue. No definitive answer has been given yet whether this assumption holds for surface EMG, although it has been made plausible that for high frequencies (over several kHz), as are present in EMG recorded with needle or wire electrodes, a pertinent influence of capacitive aspects of the muscle tissue has to be expected [34,65,76]. As far as we know, it has not been clarified if these influences, observed extracellularly close to the source, also influence the surface EMG noticeably.

8.5 ELEMENTARY SOURCES OF BIOELECTRIC MUSCLE ACTIVITY

The sources of EMG can be considered at different levels, ranging from the membrane channel phenomena to the single-fiber depolarized zone to the MUAP. Only the last two are used in existing models.

8.5.1 The Lowest Level: Intracellular Muscle-Fiber Action Potentials

The wave shapes of intracellular muscle-fiber action potentials (IAPs), taken from experimental or from theoretical work are often used as a basis for surface EMG modeling. The initiation, propagation, and extinction of membrane activity along the muscle fiber's external membrane (the sarcolemma) is usually the "lowest" level at which the source of EMG is considered in EMG models. This can be regarded as neglecting a large part of the knowledge behind EMG generation.

The complex function of transmembrane ion channels, and their structure in nervous and muscle tissues, has triggered many investigations, mostly because the (dys)function of such channels is closely linked to a large set of neurological and neuromuscular disorders [2]. Mathematical modeling of the cell membrane electrophysiology is an accepted macroscopic way to incorporate knowledge at lower levels [77]. Although models describing ion channel and T-system phenomena are being investigated, models that integrate the whole range of electrophysiological processes between the ion channel and the EMG have not been described yet. Whether the integration of models at largely diverging levels will increase our understanding can be questioned both in general and in this specific case. We think that it is scientifically valid to define the intracellular muscle-fiber action potential profile as a useful basic input for surface EMG modeling, without further investigation of the mechanisms producing such profile. In turn, detailed models of this basic input function can be developed separately, and their results may lead to different representations of such source.

8.5.2 The Highest Level: MU Action Potentials

Extending the preceding reasoning, it can be stated that at least knowledge on MUAP wave shapes as the sources of activity at the muscle level should be available in a structural EMG model, since MUAP wave shapes largely determine the EMG of an active muscle. MUAPs can be predicted starting from the level of IAPs. Although not trivial, for needle-EMG, MUAPs can be separated experimentally with special algorithms (for a review, see [71] and Chapter 3). Motor unit potentials can be isolated from the sEMG by so-called spike-triggered averaging using a needle or a highly selective electrode for detecting the firing events [17,19]. Recently the direct decomposition in single MU contributions from an sEMG signal has been presented [17,48,79] (see also Chapter 7). Predicted or experimentally isolated MUAP wave shapes can be used as building blocks of the interference EMG pattern by taking recruitment strategies, firing patterns, and synchronization levels as additional model parameters.

8.6 FIBER MEMBRANE ACTIVITY PROFILES, THEIR GENERATION, PROPAGATION, AND EXTINCTION

To describe the extracellular field, either the membrane voltage or the membrane current density may be considered to be the field source. A mathematical description of the action potential $V_m(z)$ was suggested by Rosenfalck [64]. A modified version of this description, with the added scaling factor λ is reported below in Eq. (1) together with the membrane current distribution that is proportional to the second derivative of $V_m(z)$ (see Chapter 4 and [1]). For detection electrodes at a sufficient distance from the source (as in surface EMG), this model can be further approximated as a current tripole, which has interesting properties. If the three current point sources are placed in the centroid of the three phases and have intensities proportional to the areas of such phases, their sum and the sum of their moments with respect to any of the three application points are zero. This approximation is described in Figure 8.2 and the tripole equations connected to Rosenfalck's model are reported in Eqs. (2) and (3):

$$Vm(z) = Az^3e^{-\lambda z} - B \quad \text{and} \quad I_m(z) = \frac{\sigma_i a}{2}\frac{\partial^2 Vm(z)}{\partial z^2} \propto \lambda z(6 - 6\lambda z + \lambda^2 z^2)e^{-\lambda z} \tag{1}$$

$$I_1 + I_2 + I_3 = 0 \tag{2}$$

$$I_2\, a + I_3\, b = 0 \tag{3}$$

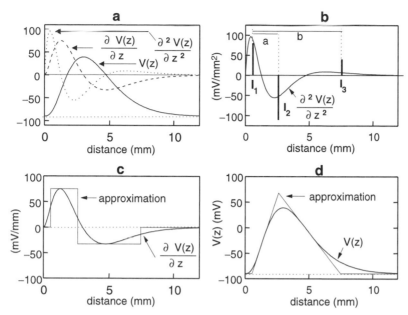

Figure 8.2. (a) Plot of the membrane potential and of its first and second derivatives (normalized values) according to the model defined by Eq. (1); (b) simplified current tripole (I_1, I_2, I_3) representation of the transmembrane current; (c) first derivative of the membrane voltage and its approximation; (d) membrane voltage and its approximation. The horizontal axis denotes the position along the fiber's long axis. Division of this scale by the conduction velocity provides plots of the same variables versus time.

where z is the coordinate along the fiber, V_m is the membrane voltage, A, B, and λ are appropriate coefficients, a is the fiber diameter and σ_i is the intracellular medium conductivity. See Figure 8.2 and Figure 8.4 for the definition of I_1, I_2, I_3, a, and b. This approach allows describing each source with only three numbers.

Remarkably enough, these simple relations between transmembrane and intracellular wave shapes could not be confirmed by direct transmembrane current measurements [4]. More complex descriptions of the intracellular action potential, for example, including the small and slow repolarization tail, have been proposed [13,55].

The initiation (at the neuromuscular junction) and the extinction (at the tendon) of the single fiber action potential (SFAP) present a particular challenge. The exact mechanisms are not well known, but a valid and straightforward description in terms of the sources appears to be possible (e.g., [35]).

If the tripole simplification is adopted, the three poles are supposed to be coincident at the neuromuscular junction at the moment of excitation. The first pole then moves in discrete steps for distance a, the second then moves with the first, and finally the entire tripole is formed and moves as a rigid body. The same happens in a mirrorlike fashion in the other direction. At the muscle-tendon junction the opposite happens: pole 1 stops and poles 2 and 3 approach it until the tripole becomes a dipole, the dipole then narrows until the three poles overlap and cancel out. This approach in discrete space and time [3,57], may be replaced by a faster and more straightforward one where the entire current distribution is considered as the input to a filter and convolved with its impulse response. Moreover the same operation can be replaced by a multiplication of the Fourier transforms of

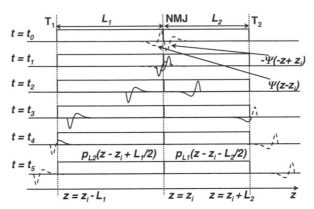

Figure 8.3. First derivative of the two intracellular action potentials propagating in opposite directions along the muscle fiber as a function of space for six different instants of time corresponding to the pre-generation situation (t_0), the generation (t_1), the propagation (t_2), the extinction of the first source (t_3), the extinction of the second source (t_4), and the final situation (t_5). Dotted lines represent the parts of the sources not included in the calculation of the surface signal. T_1 and T_2 represent the tendons. z_i is the position of the NMJ, L_1 and L_2 are the lengths of the fiber from the NMJ to tendon T_1 and T_2, respectively. (Reproduced from [25] with permission)

the two functions, further increasing computational efficiency. Dimitrov and Dimitrova [14] assumed a progressive appearance of the (continuous) first derivative of the action potential instead of the second. Farina and Merletti [25] adopted a similar approach (see Chapter 4). This way a current density source traveling at velocity v along a finite length fiber can be analytically described in space and time domain as (see Fig. 8.3 and Chapter 4)

$$i(z,t) = \frac{d}{dz}\left[\psi(z - z_i - vt)p_{L_1}\left(z - z_i - \frac{L_1}{2}\right) - \psi(-z + z_i - vt)p_{L_2}\left(z - z_i + \frac{L_2}{2}\right)\right] \quad (4)$$

where $i(z,t)$ is the current density source, $\psi(z) = dV_m(z)/dz$, $p_L(z)$ a function which takes value 1 for $-L_1/2 \leq z \leq L_2/2$ and 0 otherwise, z_i the position of the end plate, L_1 and L_2 the length of the fiber from the end plate to the right and to the left tendon, respectively. Equation (4) represents the derivative of the first derivative of the two IAPs, which are windowed to simulate the initiation and extinction at the neuromuscular junction and tendon regions.

Two main approaches may be adopted to simulate the propagation of each source and to compute its contribution to the surface potential. According to the first approach, the source may be displaced by an amount corresponding to an elementary time interval. Then its contribution to the superficial potential distribution is calculated and added to that of the other sources. The process is repeated for each elementary time increment to obtain the surface potential distribution in space and time [3,57]. Alternatively, the Fourier transforms of the source and of the transfer function of the tissue filter may be multiplied to obtain the 2-D Fourier transform of the surface potential. This approach operates in the spatial frequency domain and allows a considerable acceleration by considering that only a section of the 2-D surface potential distribution is of interest: that is the section including the point electrodes. Such a section may be obtained by integration of the 2-D Fourier transform in a particular direction and subsequent 1-D inverse Fourier transform [25]. On the other hand, the second approach does not allow simulation of fibers that are not parallel to the skin plane while the first one does (with approximations due to the direction of anisotropy, as discussed below).

8.7 STRUCTURE OF THE MOTOR UNIT

8.7.1 General Considerations

The previous sections describe the components of a single muscle fiber action potential as electrical source. This conceptualization has to be followed by a description of the collective activity of muscle fibers in a MU. This step from single fiber to MUAP is essential and not trivial because of the variation of position of single fibers within the territory of the MU they belong to. In other words, the MUAP of a MU with N fibers is not simply the SFAP wave shape with an N times larger amplitude. Rather, it is the summation of filtered versions of the individual source inputs, each applied to a different filter.

In general, MU are assumed to be made by parallel fibers, scattered in the MU territory, with the same conduction velocity, with neuromuscular junctions scattered within the width of the innervation zone and terminations scattered within the width of the termination zone, as indicated in Figure 8.4. However, it is possible to incorporate in the MU model, fibers with different orientation, and conduction velocity. If the fibers of one motor unit or those of different motor units are not parallel to each other or are grouped in different directions, as is the case in a pennated muscle, the problem of a complex anisotropy arises. This is because the muscle can no longer be considered as a medium with only two conductivities in the longitudinal and radial (or transversal) direction, respectively. Analytical models available at this time do not account for this possibility that must be addressed by finite element models.

The differences between individual fibers have consequences for the spatial and temporal properties of their SFAP contributions. In this context, anatomical regions such as (1) end plate (or neuromuscular junction) location, (2) fiber endings, (3) relative positions of muscle fibers within a MU, and (4) the location of MUs within a muscle are all significant. These anatomical features in conjunction with propagation in the terminal motor axon branches and the (often neglected) distribution of propagation velocities of fibers within a MU have an effect on the MUAP shape and should be investigated. Various models with different degrees of complexity have been suggested along these lines of reasoning [3,5,35,37,38,57,58,61].

8.7.2 Inclusion of Force in Motor Unit Modeling

The prediction of the force added by a MU or by a single muscle fiber is functionally the final property that may be included in a muscle model [32,70]. Although many theories and models regarding the muscle's mechanical properties have been described, it is difficult to acknowledge the state of the art in both the EMG and muscle force domains in a single comprehensive model [45,80]. Since most information related to MUs cannot be gained by methods other than the interpretation of EMG recordings, it is especially difficult to obtain independent confirmation for the modeling choices made at the single MU level. Only a limited number of animal experiments [38], advanced electrophysiological techniques using multi-electrodes [7], and scanning EMG [36,69] are available to support

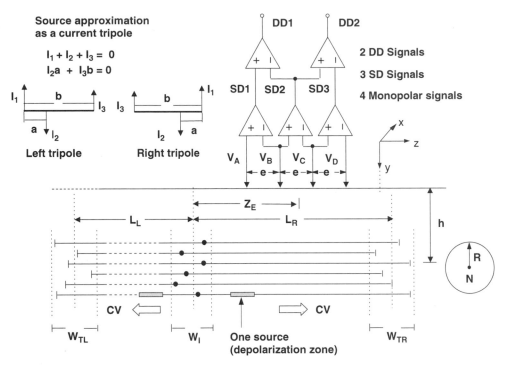

Figure 8.4. A possible model of one motor unit (MU) and the detection system. The electrode system may be at an angle with respect to the z axis. The MU has *N* fibers uniformly distributed in a cylinder of radius *R* at depth *h*. The axis of this cylinder may have an angle, with respect to the skin plane and another angle with respect to the z axis, to allow simulation of pennated muscles. However, since muscle anisotropy is considered only in the directions of the coordinate system, an error is introduced if these angles are too large. The neuromuscular junctions are uniformly distributed in a region W_I and the fiber-tendon terminations are uniformly distributed in two regions W_{TR} and W_{TL}. A right and a left current tripole originate from each neuromuscular junction and propagate to the fiber-tendon termination, where they become extinguished. The conduction velocity is the same in both directions and for all fibers of a MU but may be different in different MUs. Each of the voltages V_A, V_B, V_C, and V_D is the summation of the contributions from all the sources. (Reproduced from [57] with permission)

any modeling choice at a MU level. The issue of EMG–force relationship is partially addressed in Chapter 4 and is still open.

8.8 VOLUME CONDUCTION

8.8.1 General Considerations

Tissue acts as a "volume conductor" of electric (ion flow) current, as is elucidated in Chapter 4. Volume conduction is the reason why extracellular electrophysiological measurements are possible at some distance from the actual source. However, volume conduction is a complex and often a counterintuitive phenomenon, which can hardly be understood quantitatively without modeling. For example, consider the following questions: Is a more conductive skin going to increase or decrease the amplitude of the surface EMG signal? Is it going to increase or decrease crosstalk? The answers to such questions often appear to be counterintuitive. Moreover the volume conduction mechanism is not the type of problem that the average neurologist or neurophysiologist is investigating with priority. Clinical investigators are interested in neurophysiological processes, not in the laws of electrodynamics or electrostatics. Nevertheless, many biophysical questions that might be answered by observing electrophysiological signals remain unanswered without proper knowledge of volume conduction and its implementation in models. The lack of intuition for volume conduction phenomena might well be the main reason for misinterpretation of the surface EMG signal and, thus, the main reason for model developments in EMG. We therefore will elaborate on this aspect in some detail expanding the basic biophysical concepts provided in Chapter 4.

8.8.2 Basics Concepts

The basic physical laws describing the potential distribution in an infinite linear volume conductor, where all phenomena are quasi-static and the principle of superposition applies, are Poisson's and Ohm's equations, given in Chapter 4.

Many investigators have assumed a hemispace isotropic medium (air in the other hemispace), others have assumed an anisotropic medium extending to the boundary skin plane (separating tissue from air). Both approximations neglect the presence of layers of subcutaneous and skin tissues which have different characteristics and likely are isotropic (although not very homogeneous, as observed by Schneider et al. [66]). The effect of these layers has been investigated and accounted for in a number of recent works both in rectangular [25,28] and in cylindrical coordinates [3,35]. The boundary conditions to be derived from these physical constrains allow the solution of Poisson's equation in the different media. More refined models account for the finite volume conductor resulting from the finite dimensions of the limb [3] and the presence of the bone.

At this time most EMG models are analytical models for which a relatively simple coordinate system must be defined (e.g., Cartesian or cylindrical), which means that they are based on analytical mathematical expressions (e.g., Eqs. (1) and (4)) which can be evaluated with computers. Alternatively, purely numerical methods can be used, for which the volumes or their boundary surfaces are covered with the necessary precision with a grid of points where the electric potential will be calculated (finite elements method, boundary elements method) [53]. More complex source and/or volume conductor configurations can only be dealt with using the latter type of approach.

Despite the increasing power of computers, it is still wise to consider the computational load. This is of importance especially in inverse model applications where the most frequently applied method is a repeated use of a forward model with advanced parameter optimization algorithms [75]. Care should be taken that calculations are made in an efficient order, so as to avoid unnecessary repetitions. Often such considerations of efficiency increase insight in the problem to solve.

8.9 MODELING EMG DETECTION SYSTEMS

The basic concepts concerning the EMG detection methods, electrode montages and the effect of electrode size and location have been introduced in Chapter 4 and in Chapter 7.

These concepts are now expanded and considered from the point of view of a modeling approach.

8.9.1 Electrode Configuration

Different detection systems [16] perform spatial filtering of the surface EMG signal by linear combination of the signals detected by the different individual electrodes. To simulate such systems, it is necessary to generate and combine the signals detected at the different electrodes as indicated in Chapters 4 and 7. A considerable simplification derives from the assumption that very small electrodes are point like. When this is not possible, as in the case of a large or a ring-shaped electrode, it is assumed that the potential detected is the average of the potentials present on all points under the electrode surface (see below). A computationally faster approach is to incorporate the spatial filter introduced by the electrodes with the volume conductor transfer function in a spatial frequency approach. In general, the weighted average of the signals detected by a 1-D or 2-D array of point electrodes can be viewed as the application to the potential in spatial domain of a 1-D or 2-D finite impulse response (FIR) filter [16]. The spatial transfer function of such systems depends on the number of electrodes, interelectrode distances, and weights given to the electrodes. The transfer function is thus expressed as

$$H_{sf}(k_x, k_z) = \sum_{i=-m}^{n-1} \sum_{r=-w}^{h-1} a_{ir} e^{-jk_x i d_x} e^{-jk_z r d_z} \tag{5}$$

where $(w + h)$ is the number of rows of the matrix, $(m + n)$ the number of columns, a_{ir} the weights given to the electrodes, d_x the interelectrode distance in the x direction and d_z the interelectrode distance in the z direction, and k_x and k_z are the spatial angular frequencies. The case of nonconstant interelectrode distance in the two directions is a simple extension of Eq. (5). The Fourier transform of the potential detected over the skin by a detection system with transfer function given by Eq. (5) is simulated by multiplying the Fourier transform of the source by the transfer function of the volume conductor and by the transfer function of the detection system.

8.9.2 Physical Dimensions of the Electrodes

The effect of electrode physical dimensions on surface EMG has been described in modeling studies as an integration of the potential over the area under the electrode [40]. With this assumption, which neglects the effect of the electrode on the potential distribution in the area surrounding the electrode, Fuglevand [33] approximated the integral operation by the average of the potentials detected under the electrode area by closely spaced point electrodes (numerical integration). The same approach was suggested by other authors, although it is time-consuming and approximated. Helal and Bouissou [40] addressed the problem for the first time in frequency domain by deriving an equivalent transfer function describing the effect of electrode area. They limited their analysis to a circular electrode and obtained a 1-D spatial transfer function (in the direction of the fibers). Their contribution was important in understanding that spatial integration can be modeled by a linear filtering operation. The main limitation of their approach (which led to theoretical contradictions) was that their analysis was done in only one spatial dimension, while a circular electrode affects the spectral content of a 2-D potential in both directions. Dimitrov and Dimitrova [14] included the effect of electrode size in their model but limited the analysis to rectangular electrodes. Their approach, already proposed by Griep et al. [37], does not introduce approximations and is based on direct analytical integration of the potential over the skin in the two dimensions.

Farina and Merletti [25] proposed a model for surface EMG simulation based on the description of all the spatial and temporal phenomena in the generation of the surface MUAPs as multi-dimensional filtering operations in space and time domain. They describe the effect of electrode size by a 2-D spatial transfer function, as is done for the volume conductor and the combination of the signals detected by different electrodes. It has been shown [20] that the same approach can be used to theoretically describe new classes of spatial filters based on electrode physical dimensions.

With the 2-D spatial filtering approach it is possible to simulate the effect of any electrode shape with low computational cost. It is sufficient to recognize that the integration operation can be viewed as the 2-D convolution of the potential in space domain with a spatial filter with impulse response given by

$$h_{\text{size}}(x, z) = \begin{cases} \dfrac{1}{S} & f(-x, -z) \leq 0 \\ 0 & f(-x, -z) > 0 \end{cases} \tag{6}$$

where S is the electrode area and $f(x, z) < 0$ the mathematical expression which defines the electrode shape.

From Eq. (6) the corresponding transfer function is obtained analytically, when possible, or by numerical computation of the 2-D Fourier transform $H_{\text{size}}(k_x, k_z)$ of $h_{\text{size}}(x, z)$. This way the effect of the electrode size and shape is simulated with a 2-D spatial filter in the same way as the volume conductor and the electrode configuration. For example, in case of rectangular electrode with dimensions a and b in the x and z directions, respectively, we obtain:

$$H_{\text{size}}(k_x, k_z) = \frac{\sin(k_x a/2)}{k_x a/2} \frac{\sin(k_z b/2)}{k_z b/2} \tag{7}$$

In the case of circular electrode with radius r (Fig. 8.7):

$$H_{\text{size}}(k_x, k_z) = \frac{2}{r} \frac{J_1(r\sqrt{k_x^2 + k_z^2})}{\sqrt{k_x^2 + k_z^2}} \tag{8}$$

where $J_1(k)$ is the Bessel function of the first order. And, in the case of elliptic electrode with axes $2a$ and $2b$ in the x and z directions, respectively:

$$H_{\text{size}}(k_x, k_z) = 2 \frac{J_1(\sqrt{(ak_x)^2 + (bk_z)^2})}{\sqrt{(ak_x)^2 + (bk_z)^2}} \tag{9}$$

Note that, as expected, the circular electrode has a transfer function that is invariant to rotations (isotropic).

All the expressions above are approximated because they do not account for the changes of potential distribution (outside the electrode area) consequent to the presence of the electrode itself. In the practical situation of surface EMG, such influence is rather limited because of the relatively (compared to the tissue impedances) high electrode–skin impedance and the location of the electrode at the border of the volume conductor. In the case of an intramuscular needle electrode, its presence as a inhomogeneity in the volume conductor may have a substantial influence [72]. Figure 8.5a shows the magnitude of the transfer function of a large circular electrode. The low pass filtering effect is evident. When applied to a simulated EMG signal produced by a single fiber, this transfer function leads to the two-dimensional amplitude spectrum depicted in Figure 8.5b (monopolar detection). Figure 8.5c shows the amplitude spectrum for different lateral displacements between electrode and fiber and for positive spatial frequencies in the z direction.

In case of detection with a 1-D or 2-D array of physical electrodes the total transfer function of the detection system is obtained by combining Eq. (6) and Eq. (5):

$$H_{\text{ele}}(k_x, k_z) = \sum_{i=-m}^{n-1} \sum_{r=-w}^{h-1} a_{ir} H_{\text{size}}^{ir}(k_x, k_z) e^{-jk_x i d_x} e^{-jk_z r d_z} \tag{10}$$

where $H_{\text{size}}^{ir}(k_x, k_z)$ is the transfer function related to size and shape of the electrode located in position ir of the matrix. Equation (10) describes the very general case in which each of the electrodes of the matrix can have a different shape. Usually all the electrodes of the matrix have the same shape and $H_{\text{ele}}(k_x, k_z)$ is obtained by the multiplication of $H_{\text{size}}(k_x, k_z)$ and the spatial filter transfer function $H_{sf}(k_x, k_z)$ given by Eq. (5). In the frequency domain the transfer function defined by Eq. (10) can be applied as a multiplicative factor to the 2-D transfer function of the volume conductor. If the volume conductor is investigated in the spatial domain, the impulse response corresponding to transfer function (10) must be computed and convolved with the impulse response of the tissue.

The relative inclination between the fibers and the detection system can be modeled by rotating the detection system. Since the detection system is described by the equivalent transfer function $H_{\text{ele}}(k_x, k_z)$, the problem is to find the relationship between the two expressions of $H_{\text{ele}}(k_x, k_z)$ in case of alignment and in case of inclination of the detection system with respect to the fiber direction. Rotating the detection system implies rotating its equivalent impulse response. The impulse response of the spatial filter is a grid of delta functions with proper weights centered at the electrode locations (Chapter 7), while the equivalent impulse response of the shape of the electrodes is a constant 2-D function in the area of the electrodes, Eq. (6). Since the 2-D Fourier transform of the rotation of a 2-D function is the rotation of the Fourier transform of the function, the inclination of the fibers can be included by rotating the transfer function $H_{\text{ele}}(k_x, k_z)$:

$$H_{\text{ele}}(k_x, k_z, \theta) = H_{\text{ele}}(k_x \cos\theta - k_z \sin\theta, k_x \sin\theta + k_z \cos\theta) \tag{11}$$

where θ is the angle of inclination.

8.10 MODELING MOTOR UNIT RECRUITMENT AND FIRING BEHAVIOR

There is a wide variability between muscles with MU recruitment and with MU firing rate coding in obtaining the required muscle force. These differences are related to the specific tasks of the muscle (e.g., fine or coarse motor tasks) and thus to its size and composition. The recruitment and firing behavior of MUs within the muscle must be defined with a model describing the interference pattern. The "classical" paradigm is the so-called Henneman's size principle [32,41,59], which states that with increasing muscle force progressively larger MUs are recruited. Recent studies suggest that this functionally convincing principle is not generally valid. In the arm muscles, for example, the order of

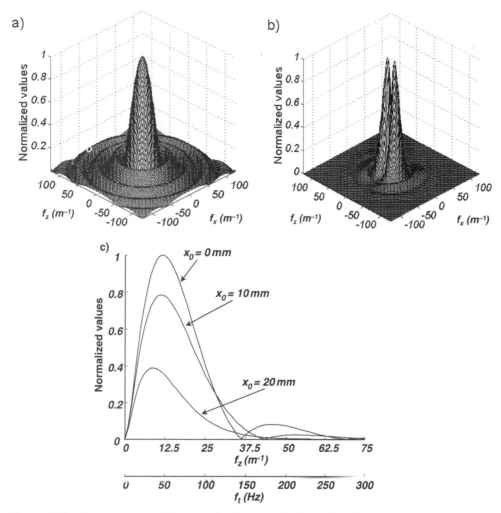

Figure 8.5. Absolute value of the transfer function of a large (for demonstrative purposes) circular electrode with radius equal to 15 mm (*a*), and absolute value of the 2-D Fourier transform of the potential detected over the skin with such a system and generated by a single IAP, approximated with Rosenfalck's analytical expression, at a depth of 2 mm in the muscle directly below the electrode (*b*). The spectrum of the surface signal is modified by a transversal displacement between the electrode and the fiber as indicated in (*c*), where spectra are depicted for lateral distances of the fiber x_0 of 0 mm, 10 mm, and 20 mm from the center of the electrode (same depth y_0 = 2 mm). The time frequency axis in case of CV = 4 m/s is also shown.

recruitment appears strongly dependent on the specific task requirements, especially movement speed and direction [74]. This complex behavior is a challenge for surface EMG analysis and for (inverse) models unraveling the surface EMG interference pattern into the contributions of the individual MUAP trains. A near endless set of activation patterns is possible. The recruitment of each new MU and a change in its firing frequency also has consequences for the mechanical output of the muscle [32,39,80]. Here the Henneman principle is an important modeling factor and, again, with some risk of oversimplification.

In the context of this overview on EMG modeling, only the modeling of the firing patterns in a stationary condition (constant isometric load), assuming a constant activation of the MU pool, will be discussed. The MU firing processes (statistics and mutual dependency of firing events) can then be described by the following parameters:

- Distribution of the interpulse interval statistics of a single MU
- Distribution of the interpulse interval statistics among different mutually active MUs
- Interdependency of the firing instants of the different MUs (synchronization, common drive)

8.10.1 MU Interpulse Intervals

The general principle describing a MU firing process in terms of consecutive interpulse intervals (IPIs) is to consider them as independent samples of a random variable. In the literature various distribution functions have been selected to most appropriately fit the IPI histograms derived from experimental data. In most cases a Gaussian distribution function was confirmed [8,30,31,50,60,63], but also a Weibull distribution [10,12], a Poisson distribution [6], and a gamma distribution [67] were proposed.

Person and Kudina [63] and Kranz and Baumgartner [50] found that at low firing rates the IPI histograms were slightly skewed and became more symmetrical at higher firing rates. Based on the observation that the character of the distribution only has minor influences on the sEMG spectral content [51,78], a Gaussian distribution appears justifiable to model the consecutive IPIs of a MU.

With respect to the variability of the firings, often constant ranges for the variation are chosen, irrespective of the mean value. Typical values for the coefficient of variation reported range from 0.1 to 0.33 (corresponding to a st. dev. of 10–30 ms at 10 Hz).

Sometimes (e.g., [18]) a refinement is implemented in the sense that the standard deviation is related to the mean interpulse interval. This is based on the experimental finding [8] that when a MU is firing at a higher rate, its frequency is more stable.

8.10.2 Mean Interpulse Intervals across Motor Units

Differences in mean IPI of the MUs reflect the differences in level of activity, and consequently the force output of the MUs. Once a MU has been recruited, its firing rate will increase with increasing force. It has been shown that MUs recruited at a higher force level show a higher initial firing rate, although those MUs usually fire at a lower rate than the earlier recruited MUs at the same force level [11]. Roughly, active MUs tend to increase their firing rate approximately linearly with the force output of the muscle.

In contrast to the IPI within a single MUAP train, no systematic assessment is found in the literature to describe the distribution of mean IPIs of the whole pool of active MUs. However, visual inspection of some figures in published articles shows that firing rates of simultaneously active MUs never differ by more than 10 Hz. The variability seems to decrease with increasing force level [11]. At this moment there still is no reliable experimental method available to study the firing patterns of the majority of the MUs in a muscle.

8.10.3 Synchronization

Strongly related to the previous aspects of the firing behavior of the individual motor units is the synchronization of firing instants between the different motor units. After many years of discussion and partly contradicting experimental evidence, the general consensus is that synchronized firing does exist, but the extent of this phenomenon is not clear. The concept of synchronization is related to the likelihood that one or more MU would fire within a specific time window from another MU. This concept may be described by the number of firings that are synchronized and by the number of MU that are involved in each synchronization event [23,79].

Hermens Model. One way of simulating synchronization of the firing events between different MUs is by generating the firing instants of a first train and then linking the firing instants of consecutive trains to them. In early studies [51,62] a constant delay was used, which is mathematically convenient but not very realistic for a biophysical process. Weytjens and Steenberghe [78] and later Hermens [42] used a Gaussian distribution to model firing synchronicity. This distribution is placed on the firings of a first train. Firings of synchronised MUs are then drawn from this distribution function. This choice is based on the experimental work of Kirkwood and Sears [47], who investigated changes in the state of depolarization of one motoneuron at the moment other motoneurons of the same pool were activated.

Figure 8.6 illustrates the effect of synchronized motor unit activity on the median frequency of the simulated surface EMG signal, using the Gaussian distribution function placed around each firing moment of a first (reference) train. When all MUAP trains are involved in synchronous firing ($N_S = 100\%$), an increase of the standard deviation of the distribution function, σ_S, from zero lowers the median frequency (MDF) substantially. However, above a certain value of σ_S, MDF is found to increase again up to its original value. This means that in case of perfect synchronization, such as during electrical stimulation of the muscle, MDF has approximately the same value as without synchronization. Realistically, a temporal dispersion of arrival times will also happen in the case of electrical stimulation of the muscle due to the different delays in the motor axons and muscle fibers, depending on the site of stimulation.

The simulation also predicts that when only part of the motor unit pool is assumed to fire synchronously, not only the effect on MDF is less but also the curve changes its shape; with lower values of N_s the minimum of MDF is reached at lower values of σ_S.

Yao/Fuglevand Model. Recently Yao et al. [80] proposed an alternative method to introduce MU synchronization based on the shift of selected MU firings. By this method, first, the firing patterns are generated independently; then some firings are synchronized in order to minimize the effect of the shift on the statistical properties of the firing patterns. This model has been used to assess the effect of synchronization on surface EMG amplitude and steadiness of force produced [80]. The same approach was recently adopted to analyze the effect of MU synchronization on the features of surface EMG signals extracted by linear and nonlinear analysis techniques [23].

Matthews/Kleine Model. The synaptic input to a particular motoneuron (MN) comes from many different sources that could have different firing distributions within the MN pool. One extreme would be an input that is unique for a particular MN and does not reach the other MNs. The other extreme corresponds to a branching of the axons of the upper MNs to form an evenly distributed projection to the whole MN pool. These two types of inputs lead to the extension of Matthews's model [54] with synchronization induced by splitting the noisy excitatory post synaptic potential input into a "common" and an "individual" component (Fig. 8.7). The common noise represents the input that is shared by all MNs of the pool. Peaks in the common noise will discharge all MNs of the pool that are close to threshold and will therefore tend to synchronize the firings. The indi-

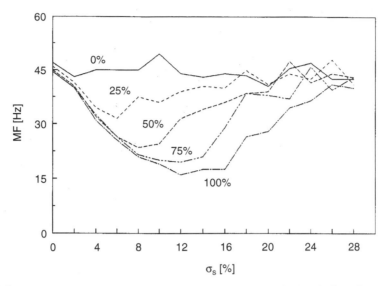

<u>Figure 8.6.</u> The dependence of the median frequency (MF) on the level of synchronization in a MU pool (σ_s is the standard deviation of the "freedom of firing" of the MUs with respect to each other in percentages of the mean firing interval). For $\sigma_s = 0$, a perfect synchronization of MUs is present, corresponding to the case of electrical stimulation. Large σ_s values denote independent firing behavior between the MUs. Note that the median frequency increases for low and for high levels of synchronization. The percentage indicated at the different curves denotes the fraction N_s of MUs being "time connected." (From [44] with permission)

vidual component of noise varies between MNs and represents the input that is unique to a particular MN.

8.11 INVERSE MODELING

In general, the most practically relevant type of modeling is the use of the inverse approach, namely the estimation of the values of model parameters from experimental results. Although the use of inverse modeling in EMG research is still limited, some examples are given in the literature. One of these examples follows from [57,58]. During low-level voluntary isometric contractions individual MUAPs may be observed using linear electrode arrays and classified as belonging to the same MU. The multi-channel patterns, indicated in Figure 8.8a for monopolar detection and in Figure 8.8b for single differential detection from different biceps brachii muscles, may be simulated by searching for the set of MU model parameters that would provide a good fit of the experimental results according to a specific model. By this inverse modeling approach, possible (not necessarily unique) structures of the MU, compatible with the experimental findings, may be identified.

8.12 MODELING OF MUSCLE FATIGUE

8.12.1 Myoelectric Manifestations of Muscle Fatigue during Voluntary Contractions

Myoelectric manifestations of muscle fatigue (described in Chapter 9) are associated with a number of changes that can be observed in the surface EMG signal. These changes can

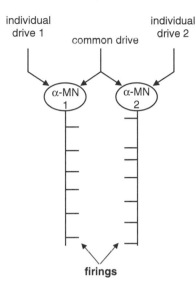

Figure 8.7. Schematic presentation of the model of Kleine et al. [49] (with permission) on the generation of synchronized MU firings. The higher level drive to each α-motoneuron is the combined activity from synapses with a component common to both motoneurons and individual input. The α-motoneuron behavior is based on the concept of Matthews [54] in which the after hyperpolarization following each depolarization of the neuron is essential.

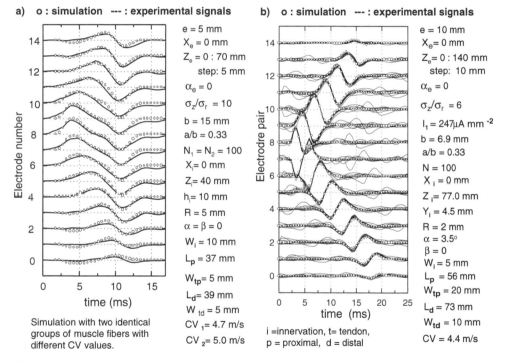

a)   o : simulation   --- : experimental signals

$e = 5$ mm
$X_e = 0$ mm
$Z_e = 0 : 70$ mm
     step: 5 mm
$\alpha_e = 0$
$\sigma_z/\sigma_r = 10$
$b = 15$ mm
$a/b = 0.33$
$N_1 = N_2 = 100$
$X_i = 0$ mm
$Z_i = 40$ mm
$h_i = 10$ mm
$R = 5$ mm
$\alpha = \beta = 0$
$W_i = 10$ mm
$L_p = 37$ mm
$W_{tp} = 5$ mm
$L_d = 39$ mm
$W_{td} = 5$ mm
$CV_1 = 4.7$ m/s
$CV_2 = 5.0$ m/s

Simulation with two identical groups of muscle fibers with different CV values.

b)   o : simulation   --- : experimental signals

$e = 10$ mm
$X_e = 0$ mm
$Z_e = 0 : 140$ mm
     step: 10 mm
$\alpha_e = 0$
$\sigma_z/\sigma_r = 6$
$I_1 = 247\mu A$ mm^{-2}
$b = 6.9$ mm
$a/b = 0.33$
$N = 100$
$X_i = 0$ mm
$Z_i = 77.0$ mm
$Y_i = 4.5$ mm
$R = 2$ mm
$\alpha = 3.5°$
$\beta = 0$
$W_i = 5$ mm
$L_p = 56$ mm
$W_{tp} = 20$ mm
$L_d = 73$ mm
$W_{td} = 10$ mm
$CV = 4.4$ m/s

i = innervation, t = tendon, p = proximal, d = distal

Figure 8.8. (a) Set of monopolar signals from a single MUAP generated during a weak voluntary contraction of a biceps brachii (solid lines) and simulated signals (dots) obtained by assuming two values of CV (4.7 and 5 m/s), each associated to 100 fibers of the motor unit. A possible set of model parameters is indicated in the figure (see Fig. 8.4 for model description and explanation of symbols). Such a set may not be unique. (b) Ten superimposed occurrences of a differentially detected MUAP identified during a weak voluntary contraction of the biceps brachii (dashes) and the corresponding simulated signals (circles). A possible set of model parameters is indicated in the figure (see Fig. 8.4 for model description and explanation of symbols). The set of model parameters may not be unique. See also Figure 8.4. (Reproduced from [58] with permission)

be simulated using a model that allows to control the source, the variation of muscle-fiber conduction velocity, and the changes in the firing behavior of the motor units. With respect to the last item, synchronization of motor unit firings has been mentioned in literature to explain the often found different time course of the muscle-fiber conduction velocity and the median frequency during a fatiguing contraction. Other firing aspects like the mean firing rate were shown to have negligible influence on spectral variables [42,44]. Figure 8.9 provides an illustration of the effect of the duration and shape of the MUAP on the median frequency.

In this simulation different MUAP shapes were extracted from the EMG using a needle triggered averaging of the surface EMG signal. These MUAPs were used to generate an artificial interference EMG signal. As expected, the results show an inverse proportional relation between MUAP duration and the median frequency, whose slope is determined by the shape of the MUAP. Contrary to expectations, many experimental studies show a relative decrease of the median frequency considerably larger than that of mean muscle-fiber conduction velocity. In some cases a decrease of spectral variables may be observed while conduction velocity remains constant. A model becomes extremely useful for the identification of the mechanisms leading to these unexpected behaviors and for their interpretations. As discussed above, changing levels of synchronization between motor units and changes of MUAP shapes may explain these different behaviors.

8.12.2 Myoelectric Manifestations of Muscle Fatigue during Electrically Elicited Contractions

During electrical stimulation the activated MUs behave like a single giant MU where different groups of fibers may have different behaviors. This condition reduces the number

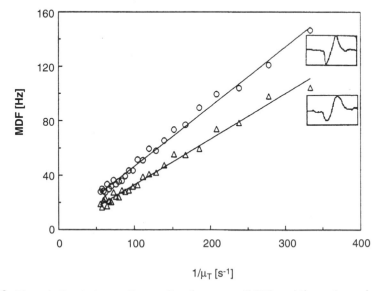

Figure 8.9. The relation between the median frequency (MDF) and the reciprocal value of the mean peak-peak time distribution (μ_T) with two different normalized MUAP shapes that have been extracted from the biceps brachii by means of spike-triggered averaging of the surface EMG signal. (From [42] with permission)

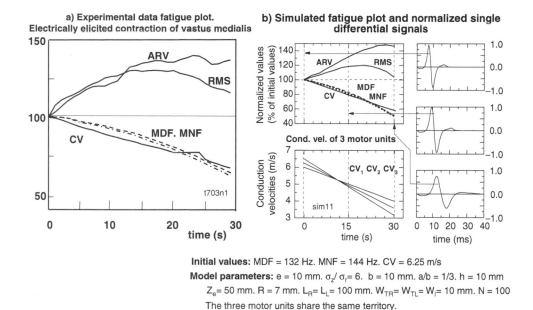

Figure 8.10. (*a*) Fatigue plot of an electrically elicited contraction of a healthy vastus medialis muscle (stimulation frequency: 30 Hz). A peculiar behavior of CV and MNF-MDF may be observed. (*b*) Simulation of the observed behavior using the model described in [57]. First converging, and then diverging, patterns of CV of different MUs provide a situation that explains the experimental results. In the simulation only three MUs are included. The solution may not be unique.

of variables and makes simulation easier. Different time patterns of myoelectric variables are observed in stimulated muscles of different subjects. This fact suggests that a model might outline the differences between such muscles. Different behaviors concerning CV and spectral variables include identical relative decrease of CV, MNF, and MDF; decrease of MNF and MDF greater than that of CV; no decrease or slight increase of CV and decrease of MNF and MDF; opposite concavities in the patterns of CV and MNF-MDF; and so on. Models may be very useful in explaining such behaviors by testing possible mechanisms. Of course, more than one mechanism may fit the experimental data and the answer may therefore be not unique. Figure 8.10 depicts the fatigue plot of an electrically stimulated contraction of the vastus medialis showing a MNF-MDF decrement that is first slower and then faster than that of CV. A possible explanation (which also fits the behavior of the amplitude variables) is provided by the hypothesis of initially converging and then diverging values of CV of different motor units. This hypothesis is tested in Figure 8.10 using the model developed by Merletti et al. [57,58], with only three MUs that represent groups of MUs with similar behaviors. The right part of Figure 8.10 also shows the simulated M-waves resulting from this hypothesis, at the beginning, mid-time, and end of the 30 s long contraction.

Figure 8.11 shows M-waves detected with a linear electrode array placed between the innervation and tendon zones of an electrically stimulated biceps, during a 10 s contraction. The single differential signals show a progressive widening with increasing time and distance from the motor point. The double differential signals (more spatially selective) show a change of M-wave shape suggesting a progressive separation of CV values in two groups of MUs. Simulation according to [57] and [58] provides a good (not necessarily

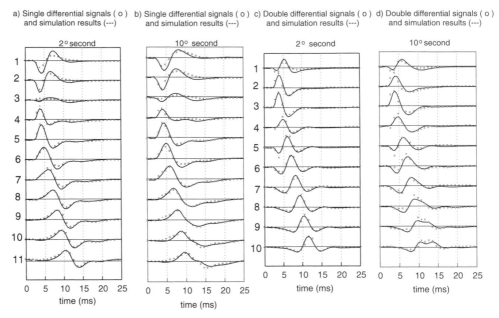

Figure 8.11. Single differential (*a*, *b*) and double differential (*c*, *d*) signals at the beginning (*a*, *c*) and end (*b*, *d*) of a 10s electrically evoked contraction of a healthy biceps brachii (stimulation frequency of 28 Hz). Detection with a 12 contact array with 5 mm interelectrode distance. Electrode pair 1 is proximal and pair 11 is distal. Observe the widening of the SD M-wave from (*a*) to (*b*) but the stability of the zero crossing and the formation of two peaks in the double differential distal traces at the 10th second. Dots are experimental signal sampled at 1024 samples/s. Solid lines are the result of simulation. The model parameters are reported in the table below. Although the set of parameters may not be unique, it is interesting to note that the experimentally observed changes of the M-waves may be explained by assuming only changes of conduction velocity values.

Parameters of the Simulation (see Fig. 8.4 for explanation)

	N	X_l	Z_l	h	R	W_l	L_d	W_{td}	L_p	W_{tp}	CV_2	CV_{10}
		mm	mm	mm	mm	mm	mm	mm	mm	mm	m/s	m/s
MU 1	30	0	14.5	12	10	3	100	40	40	30	4.5	5.4
MU 2	40	0	12.5	12	10	3	100	40	40	30	4.3	3.9
MU 3	55	0	−18.5	17	10	15	100	40	60	60	4.9	4.9

Note: CV_2 and CV_{10} are the CV values at the 2nd and 10th second of contraction. Fibers are parallel to the *z* axis. Tripole width b = 12 mm; asymmetry a/b = 0.33.

unique) fitting of the experimental data assuming two populations of MU with different CV behavior.

8.13 OTHER APPLICATIONS OF MODELING

Surface EMG models have been used in a number of studies to investigate the effects of the system parameters on the properties of the detected surface EMG signals. A number of papers in the literature address the effect of anatomical, physical, and detection-system parameters on surface EMG features.

Fuglevand et al. [33] addressed the problem of interelectrode distance and electrode size, indicating their effects on surface EMG amplitude and spectral characteristics. Fuglevand et al. [32] also addressed the relationship between surface EMG signal amplitude and developed force by EMG modeling. Farina et al. [24] indicated the serious limitations in using surface EMG spectral analysis for the investigation of MU recruitment strategies and clarified controversial experimental findings of the past. Other important issues, such as crosstalk [15,26] and the effect of MU synchronization [23,48,80] on EMG features, have been explained by modeling approaches. The model approach is particularly useful when the experimental separation of different phenomena is difficult or impossible.

Moreover structure-based models are becoming more important for testing complex signal processing techniques for information extraction from surface EMG signals. In biological systems it is of paramount importance to address the sensitivity of a technique to variations of anatomical and physiological properties of the system. A useful example is the development of techniques for the estimation of global muscle-fiber conduction velocity (see Chapter 7). Although a low-estimation variance is necessary in many applications, this aspect is not the only one characterizing the estimation method [21]. In experimental conditions there are many nonidealities that introduce a bias in the estimation and depend on the estimator applied. These aspects can only be assessed by structure-based models. Farina et al. [27] showed, by using a structure-based model, that optimal (minimal variance) CV estimation algorithms are not the best when bias is introduced, for example, by the inclination of the muscle fibers with respect to the detection systems.

An important application of modeling is in teaching and training health professionals. The availability of a "virtual" muscles that can answer questions of the type "What if . . . ?" is of great value for muscle understanding and teaching. Different models may be suitable for answering different questions. The efforts toward modeling and model dissemination promoted by the European Concerted Action, "Surface EMG for Non-invasive Assessment of Muscles" (SENIAM) [43], provided an important contribution and should be considered with great attention by professors of neurophysiology, health scientists, and EMG clinical researchers. Four different models are available, among other materials, on CD-ROM to the scientific and clinical community by SENIAM.

8.14 CONCLUSIONS

This chapter deals with the properties of structure-based models describing sEMG patterns and the assumptions behind them. Since the mid-1970s a considerable number of papers on the use and development of EMG models were published. Models are particularly helpful in obtaining insight into the basic elements of surface (and needle) EMG characteristics. In general, what is involved is the search for a uniform approach. Considering the variety of questions posed to the EMG and thus to the models, such a goal will probably never be reached in full. The present overview is meant to give some background and suggest the possibilities of sEMG models and their limitations.

REFERENCES

1. Andreassen, S., and A. Rosenfalk, "Relationship of intracellular and extracellular action potentials of skeletal muscle fibers," *CRC Crit Rev Bioeng* 267–305 (1981).

2. Ashcroft, F. M., *Ion channels and disease*, Academic Press, San Diego, 2000.

3. Blok, J. H., D. F. Stegeman, and A. van Oosterom, "A three-layer volume conductor model and software package for applications in surface electromyography," *Ann Biomed Eng* **30**, 566–577 (2002).

4. Boom, H. B. K., and W. Wallinga, "Source characteristics from inverse modelling of EMG signals," in I. Gath and G. F. Bar, eds., *Advances in processing and pattern analysis of biological systems*, Plenum Press, New York, 1996, pp. 319–338.

5. Boyd, D. C., P. D. Lawrence, and P. J. Bratty, "On modelling the single motor unit action potential," *IEEE Trans BME* **25**, 236–243 (1978).

6. Brody, G. R., N. Scott, and R. Balasubramanian, "A model for myoelectric signal generation," *Med Biol Eng* **12**, 29–41 (1974).

7. Buchthal, F., C. Guld, and P. Rosenfalck, "Multielectrode study of the territory of a motor unit," *Acta Physiol Scand* **39**, 83–104 (1959).

8. Clamann, H. P., "Activity of single motor units during isometric tension," *Neurology* **20**, 254–260 (1970).

9. Clancy, E. A., and N. Hogan, "Probability density of the surface electromyogram and its relation to amplitude detectors," *IEEE Trans BME* **46**, 730–739 (1999).

10. De Luca, C. J., "Physiology and mathematics of myoelectric signals," *IEEE Trans BME* **26**, 313–325 (1979).

11. De Luca, C. J., and Z. Erim, "Common drive of motor units in regulation of muscle force," *Trends Neurosci* **17**, 299–305 (1994).

12. De Luca, C. J., and W. J. Forrest, "Some properties of motor unit action potential trains recorded during constant force isometric contractions in man," *Kybernetik* **12**, 160–168 (1973).

13. Dimitrov, G. V., "Changes in the extracellular potentials produced by unmyelinated nerve fibre resulting from alterations in the propagation velocity or the duration of the action potential," *Electromyogr Clin Neurophysiol* **27**, 243–249 (1987).

14. Dimitrov, G. V., and N. A. Dimitrova, "Precise and fast calculation of the motor unit potentials detected by a point and rectangular plate electrode," *Med Eng Phys* **20**, 374–381 (1998).

15. Dimitrova, N. A., G. V. Dimitrov, and O. A. Nikitin, "Neither high-pass filtering nor mathematical differentiation of the EMG signals can considerably reduce cross-talk," *J Electromyogr Kinesiol* **12**, 235–246 (2002).

16. Disselhorst-Klug, C., J. Silny, and G. Rau, "Improvement of spatial resolution in surface-EMG: A theoretical and experimental comparison of different spatial filters," *IEEE Trans BME* **44**, 567–574 (1997).

17. Disselhorst-Klug, C. G. Rau, A. Schmeer, and J. Silny, "Non-invasive detection of the single motor unit action potential by averaging the spatial potential distribution triggered on a spatially filtered motor unit action potential," *J Electromyogr Kinesiol* **9**, 67–72 (1999).

18. Duchene, J., and J. Y. Hogrel, "SEMG simulation for MFCV distribution estimates assessment," in H. J. Hermens, G. Hagg, and B. Freriks, eds., *European applications on surface electromyography, Proc 2nd General SENIAM Workshop*, Roessingh Research and Development, Enschede, Netherlands, 1997, pp. 122–128.

19. Farina, D., L. Arendt-Nielsen, R. Merletti, and T. Graven-Nielsen, "A spike triggered averaging technique for high resolution assessment of single motor unit conduction velocity during fatiguing voluntary contractions," *J Neurosci Meth* **115**, 1–12 (2002).

20. Farina, D., and C. Cescon, "Concentric ring electrode systems for non-invasive detection of single motor unit activity," *IEEE Trans BME* **48**, 1326–1334 (2001).

21. Farina, D., C. Cescon, and R. Merletti, "Influence of anatomical, physical and detection-system parameters on surface EMG," *Biol Cybern* **86**, 445–456 (2002).

22. Farina, D., A. Crosetti, and R. Merletti, "A model for the generation of synthetic intramuscular EMG signals to test decomposition algorithms," *IEEE Trans BME* **48**, 66–77 (2001).

23. Farina, D., L. Fattorini, F. Felici, and G. Filligoi, "Nonlinear surface EMG analysis to detect changes of motor unit conduction velocity and syncronization," *J Appl Physiol* **93**, 1753–1763 (2002).

24. Farina, D., M. Fosci, and R. Merletti, "Motor unit recruitment strategies investigated by surface EMG variables," *J Appl Physiol* **92**, 235–247 (2002).

25. Farina, D., and R. Merletti, "A novel approach for precise simulation of the EMG signal detected by surface electrodes," *IEEE Trans BME* **48**, 637–646 (2001).

26. Farina, D., R. Merletti, B. Indino, M. Nazzaro, and M. Pozzo, "Surface EMG crosstalk between knee extensor muscles: Experimental and model results," *Muscle Nerve* **26**, 681–695 (2002).

27. Farina, D., W. Muhammad, E. Fortunato, O. Meste, R. Merletti, and H. Rix, "Estimation of single motor unit conduction velocity from surface electromyogram signals detected with linear electrode arrays," *Med Biol Eng Comput* **39**, 225–236 (2001).

28. Farina, D., and A. Rainoldi, "Compensation of the effects of subcutaneous tissue layers on surface EMG: A simulation study," *Med Eng Phys* **21**, 487–496 (1999).

29. Farmer, S. F., D. M. Halliday, B. A. Conway, J. A. Stephens, and J. R. Rosenberg, "A review of recent applications of cross-correlation methodologies to human motor unit recording," *J Neurosci Meth* **74**, 175–187 (1997).

30. Freund, H. J., H. J. Budingen, and V. Dietz, "Activity of single motor units from human forearm muscles during voluntary isometric contractions," *J Neuropsysiol* **38**, 933–946 (1975).

31. Freund, H. J., and C. W. Wita, "Computer analysis of interval statistics of single motor units in normals and patients with supraspinal motor lesions," *Arch Psychiatr Nervenkr* **214**, 56–71 (1971).

32. Fuglevand, A. J., D. A. Winter, and A. E. Patla, "Models of recruitment and rate coding organization in motor-unit pools," *J Neurophysiol* **70**, 2470–2488 (1993).

33. Fuglevand, A. J., D. A. Winter, A. E. Patla, and D. Stashuk, "Detection of motor unit action potentials with surface electrodes: influence of electrode size and spacing," *Biol Cybern* **67**, 143–153 (1992).

34. Gielen, F. L., W. Wallinga-De Jonge, and K. L. Boon, "Electrical conductivity of skeletal muscle tissue: experimental results from different muscles in vivo," *Med Biol Eng Comput* **22**, 569–577 (1984).

35. Gootzen, T. H. J. M., D. F. Stegeman, and A. Van Oosterom, "Finite limb dimensions and finite muscle length in a model for the generation of electromyographic signals," *Electroencephalogr Clin Neurophysiol* **81**, 152–162 (1991).

36. Gootzen T. H. J. M., H. J. M. Vingerhoets, and D. F. Stegeman, "A study of motor unit structure by means of scanning EMG," *Muscle Nerve* **15**, 349–357 (1992).

37. Griep, P. A., K. L. Boon, and D. F. Stegeman, "A study of the motor unit action potential by means of computer simulation," *Biol Cybern* **30**, 221–230 (1978).

38. Griep, P. A., F. L. Gielen, H. B. Boom, K. L. Boon, L. L. Hoogstraten, C. W. Pool, and W. Wallinga-De Jonge, "Calculation and registration of the same motor unit action potential," *Electroencephalogr Clin Neuroph* **53**, 388–404 (1982).

39. Halliday, D. M., B. A. Conway, S. F. Farmer, and J. R. Rosenberg, "Load-independent contributions from motor-unit synchronisation to human physiological tremor," *J Neurophysiol* **82**, 664–675 (1999).

40. Helal, J. N., and P. Bouissou, "The spatial integration effect of surface electrode detecting myoelectric signal," *IEEE Trans BME* **39**, 1161–1167 (1992).

41. Henneman, E., "Recruitment of motoneurons: The size principle. In: Motor unit types, recruitment, and plasticity in health and disease," *Progr Clin Neurophysiol* **9**, 26–60 (1981).

42. Hermens, H. J., "Surface EMG," PhD thesis, Twente University of Technology, 1991.

43. Hermens, H. J., B. Freriks, R. Merletti, D. Stegeman, J. Blok, G. Rau, C. Disselhorst-Klug, and G. Hägg, *European recommendations for surface electromyography*, Roessingh Research and Development, Enschede, Netherlands 1999.

44. Hermens, H. J., T. A. M. Van Bruggen, C. T. M. Baten, W. L. C. Rutten, and H. B. K. Boom, "The median frequency of the surface EMG power spectrum in relation to motor unit firing and action properties," *J Electromyogr Kinesiol* **2**, 15–25 (1992).

45. Hof, A. L., "EMG and muscle force: an introduction," *Hum Mov Sci* **3**, 119–153 (1984).

46. Jansen, P. H., E. M. Joosten, and H. M. Vingerhoets, "Muscle cramp: main theories as to aetiology," *Eur Arch Psychiatry Neurol Sci* **239**, 337–342 (1990).

47. Kirkwood, P. A., and T. A. Sears, "The synaptic connections to intercostal motoneurones as revealed by the average common excitation potential," *J Physiol* **275**, 103–134 (1978).

48. Kleine, B. U., J. H. Blok, R. Oostenveld, P. Praamstra, and D. F. Stegeman, "Magnetic stimulation-induced modulations of motor unit firings extracted from multi-channel surface EMG," *Muscle Nerve* **23**, 1005–1015 (2000).

49. Kleine, B. U., D. F. Stegeman, D. Mund, and C. Anders, "Influence of motoneuron firing synchronization on SEMG characteristics in dependence of electrode position," *J Appl Physiol* **91**, 1588–1599 (2001).

50. Kranz, H., and G. Baumgartner, "Human alpha motorneurone discharge, a statistical analysis," *Brain Res* **67**, 324–329 (1974).

51. Lago, P. J., and N. B. Jones, "Effect of motor unit firing time statistics on EMG spectra," *Med Biol Eng Comp* **15**, 648–655 (1977).

52. Lo Conte, L., and R. Merletti, "Advances in signal processing of surface myoelectric signals. Part 2," *Med Biol Eng Comput* **33**, 372–384 (1995).

53. Lowery, M. M., N. S. Stoykov, A. Taflove, and T. A. Kuiken, "A multiple-layer finite-element model of the surface EMG signal," *IEEE Trans BME* **49**, 446–454 (2002).

54. Matthews, P. B. "Relationship of firing intervals of human motor units to the trajectory of post-spike after-hyperpolarization and synaptic noise," *J Physiol* **492**, 597–628 (1996).

55. McGill, K. C., and Z. C. Lateva, "Slow repolarization phase of the intracellular action potential influences the motor unit action potential," *Muscle Nerve.* **23**, 826–828 (2000).

56. Merletti, R., and L. Lo Conte, "Advances in signal processing of surface myoelectric signals. Part 1," *Med Biol Eng Comput* **33**, 362–372 (1995).

57. Merletti, R., L. Lo Conte, E. Avignone, and P. Guglielminotti, "Modelling of surface myoelectric signals. Part I: Model implementation," *IEEE Trans BME* **46**, 810–820 (1999).

58. Merletti, R., S. H. Roy, E. Kupa, S. Roatta, and A. Granata, "Modelling of surface myoelectric signals. Part II: Model based signal interpretation," *IEEE Trans BME* **46**, 821–829 (1999).

59. Milner-Brown, H. S., R. B. Stein, and R. Yemm, "The orderly recruitment of human motor units during voluntary isometric contractions," *J Physiol* **230**, 359–370 (1973).

60. Mori, S., "Discharge patterns of soleus motor units with associated changes in force exerted by foot during quiet stance in man," *J Neurophysiol* **36**, 458–471 (1973).

61. Nandedkar, S. D., D. B. Sanders, E. V. Stalberg, and S. Andreassen, "Simulation techniques in electromyography," *IEEE Trans BME* **31**, 775–785 (1985).

62. Person, R. S., and M. S. Libkind, "Simulation of electromyograms showing interference patterns," *Electroenceph Clin Neurophysiol* **28**, 625–632 (1970).

63. Person, R. S., and L. P. Kudina, "Discharge frequency and discharge pattern in human motor units during voluntary contractions of muscle," *Electroenceph Clin Neurophysiol* **32**, 47–83 (1972).

64. Rosenfalck, P., "Intra- and extracellular potential fields of active nerve and muscle fibers: A physico-mathematical analysis of different models," *Acta Physiol Scand* **321** (suppl), 1–168 (1969).

65. Roth, B. J., F. L. H. Gielen, and J. P. Wikswo, "Spatial and temporal frequency-dependent conductivities in volume conduction calculations of skeletal muscle," *Math Biosci* **88**, 159–189 (1988).

66. Schneider, J., J. Silny, and G. Rau, " Influence of tissue inhomogeneities on noninvasive muscle fiber conduction velocity measurements—Investigated by physical and numerical modeling," *IEEE Trans BME* **38**, 851–860 (1991).

67. Shiavi, R., and M. Negin, "Stochastic properties of motoneuron activity and the effect of muscular length," *Biol Cybern* **19**, 231–237 (1975).

68. Shwedyk, E., R. Balasubramanian, and R. Scott, "A non-stationary model of the electromyogram," *IEEE Trans BME* **24**, 417–424 (1977).

69. Stalberg, E., and L. Antoni, "Electrophysiological cross section of the motor unit," *J Neurol Neurosurg Psychiatry* **43**, 469–474 (1980).

70. Stashuk, D. W., "Simulation of electromyographic signals," *J Electromyogr Kinesiol* **3**, 157–173 (1993).

71. Stashuk, D., "EMG signal decomposition: how can it be accomplished and used?" *J Electromyogr Kinesiol* **11**, 151–173 (2001).

72. Stegeman, D. F., T. H. Gootzen, M. M. Theeuwen, and H. J. Vingerhoets, "Intramuscular potential changes caused by the presence of the recording EMG needle electrode," *Electroencephalogr Clin Neurophysiol* **93**, 81–90 (1994).

73. Tegtmeyer, K., L. Ibsen, and B. Goldstein, "Computer-assisted learning in critical care: From ENIAC to HAL," *Crit Care Med* **29**, 177–182 (2001).

74. Van Bolhuis, B. M., W. P. Medendorp, and C. C. A. M. Gielen, "Motor-unit firing behaviour in human arm flexor muscles during sinusoidal isometric contractions and movements," *Brain Res* **117**, 120–130 (1997).

75. Van Oosterom, A., "History and evolution of methods for solving the inverse problem," *J Clin Neurophysiol* **8**, 371–380 (1991).

76. Van Veen, B. K., N. J. Rijkhoff, W. L. Rutten, W. Wallinga, and H. B. Boom, "Potential distribution and single-fibre action potentials in a radially bounded muscle model," *Med Biol Eng Comput* **30**, 303–310 (1992).

77. Wallinga, W., S. L. Meijer, M. J. Alberink, M. Vliek, E. D. Wienk, and D. L. Ypey, "Modelling action potentials and membrane currents of mammalian skeletal muscle fibers in coherence with potassium concentration changes in the T-tubular system," *Eur Biophys J* **28**, 317–329 (1999).

78. Weytjens, J. L. F., and D. Van Steenberghe, "The effects of motor unit synchronisation on the power spectrum of the electromyogram," *Biol Cybern* **51**, 71–77 (1984).

79. Wood, S. M., J. A. Jarratt, A. T. Barker, and B. H. Brown, "Surface electromyography using electrode arrays: A study of motor neuron disease," *Muscle Nerve* **24**, 223–230 (2001).

80. Yao, W., R. J. Fuglevand and R. M. Enoka, "Motor-unit synchronisation increases EMG amplitude and decreases force steadiness of simulated contractions," *J Neurophysiol* **83**, 441–452 (2000).

MYOELECTRIC MANIFESTATIONS OF MUSCLE FATIGUE

R. Merletti

Laboratory for Engineering of the Neuromuscular System
Department of Electronics
Politecnico di Torino, Italy

A. Rainoldi

Laboratory for Engineering of the Neuromuscular System
Department of Electronics
Politecnico di Torino, Italy
University of Tor Vergata, Roma, Italy

D. Farina

Laboratory for Engineering of the Neuromuscular System
Department of Electronics
Politecnico di Torino, Italy

9.1 INTRODUCTION

Fatigue is an experience of our daily life, but its definition is very complex, not unique and controversial. In common language, fatigue may be described as a feeling or sensation of weakness or muscle pain or a decrement of performance, not easily suitable for quantification or measurement. In this chapter a possible approach to quantitative measurement of fatigue is described. The approach is based on the analysis of the surface EMG signal in the time and spectral domains. Indexes of fatigue are defined on the basis of the time evolution of the surface EMG signal features during the contraction. In this way fatigue can be assessed since the very beginning of a muscle effort. The indication pro-

Electromyography: Physiology, Engineering, and Noninvasive Applications, edited by Roberto Merletti and Philip Parker.
ISBN 0-471-67580-6   Copyright © 2004 Institute for Electrical and Electronics Engineers, Inc.

vided is often related to a large number of motor units rather than specific for the single contributions to the surface EMG signal. The mathematical details for the extraction of amplitude and spectral variables from the surface EMG signal are not addressed in this chapter, since they are described extensively in Chapter 6. Rather, here we describe applications of the assessment of muscle fatigue by the surface EMG signal analysis, we indicate the potentials and limitations of this technique, and we describe future research topics to be addressed for its full validation.

9.2 DEFINITIONS AND SITES OF NEUROMUSCULAR FATIGUE

A quantitative approach to fatigue is often associated to an event, or to the time instant corresponding to an event, such as the inability to further perform a task or sustain an effort, and therefore is somehow related to mechanical performance. Another description deals with the inability to reach the same initial level of maximal voluntary contraction (MVC) force (the force generation capacity), again related to an event or time instant associated to the inability to produce a specified mechanical performance. These definitions indirectly imply that there is no fatigue before a time instant or event, whereas there is fatigue after it.

The engineering approach to fatigue of a material or a mechanism is different. In this case fatigue is defined as a process developing in time and progressively changing the characteristics of the material or the mechanism without evident changes of performance up to the time (or point) of deformation or rupture. This definition can be adapted to a muscle (where inability to perform is the analogue of deformation or rupture but does not imply any irreversible process in the muscle) where the fatigue process accounts for all the physiological changes taking place in a muscle, before the inability to perform sets in as "mechanical manifestation of muscle fatigue." This different definition is clearly explained in [3], and supports the concept of fatigue as an analogue function of time, which starts evolving from the beginning of the contraction.

The evolution may be fast or slow, depending on the effort performed, and lead sooner or later to mechanically detectable changes of performance. Many factors contributing to this evolution affect the surface EMG signal, and can be detected through it, leading to what is commonly defined as "myoelectric manifestations of muscle fatigue." Unless otherwise specified, this is the concept of fatigue that will be adopted in this chapter.

There are many potential sites of fatigue in the neuromuscular system: the motor cortex, the excitatory drive, the control strategies of the spinal (upper) and the α (lower) motoneurons, the motoneuron conduction properties, the neuromuscular transmission, the sarcolemmal excitability and conduction properties, the excitation-contraction coupling, the metabolic energy supply, and the contraction mechanisms. They can be grouped under the headings of (1) central fatigue, (2) fatigue of the neuromuscular junction, and (3) muscle fatigue. All these factors directly or indirectly affect the EMG signal in ways that are very difficult to unscramble, especially because the information obtained from the surface EMG signal is usually related to a large group of MUs (see Chapter 6).

To reduce the difficulty of the problem and the number of factors affecting the EMG signal, most past research focused on myoelectric manifestations of muscle fatigue during isometric, constant force conditions. It is clear that such easy to study conditions do not reflect the muscle function in daily life. It is very likely that fatigue in highly dynamic conditions could be estimated in the coming years, as anticipated by the special issue of the *IEEE Engineering in Medicine and Biology* magazine (November–December 2001)

[9]. At this time such estimates are questionable because dynamic conditions introduce changes in the geometry (relative location of the electrodes with respect to the source) that may easily lead to incorrect conclusions [30] (see Section 9.7). The situation is analogue to testing the engine of a vehicle whose workings are unknown: certain information can be more easily extracted during a bench test in controlled conditions, such as constant torque, speed, or power, in a laboratory environment rather than on the road, where many additional confounding factors may cause interpretation errors.

9.3 ASSESSMENT OF MUSCLE FATIGUE

Fatigue itself is not a physical variable. Its assessment requires the definition of indexes based on physical variables that can be measured, such as force or torque (current value or MVC value), power, angular velocity of a joint, or variables associated to the single motor units (MUs), such as firing rates, conduction velocity, degree of synchronization, and intermittent activation, or variables associated to the EMG signal, such as amplitude and spectral estimates or global (as opposed to single MU) frequency and amplitude or conduction velocity estimates (see Chapters 6 and 7).

The association between mechanical and myoelectric variables requires great caution and awareness of the different phenomena affecting the two sets of variables. In most cases voluntary efforts activate muscle compartments, not individual muscles. They may even involve coactivation of antagonist muscles, and it is possible that the net force acting over a joint is near zero while agonist and antagonist muscles are both active and "fatiguing." While the resulting force or torque is the algebraic summation of the contributions of different muscles, the detected EMG signal predominantly reflects the activity of muscle(s) underneath the electrodes.

A variation of the sharing of force contributions among synergic and antagonist muscles may leave the net force or torque unchanged but may redistribute the EMG signals, decreasing those of some muscles and increasing those of others, thereby changing their myoelectric manifestations of fatigue. Furthermore, even within the same muscle, the MU pool may not remain constant during a constant force sustained contraction. New MUs may be recruited to replace failing ones (see Fig. 7.9 and [37]), therefore altering mechanical or myoelectric manifestations of fatigue. As a consequence detection of this phenomenon becomes important in order to describe and understand what is going on within the muscle and its controlling system. Because of the difficulties in isolating so many factors influencing the surface EMG signal features during fatigue, particular experimental protocols have been designed to reduce the parameters of the neuromuscular system that are altered during the contraction.

The main alternative way to investigate fatigue implies the use of selective electrical stimulation of a nerve branch or of the motor point of a muscle [66]. The purpose of this approach is to "disconnect" the muscles from the CNS, and activate only one (or a portion of one) muscle at a time at a controlled frequency and with a MU pool that is more likely to be stable. Again, this may be considered a highly unphysiological approach, and the same considerations given at the end of the previous section still apply; we should not forget that our knowledge of the system under investigation is very limited and can be increased only by testing it in controlled conditions designed and focused toward the goal of answering a question at a time. Any attempt to study it in its full operational mode, as it would be clinically desirable, is doomed to failure. During electrically elicited contractions we can avoid the factors of variability due to the central control strategies and many methodological issues, such as crosstalk (see Section 9.7).

9.4 HOW FATIGUE IS REFLECTED IN SURFACE EMG VARIABLES

In 1912 Piper [73] observed a progressive "slowing" of the EMG during isometric voluntary sustained contractions. Given the random nature of voluntary EMG, this "slowing" cannot easily be quantified in the time domain. It is easier to describe it in the frequency domain using spectral characteristic frequencies, such as the mean or median frequencies (MNF and MDF) of the power spectral density function (indicated in the following simply as power spectrum), as suggested in the early work of Chaffin, Lindstrom, Kadefors and De Luca [12,15,52,54]. This approach applies equally well to the almost stochastic signals generated during voluntary contractions and to the almost deterministic signals generated during electrically elicited contractions (see Chapter 6).

Global estimates of muscle fiber conduction velocity (CV) decrease during sustained isometric constant force contractions [2,15,67] and, as a consequence, the cross-correlation function between two signals detected from adjacent electrode pairs shifts to the right and widens while power spectrum compresses to the left. However, the cross-correlation function may also change shape while the power spectrum is scaled by a factor usually greater than that predictable by the changes of CV. These observations strongly suggest that myoelectric manifestations of muscle fatigue cannot solely be attributed to CV decrements and other factors must be considered [67]. Myoelectric manifestations of muscle fatigue appear to be a multifactorial phenomenon, involving different physiological processes which evolve simultaneously.

Consider a segment (epoch) of the signal generated at time $t = 0$ by a single MU as $x_1(\theta)$ (where θ is the local time variable) and a second signal epoch $x_2(\theta) = hx_1(k\theta)$ generated at some later time t by the same MU and scaled in amplitude by a factor h and in time by a factor k, as a consequence of the change of CV by a factor k ($k < 1$ means slowing). The Fourier transform, power spectrum, and autocorrelation function of $x_2(\theta)$ are associated to those of $x_1(\theta)$ by the following relations:

$$X_2(f) = \frac{h}{k} X_1\left(\frac{f}{k}\right)$$

$$P_2(f) = \frac{h^2}{k^2} P_1\left(\frac{f}{k}\right)$$

$$\Phi_{22}(\tau) = \frac{h^2}{k} \Phi_{11}(k\tau) \tag{1}$$

As a consequence the EMG "variables," f_{med} (MDF), f_{mean} (MNF), lag of the first zero of the autocorrelation function τ_0, average rectified value (ARV), and root mean square value (RMS) (all these variables are defined in Chapter 6) of $x_2(\theta)$ (at time t), are related to those of $x_1(\theta)$ (at time 0) by the relations:

$$\frac{f_{\text{med}}(t)}{f_{\text{med}}(0)} = \frac{f_{\text{mean}}(t)}{f_{\text{mean}}(0)} = k$$

$$\frac{\tau_0(t)}{\tau_0(0)} = \frac{1}{k}, \quad \frac{\text{ARV}(t)}{\text{ARV}(0)} = \frac{h}{k}, \quad \frac{\text{RMS}(t)}{\text{RMS}(0)} = \frac{h}{\sqrt{k}} \tag{2}$$

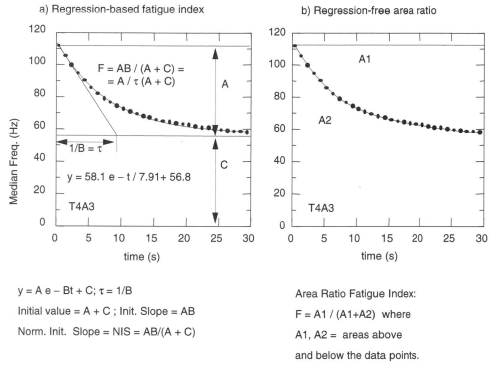

Figure 9.1. (*a*) Median frequency as a function of time computed during a stimulated muscle contraction with the correspondent exponential regression. (*b*) The definition of the regression-free "area ratio" fatigue index for the same data depicted in (*a*).

During a sustained isometric contraction the value k changes, in subsequent epochs, as a function of time, starting from the initial value equal to 1. This value can therefore be chosen as a fatigue index. Unfortunately, not all the MUs have the same value of k. In addition their MUAPs may change shape and other phenomena, such as a change of MU pool, may further affect the interference EMG signal. As a consequence different values of k are obtained, depending on the variable used for the estimate (CV, MNF, MDF, ARV, RMS). The plots of these variables, normalized with respect to their initial values, versus time, is referred to as the "fatigue plot" [64]. This plot allows to compare percent changes of the different variables during an isometric constant force contraction. The association of this plot with physiological events is not trivial, but the approach is very useful to outline differences among muscles.

Although the fatigue plot provides considerable information concerning the evolving situation of the neuromuscular system, a more compact representation is desirable to facilitate comparisons. If the scatter plot of a variable shows a linear pattern, the slope of the regression line will provide an intuitive fatigue index. If a pattern is curvilinear, the parameters of an exponential or polynomial regression can be used as indexes. This implies the choice of a regression curve, a fact that introduces a subjective factor. In addition a negative slope (e.g., of CV or MNF) would be associated to "positive" fatigue. A way to solve this problem was proposed by Merletti et al. [64] who suggested the use of the regression-free "area ratio," defined in Figure 9.1, that provides a positive fatigue index between 0 and 1 for decreasing patterns and a negative index for increasing patterns. This index may either be defined for the entire contraction or as a variable function of time. While in the case of linear patterns this approach does not provide significant advantages, in the case of curvilinear behavior, it avoids dealing with a wealth of coefficients of regression functions. Of course, an "area ratio" index may be computed for each EMG variable, therefore providing a "fatigue vector" and allowing clustering of behaviors on the basis of the length and direction of such vector in the space defined by the EMG variables considered.

Other indexes of fatigue have been proposed by Lo Conte and Merletti [58]. In particular, the one based on the cumulative power functions of two EMG signals (the integral of the normalized power density function) is of interest because it provides a value for the spectral compression and an additional number indicating the degree of change of shape of the power spectrum. A band, similar to a confidence interval, can then be associated to the fatigue curve describing spectral compression. Finally fatigue indexes of single MUs have been recently defined and are based on more advanced methods of signal detection and processing [28,34], described in Chapter 7.

Isometric constant force contractions are performed to observe differences between muscles, or between tests performed on the same muscle at different times. The force or torque level is sustained for some time at a fraction of the MVC value. In this way it is believed that the results obtained from different individuals performing similar relative efforts could be compared. Indeed, the same force or torque would imply a rather different relative effort from a weight lifter and an elderly woman. However, this approach has a major drawback: fatigue depends on the rate of metabolite removal, which in turn depends on blood flow. At a certain level of contraction, blood flow is stopped by intramuscular pressure and the muscle becomes ischemic. Myoelectric manifestations of muscle fatigue are affected by this event [68], which probably depends on absolute, rather than relative, contraction level.

9.5 MYOELECTRIC MANIFESTATIONS OF MUSCLE FATIGUE IN ISOMETRIC VOLUNTARY CONTRACTIONS

Figure 9.2 shows an example of fatigue plot observed during an isometric contraction of a healthy tibialis anterior muscle sustained for 100 seconds at 60% of the maximal voluntary contraction level (MVC). It is evident that this level can be maintained for 60 seconds and mechanical manifestations of muscle fatigue become evident after this time. On the contrary, myoelectric manifestations of muscle fatigue begin at the very beginning of the contraction. Short (0.2 s) signal epochs at $t = 1, 30, 60, 90$ s are shown together with the corresponding power spectra normalized with respect to the highest power density value. Epochs of 1 second were used for the calculation of the variables, and one value every three is depicted to increase clarity. The initial values of the depicted variables are also provided in order to de-normalize the plot. The progressive slowing of the EMG signal and compression of its power spectrum are evident. Of course, while at time near zero the contraction level is indeed 60% MVC, after one minute it is 100% of the MVC at that time. It is therefore likely that new MUs have been recruited, thereby reducing the global manifestations of muscle fatigue that would have been more marked had the MU pool been stable. It should be taken into consideration that different muscles recruit their MUs with different strategies: some recruit all the available MUs at low force levels and then increase force by increasing their firing rates; others keep recruiting MUs up to 80% MVC [3].

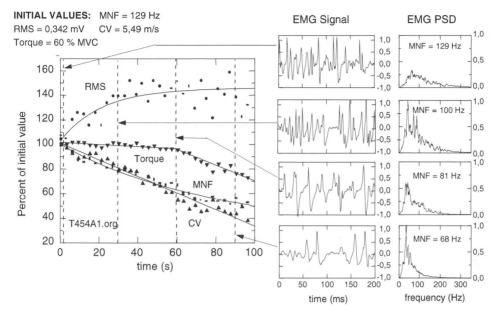

Healthy tibialis anterior: voluntary contraction at 60% MVC

Figure 9.2. Fatigue plot of a voluntary isometric contraction of the tibialis anterior muscle, sustained for 100 s. The torque is maintained constant for 60 s after which the subject cannot further sustain the torque level, that progressively decreases. Four raw surface EMG signal epochs with the respective power spectra are also presented.

Figure 9.3 shows the pattern of torque and MDF of a tibialis anterior muscle sustaining a contraction at 50% MVC and at 70% MVC (in two different days) up to beyond the endurance time [63]. It can be observed that the initial and asymptotic value are similar, but the time constant of the exponential regression is different. In both cases, at the limit of endurance, MDF dropped by 35%.

Myoelectric manifestations of muscle fatigue allow to differentiate between muscles with different characteristics or adopting different strategies. Figure 9.4 depicts the behavior of the tibialis anterior muscles of two subjects performing the same task. Subject A sustains the 50% MVC effort for 75 seconds, and shows equal percentage decrements of CV and MDF and a large increment of EMG amplitude up to the endurance time limit. Subject B sustains the effort for 120 seconds, shows no decrement of CV, smaller decrement of MDF, and smaller increment of amplitude. The two different behaviors are clearly detected by the technique. Subject B shows less evident manifestations of muscle fatigue (changes in all the variables are smaller than for subject A), perhaps attributable to a different distribution of fatigable and fatigue resistant MUs in the two muscles. Figure 9.5 shows examples of fatigue plots representative of different groups of subjects or muscle conditions. Figure 9.5*a* shows myoelectric manifestations of fatigue of a biceps contracting at 80% MVC in an isometric brace fixed to a table (a1) or hanging from the ceiling with elastic bands (a2) [35]. Figure 9.5*b* shows the fatigue plots of the sternocleidomastoid of two subjects, one healthy (b1) and one affected by chronic neck pain (b2) [22]. Figure 9.5*c* shows the fatigue plot of the biceps brachii of a young (c1) and an elderly

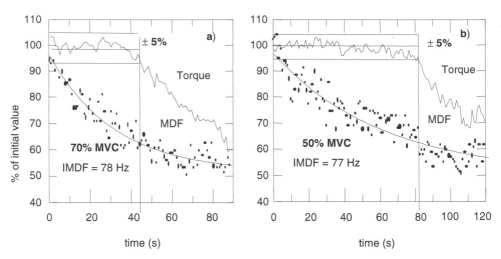

Figure 9.3. Median frequency (MDF) and torque during sustained isometric contractions of the tibialis anterior muscle at 70% MVC and 50% MVC in the same healthy subject. The torque is maintained constant for 45s at 70% MVC and 82s at 50% MVC (endurance values). The starting values (IMDF) and total decrement of MDF are the same but the time constants of the decay are different.

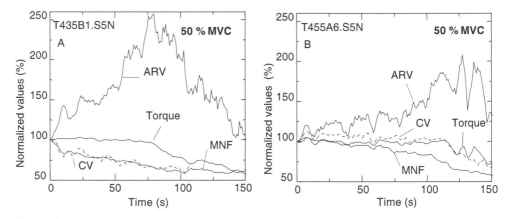

Figure 9.4. Fatigue plots of two healthy subjects requested to perform isometric contractions of the tibialis anterior muscle sustained for 150s at 50% MVC. The first sustains the contraction level for 75s, the second for 120s. Myoelectric manifestations of muscle fatigue are more marked in the first subject.

(c2) subject [69]. These individual plots are reported with the objective of showing the possibility of detecting different behaviors. Indications about repeatability and statistical significance of the differences observed between groups are reported in the cited literature.

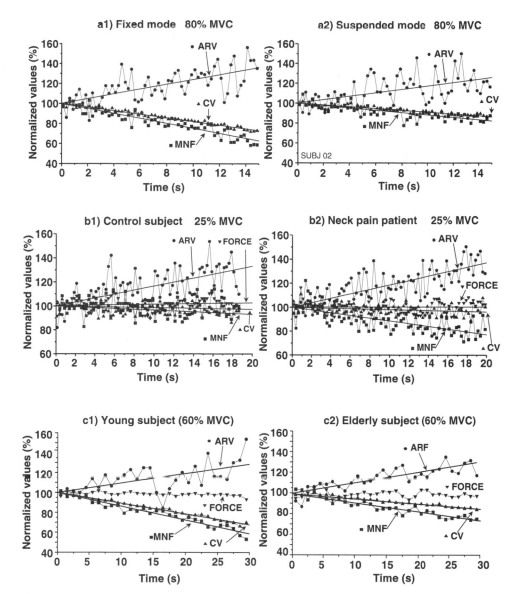

Figure 9.5. Examples of fatigue plots observed in different conditions or subjects. (a1, a2): Comparison between the fatigue development of the biceps brachii in the same subject when the elbow brace is fixed to a table and when it is hanging from the ceiling and free to move (from [35] with permission). (b1, b2): Fatigue patterns of the sternocleidomastoid of a healthy subject and of a subject affected by neck pain (from [22] with permission). (c1, c2): Fatigue in a young and an elderly subject (muscle biceps brachii) (from [69] with permission). See the referenced literature for statistical results.

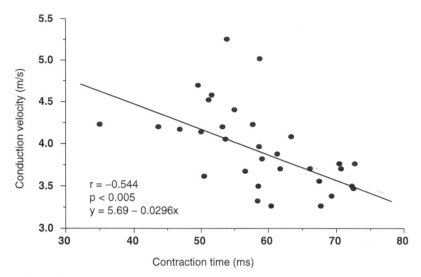

<u>Figure 9.6.</u> Muscle-fiber conduction velocity plotted against the corresponding time to peak from slow and fast motor units during electrically evoked single twitches. Motor units with greater contraction time from the biceps brachii muscle (i.e., type I, slow oxidative) show lower conduction velocity with respect to motor unit with lower contraction time (type II, fast glycolytic). (Redrawn from [41] with permission)

9.6 FIBER TYPING AND MYOELECTRIC MANIFESTATIONS OF MUSCLE FATIGUE

Different motor unit types are distinguished on the basis of their mechanical response. Motor units with fast fibers produce force twitches characterized by short time to peak, while motor units made of slow fibers need more time to produce the maximum. In general, slow twitch motor units are comprised of fibers with slower CV, whereas fast twitch motor units are comprised of fibers with higher CV. Probably the first work aimed to assess this correlation was done by Hopf et al. [41]. They electrically evoked single twitches in human biceps brachii muscle estimating both the contraction times (defined as the time from the onset of the deflection to the peak) and the muscle fiber CV (using invasive needle technique). A negative correlation between contraction times of the two fiber types and muscle fiber CV was found (Fig. 9.6).

Fiber composition is usually investigated by biopsy and histochemical analysis. Johnson et al. [43], analyzed 36 muscles during an autopsy study on six male cadavers (aged 17–30 years) and provided for these muscles the percentages of type I and type II fibers. Other authors focused their efforts on fewer muscles and a larger number of subjects, producing more reliable data [36,39,83,92]. The high costs and ethical problems of biopsies limit the use of such techniques. Moreover the information obtained from the specimens are not representative of the muscle as a whole, and this constitutes a serious drawback, supporting the development of alternative noninvasive methods.

The variables potentially useful for this purpose are the characteristic spectral frequencies (e.g., MNF and MDF) and the estimated muscle fiber conduction velocity (CV). They are related (although not exclusively; see Section 9.7) to the pH decrease due to the increment of metabolites produced during fatiguing contractions. Lactic acid accumula-

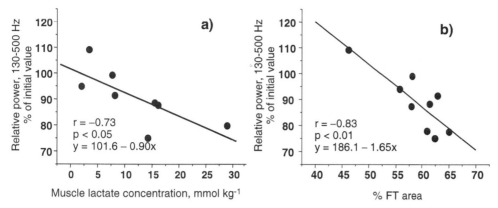

Figure 9.7. (a) Relationship between the power in the highest bandwidth (130–500 Hz) of the EMG spectrum at the end of the exercise (120 isokinetic contractions at 1.5 rad/s), expressed as percentage of initial value, and the muscle lactate concentration (N = 8 subjects, vastus lateralis muscle). (b) The same variable is depicted in the right diagram plotted with respect to the percentage area of type II fiber (fast fibers) estimated by means of biopsies. (Redrawn from [84] with permission)

tion in fact determines a decrease of muscle fiber CV (and as a consequence of MNF) [11]. Findings in the opposite direction are albeit shown in the work of Linssen et al. [57] on McArdle's disease patients, where a CV reduction was observed during biceps brachii contraction at 80% MVC, although, because of the pathology, muscles could not produce lactic acid. This finding confirms that pH modification is not the only cause of the myoelectric manifestations of fatigue. Since the percentage of type II fibers were shown to be positively correlated to the lactate accumulation and both were negatively correlated with MNF initial values [84] (Figs. 9.7 and 9.8), this variable can be considered as a good noninvasive tool to estimate muscle fiber composition.

Enoka and Fuglevand [20], in their review, clarified the issue of the innervation ratios within a given muscle, that is, the number of muscle fibers innervated by a single motoneuron and therefore the proportions of the motor neuron pool that innervate the different muscle fiber types. It seems that consistently across muscles and species, motor unit innervation number has actually a skewed distribution among different fibers, toward type I. In first dorsal interosseus muscle, for instance, that comprises equal number of type I and II fibers, about 84% of the motor units are type I and only 16% are type II. In the same way, in the triceps brachii muscle, 75% of motor neurons innervate type I fibers, which represent only the 33% of the whole fiber population. Hence age-related remodeling [69] (see Chapter 17), pathologies and special exercise training, able to modify type II fiber (their number or their diameter), can actually act on a very small number of large motor units of the whole pool. The effect of such a motor unit organization asymmetry on EMG signal pattern should be considered as an important factor in the fiber type composition evaluation.

A variety of experiments were carried out both on animals and human models. The work by Kupa et al. [49] can be considered the most evident demonstration, within the animal muscle, of the positive correlation between variations of the EMG signal power spectrum and the percentage of fiber types activated during a contraction. EMG signals were recorded from rat soleus, extensor digitorum longus, and diaphragm muscles during

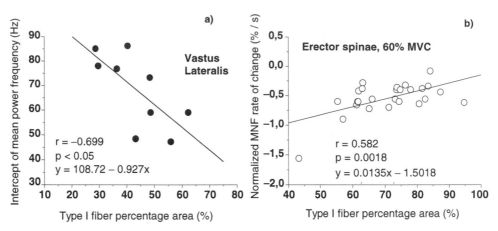

Figure 9.8. (*a*) Relationship between the percentage of type I fiber area and the intercept of mean power frequency in human vastus lateralis muscle. The percentage of slow fibers (type I) negatively correlates with the mean frequency content of the EMG power spectrum (redrawn from [39] with permission). (*b*) Correlation between relative area of the muscle occupied by type I fiber and the rate of change of MNF during a 60% MVC contraction recorded from the erector spinae in 31 subjects (Redrawn from [61] with permission).

electrically elicited contractions applied to an in vitro neuromuscular preparation. Fibers from the three rat muscles were typed as slow oxidative, fast oxidative glycolytic, and fast glycolytic, respectively. MNF initial values and rates of change during the contraction were positively correlated with the percentage of fast glycolytic fiber within the muscle (Fig. 9.9). Findings from the work by Kupa et al. [49] were obtained in a very particular laboratory condition, where the whole muscle was extracted with a nerve segment, free of subcutaneous layers, entirely stimulated by electrical current, and the electrodes were placed directly on its surface. Thus the strong correlation found may not be observable with noninvasive methods because of several factors that modify the surface EMG variables.

In the same direction Sadoyama et al. [79] showed a strong correlation between different fiber type compositions and conduction velocities estimated during voluntary contractions from two different groups of athletes (sprinters vs. distance runners; see Fig. 9.10). Findings similar to those of Kupa et al. were reported in humans by Gerdle et al. [39] and by Mannion et al. [61]. The former showed a negative correlation between initial values of MNF and the percentage of type I fibers found in the vastus lateralis muscle. The increase of aerobic type I fibers shifts the power spectrum to lower frequency components determining a lower MNF value. The latter work indicates a correlation between the rate of change of MNF and the percentage of type I fibers in the erector spinae muscle (Fig. 9.8).

Modifications of fiber type composition due to diseases and/or natural adaptation can be monitored by the EMG spectral modification, as shown in several recent works. In the very particular case of the congenital myopathy, for instance, where a generalized congenital predominance (95–100%) of fiber I is found in all skeletal muscles, initial values and rate of change of CV, estimated from vastus lateralis muscle, were found significantly lower compared to the control group of healthy subjects. According to these findings, the

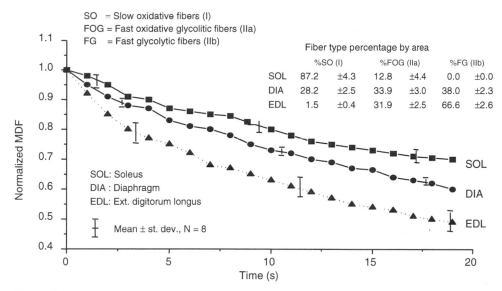

Figure 9.9. Normalized values of MDF plotted for the soleus (SOL), diaphragm (DIA) and extensor digitorum longus (EDL) muscles of 8 Wistar rats (mean ± st. dev.). The neuromuscular preparations were stimulated at 40 Hz for 20 s. Data for each muscle groups are normalized with respect to their initial values. The rate of change of MDF in the three muscles is correlated with the percentage of fast glycolytic fiber. SO = slow oxidative fibers (type I); FOG = fast oxidative glycolitic fibers (type IIa); FG = fast glycolytic fibers (type IIb). (Redrawn from [49] with permission)

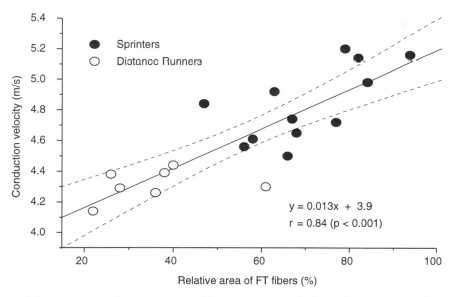

Figure 9.10. Correlation between muscle-fiber conduction velocity and the portion of fast type fiber areas in the vastus lateralis. Filled circles are for sprinters, and open circles are for distance runners. The two separate clusters confirm the hypothesis that the two groups have opposite fiber type distributions and suggest that conduction velocity is a suitable variable for the non-invasive estimate of the fast type fiber percentage in the studied muscle. (Redrawn from [79] with permission)

lactic acid production was found lower in pathological subjects than in the control group [56]. In subjects with peripheral vascular disease CV initial values from the tibialis anterior muscle were found to correlate with type II fiber diameter, while MNF initial values correlated both with type II fiber diameter and with their percentage in the muscle [72].

With their findings, Linssen, Pedrinelli, and Mannion, provided a further confirmation of the hypothesis of Sadoyama [79], indicating that myoelectric manifestations of fatigue are related to the percentage area of type II fibers per cross sectional unit area, regardless of the muscle-fiber diameter. Uhlig et al. [86] demonstrated the effect of pathologies in modulating the transformation of transitional fiber (IIa). Biopsies from both ventral and dorsal neck muscles were taken from patients with rheumatoid arthritis or whiplash injury. Results showed the occurrence of transformations strongly correlated with the duration of symptoms proceeding in the direction from "slow oxidative" to "fast glycolytic." These bioptical findings had a surface EMG confirmation in the work of Falla et al. [22], where the myoelectric manifestations of fatigue were found to be greater in the whiplash patient group with respect to the control group, suggesting the possibility that the patient group had a greater percentage of more fatigable type II fibers.

Further work is necessary to develop a noninvasive tool, alternative to the biopsy technique, for noninvasive fiber type estimation. The information provided by the myoelectric manifestations of fatigue seem, to date, a promising source to try to win such a challenge.

9.7 FACTORS AFFECTING SURFACE EMG VARIABLES

9.7.1 Isometric Contractions

As indicated in Chapter 6, most of the variables extracted from the surface EMG signal are not directly related to any physiological quantity. For example, the mean power spectral frequency does not reflect any frequency physiologically generated in the signal. Rather, these variables are connected to physiological phenomena and are sensitive to some of them (usually to many of them at the same time). Thus, apart from the efforts done in the past for the proper estimation of surface EMG variables, the most important challenge has been (and still is) the analysis of the relationships between surface EMG variables and the physiological phenomena under study. Because of the high complexity of the EMG generation and detection system (see Chapter 8), predictions were usually based on simplifications of the problem. Some predictions based on simple assumptions were validated by experimental evidence and are now well established. For example, it was proved that the rates of change of spectral variables and global conduction velocity during sustained isometric constant force contractions are indicative of muscle fatigue and may be related to MU type (see Section 9.6). It was also shown, both theoretically [53,82] and experimentally [2], that during fatiguing isometric constant force contractions, CV and MNF or MDF of the surface EMG signal are highly correlated, with MNF and MDF reflecting mainly the changes in CV of the active MUs.

These considerations are based on the following simplifications: (1) all the motor units have the same CV, (2) only the MUAP shape is concurring to the global power spectrum (i.e., MU firing rates do not play any role in modifying the power spectral density), (3) the intracellular action potential is not changing shape during the contraction, and (4) there is no re-derecruitment of MUs during fatigue. Indeed, none of the previous assumptions are valid in real cases, even for the very simple case of isometric constant force contractions.

Physiological Phenomena Affecting EMG Signal Features: Fatigue Assessment

VOLUNTARY CONTRACTIONS. In case of voluntary contractions, the MUs have a CV distribution with a mean of 3 to 5 m/s and a standard deviation of approximately 0.3 to 0.4 m/s [85]. Both mean and standard deviation may change due to fatigue. Moreover the distribution of signal energy among MUAPs with different CVs is not uniform, since larger MUs usually have higher CVs [1]. Thus, considering a random location of the MUs in the detection volume, it is more likely that high CV MUs generate large surface potentials. This implies that the global mean CV estimates is biased toward higher values [93].

The MU firing rates have an influence on the low frequency region of the surface EMG power spectrum. MU firing rates are changing with fatigue [6] and as a consequence of ageing or pathological conditions [45,46]. MU firing rates also depend on the MU type, with fast-twitch units resulting in lower firing rates than slow-twitch units. The choice of different percentile frequencies to characterize the EMG power spectrum indicated that some of them are more sensitive than others to the mean MU firing rate and thus provide slightly different indications on muscle fatigue [59]. In case of dependent firing patterns the degree of dependency is also affecting the spectrum, increasing the power at low frequencies [50,87].

The intracellular action potential is probably changing during time, both in shape and length, and this is affecting spectral characteristics [17]. The effect of the intracellular action potential length and shape on the characteristic spectral frequencies is limited to superficial MUs but has to be taken into account especially for muscles covered by a thin fat layer.

Finally it has been shown that the MU recruitment threshold decreases with fatigue. As a consequence, during contractions at force levels below the end of the recruitment, additional MUs, not active at the beginning of the contraction, are activated during time. This recruitment process seems to follow Henneman's principle with the additionally recruited MUs having amplitude and CV progressively larger [37]. This phenomenon is affecting EMG amplitude, spectral variables, and global CV estimates. EMG amplitude increases more than predicted by the change of CV and may change slope during time (i.e., its behavior can considerably deviate from the linearly increasing one, often seen at high force level contractions).

Global CV tends to increase as additional, not fatigued, fast MUs are recruited. The power spectrum changes in a manner difficult to predict as a consequence of this phenomenon. High CV MUs tend to have a short signal duration, and thus to bias characteristic spectral frequencies toward higher values. On the other hand, MUAP duration is also highly affected by the distance of the source from the detection point; thus deep MUs recruited after the beginning of the contraction may bias the power spectrum toward lower frequencies, even if the recruited MUs are of the fast type. Conditions in which MU recruitment may be the factor mostly affecting signal features are very low force level, long duration contractions. In these cases spectral variables and CV may show an initial decreasing pattern followed by an increase [37]. Moreover, in the part of the contraction with substantial MU recruitment, the relationship between spectral variables and CV can no longer be predicted analytically because of the masking effect of the volume conductor and MU location in the muscle [29].

De-recruitment of MUs has been suggested as a compensating effect against MU over-load [91]. This phenomenon is controversial and no clear indications on its real occur-rence are provided in the literature. As recruitment, de-recruitment would determine similar, unpredictable effects on the EMG variables. However, since the de-recruited MUs should be the smallest ones, their influence on the interference EMG signal would also be much lower than recruitment of larger units.

ELECTRICALLY ELICITED CONTRACTIONS. During electrically elicited contractions the same physiological effects, as those described for voluntary contractions, occur, except for those related to MU firing rate modulation. In electrically elicited contractions the MU CV spread is playing a much more important role on spectral variables than in voluntary contractions [23]. In this case the spread of CVs implies that the single MU action poten-tials contributing to the M-wave become desynchronized as they propagate along the muscle fibers so that the M-wave becomes progressively wider and smaller (the effect is equivalent to that of a low-pass filter).

The changes of intracellular action potential shape and duration have similar effects on the signal power spectrum as for the voluntary contractions. De-recruitment may occur also in stimulated contractions when the sarcolemma excitability is no longer possible, because of repeated stimulations. In some contraction modalities and pathologies the lat-ter effect may be the dominant one [13], resulting in amplitude decrease and changes in spectral frequencies and CV (difficult to predict analytically), as in the case of voluntary contractions.

Central Nervous System Control Strategy Assessment. Many past works applied frequency analysis of the surface EMG signal as indicator of the recruitment strategies of the central nervous system [4,5,7,38,88,90] which may, however, be critical (see Chapter 4).

SURFACE EMG VARIABLES AND FORCE LEVEL. The theory of volume conductor indi-cates that any relationship between recruitment and spectral features may be masked by anatomical or geometrical factors. Indeed, when a deep MU is recruited it may contribute to the low-frequency region of the EMG power spectrum even if its CV is higher than the mean CV of the previously recruited MUs. Thus a recruitment order from the slow to the fast MUs does not necessarily imply an increase of characteristic spectral frequencies. As a consequence the behavior of MNF or MDF with force may change significantly between different subjects and muscles. This has been shown both experimentally and with simu-lated signals by Farina et al. [29], who pointed out the need to consider carefully the estab-lishment of a relationship between force level and spectral EMG variables (Fig. 9.11). This relationship may occur in particular conditions but, in general, may be masked by the random location of the MUs in the muscle. Regarding the other EMG variables, the ampli-tude variables increase monotonically with force independently on the muscle recruitment strategy, since both an increase of firing rate and an increase of the number of active MUs increases signal amplitude. CV was shown to monotonically increase with the contrac-tion level, since its estimation is less affected by MU depth than the spectral variables. However, CV increases monotonically also when muscle force increases as a consequence of rate modulation and not of recruitment, since the last activated MUs have surface poten-tials in average with larger energy and their relative weight on the surface signal is larger when firing rate increases. In addition muscle fiber CV increases slightly with firing rate.

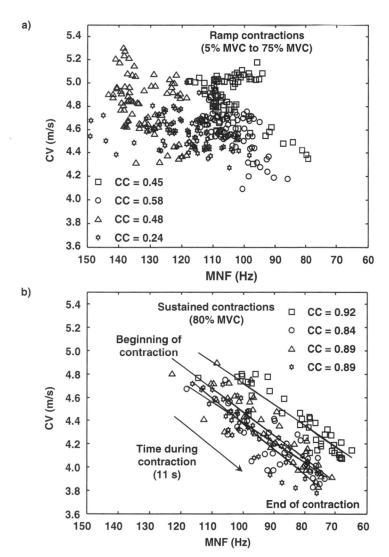

Figure 9.11. Scatter plots of CV versus MNF for four contractions performed in two different days by a healthy subject during (*a*) a ramp and (*b*) a constant torque contraction of the biceps brachii. Each point in the plot has coordinates corresponding to MNF and CV estimated from the same signal epoch. Correlation coefficients between CV and MNF are also reported. The four symbols indicate the four contractions. Analysis window of 250 ms without overlapping in case of sustained torque part, overlapping of 511 samples in case of ramp torque part (only one value every 64 is shown for clarity). (From [29] with permission)

Anatomical, Physical, and Detection System Parameters Affecting EMG Variables. The EMG variables are sensitive to a number of anatomical, physical, and methodological factors involved in their estimation. The most important of these factors are (1) the thickness of the subcutaneous tissue layer, (2) the depth of the sources within the muscle, (3) the inclination of the detection system with respect to the muscle fiber orientation, (4) the length of the fibers, (5) the location of the electrodes over the muscle, (6) the spatial filter used for signal detection, (7) the interelectrode distance, (8) the electrode size and shape, (9) crosstalk among nearby muscles, and (10) the estimators adopted for EMG variable computation. The influence of all these factors on surface EMG features has been recently discussed in simulation and in experimental signals [18,19,24,25,27,30,31,32,33,42,52,76,78,80].

The thickness of the subcutaneous tissue layer decreases signal amplitude and frequency content (it acts as a low-pass spatial filter) [26]. CV estimates can be biased toward higher values with increasing subcutaneous tissue because the end-of-fiber components may increase their weight on the signal. The depth of the sources within the muscle has a similar effect as the thickness of subcutaneous layers. The effect of the inclination of the muscle fibers with respect to the detection system is very difficult to predict because it depends on the spatial filter used, electrode size and shape, and many other factors (refer to [27]). The length of the fibers determines the amplitude of the end-of-fiber components, with larger end-of-fiber potentials for shorter fibers. The effect of electrode location, in particular, with respect to the tendon and innervation zones, has been discussed by many authors [30,42,76,78]. It is well established that for getting the best estimates of amplitude, spectral variables, and CV, the location should be selected between the innervation zone and the tendon. The innervation zone and tendon regions determine positive bias in spectral frequency estimates and negative bias in amplitude values. The estimation of CV from regions close to the innervation zone or tendon locations may be completely wrong, due to the absence of potentials traveling in space with similar shape.

The spatial filter used for signal detection determines the attenuation of the end-of-fiber components and performs a high-pass filtering of the traveling components. As a consequence the amplitude is reduced and the frequency content is increased when high selective spatial filters are used. Also CV presents a bias depending on the spatial filter, as shown experimentally (biceps brachii and upper trapezius) [31,80]. Interelectrode distance changes the spatial transfer function of the detection system, and thus affects the signal properties [32]. Finally electrode size has the effect of decreasing amplitude and frequency content [25,27] and of affecting CV estimates when the detection system is not perfectly aligned with respect to the fiber orientation.

Crosstalk depends on many anatomical factors, such as the subcutaneous layer thickness and the length of the fibers, and on many detection system parameters, such as the spatial filter used, the interelectrode distance, the size of the electrodes, and the electrode location [33].

Many estimators have been proposed for computing EMG variables. Spectral variables can be computed from Fourier-based approaches, autoregressive methods, and other time frequency analysis (for details refer to Chapters 6 and 10). Amplitude can be estimated by simple RMS or ARV analysis on raw data, after whitening of the signal (see Chapter 6), or as the envelope, and so on. CV can be computed as the peak distance between waveforms detected in two locations of the muscle [1], with the distribution function method [77], by two-channel maximum likelihood approaches [62], by multi-channel channel MLE methods [34], by modified beamforming approaches [28], and so on. The choice of the estimator determine the variance and the bias of the estimate of the "true" variable value.[1]

---

[1] The "true" value of an EMG variable is sometimes difficult to define. A typical example is CV estimation. Since CV is not a single value but rather a distribution of values, the "true" global CV cannot be defined in a unique way. It can be, for example, the mean of the distribution weighed by the potential energy or by MUAP occurrences, and so on.

9.7.2 Dynamic Contractions

In case of dynamic contractions many other factors affect the results in addition to those encountered with isometric contractions. One of the main problems when dealing with dynamic muscle contractions is that of the high degree of nonstationarity of the signal. EMG signals may suddenly change their spectral properties during a dynamic task, and this may be difficult to investigate with classic spectral techniques. Recently methods for time-frequency analysis of surface EMG signals detected during dynamic conditions have been proposed. Most of them are based on the Cohen's class of time-frequency distributions [9] or wavelets [47]. These methods are believed to provide better results than classic short-time Fourier transform, although there are still no modeling studies showing real improvements in practical applications.

Apart from the spectral estimation problem, many issues related to the dynamic condition should be considered in addition to the factors described above that affect results in isometric contractions. During movement, the muscle changes its size and moves with respect to the electrodes located over the skin. This phenomenon depends on the specific muscle. The relative shift of the muscle with respect to the skin during movements, almost covering the dynamic range of the joint, ranges from a few millimeters, as in the upper trapezius muscle or gastrocnemii muscles [30,32], up to 3 centimeters, as in semitendinosus muscle [30]. Due to the large sensitivity of EMG variables to electrode location with respect to the innervation zone and tendon regions, a shift of the muscle may result in large geometrical artifacts and forms of nonstationarity. Measurements during cyclic contractions [10] may partially overcome this problem.

Apart from geometrical factors, during dynamic contractions consistent rederecrutiment of MUs occurs, and thus interpretation of changes of spectral EMG variables may be critical as explained for the case of nonconstant force isometric contractions.

9.8 REPEATABILITY OF ESTIMATES OF EMG VARIABLES AND FATIGUE INDEXES

Assessment of the repeatability of EMG variables is of considerable relevance for the clinical daily use of surface EMG. To reach this goal, it is necessary (1) to ensure that data recorded in different laboratories and by different operators are comparable and (2) to define the minimum change of the observed variables that may be related to physiological variations and not to random fluctuations. Such an assessment is a precondition to demonstrate the potential clinical usefulness of the technique.

Several works on repeatability of surface EMG variables on many different muscles, such as elbow flexors, quadriceps, back, and respiratory muscles, can be found in the literature of the last 10 years [8,16,48,55,70,71,74,75]. If associated with the problem of proper electrode positioning and with the correct use of statistical tools, the issue of repeatability assumes a dramatic importance in the validation process of the surface EMG methodology. The efforts toward standardization promoted by the European Concerted Action "Surface EMG for Noninvasive Assessment of Muscles" (SENIAM) [40] move in this direction and should be considered with great attention by EMG clinical researchers.

The issue of repeatability can be addressed considering two different and complementary aspects. The first aspect concerns the reliability (within days) and constancy (between days) of the repeated measure [89], which are important in order to assess the measure precision (e.g., described by the normalized standard error of the mean, NSEM)

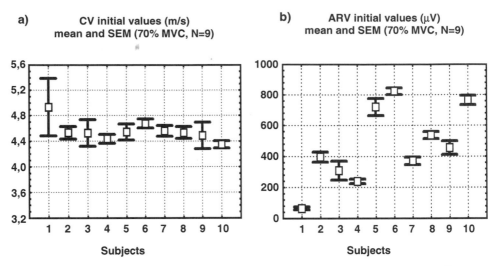

Figure 9.12. (*a*) Within subject variance is greater than between subject variance. Differences between subjects cannot be detected. ICC is meaningless. CV is not affected by individual differences (which are smaller than measurement error). Total variance is due mostly to intra-subject variability. (*b*) Within subject variance is smaller than between subject variance. Differences between subjects can be detected. ICC is meaningful. ARV is affected by individual differences (which are larger than measurement error). Total variance is mostly due to inter-subject variability. (Redrawn from [75] with permission)

and the variability due to repeated trials and electrode repositioning. The second concerns the capability of the variables investigated in distinguishing between different subjects. This capability is in general assessed using the Intraclass Correlation Coefficient (ICC) or the Fisher test; both lead to similar results [74]. When the between-subject variability is comparable to (or less than) the within-subject variability, the degree of repeatability defined by the ICC becomes meaningless [74,75]. When very little variation in the EMG variable estimates across subjects and trials is found, the measure might not be able to detect different muscle properties among uniform groups (e.g., healthy subjects). In this case the EMG variable might be used to provide a reference range for a uniform group of subjects (i.e., normative data), but the ICC is not an appropriate indicator of inter-subject differences, since the within-subject variance may be larger than the between-subject variance (see Fig. 9.12).

Several publications are available in the literature regarding the repeatability of surface EMG variables [8,14,16,21,48,51,60,65,71,74,75,88], among others. Different surface EMG parameters are investigated, and different experimental protocols are used. Therefore comparison of results is difficult. However, findings from some of the works cited are selected among those obtained with similar methodology and are shown in Table 9.1 to offer the reader a summary of the state of the art on the issue of repeatability of EMG variable estimates in different muscles.

9.9 CONCLUSIONS

The analysis of the surface EMG signal may provide an objective tool for the assessment of muscle fatigue, although the relationships between EMG variable changes and the

TABLE 9.1. Repeatability of Initial Values of Mean Power Spectral Frequency (MNF), Average Rectified Value (ARV), and Conduction Velocity (CV) during Isometric Voluntary and Electrically Elicited Contractions in Different Muscles

Muscle	MNF		ARV		CV		Source of Data
	ICC (%)	NSEM (%)	ICC (%)	NSEM (%)	ICC (%)	NSEM (%)	
Sternocleidomastoid (N = 9)	43.9	3.0	70.6	16.8	42.5	2.8	[21]
Anterior scalenus (N = 9)	74.3	7.15	25.7	13.6	28.0	2.6	[21]
Biceps brachii (N = 10)	71.8	4.4	77.7	18.3	<0	1.5	[75]
Longissimus[a] (N = 10)	87.0	—	73.7	—	—	—	[88]
Multifidus[a] (N = 10)	87.5	—	49.5	—	—	—	[88]
Iliocostalis[a] (N = 10)	88.5	—	71.0	—	—	—	[88]
Rectus femoris (N = 18)	88.7	—	73.7	—	—	—	[48]
Vastus lateralis[a] (N = 9)	81.5	6.8	80.1	16.3	61.4	7.3	[74]
Vastus medialis[a] (N = 9)	70.1	4.3	73.9	14.7	<0	5.3	[74]
Vastus medialis[b] (N = 9)	85.2	2.3	21.6	7.4	35.7	3.1	[70]
Tibialis anterior[b] (N = 10)	86.5	—	78.0	—	11.5	—	[65]

Note: N = number of measured subjects. The values of ICC and NSEM were reported for the three variables when available.

[a] Values were obtained as a mean between left and right findings.
[b] Results of electrically elicited contractions.

underlying physiological phenomena are very complex and not yet fully understood. This technique was proved to be able to differentiate between subject groups and muscles and to have potentially relevant applications in basic research and clinical studies. Methods for analyzing fatigue at the single motor unit level from noninvasive recordings would allow a more reliable, although limited to a small muscle portion, indication and are currently under study (see Chapter 7).

REFERENCES

1. Andreassen, S., and L. Arendt-Nielsen, "Muscle fibre conduction velocity in motor units of the human anterior tibial muscle: A new size principle parameter," *J Physiol* **391**, 561–571 (1987).
2. Arendt-Nielsen, L., and K. R. Mills, "The relationship between mean power frequency of the EMG spectrum and muscle fibre conduction velocity," *Electroenceph Clin Neurophysiol* **60**, 130–134 (1985).

3. Basmajian, J. V., and C. J. De Luca, *Muscles alive*, 5th ed., Williams and Wilkins, Philadelphia, 1985.

4. Bernardi, M., M. Solomonow, G. Nguyen, A. Smith, and R. Baratta, "Motor unit recruitment strategies changes with skill acquisition," *Eur J Appl Physiol* **74**, 52–59 (1996).

5. Bernardi, M., F. Felici, M. Marchetti, F. Montellanico, M. F. Piacentini, and M. Solomonow, "Force generation performance and motor unit recruitment strategy in muscles of contralateral limbs," *J Electromyogr Kinesiol* **9**, 121–130 (1999).

6. Bigland-Ritchie, B., R. Johansson, O. C. Lippold, S. Smith, and J. J. Woods, "Changes in motoneurone firing rates during sustained maximal voluntary contractions," *J Physiol* **340**, 335–346 (1983).

7. Bilodeau, M., A. B. Arsenault, D. Gravel, and D. Bourbonnais, "EMG power spectra of elbow extensors during ramp and step isometric contractions," *Eur J Appl Physiol* **63**, 24–28 (1991).

8. Bilodeau, M., A. B. Aesenault, D. Gravel, and D. Bourbonnais, "EMG power spectrum of elbow extensors: A reliability study," *Electrom Clin Neurophys* **34**, 149–158 (1994).

9. Bonato, P., "Recent advancements in the analysis of dynamic EMG data," *IEEE Eng Med Biol Mag* **20**, 29–32 (2001).

10. Bonato, P., S. H. Roy, M. Knaflitz, and C. J. DeLuca, "Time-frequency parameters of the surface myoelectric signal for assessing muscle fatigue during cyclic dynamic contractions," *IEEE Trans Biomed Eng* **48**, 745–753 (2001).

11. Brody, L., M. Pollock, S. Roy, C. J. De Luca, and B. Celli, "pH induced effects on median frequency and conduction velocity of the myoelectric signal," *J Appl Physiol* **71**, 1878–1885 (1991).

12. Chaffin, D. B., "Localized muscle fatigue, definition and measurements," *J Occup Med* **15**, 346–354 (1973).

13. Chisari, C., C. D'Alessandro, M. L. Manca, and B. Rossi, "Sarcolemmal excitability in myotonic dystrophy: Assessment through surface EMG," *Muscle Nerve* **21**, 543–546 (1998).

14. Daanen, H., M. Mazure, M. Holewijn, and E. Van Der Velde, "Reproducibility of the mean power frequency of the surface electromyogram," *Eur J Appl Physiol* **61**, 274–277 (1990).

15. De Luca, C. J., "Myoelectric manifestations of localized muscular fatigue in humans," *CRC Crit Rev Biomed Eng* **11**, 251–279 (1984).

16. Dedering, A., M. Roos af Hjelmsater, B. Elfving, K. arms-Ringdahl, and G. Nemeth, "Between days reliability of subjective and objective assessments of back extensor muscle fatigue in subjects without lower back pain," *J Electromyogr Kinesiol* **10**, 151–158 (2000).

17. Dimitrov, G. V., Z. C. Lateva, and N. A. Dimitrova, "Effects of changes in asymmetry, duration and propagation velocity of the intracellular potential on the power spectrum of extracellular potentials produced by an excitable fiber," *Electromyogr Clin Neurophysiol* **28**, 93–100 (1988).

18. Dimitrova, N. A., and G. V. Dimitrov, "Calculation of extracellular potentials produced by an inclined muscle fiber at a rectangular plate electrode," *Med Eng Phys* **21**, 583–588 (1999).

19. Dimitrova, N. A., G. V. Dimitrov, and V. N. Chihman, "Effect of electrode dimensions on motor unit potentials," *Med Eng Phys* **21**, 479–485 (1999).

20. Enoka, R. M., and A. J. Fuglevand, "Motor unit physiology: Some unresolved issues," *Muscle Nerve* **24**, 4–17 (2001).

21. Falla, D., P. Dall'Alba, A. Rainoldi, R. Merletti, and G. Jull, "Repeatability of surface EMG variables in the sternocleidomastoid and anterior scalene muscles," *Eur J Appl Physiol* **87**, 542–549 (2002).

22. Falla, D., A. Rainoldi, R. Merletti, and G. Jull, "Myoelectric manifestations of sternocleidomastoid and anterior scalene muscle fatigue in chronic neck pain patients," *Clin Neurophysiol* **114**, 488–495 (2003).

23. Farina, D., "Advances in myoelectric signal modelling, detection and processing in motor control studies," PhD thesis, Politecnico di Torino and Ecole Centrale de Nantes, 2002.

24. Farina, D., and R. Merletti, "Comparison of algorithms for estimation of EMG variables during voluntary isometric contractions," *J Electromyogr Kinesiol* **10**, 337–350 (2000).

25. Farina, D., and R. Merletti, "Effect of electrode shape on spectral features of surface detected motor unit action potentials," *Acta Physiol Pharmacol Bulg* **26**, 63–66 (2001).

26. Farina, D., and A. Rainoldi, "Compensation of the effect of sub-cutaneous tissue layers on surface EMG: A simulation study," *Med Eng Phys* **21**, 487–496 (1999).

27. Farina, D., C. Cescon, and R. Merletti, "Influence of anatomical, physical and detection system parameters on surface EMG," *Biol Cybern* **86**, 445–456 (2002).

28. Farina, D., E. Fortunato, and R. Merletti, "Non-invasive estimation of motor unit conduction velocity distribution using linear electrode arrays," *IEEE Trans Biomed Eng* **41**, 380–388 (2000).

29. Farina, D., M. Fosci, and R. Merletti, "Motor unit recruitment strategies investigated by surface EMG variables: An experimental and model based feasibility study," *J Appl Physiol* **92**, 235–247 (2002).

30. Farina, D., R. Merletti, M. Nazzaro, and I. Caruso, "Effect of joint angle on surface EMG variables for the muscles of the leg and thigh," *IEEE Eng Med Biol Mag* **20**, 62–71 (2001).

31. Farina, D., E. Schulte, R. Merletti, G. Rau, and C. Disselhorst-Klug, "Single motor unit analysis from spatially filtered surface EMG signals. Part 1: Spatial selectivity," *Med Biol Eng Comput* **41**, 330–337 (2003).

32. Farina, D., P. Madeleine, T. Graven-Nielsen, R. Merletti, and L. Arendt-Nielsen, "Stardardising surface electromyogram recordings for assessment of activity and fatigue in the human upper trapezius muscle," *Eur J Appl Physiol* **86**, 469–478 (2002).

33. Farina, D., R. Merletti, B. Indino, M. Nazzaro, and M. Pozzo, "Cross-talk between knee extensor muscles. Experimental and modelling results," *Muscle Nerve* **26**, 681–695 (2002).

34. Farina, D., W. Muhammad, E. Fortunato, O. Meste, R. Merletti, and H. Rix, "Estimation of single motor unit conduction velocity from surface electromyogram signals detected with linear electrode arrays," *Med Biol Eng Comput* **39**, 225–236 (2001).

35. Farina D., R. Merletti, A. Rainoldi, M. Buonocore, and R. Casale, "Two methods for the measurement of voluntary contraction torque in the biceps brachii muscle," *Med Eng Phys* **21**, 533–540 (1999).

36. Froese, E. A., and M. E. Houston, "Torque-velocity characteristics and muscle fiber type in human vastus lateralis," *J Appl Physiol* **59**, 309–314 (1985).

37. Gazzoni, M., D. Farina, and R. Merletti, "Motor unit recruitment during constant low force and long duration muscle contractions investigated by surface electromyography," *Acta Physiol Pharmacol Bulg* **26**, 67–71 (2001).

38. Gerdle, B., N. E. Eriksson, and L. Brundin, "The behaviour of the mean power frequency of the surface electromyogram in biceps brachii with increasing force and during fatigue: With special regard to the electrode distance," *Electromyogr Clin Neurophysiol* **30**, 483–489 (1990).

39. Gerdle, B., K. Henriksson-Larsen, R. Lorentzon, and M. L. Wretling, "Dependence of the mean power frequency of the electromyogram on muscle force and fibre type," *Acta Physiol Scand* **142**, 457–465 (1991).

40. Hermens, H. J., B. Freriks, C. Disselhorst-Klug, and G. Rau, "Development of recommendations for SEMG sensor and sensor placement procedures," *J Electromyogr Kinesiol* **10**, 361–374 (2000).

41. Hopf, H. C., R. L. Herbort, M. Gnass, H. Günther, and K. Lowitzsch, "Fast and slow contraction times associated with fast and slow spike conduction of skeletal muscle fibers in normal subject and in spastic hemiparesis," *Z Neurol* **206**, 193–202 (1974).

42. Jensen, C., O. Vasseljen, and R. H. Westgaard, "The influence of electrode position on bipolar surface electromyogram recordings of the upper trapezius muscle," *Eur J Appl Physiol* **67**, 266–273 (1993).

43. Johnson, M. A., J. Polgar, D. Weightman, and D. Appleton, "Data on the distribution of fiber types in thirty-six human muscles: An autopsy study," *J Neurol Sci* **18**, 111–129 (1973).

44. Kadefors, R., E. Kaiser, and I. Petersen, "Dynamic spectrum analysis of myo-potentials and with special reference to muscle fatigue," *Electromyography*, **8**, 39–74 (1968).

45. Kamen, G., and C. J. DeLuca, "Unusual motor firing behavior in older adults," *Brain Res* **482**, 136–140 (1989).

46. Kamen, G., S. V. Sison, C. C. Du, and C. Patten, "Motor unit discharge behavior in older adults during maximal-effort contractions," *J Appl Physiol* **79**, 1908–1913 (1995).

47. Karlsson, S., J. Yu, and M. Akay, "Time-frequency analysis of myoelectric signals during dynamic contractions: A comparative study," *IEEE Trans BME* **47**, 228–238 (2000).

48. Kollmitzer, J., G. R. Ebenbichler, and A. Kopf, "Reliability of surface electromyographic measurements," *Clin Neurophysiol* **110**, 725–734 (1999).

49. Kupa, E. J., S. H. Roy, S. C. Kandarian, and C. J. De Luca, "Effects of muscle fiber type and size on EMG median frequency and conduction velocity," *J Appl Physiol* **79**, 23–32 (1995).

50. Lago, P., and N. B. Jones, "Effect of motor unit firing time statistics on EMG spectra," *Med Biol Eng Comput* **15**, 648–655 (1977).

51. Lariviere, C., A. B. Arsenault, D. Gravel, D. Gagnon, and P. Loisel, "Evaluation of measurement strategies to increase the reliability of EMG indices to assess back muscle fatigue and recovery," *J Electromyogr Kinesiol* **12**, 91–102 (2002).

52. Li, W., and K. Sakamoto, "The influence of location of electrode on muscle fiber conduction velocity and EMG power spectrum during voluntary isometric contractions measured with surface array electrodes," *Appl Human Sci* **15**, 25–32 (1996).

53. Lindstrom, L., and R. Magnusson, "Interpretation of myoelectric power spectra: A model and its applications," *Proc IEEE* **65**, 653–662 (1977).

54. Lindstrom, L., R. Magnusson, and R. Petersen, "Muscle fatigue and action potential conduction velocity changes studied with frequency analysis of EMG signals," *Electrom Clin Neurophysol* **10**, 341–356 (1970).

55. Linssen, W., D. Stegeman, E. Joosten, M. van't Hof, R. Binkhorst, and S. Notermans, "Variability and interrelationships of surface EMG parameters during local muscle fatigue," *Muscle Nerve* **16**, 849–856 (1993).

56. Linssen, W. H., D. F. Stegeman, E. M. Joosten, R. A. Binkhorst, M. J. Merks, H. J. ter Laak, and S. L. Notermans, "Fatigue in type I fiber predominance: A muscle force and surface EMG study on the relative role of type I and type II muscle fibers," *Muscle Nerve* **14**, 829–837 (1991).

57. Linssen, W. H., M. Jacobs, D. F. Stegeman, E. M. Joosten, and J. Moleman, "Muscle fatigue in McArdle's disease: Muscle fibre conduction velocity and surface EMG frequency spectrum during ischaemic exercise," *Brain* **113**, 1779–1793 (1990).

58. Lo Conte, L., and R. Merletti, "Advances in processing of surface myoelectric signals. Part 2," *Med Biol Eng Comput* **33**, 373–384 (1995).

59. Lowery, M. M., C. L. Vaughan, P. J. Nolan, and M. J. O'Malley, "Spectral compression of the electromyographic signal due to decreasing muscle fiber conduction velocity," *IEEE Trans Rehabil Eng* **8**, 353–361 (2000).

60. Maarsingh, E. J., L. A. van Eykern, A. B. Sprikkelman, M. O. Hoekstra, and W. M. van Aalderen, "Respiratory muscle activity measured with a noninvasive EMG technique: Technical aspects and reproducibility," *J Appl Physiol* **88**, 1955–1961 (2000).

61. Mannion, A. F., G. A. Dumas, J. M. Stevenson, and R. G. Cooper, "The influence of muscle fiber size and type distribution on electromyographic measures of back muscle fatigability," *Spine* **23**, 576–584 (1998).

62. McGill, K. C., and L. J. Dorfman, "High resolution alignment of sampled waveforms," *IEEE Trans Biomed Eng* **31**, 462–470 (1984).

63. Merletti, R., and S. Roy, "Myoelectric and mechanical manifestations of muscle fatigue in voluntary contractions," *J Orthop Sports Phys Ther* **24**, 342–353 (1996).

64. Merletti, R., L. R. Lo Conte, and C. Orizio, "Indices of muscle fatigue," *J Electromyogr Kinesiol* **1**, 20–33 (1991).

65. Merletti, R., L. R. Lo Conte, and D. Sathyan, "Repeatability of electrically-evoked myoelectric signals in the human tibialis anterior," *J Electromyogr Kinesiol* **5**, 67–80 (1995).

66. Merletti, R., M. Knaflitz, and C. J. De Luca, "Electrically evoked myoelectric signals," *CRC Rev Biomed Eng* **19**, 293–340 (1992).

67. Merletti, R., M. Knaflitz, and C. J. De Luca, "Myoelectric manifestations of fatigue in voluntary and electrically elicited contractions," *J Appl Physiol* **69**, 1810–1820 (1990).

68. Merletti, R., M. Sabbahi, and C. J. De Luca, "Median frequency of the myoelectric signal: Effects of muscle ischemia and cooling," *Eur J Appl Physiol* **52**, 258–265 (1984).

69. Merletti, R., D. Farina, M. Gazzoni, and M. P. Schieroni, "Effect of age on muscle functions investigated with surface electromyography," *Muscle Nerve* **25**, 65–76 (2002).

70. Merletti, R., A. Fiorito, L. R. Lo Conte, and C. Cisari, "Repeatability of electrically evoked EMG signals in the human vastus medialis muscle," *Muscle Nerve* **21**, 184–193 (1998).

71. Ng, J. K., and C. A. Richardson, "Reliability of electromyographic power spectral analysis of back muscle endurance in healthy subjects," *Arch Phys Med Rehabil* **77**, 259–264 (1996).

72. Pedrinelli, R., L. Marino, G. Dell'Omo, G. Siciliano, and B. Rossi, "Altered surface myoelectric signals in peripheral vascular disease: Correlations with muscle fiber composition," *Muscle Nerve* **21**, 201–210 (1998).

73. Piper, H., *Electrophysiologie, Menschlicher Muskeln*, Springer Verlag, Berlin, 1912.

74. Rainoldi, A., J. E. Bullock-Saxton, F. Cavarretta, and N. Hogan, "Repeatability of maximal voluntary force and of surface EMG variables during voluntary contraction of quadriceps muscles in healthy subjects," *J Electromyogr Kinesiol* **11**, 25–438 (2001).

75. Rainoldi, A., G. Galardi, L. Maderna, G. Comi, L. R. Lo Conte, and R. Merletti, "Repeatability of surface EMG variables during voluntary isometric contractions of the biceps brachii muscle," *J Electromyogr Kinesiol* **9**, 105–119 (1999).

76. Rainoldi, A., M. Nazzaro, R. Merletti, D. Farina, I. Caruso, and S. Gaudenti, "Geometrical factors in surface EMG of the vastus medialis and lateralis," *J Electromyogr Kinesiol* **10**, 327–336 (2000).

77. Rix, H., and J. P. Malengé, "Detecting small variation in shape," *IEEE Trans Syst Man Cybern* **10**, 90–96 (1980).

78. Roy, S. H., C. J. DeLuca, and J. Schneider, "Effects of electrode location on myoelectric conduction velocity and median frequency estimates," *J Appl Physiol* **61**, 1510–1517 (1986).

79. Sadoyama, T., T. Masuda, H. Miyata, and S. Katsuta, "Fiber conduction velocity and fiber composition in human vastus lateralis," *Eur J Appl Physiol* **57**, 767–771 (1988).

80. Schulte, E., D. Farina, G. Rau, R. Merletti, and C. Disselhorst-Klug, "Single motor unit analysis from spatially filtered surface EMG signals. Part 2: Conduction velocity estimation," *Med Biol Eng Comput* **41**, 338–345 (2003).

81. Solomonow, M., C. Baten, J. Smith, R. Baratta, H. Hermens, R. D'Ambrosia, and H. Shoji, "Electromyogram power spectra frequencies associated with motor unit recruitment strategies," *J Appl Physiol* **68**, 1177–1185 (1990).

82. Stulen, F. B., and C. J. DeLuca, "Frequency parameters of the myoelectric signals as a measure of muscle conduction velocity," *IEEE Trans BME* **28**, 515–523 (1981).

83. Suter, E., W. Herzog, J. Sokolosky, J. P. Wiley, and B. R. Macintosh, "Muscle fiber type distribution as estimated by Cybex testing and by muscle biopsy," *Med Sci Sports Exer* **25**, 363–370 (1993).

84. Tesch, P. A., P.-V. Komi, I. Jacobs, J. Karlsson, and J. T. Viitasalo, "Influence of lactate accumulation of EMG frequency spectrum during repeated concentric contractions," *Acta Physiol Scand* **119**, 61–67 (1983).

85. Troni, W., R. Cantello, and I. Rainero, "Conduction velocity along human muscle fibers in situ," *Neurology* **33**, 1453–1459 (1983).

86. Uhlig, Y., B. R. Weber, D. Grob, and M. Muntener, "Fiber composition and fiber transformations in neck muscles of patients with dysfunction of the cervical spine," *J Orthop Res* **13**, 240–249 (1995).

87. van Boxtel, A., and L. R. B. Shomaker, "Influence of motor unit firing statistics on the median frequency of the EMG power spectrum," *Eur J Appl Physiol* **52**, 207–213 (1984).

88. van Dieen, J. H., and P. Heijblom, "Reproducibility of isometric trunk extension torque, trunk extensor endurance, and related electromyographic parameters in the context of their clinical applicability," *J Orthop Res* **14**, 139–143 (1996).

89. Viitasalo, J. H. T., and P. Komi, "Signal characteristics of EMG with special reference to reproducibility of measurements," *Acta Physiol Scand* **93**, 531–539 (1975).

90. Viitasalo, J. T., and P. V. Komi, "Interrelationships of EMG signal characteristics at different levels of muscle tension during fatigue," *Electromyogr Clin Neurophysiol* **18**, 167–178 (1978).

91. Westgaard, R. H., and C. J. DeLuca, "Motor unit substitution in long-duration contractions of the human trapezius muscle," *J Neurophysiol* **82**, 501–504 (1999).

92. Zijdewind, I., and D. Kernell, "Fatigue associated EMG behavior of the first dorsal interosseous and adductor pollicis muscles in different groups of subjects," *Muscle Nerve* **17**, 1044–1054 (1994).

93. Zwarts M. J., "Evaluation of the estimation of muscle fiber conduction velocity: Surface versus needle method," *Electroencephalogr Clin Neurophysiol* **73**, 544–548 (1989).

10

ADVANCED SIGNAL PROCESSING TECHNIQUES

D. Zazula

Faculty of Electrical Engineering and Computer Science
University of Maribor, Maribor, Slovenia

S. Karlsson

University Hospital
Department of Biomedical Engineering and Informatics, Umea, Sweden

C. Doncarli

IRCCyN
Ecole Centrale de Nantes, Nantes, France

10.1 INTRODUCTION

In many different application fields knowledge on a physical phenomenon is obtained through the recording of a signal. The processing of this signal can be aimed at description, analysis, and decision.

To shape the experimental results, the user must already be familiar with the studied phenomenon. Training is critical because experience is needed to abstract relevant information contained in the signals. However, when an unfamiliar recorded signal appears, what is the right decision? The experienced researcher will diagnose the presence of an anomaly (binary decision or detection), or assign the signal to a class corresponding to a particular behavior (classification)—with and without rejection decisions. Such a decision method requires one to represent the signal by a limited number of features known as "dis-

Electromyography: Physiology, Engineering, and Noninvasive Applications, edited by Roberto Merletti and Philip Parker.
ISBN 0-471-67580-6   Copyright © 2004 Institute for Electrical and Electronics Engineers, Inc.

criminating characters." To do so, one must define a representation space in which the features differentiating the classes are present in distinguishable clusters.

Many studies have tackled stationary signals from two points of view: One way is through a mathematical (statistical) model, using assumed parameters to represent the signal. The model is used to calculate the conditional probability of each class (the measurements and the training set being known), which is a basis for an optimization methodology. Often required are the use of computationally heavy numerical methods (e.g., Monte Carlo method and Markov chains), and their success depends on the fit of the model to real data. Another way, when the modeling method is too difficult, is to define nonparametric descriptors, from which it is possible to derive ad hoc decision algorithms.

For nonstationary signals the problem is made of choosing the right model becomes more difficult, and this is why a number of other approaches—based on the time-frequency or time-scale properties—have been developed.

10.1.1 Parametric Context

Bayesian decision theory, which can be used when the relevant statistical parameters are known (parametric context), has received much attention in the pattern recognition literature. This is because a Bayes classifier is optimal in terms of minimum average error probability. To implement the Bayes classifier, one assumes that the class conditional distributions and the a priori probabilities of the classes are known. Unfortunately, these probabilities are unknown in many practical applications and have to be estimated from a learning set, which can be troublesome. One must then consider two cases: using classical hypotheses, where each class-conditional density is defined by an unknown but constant parameter vector, or using hierarchical hypotheses, where the probability density function of each individual (item), inside a given class, depends on a random parameter vector, distributed by a law that depends on an unknown constant vector that characterizes the class. The problem with this Bayesian learning procedure is that it requires the computation of multidimensional integrals, which do not admit any closed-form expressions. Because the integrals represent mathematical expectations, a Monte Carlo method is used to obtain the accurate numerical evaluation of these integrals. However, the relevant samples must be generated by running an ergodic Markov chain (e.g., the Metropolis-Hastings algorithm).

Second-order statistics are sufficient to describe many real signals. However, higher order moments are used when the Gaussian hypothesis cannot be applied or when the discriminating characters cannot be seen in the second-order properties. For example, source separation (the decomposition of superimposed signals) calls for the analysis of at least third-order moments.

10.1.2 Nonparametric Context

In many situations the size of the learning set is too small to permit correct parameter estimation, or worse, no suitable model structure can be assumed. It is then necessary to find a model-free representation space in which the differences between the classes are emphasized and the similarities are attenuated. A variety of approaches are based on wavelet or wavelet packet decomposition and selection of the best basis of packets. Only the discriminant packets are selected, and the representation space dimension is then reduced. Also a signal-dependant mother wavelet can be chosen to improve the detector or classifier's accuracy. Other methods use time-frequency representations, and these are interesting for several reasons. First, for the detection of Gaussian signals in Gaussian noise, the optimal (Bayesian) classifier admits an equivalent closed-form expression in the time-frequency plane. Second, energy location (in time and frequency) is often a discriminant character. Third, the representation kernel can be adapted to the learning classes to improve the classifier's performance.

10.1.3 Conclusion

Parametric and nonparametric approaches can be both used in practical decision for classification problems. Nonparametric representations (time-scale or time-frequency representations) sometimes outline features that can help the user in the choice of a parametric model. Nonparametric representations are directly related to the data, and the results deducted are thus independent of a possible error in the model choice. However, model-based results (e.g., Bayesian learning) can be optimal, so the parametric approach should be used whenever possible. Nevertheless, we should keep in mind that if there is an error in the model form, the results of the parametric approach can be entirely meaningless. Because EMGs models depend on experimental conditions (level of force, static or dynamic task, etc.), beginning with nonparametric methods seems to be a good approach.

10.2 THEORETICAL BACKGROUND

10.2.1 Multichannel Models of Compound Signals

Signal theory involves a system transfer function that, in a mathematical sense, describes a transform relating two functions (i.e., two signals). Take a linear, shift-invariant, stable system whose input is excited by the time-discrete input signal, $x(n)$, while the time-discrete output $y(n)$ depends on a so-called system unit-sample response, $h(n)$. These are interconnected by linear convolution:

$$y(n) = \sum_{i=0}^{\infty} x(i)h(n-i) \tag{1}$$

Such a model has only one transfer channel with a single input and single output (SISO). It can be applied when a system is excited only by a single signal source and its response is obtained from the observation of a single signal realization or measurement. However, physiological phenomena tend to be multivariate. Recall the discussion of EMG generation and modeling in Chapters 1, 4, and 8. It is evident that the EMG belongs to the class of compound signals. Several motor units are excited by different firing patterns caused by the discharges that trigger the potentials traveling along the fibers and their diffusion through the volume conductor. These potentials may be detected in a form of action potentials (AP) (see Chapters 2 and 5). If a single-fiber action potential (SFAP) can be detected, the SISO modeling can be applied. A similar situation appears when the response of only one motor unit, namely a single motor unit action potential (MUAP), is measured. However, such cases with only one innervation source and only one response captured by the measurement are very limited in their clinical relevance.

EMG measuring approaches use concentric needle or surface electrodes. Both techniques detect MUAPs of several motor units at the same time, which means that the meas-

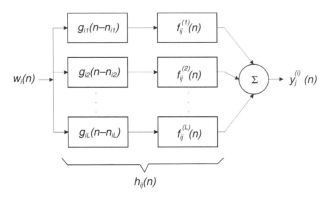

Figure 10.1. Contribution of the ith motor unit to the jth EMG measurement. $w_i(n)$ stands for the motor-unit innervation train, $g_{il}(n - n_{il})$ for the lth fiber impulse response, $f_{ij}^{(l)}(n)$ for a spatial filtering characteristic of the tissue, fat, and possibly skin between a fiber and the measurement point, and $y_j^{(i)}(n)$ for the measurement contributed by the ith motor unit at the jth pickup.

ured signal is influenced by several sources. Such compound signals may be modeled with multiple inputs to a system whose output configuration depends on the number of measuring electrodes. For a concentric needle electrode or either unipolar or bipolar surface detection systems, a multiple-input, single-output model (MISO) suits the needs. Measurements with more surface electrodes, regardless of their placement, lead to multiple-input, multiple-output models (MIMO).

Suppose we denote the ith innervation pulse train by $w_i(n)$, and let it trigger L muscle fibers that belong to the same motor unit (MU). Because of the scattered end plates in the innervation zone, the individual fiber firings will be differently delayed, in this case the lth fiber by n_{il} samples. Figure 10.1 depicts fiber APs as fiber responses $g_{il}(n)$, where indexes i and l denote the ith MU and its lth fiber, respectively.

However, the EMG measurement electrodes capture APs after they have been transferred and modified through tissue layers, subcutaneous fat, and for surface measurements, also the skin. This effect may be observed as an additional filtering. Each single-fiber AP propagates through a spatial filter $f_{ij}^{(l)}(n)$, where indexes i and j stand for the ith MU and the jth pickup, respectively, while subscript l denotes the MU's lth fiber. As Figure 10.1 shows the ith innervation train generates the following signal at the jth electrode:

$$y_j^{(i)}(n) = \left\{ \sum_{l=1}^{L} [g_{il}(n - n_{il}) * f_{ij}^{(l)}(n)] \right\} * w_i(n) = h_{ij}(n) * w_i(n) \tag{2}$$

where $*$ means convolution and $h_{ij}(n)$ stands for the shape of the ith MUAP as it appears at the jth pickup.

In all cases, except single-fiber AP measurements, electrodes detect superimposed contributions of several signals. Now suppose that K motor units are active at the same time; then the jth pickup measures the following signal:

$$y_j(n) = \sum_{i=1}^{K} y_j^{(i)}(n) = \sum_{i=1}^{K} h_{ij}(n) * w_i(n) \tag{3}$$

An obvious extension describes the MIMO model depicted in Figure 10.2:

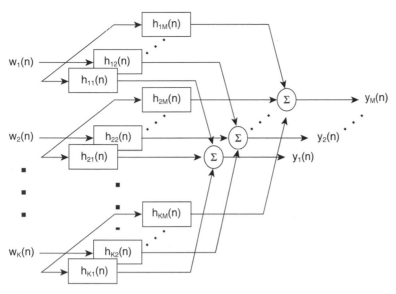

Figure 10.2. Multiple-input, multiple-output (MIMO) model. $w_i(n)$ denotes the input excitations, $h_{ij}(n)$ the model channel impulse responses, and $y_j(n)$ the model outputs.

$$\mathbf{y}(n) = \sum_{i=0}^{\infty} \mathbf{H}^T(i)\mathbf{w}(n-i) \qquad (4)$$

where

$$\mathbf{y}(n) = [y_1(n), y_2(n), \ldots, y_M(n)]^T$$

is a vector of M output measurements (M-electrodes), and

$$\mathbf{H}(n) = \begin{bmatrix} h_{11}(n), h_{12}(n), \ldots, h_{1M}(n) \\ h_{21}(n), h_{22}(n), \ldots, h_{2M}(n) \\ \vdots \\ h_{K1}(n), h_{K2}(n), \ldots, h_{KM}(n) \end{bmatrix},$$

$$\mathbf{w}(n) = [w_1(n), w_2(n), \ldots, w_K(n)]^T$$

with $h_{ij}(n)$ denoting the MUAP originating from the ith source (MU) and measured by pickup j, and $w_i(n)$ the innervation pulse train at the ith MU.

The proposed models explain the synthesis of MUAPs based on the incorporated SFAPs and the single- and multiple-electrode EMG measurements (the MISO and MIMO models, respectively). On the other hand, the same model may serve as a decomposition vehicle for the appropriate EMG decomposition methods. The resulting shapes of the basic EMG components, such as MUAPs, SFAPs, or the innervation trains, have significant clinical importance. Some advanced EMG decomposition methods based on time scale representations and on higher order statistics will be derived in the sequel.

10.2.2 Stochastic Processes

Events that may be influenced by processes, such as atmospheric and weather changes, that have unpredictable outcomes or behavior are treated as random (stochastic) processes. A random process is evolving in time. Its random nature requires several long observations if we want to recognize the inherent features of the process. We can think of these observations as several measurements of a system's response. After the measured values are taken over time, a so-called random sequence appears. A single value can be taken out of every measurement regardless of time, and this set of samples constitutes a so-called random variable. Two important properties characterize the random process. The first is stationarity, meaning that all random variables across time exhibit the same statistical behavior (e.g., the same moments). The second is ergodicity: the random process is ergodic if it shows the same statistical properties regardless of whether they are determined along random sequences or with the help of random variables [1]. Ergodicity plays an important role in practical implementations. Let us demonstrate this fact using the EMG measurements. If the statistical properties of EMG are varied with time, no stable and consistent statistical measure can be obtained from a single temporal EMG recording. Instead, we should repeat the measurement session under the same conditions several times, and then combine the recorded values from all the measurements at certain time instants in order to obtain random variables. These variables can separately serve further statistical processing.

Statistical processing of random variables opens some important insights into their features and extracts the information buried in their random nature. To understand how different statistics can help EMG analysis, we have to first get acquainted with the most appropriate statistical measures [27,39].

Statistical Signal Processing: Moments and Cumulants. Let $X = \{x_1, x_2, \ldots, x_N\}$ be a random variable. Then a well-known relationship [27,43] generates the moments, m_k, of X:

$$m_k = E(X^k) \tag{5}$$

where E stands for mathematical expectation. First- and second-order moments are the most often applied statistical measures because they correspond to the mean (average) and variance (deviation) of a random variable.

Combinations of certain moments possess the property of linearity. These combinations are called cumulants. A cumulant of order k, C_k, enters the following relationship with the moments [31].

$$\sum_{k=1}^{\infty} \frac{C_k}{k!} t^k = \sum_{i=1}^{\infty} \frac{(-1)^{i+1}}{i} \left(\sum_{j=1}^{\infty} \frac{m_j}{j!} t^j \right)^i \tag{6}$$

The calculation of C_k, suggested by the Taylor expansion of Eq. (6), turns into a summation of all right-hand terms whose exponent of a small real variable t (i.e., $j \cdot i$) equals k. Hence the following outcomes are obtained from Eq. (6), where $k = j \cdot i$, such that for $k = 1$ the only possibility is $i = j = 1$, for $k = 2$ two index combinations with $i = 1, j = 2$, and $i = 2, j = 1$, and so on:

$$C_1 = \operatorname{cum}(X) = m_1$$

$$C_2 = \operatorname{cum}(X, X) = m_2 - m_1^2$$

$$C_3 = \operatorname{cum}(X, X, X) = m_3 - 3m_2 m_1 + 2m_1^3$$

$$C_4 = \operatorname{cum}(X, X, X, X) = m_4 - 4m_3 m_1 - 3m_2^2 + 12 m_2 m_1^2 - 6 m_1^4, \dots \quad (7)$$

where "cum" denotes the cumulant calculation. In theory, the moments, and consequently the cumulants (see Eqs. (7) and (5)), are computed using mathematical expectation. In practice, the expectation is replaced by the sample statistics; for example,

$$\hat{m}_1 = \frac{1}{N} \sum_{i=1}^{N} x_i, \quad \hat{m}_2 = \frac{1}{N} \sum_{i=1}^{N} x_i^2 \quad (8)$$

where $\wedge$ designates estimate.

Multivariable Approach. Let $X_1, X_2, \dots, X_n$ be a set of random variables. The notation introduced in Eq. (7) can be extended in such a way that the cumulant C is subscribed by a set of n indexes, each of them standing for one of the observed variables. The value of an index defines the order (the number of times) of appearance for a certain variable; for example, C_{201} indicates the cumulant of three variables X_1, X_2, and X_3 in such a way that only X_1 and X_3 are included as follows:

$$C_{201} = \operatorname{cum}(X_1, X_1, X_3) = E(X_1 X_1 X_3)$$

If the variables are real with zero mean, that is, $E(X_i) = m_i = 0$, the following results from Eq. (7):

$$C_{10\dots0} = \operatorname{cum}(X_1) = E(X_1) = 0$$

$$C_{20\dots0} = \operatorname{cum}(X_1, X_1) = E(X_1^2)$$

$$C_{30\dots0} = \operatorname{cum}(X_1, X_1, X_1) = E(X_1^3)$$

$$C_{40\dots0} = \operatorname{cum}(X_1, X_1, X_1, X_1) = E(X_1^4) - 3[E(X_1^2)]^2, \dots$$

For combinations of more random variables Eq. (7) yields

$$C_{110\dots0} = \operatorname{cum}(X_1, X_2) = E(X_1 X_2)$$

$$C_{1110\dots0} = \operatorname{cum}(X_1, X_2, X_3) = E(X_1 X_2 X_3)$$

$$C_{11110\dots0} = \operatorname{cum}(X_1, X_2, X_3, X_4)$$

$$= E(X_1 X_2 X_3 X_4) - E(X_1 X_2)E(X_3 X_4) - E(X_1 X_3)E(X_2 X_4) - E(X_1 X_4)E(X_2 X_3)$$

For further details refer to [30].

Some Properties of Cumulants. Cumulants have some nice properties that allow many practical computations to be considerably simplified. The most important [31] are summarized here in order to set the basis for the introduction of system identification based on higher-order statistics, namely the cumulants.

Property 1. If λ_i are constants and X_i; $i = 1, \ldots, n$ random variables, then

$$\text{cum}(\lambda_1 X_1, \lambda_2 X_2, \ldots, \lambda_n X_n) = \left(\prod_{i=1}^{n} \lambda_i \right) \text{cum}(X_1, X_2, \ldots, X_n) \qquad (9)$$

Property 2. The computation of cumulants is symmetric:

$$\text{cum}(X_1, X_2, \ldots, X_n) = \text{cum}(X_{i_1}, X_{i_2}, \ldots, X_{i_n}) \qquad (10)$$

where $(i_1, \ldots, i_n)$ is an arbitrary permutation of $(1, \ldots, n)$.

Property 3. If X_0, Y_0, and Z_i random variables, there is additivity in arguments:

$$\text{cum}(X_0 + Y_0, Z_1, \ldots, Z_n) = \text{cum}(X_0, Z_1, \ldots, Z_n) + \text{cum}(Y_0, Z_1, \ldots, Z_n) \qquad (11)$$

Property 4. If α equals a constant, then

$$\text{cum}(\alpha + Z_1, \ldots, Z_n) = \text{cum}(Z_1, \ldots, Z_n) \qquad (12)$$

This property suggests how to create a zero-mean sequence, if necessary, without changing the cumulants' values, for examples, $y(n) - \text{mean}(y)$.

Property 5. If variables X_i and Y_i; $i = 1, 2, \ldots, n$, are statistically independent, then

$$\text{cum}(X_1 + Y_1, \ldots, X_n + Y_n) = \text{cum}(X_1, \ldots, X_n) + \text{cum}(Y_1, \ldots, Y_n) \qquad (13)$$

It is evident that the linearity property is ensured by properties 1 and 3.

Cumulants of Random Processes. If $\xi(t)$ is a stationary random process, the kth order cumulant, $C_{k,x}(\tau_1, \tau_2, \ldots, \tau_{k-1})$, is defined over random variables $X(t)$, $X(t + \tau_1)$, $\ldots$, $X(t + \tau_{k-1})$:

$$C_{k,x}(\tau_1, \tau_2, \ldots, \tau_{k-1}) = \text{cum}(X(t), X(t + \tau_1), \ldots, X(t + \tau_{k-1})) \qquad (14)$$

The calculation in Eq. (14) depends only on the shifts $\tau_1, \ldots, \tau_{k-1}$ because a wide-sense stationarity condition is assumed. We calculate the cumulant for one or for the multivariable case as a scalar. A random process may be viewed as a multivariable case; the sequences beginning at the chosen time instants τ_1, τ_2, etc., are considered values of different random variables. A single scalar cumulant value is determined for any combination of indexes; in going through an interval of values τ_1, τ_2, etc., a multidimensional matrix is obtained.

Let us illustrate the symmetry property in calculation of a third-order cumulant. This cumulant is two-dimensional, but it consists of six equal-valued regions in the (τ_1, τ_2)-plane:

$$
\begin{aligned}
C_{3,x}(\tau_1, \tau_2) &= C_{3,x}(\tau_2, \tau_1) \\
&= C_{3,x}(-\tau_1, \tau_2 - \tau_1) \\
&= C_{3,x}(-\tau_2, \tau_1 - \tau_2) \\
&= C_{3,x}(\tau_1 - \tau_2, -\tau_2) \\
&= C_{3,x}(\tau_2 - \tau_1, -\tau_1)
\end{aligned}
$$

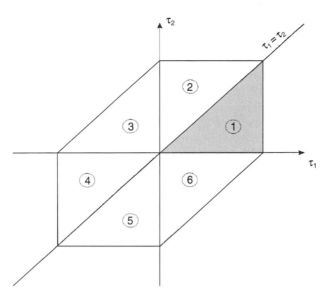

Figure 10.3. Six identical regions of third-order cumulants.

A transition from the first to the second and third lines in the symmetrical relationship above is easily understood if we imagine three variables in Eq. (14) with their arguments being t, $t + \tau_1$, and $t + \tau_2$. After substituting the second and the third variable argument with a new compact argument, the second and the third lines of symmetry become as follows: for $v = t + \tau_1$, the variable arguments are $v - \tau_1$, v, and $v + \tau_2 - \tau_1$ and analogously for $v = t + \tau_2$. The rest comes from the cumulant property 2.

Figure 10.3 shows the third-order cumulant for variable X of finite length q and six identical regions. Two are in the first and third quadrant, and one in the second and fourth quadrant.

System Identification Using Cumulants. Recall the SISO system characterized by Eq. (1). Now consider its input excitation to be in the form of non-Gaussian, white, zero-mean, independent identically distributed (i.i.d.) random noise $w(n)$. The system's output cumulant can be computed using Eq. (14). Therefore, for a discrete, sampled output $y(n)$, we have

$$C_{k,y}(\tau_1, \tau_2, \ldots, \tau_{k-1}) = \text{cum}[y(n), y(n + \tau_1), \ldots, y(n + \tau_{k-1})] =$$

$$\text{cum}\left[\sum_{i_0=0}^{\infty} w(i_0)h(n - i_0), \sum_{i_1=0}^{\infty} w(i_1)h(n + \tau_1 - i_1), \ldots, \sum_{i_{k-1}=0}^{\infty} w(i_{k-1})h(n + \tau_{k-1} - i_{k-1})\right] \quad (15)$$

Equation (15) can further be modified according to the cumulant properties 1 and 3,

$$\sum_{i_0}\sum_{i_1} \cdots \sum_{i_{k-1}} h(n - i_0)h(n + \tau_1 - i_1)\ldots h(n + \tau_{k-1} - i_{k-1}) \cdot \text{cum}[w(i_0), w(i_1), \ldots, w(i_{k-1})]$$

and according to the fact that the cumulant of the system's non-Gaussian random noise excitation differs from zero only if $i_0 = i_1 = \ldots = i_{k-1}$. Hence the final form of (15) yields

$$C_{k,y}(\tau_1, \tau_2, \ldots, \tau_{k-1}) = \gamma_{k,w} \sum_{n=0}^{\infty} h(n) \cdot h(n + \tau_1) \ldots h(n + \tau_{k-1}) \qquad (16)$$

which is known as the Barlett-Brillinger-Rosenblatt equation; $\gamma_{k,w}$ stands for the kth order cumulant of the system input excitation noise (e.g., for exponentially distributed noise $\gamma_{3,w}$ = 2, while for Gaussian distributed noise it equals 0 for all cumulant orders above 2 [31]).

It is clear that the cumulants of a system's output signal extract the information on the system's response, $h(n)$, when the system is stimulated by non-Gaussian, zero-mean, i.i.d random noise. The simplest theoretical approach for determining the system's response using cumulants is based on the so-called $C(q, k)$ formula [31]. The idea can easily be demonstrated using third-order cumulants as

$$C_{3,y}(\tau_1, \tau_2) = \gamma_{3,y} \sum_{n=0}^{\infty} h(n)h(n + \tau_1)h(n + \tau_2) \qquad (17)$$

Now consider a moving-average (MA) system response of order q and consider $\tau_1 = q$, while $\tau_2 = k$:

$$C_{3,y}(q, k) = \gamma_{3,w} h(0)h(q)h(k) \qquad (18)$$

Equation (18) is obtained from Eq. (17), since $h(n) = 0$ for all $n > q$. An additional combination of $\tau_1 = q$ and $\tau_2 = 0$ gives

$$C_{3,y}(q, 0) = \gamma_{3,w} h^2(0)h(q) \qquad (19)$$

If $h(n)$ is assumed to be normalized, so that $h(0) = 1$, then Eq. (18) divided by Eq. (19) leads to the desired system response:

$$h(k) = \frac{C_{3,y}(q, k)}{C_{3,y}(q, 0)}, \qquad k = 0, 1, \ldots, q \qquad (20)$$

Equation (20), which is called the $C(q, k)$ formula, is applicable only to systems that can be represented by MA models. Despite its theoretical importance, its behavior in practice is rather unreliable. However, a large number of better identification methods based on higher order statistics have been developed [35]. Some of these models are applicable only to MA; others, to autoregressive (AR) and autoregressive moving-average (ARMA) models. The majority can handle only the SISO cases, but there have also been approaches that satisfy the needs of the MISO and MIMO decomposition [35]. The latter will be discussed in Section 10.3.2 and applied to the EMG decomposition problem.

Let us now look at the difference between the second-order and higher order statistical approaches. The former comprise a large arsenal of methods and techniques that are well known and have often been successfully applied over a long time [27,43]. However, except for the cross-statistics in the MIMO cases, the second-order statistics are phase blind, which means that only the magnitude of the system response may be obtained in a system identification. In general, to reconstruct the entire response, including the signal phase, the application of higher order statistical approaches is mandatory. While second-order statistics work for any type of random variable distribution, and are very often used in conjunction with Gaussian processes, the cumulants of orders higher than 2 are blind in Gaussian processes [31]. This is why the non-Gaussian property is always stressed when higher order statistics are discussed. The EMG characteristics conform to a non-Gaussian condition because the AP's firing instants are close to a Bernoulli distribution [53]. Actually the innervation pulse trains may be considered non-Gaussian, white, and practically i.i.d. random processes, which makes higher order statistical processing of the EMG signals advisable.

10.2.3 Time-Frequency Representations

The Limitations of the Fourier Transform. The frequency content of a stationary signal $x(t)$ is classically derived from the Fourier transform (FT), defined by

$$FT_x(f) = \int x(t)\exp(-j2\pi ft)\,dt$$

This integral can be seen as a scalar product between the signal $x(t)$ and a family of monochromatic waves of frequency f. The result $FT_x(f)$ can be considered the component of $x(t)$ on the basis of vectors $\exp(j2\pi ft)$.

The Fourier integral can also be treated as a convolution product between the signal $x(t)$ and the impulse response of a monochromatic bandpass filter centered on the frequency f. The result $FT_x(f)$ can then be considered the (demodulated) output of this filter. It must be emphasized that as the analyzing waves are of infinite length, the analyzing filters are also of infinite length impulse response, meaning of infinite frequency concentration. An important result (Parseval's theorem) deals with finite energy/power signals: for a signal $x(t)$ of energy E

$$E_x = \int |x(t)|^2\,dt = \int |FT_x(f)|^2\,df$$

where $|x(t)|^2$ is the energy distribution in time and $|FT_x(f)|^2$ is the energy distribution in frequency.

The Fourier transform can be applied to nonstationary signals. For example, consider an infinite length monochromatic wave clipped by a rectangular window. The resulting FT shows the convolution between the FTs of the wave (Dirac impulse) and of the window (sinc function) resulting in a translated sinc function. The original truncated wave can then be described in a stationary context by the contribution of an infinite number of monochromatic waves, the amplitudes being distributed (along the frequencies) by the sinc function. This possible interpretation is not, of course, a very good way to describe a monochromatic wave present only from the beginning to the end of the window. Another example concerns musical scores. A tune is made of frequencies appearing and disappearing at very precise instants, and the musicians read this information on the score, which can be regarded as the description of the time-varying frequency content of the tune. These examples show the necessity of using specific tools for nonstationary signals, called time-frequency representations (TFRs). However, applying FT to nonstationary signals is not wrong or incorrect: it simply does not detect transient events that take place within the observation time.

The Short-Term Fourier Transform and the Spectrogram. The short-term Fourier transform, $STFT_{hx}(t, f)$ of a signal $x(t)$, is a function of time t and frequency f, depending on a window $h(t)$, and defined by

$$STFT_{hx}(t, f) = \int x(u)h*(u-t)\exp(-j2\pi fu)\,du$$

with $\int |h(t)|^2\,dt = 1$ (normalized window).

This definition can be seen to have the following components:

1. A standard FT of the signal $x(u)$ clipped by the window $h(u)$, (centered at current analysis time t), assuming that the signal is stationary inside the window.
2. A decomposition (scalar product) on a basis of clipped (by the window h) monochromatic waves of frequency f, centered around time t (here the window is independent of the frequency, as opposed to the continuous wavelets transform case).
3. Filtering (convolution product) the signal by a bank of finite length impulse response filters (narrowband filters) centered on the current analysis time and frequency.

For a finite energy signal (and a normalized window), Parseval's theorem can be easily extended to the STFT:

$$E_x = \int\int |STFT_{hx}(t, f)|^2\,dt\,df$$

The spectrogram can be defined as

$$SP_{hx}(t, f) = |STFT_{hx}(t, f)|^2$$

so that the spectrogram represents the distribution of the signal energy in the time-frequency plane. The spectrogram is a particular case of bilinear representations. Using standard integral properties, it can be written as

$$SP_{hx}(t, f) = \int\int h\left(t-u+\frac{\tau}{2}\right)h*\left(t-u-\frac{\tau}{2}\right)x\left(u+\frac{\tau}{2}\right)x*\left(u-\frac{\tau}{2}\right)\exp(-j2\pi f\tau)\,du\,d\tau$$

We can define the instantaneous product as

$$p(t, \tau) = x\left(t+\frac{\tau}{2}\right)x*\left(t-\frac{\tau}{2}\right)$$

where t is the time instant for analysis and τ is the delay. We can define the time-delay smoothing operator as

$$\Phi_{tr}(t, \tau) = h\left(t+\frac{\tau}{2}\right)h*\left(t-\frac{\tau}{2}\right)$$

So the spectrogram can be seen as the FT of the instantaneous product $p(t, \tau)$ smoothed by the operator $\Phi_{tr}(t, \tau)$. We can proceed to compute the time-frequency expression of the smoothing operator, by computing the FT (delay to frequency) of $\Phi_{tr}(t, \tau)$:

$$\Phi_{tf}(t, f) = \int \Phi_{tr}(t, \tau)\exp(-j2\pi f\tau)\,d\tau$$

With FT thus defined, $WV_x(t, f)$, the delay to frequency of the instantaneous product $p(t, \tau)$, gives the Wigner-Ville transform of $x(t)$:

$$WV_x(t, f) = \int x\left(t + \frac{\tau}{2}\right)x^*\left(t - \frac{\tau}{2}\right)\exp(-j2\pi f\tau)d\tau$$

Now the spectrogram can then be written as

$$SP_{hx}(t, f) = \int \Phi_{tf}(u - t, v - f)WV_x(u, v)dudv$$

The spectrogram can be seen as the result of a double smoothing (in time and frequency) of the Wigner-Ville transform of $x(t)$. The smoothing kernel has a particular form and can be defined as the Wigner-Ville transform of the window h in terms of the STFT:

$$\Phi_{tf}(t, f) = \int h\left(t + \frac{\tau}{2}\right)h^*\left(t - \frac{\tau}{2}\right)\exp(-j2\pi f\tau)d\tau$$

For random signals all the mathematical expectations must be treated as integrals over the statistical distributions. This is because the nonstationarity forbids any statistical interpretation of temporal integrals (in particular, no extension of the ergodic formula can be written). Nevertheless, the expectation of the instantaneous product defines the (time-dependant) autocorrelation function, and its FT (delay to frequency) is called the Wigner-Ville spectrum.

The Bilinear Time-Frequency Representations (TFRs) of Cohen's Class. A general class of bilinear time-frequency representations can be derived from the preceding interpretation of the sprectrogram: Cohen's class of TFRs. A TFR belonging to Cohen's class is the result of a double smoothing (time and frequency) of the Wigner-Ville transform of the signal by a unitary smoothing kernel $\Phi_{tf}(t, f)$:

$$TFR_{\phi x}(t, f) = \int\int \Phi_{tf}(u - t, v - f)WV_x(u, v)dudv$$

with $\int\int \Phi_{tf}(t, f)dtdf = 1$ to ensure that $E_x = \int\int TFR_{\phi x}(t, f)dtdf$.

The main property of TFRs of Cohen's class is the respect of time translation and frequency modulation. That is, if $y(t) = x(t - t_1)\exp(j2\pi f_1 t)$, then $TFR_{\phi y}(t, f) = TFR_{\phi x}(t - t_1, f - f_1)$. This property is very important for signals recognition and classification. Additionally one can apply the FTs to the TFRs and represent the energy distribution in the following equivalent domains:

- Time-delay domain (instantaneous product defined previously).
- Doppler-delay domain (also called ambiguity domain), where the doppler is the FT of the time (as frequency is the FT of the delay).

The Doppler-delay domain is related to the well-known Doppler effect, where a stretching of the time (caused by a target's motion) causes a frequency modulation. In the ambiguity domain the double smoothing operation (convolution products) of the time-frequency domain reduces to a simple product (as a property of the FT). The design of the kernel $\Phi_{tf}(t,f)$ (see below) is usually done in the ambiguity domain because the effects of a simple product are directly seen in the ambiguity function of the signal.

Properties of the Bilinear TFRs of the Cohen Class. If a signal is made up of two (or more) components (represented by lines in the time-frequency plane), the resulting bilinear TFR is not the sum of the individual TFRs of each component. Additional terms, called interference terms, appear and reduce the readability of the TFR. This is because we are dealing with power or energy spectra where the components are summed and squared, generating cross-terms. Since the cross-terms generally cause high-frequency patterns (in the time-frequency plane), the smoothing operation results in an improvement of readability. It also results, however, in an increased uncertainty in the energy location. Thus a variety of kernels were proposed in order to realize a trade-off between interference term attenuation and energy concentration. Examples and results can be found in the tutorial part of the MATLAB "time-frequency toolbox."

The properties of the TFRs (positiveness, marginals, linear operation compatibility) are directly related to the kernel shape. Beside standard kernels (spectrogram, pseudo-smooth Wigner, Choi-Williams, etc.), some signal-dependant kernels can be defined on the basis of a priori knowledge about the signal and of the desired properties, for example, the radial Gaussian kernel. It should be noted that readability is not always the aim of a TFR. For instance, if a TFR is used for signal classification the interference terms are also informative terms, and may contain some discriminant information. The kernel should then be chosen for its classification ability rather than for its readability improvement.

10.2.4 Wavelet Transform

Considering the STFT and the associated bandpass filters, it appears that the time-frequency resolution depends on only the size of the window: a short window leads to a high resolution in time but a low-frequency resolution (time and frequency resolution are dependant, according to the *Heisenberg uncertainty principle*[1]). This resolution problem suggests that the design be of variable length in analyzing windows: short ones for high frequencies and long ones for low frequencies. This choice corresponds to the wavelet transform and the related time-scale analysis.

Wavelet Functions. The wavelet transform (WT) uses *basis functions*, meaning wavelet functions, that have time widths adapted to each frequency band. The wavelet is a smooth, oscillating function showing a bandpass character with good localization both in time and in frequency. The idea of relative time-frequency resolution (opposed to the fixed time-frequency resolution of STFT) allows the time-scale atoms to be considered as related by a time translation and a time stretch (opposed to a time translation and a frequency shift in the STFT).

[1] The principle applies to a pair of operators, such as position, **P**, and momentum, **Q**, with the property $\mathbf{QP} - \mathbf{PQ} = \mathbf{I}$. Then the Schwarz inequality gives $\|x\|^2 = x * (\mathbf{QP} - \mathbf{PQ})x \leq 2 \|\mathbf{Q}x\| \|\mathbf{P}x\|$. Application to Fourier analysis gives $\|f(t)\|^2 \leq 2 \|tf(t)\| \|\omega F(\omega)\| = 2\Delta t \Delta \omega$ [37].

A wavelet family consists of members, $\psi_{a,\tau}$, obtained by dilations (or scaling) and translations (i.e., time shifting) of one unique *prototype* or *mother wavelet*, $\psi(t)$:

$$\psi_{a,\tau}(t) = \frac{1}{\sqrt{a}}\psi\left(\frac{t-\tau}{a}\right), \tag{21}$$

where $a \in \Re^+$ represents the scale parameter, $\tau \in \Re$ represents the translation parameter. When a becomes large, the basis function $\psi_{a,\tau}$ becomes a stretched version of the prototype, which emphasizes the low-frequency components, whereas a small a contracts the basis function $\psi_{a,\tau}$ and stresses the high-frequency components. However, the shape of the basis function will always remain unchanged.

Since $\psi(t)$ can be implemented as a bandpass filter whose center frequency can change, at a given scale, the filter yields wider or narrower frequency-response changes depending on the center's frequency. This time-scale expression has an equivalent time-frequency expression. Since wavelets are well localized around a nonzero frequency f_0, at a scale $a = 1$ (i.e., the mother wavelet), there is an inversely proportional relationship between scale and frequency, given by $a = f_0/f$. Note that the factor $1/\sqrt{a}$ in Eq. (21) is introduced to guarantee energy preservation, that is, to normalize the wavelet so that it has unit energy.

Continuous Wavelet Transform. For a given input signal, $x(t)$, which is supposed to be square integrable $\psi-x(t) \in L^2(\Re)$, the different correlations (or the convolution) between wavelets and the signal can be defined as the continuous wavelet transform (CWT):

$$CWT_{\psi x}(a, \tau) = \int x(t)\psi_{a,\tau}^*(t)dt \tag{22}$$

where the asterisk stands for complex conjugation. Equation (22) is often written in the form of scalar product:

$$CWT_{\psi x}(a, \tau) = \langle x, \psi_{a,\tau} \rangle$$

where

$$\psi_{a,\tau}(t) = \frac{1}{\sqrt{a}}\psi\left(\frac{t-\tau}{a}\right), \qquad a \in \Re^+, \tau \in \Re$$

and $\psi(t) \in L^2(\Re)$ is the mother wavelet.

ADMISSIBILITY CONDITION. Wavelets $\psi(t)$ fulfill the so-called admissibility condition, in order to guarantee the inversion CWT:

$$C_\psi = \int_0^{+\infty} \frac{|\Psi(f)|^2}{f}df < +\infty \tag{23}$$

where $\Psi(f)$ is the Fourier transform of $\psi(t)$. This condition is equivalent to $\Psi(0) = 0 \Rightarrow \int_{-\infty}^{+\infty}\psi(t)dt = 0$, meaning that $|\Psi(f)|$ must decay faster than $1/\sqrt{f}$ when $f \to \infty$.

MAIN PROPERTIES of CWT. Suppose that the admissibility condition, Eq. (23), is fulfilled. Then

1. For a translated and scaled version of signal $x(t)$, the CWT yields

$$y(t) = \frac{1}{\sqrt{a_0}} x\left(\frac{t - \tau_0}{a_0}\right) \Rightarrow CWT_{\psi y}(a, \tau) = CWT_{\psi x}\left(\frac{a}{a_0}, \frac{t - \tau_0}{a_0}\right) \tag{24}$$

2. Energy conservation also holds:

$$\int_{-\infty}^{+\infty} |x(t)|^2 dt = \frac{1}{C_{\psi x}} \int_0^{+\infty} \int_{-\infty}^{+\infty} |CWT_{\psi x}(a, \tau)|^2 d\tau \frac{da}{a^2} \tag{25}$$

THE SCALOGRAM. From the previous property, we can define the energy distribution (along time and scale), called a scalogram (similar to spectrogram in the time-frequency plane), by

$$Pw_{\psi x}(a, \tau) = |CWT_{\psi x}(a, \tau)|^2 \tag{26}$$

INVERSION FORMULA. Under the admissibility condition, the signal $x(t)$ can be reconstructed from its CWT by

$$x(t) = \frac{1}{C_{\psi}} \int_0^{+\infty} \int_{-\infty}^{+\infty} CWT_{\psi x}(a, \tau) \psi_{a,\tau}^*(t) d\tau \frac{da}{a^2} \tag{27}$$

Sampling of the CWT. The Nyquist sampling theorem [36] shows a heavy redundancy of continuous representations of signals and their transforms. At the same time any digital processing calls for discrete representations and, thus, for sampling of the continuous signals. In the case of CWT there are two continuous variables that may be sampled: scale a and time-shift τ. If we introduce a discrete set of times and scales $(a_j, \tau_k)_{(j,k) \in Z^2}$, the sampled version of the transform is

$$CWT_{\psi x}(a_j, \tau_k) = \int x(t) \psi_{jk}^*(t) dt = \langle x, \psi_{jk} \rangle \tag{28}$$

where

$$\psi_{jk}(t) = \frac{1}{\sqrt{a_j}} \psi\left(\frac{t - \tau_k}{a_j}\right), (j, k) \in Z^2$$

This sampling of the time-scale plane must be fine enough to contain the same information as the signal $x(t)$. More precisely, it is possible to reconstruct the signal (inversion property) from the sampled transform by using the dual frame $\{\tilde{\psi}_{jk}\}$:

$$x(t) = \sum_{j,k} CWT_{\psi x}(a_j, \tau_k) \tilde{\psi}_{jk}(t) \tag{29}$$

It should be noted that this dual frame is not unique; it is not simple to determine and is not necessarily a frame of wavelets. So two questions must now be considered:

- What kind of sampling of the time-scale plane guarantees the inversion conditions?
- Which class of wavelets leads to a simple relationship between the primary and the dual frame of wavelets?

The common approach is to sample on a dyadic grid in the time-scale plane, meaning $a = 2^j$ and $\tau = k2^j$, which leads to

$$d_{j,k} = CWT_{\psi x}(2^j, k2^j) = \int x(t)\psi_{j,k}^*(t)dt, \qquad j, k \in Z \tag{30}$$

Equation (30) is known as a *wavelet series expansion* (WSE). The *analysis wavelets* are, in this case,

$$\psi_{j,k}(t) = 2^{-j/2}\psi(2^{-j}t - k) \tag{31}$$

The original signal can be recovered through the following expression:

$$x(t) = \sum_{j \in Z}\sum_{k \in Z} d_{j,k}\tilde{\psi}_{j,k}(t) \tag{32}$$

where the *synthesis wavelets* $\tilde{\psi}_{j,k}(t)$ are of the form (31). Note that in the case of the *orthogonal* wavelet transform, the same wavelet prototype is used to perform analysis (decomposition) and synthesis (reconstruction) [10,30]. This is not a requirement of the method. In fact there are bases of wavelets, called *biorthogonal wavelet bases*, in which the synthesis wavelet is not the same as the analysis wavelet [10].

Discrete Wavelet Transform. Discrete wavelet transform (DWT) is very similar to WSE. The basic difference is that DWT only applies to discrete sequences $\{x[n], n \in Z\}$, which means that also the time is considered discretised. The wavelets used in DWT are discrete versions of the continuous wavelets used in CWT or WSE [1]. The main issue here is that the discretization of wavelets partially depends on the algorithm chosen to perform the transformation. However, we don't need to explicitly calculate a digitized version of the mother wavelet, $\psi(t)$, because the wavelet transform could be well approximated by digital filter banks (see Fig. 10.4).

The output from the low-pass filter is a smoothed version of the input signal whose the high-frequency components of the signal are removed. The high-pass filter removes the low-frequency components, and the result is a signal containing details of the input signal. The problem is that we now have two signals with the same length as the input signal; that is the information is doubled. The solution is to *down-sample* (decimate) the filtered sequences by a factor of two (designated by ↓ in Fig. 10.4). This operation decreases the usable frequency contents of the sampled signal by two as well. However, correct reconstruction may be performed using a reversed version of the filter bank from the decimation step. By first *up-sampling* (zero-padding) and then filtering by the reversed filter bank, it is possible to thoroughly reconstruct the original signal.

Mallat's remarkably fast pyramid [30] with complexity $O(N)$ involves use of low-pass and high-pass filters along with a down-sampling or up-sampling operator. Using Mallat's

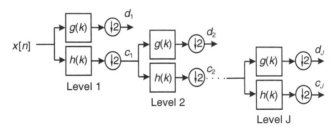

Figure 10.4. Analysis filter bank using the DWT pyramid algorithm, where the input signal is filtered recursively with pairs of low-pass, $h(k)$, and high-pass, $g(k)$, filters. The output from each filter is subjected to down-sampling.

pyramid, we are able to compute the DWT and inverse DWT for the sampled signals much faster than by FFT.

MULTIRESOLUTION ANALYSIS. Multiresolution theory was developed by Mallat, Daubechies, and Meyer. They usesd the ideas similar to sub-band decomposition and coding. A signal is divided into a set of frequency bands, which is here introduced by sampling on a dyadic grid. This, however, means that the frequency domain is divided into octave subbands.

Given a discrete signal $x[n]$; $n = 1, \ldots , N$, its DWT up to a level of J depth maps the vector $(x[1], \ldots , x[N])$ to a set of N wavelet coefficients containing $c_{J,k}$ and $d_{j,k}$; $j = 1, 2, \ldots , J$, of the wavelet series approximation. Such a *multiresolution analysis (MRA)* derivation becomes clearer if a second function, called a *scaling function*, is put along side the wavelet functions.

SCALING FUNCTION. A scaling function $\varphi(t)$ gives a set of approximations of the signal as a set of resolution levels j, by projecting it on a set of subspaces V_j. The nested subspaces V_j are generated by stretched and translated versions of $\varphi(t)$. For a given level j, the subset V_j is spanned by the base of scaling functions:

$$\varphi_{jk}(t) = \frac{1}{\sqrt{2^j}} \varphi\left(\frac{t - k2^j}{2^j}\right), \qquad k \in Z$$

The approximations coefficients yield

$$c_{j,k} = \langle x, \varphi_{jk} \rangle$$

By setting stretching factor equal to 2 (dyadic grid), we have $\varphi(t)$ corresponding to

$$\varphi(t) = \sqrt{2} \sum_k h[k] \varphi(2t - k) \tag{33}$$

where

$$h[n] = \left\langle \frac{1}{\sqrt{2}} \varphi(t), \varphi(2t - n) \right\rangle \tag{34}$$

The MRA (i.e., the decomposition of a signal on the nested subspaces V_j) is completely defined by the function $\varphi(t)$ or equivalently by the sequence $h[n]$ from Eq. (34). The signal is decomposed, that is band-pass filtered, to the frequency subbands, where the scale function always defines the low-frequency signal components. The complementary high-frequency band is, however, obtainable by the wavelets.

WAVELET FUNCTION. The information lost between two successive approximations is called a detail. It is obtained by projecting the signal on the complement of V_j, denoted by W_j, $(W_j + V_j = V_{j-1})$, and defined by the wavelet functions

$$\psi_{jk}(t) = \frac{1}{\sqrt{2^j}} \psi\left(\frac{t - k2^j}{2^j}\right), \qquad k \in Z \qquad (35)$$

where $\psi(t)$ is the mother wavelet

$$\psi(t) = \sqrt{2} \sum_k g[k]\, \varphi(2t - k)$$

with

$$g[n] = \left\langle \frac{1}{\sqrt{2}} \psi(t), \varphi(2t - n) \right\rangle \qquad (36)$$

The detail coefficients are computed as follows:

$$d_{j,k} = \langle x, \psi_{jk} \rangle$$

and they correspond to the discrete wavelet transform as generated by the MRA:

$$d_{j,k} = DWT_{\psi x}(j, k) = \langle x, \psi_{jk} \rangle$$

Therefore the practical approach to this calculation comes from the principles explained in paragraph on discrete wavelet transform.

The signal can be reconstructed from all the detail coefficients or from the approximation of level J and the details of lower levels:

$$x(t) = \sum_k c_{J,k}\, \tilde{\varphi}_{Jk}(t) + \sum_{j \le J} \sum_k d_{j,k}\, \tilde{\psi}_{jk}(t) \qquad (37)$$

When the subspaces W_j and V_j are orthogonal, the corresponding MRA, is called orthogonal MRA, and the functions $\{\varphi_{jk}(t)\}_{k \in Z}$ define the orthogonal base vectors of V_j, and $\{\psi_{jk}(t)\}_{k \in Z}$ the base of W_j. The sequence $h[n]$ is then the impulse response of a *conjugate mirror filter* (CMF), while the sequence $g[n]$ is defined by

$$g[n] = (-1)^{1-n} h[1 - n] \qquad (38)$$

The design of discrete orthogonal bases of wavelets is thus reduced to choosing either a scale function $\varphi(t)$ or a discrete sequence $h[n]$.

Wavelet Packets. The MRA in Figure 10.4 always splits the input spectrum at each stage into two bands. The higher band is taken as one of the transform outputs, whereas the lower band is again split into two narrower bands. This scheme is rather restricted. If a more general option is considered, each band, of both the lower and higher frequencies, may be split into several bands at a time. The wavelet interpretation of such generalized MRA is that the outputs produced are *wavelet packets* (*WP*). Wavelet packets can apply arbitrary band splitting. Therefore they are not bound to octave frequency resolution. The most suitable resolutions may be chosen for a particular signal, giving the option of an adaptive system.

The *general wavelet packets* are the functions

$$W_{j,b,k}(t) = 2^{-j/2} W_b(2^{-j} t - k) \qquad b \in N, \, j, k \in Z \qquad (39)$$

Each function is determined by a scaling parameter j, a localization parameter k, and an oscillation parameter b. The function $W_b(2^{-j}t - k)$ is roughly centered at $2^j k$, has support of size $\approx 2^j$, and oscillates $\approx b$ times.

These outputs are much too numerous to form an orthonormal basis. In fact we can extract several different orthonormal bases from the collection. The choices $j = 0$, $b \in Z^+$, and $k \in Z$ lead to the basic wavelet packet, while the choices $b = 1$ and $j,k \in Z$ lead to the orthonormal wavelet basis described in the previous subsection. Any collection of indexes (j, b) such that $I(j, b)$ (defined by $2^{-j}b \leq \omega \leq 2^{-j}(b + 1)$) form a disjoint cover of $[0, \infty)$ gives rise to an orthonormal basis of $L^2(\Re)$.

Similar to the wavelet transform, the WP can be approximated by digital filter banks. One way to accomplish this is by repeatedly filtering the signal from the high-pass filters (i.e., the wavelet coefficients). This gives us a tree structure of coefficients at different levels as depicted in Figure 10.5. This tree is called the wavelet packet tree, where each node has a set of coefficients that one can choose to further split or not.

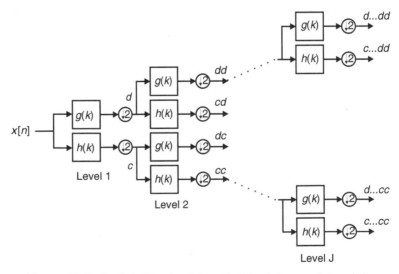

Figure 10.5. Analysis filter bank tree that leads to wavelet packets.

There are many different methods used to decide the best subset of the bases family, and that can be adapted to suit a particular application. For example, a best basis selection is used to choose the basis, among all possible bases, that has the smallest information cost [8].

PSD ESTIMATION USING WAVELET PACKETS. An alternative to Fourier and autoregressive methods is to use the segment-by-segment estimation technique with wavelet packets. By choosing a suitable resolution level, j_0 $(1 \leq j_0 < J)$, and by obtaining the energy of the WP coefficients in the (j_0, b)-th block of the wavelet packet table, one can make a better analysis of the spectrum than by using the wavelet scalogram. Thus the *WP periodogram* at level j_0 (for a segment of length 2^J) with respect to ψ can be defined as

$$\hat{P}_{WP}^{\psi}(j_0, b) = \sum_{k=1}^{2^{J-j_0}} \hat{P}_{WP}^{\psi}(j, b, k), \qquad b = 0, 1, \dots, 2^{j_0} - 1 \tag{40}$$

The oscillation parameter b in Eq. (40) gives us the opportunity to analyze the frequency components more flexibly:

$$\hat{P}_{WP}^{\psi}(j, b, k) = \left| \sum_{n-0}^{N-1} x[n] W_{j,b,k}[n] \right|^2 \tag{41}$$

In general, the WP periodograms perform similarly to the FFT periodograms. However, WP gives us some advantage over traditional methods (e.g., Fourier, AR, and ARMA). By using the library of orthonormal bases of WP, arbitrary time-frequency resolution can be obtained[2]. The WP is a generalized multiresolution method that represents the entire family of subband coded decompositions. From this family of bases, the best basis can be chosen to match the nonstationarity of the signal. An adaptive filtering of the signal can be performed by using an entropy criterion and choosing among parents and children in the tree. In sum, the WP method offers a more flexible method to arbitrary and adaptive select frequency resolution depending on the EMG application.

10.2.5 Improving the PSD Estimation Using Wavelet Shrinkage

The periodogram, both Fourier and wavelet, is not a consistent spectral estimate, since the variance does not tend to zero as the data length tends to infinity. To smooth the periodogram (i.e., to reduce the variance), there are basically three types of averaging schemes:

1. The Daniell method, which smoothes the periodogram by averaging over adjacent frequency bins.
2. The Bartlett method, which averages multiple periodograms and is produced from nonoverlapping segments of the original data sequence.
3. The Welch method, which is an extension of the Bartlett's method, uses data windows, and allows the segments to overlap.

---

[2] Multiresolution is also possible with the STFT by simply taking multiples FTs with different window lengths. However, the method is not as flexible as WP, since for every additional time-frequency tiling the complete spectra must be recalculated with respect to the selected window length.

The averaging methods reduce variance but increase bias and decrease the frequency resolution. The amount of segmentation involves a trade-off between the degree of smoothing and the frequency resolution.

An alternative method to smoothing the periodogram has been proposed [16,34]. That method rests on the principle of *wavelet shrinkage*. Wavelet smoothing techniques make use of the different properties of signal and noise wavelet coefficients. The basic idea, which is more subtle in approach than classical filtering, is called *thresholding* [11,12]. That method involves discarding the portion of the DWT detail coefficients less than a certain threshold. The advantage, as compared with classical filtering, is that high-frequency information not regarded as noise will not be suppressed.

To improve the PSD, the procedure for wavelet shrinkage of a log-periodogram can be described as follows:

1. Decomposition of the log-periodogram to get the empirical wavelet coefficients.
2. Shrinkage of the empirical wavelet coefficients by thresholding (all wavelet coefficients of magnitude less than the threshold are set to zero).
3. Reconstruction of the thresholded wavelet transform coefficients to get the smoothed log-periodogram.

A level-dependent threshold using a soft shrinkage function has been proposed by Gao, and gives as nearly a noise-free log-spectrum as possible [16]. For more details regarding wavelet shrinkage see [5,16,47].

10.2.6 Spectral Shape Indicators

The spectral shape indicators, such as mean and median frequency, and the higher order central moments defined in Chapter 6, may be redefined to include the time dependence of the signal's frequency content [24]. The time-dependent or instantaneous standard and central moments of order k are

$$IM_k(t) = \frac{\int_0^\infty \omega^k P(t,\omega)d\omega}{\int_0^\infty P(t,\omega)d\omega} \tag{42}$$

$$IM_{Ck}(t) = \frac{\int_0^\infty (\omega - IM_1(t))^k P(t,\omega)d\omega}{\int_0^\infty P(t,\omega)d\omega} \tag{43}$$

Thus the chosen indicators of spectral change are the instantaneous mean frequency $IM_1(t)$, the dispersion index (variance of the normalized PSD intended as a p.d.f.) $\sigma^2(t) = IM_{C2}(t)$, the skewness index $\gamma_1(t) = IM_{C3}(t)/\sigma^3(t)$, and the kurtosis index $\gamma_2(t) = IM_{C4}(t)/\sigma^4(t)$. In addition Roy et al. [40] introduced the instantaneous median frequency (IMDF):

$$\int_0^{IMDF(t)} P(t,\omega)d\omega = \int_{IMDF(t)}^\infty P(t,\omega)d\omega \tag{44}$$

10.3 DECOMPOSITION OF EMG SIGNALS

As shown in the preceding chapters, EMG is a compound signal that usually contains a high number of superimpositions of the basic waveforms, especially when the recordings are taken with high contraction forces. Decomposition of such a signal is not an easy task. The challenge in detecting different MUAPs and their firing patterns is owing to the great diagnostic and clinical importance of these basic EMG features. Some of "classical" decomposition techniques are discussed in Chapter 3; in this section we present some novel decomposition approaches.

10.3.1 Parametric Decomposition of EMG Signals Using Wavelet Transform

The dipole and tripole models for describing the SFAP shapes incorporate geometrical parameters, such as the distance between the motor unit innervation zone and the measuring electrode position, the depth of the motor unit under the electrode, and the conduction velocity (see Chapters 1, 4, and 8). We learned from the MIMO models in Figures 10.1 and 10.2 that the measured EMG signals combine MUAPs whose shapes are modified by low-pass filters. Suppose that an MUAP is described as the superposition of SFAPs, $g_{il}(n)$, and the filtering characteristics $f_{ij}^{(l)}(n)$ depend only on the geometrical distance between the measuring electrode and the fiber. The samples of $f_{ij}^{(l)}(n)$ reflect for every n the conditions of the path, and its length, from the spatial point in the fiber where the traveling AP is located at time instant n to the electrode. In the sequel we will call the shortest distance between a fiber and a measuring electrode the depth of the muscle fiber, d [54]. In a deterministic approach called multimodal parametric search (MPS) [55], when the SFAP characteristics are assumed known and the conduction velocity is supposed constant (its influence is carried by the SFAP shape), what remains unknown are the depths of signal sources, d, and their time delays n_{il}.

Multimodal Parametric Search. The MPS maps a compound signal from the signal space (where each sample is a dimension) to parametric space (where each signal "parameter" is a dimension). There are two basic conditions for implementing such a method: a set of signal components from which the compound signal is built of must be known a priori, and they must contribute to the signal in a linear superimposition. The MISO and MIMO models of needle and surface EMG meet these requirements. Hence a decomposition of EMGs using the MPS becomes theoretically feasible when SFAPs are taken as known signal building blocks whose shapes are regulated by a single parameter, namely the depth of the signal-emitting muscle fiber under the measuring electrode. This distance is taken as one parameter, while the instants of appearance of SFAPs, n_{il}, enter as another parameter. By choosing only two parameters, the parameter space shrinks to two dimensions that enables an illustrative and easy-to-understand representation of the parameters in two-dimensional plane. The dependence of the SFAP's shape on conduction velocity is neglected for this purpose. By varying the two parameters chosen, the absolute difference between the modeled and the measured EMG signal indicates the level of correlation. The best fit is searched for, using the following iteration [54]:

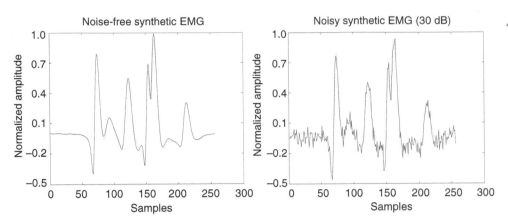

Figure 10.6. Comparison of the original (*solid line*) and the reconstructed signal after parametric search (*dotted line*); noise-free case (*left*) and noisy case (*right*). Fourteen identical sources (SFAPs) were placed at different depths and activated at different times. Depths and firing instants (delays) are estimated for SNR = infinity and SNR = 30 dB in Table 10.1.

```
repeat
      synthesize the set of basic SFAPs for all selected depths:
      the SFAP shape depends on the depth parameter only;
      for all possible delays do
             compute the error function via a distance between
      the shifted SFAP and the observed EMG;
      find the minimum of error function and the corresponding
      delay;
      with this delay do
             for all possible depths do begin
                   apply the next basic SFAP for the next
                   depth;
                   compute the new error function;
            end;
      find the minimum of this new error function and the
      corresponding depth;
      peel the SFAP obtained at the detected delay and depth off
      the observed EMG;
until the distance between the original EMG and the model output
is higher than in the previous step.
```

An example of this approach for a simple sequence of synthetic EMG signal depicted with a solid line is given in Figure 10.6 [55]. The source depths and delays used to generate the synthetic signal are shown in the upper panel of Table 10.1. The results of parametric decomposition follow for the noise-free case in the middle panel (shown dotted in the left-hand side of Fig. 10.6), and for a noisy signal with SNR of 30 dB in the bottom panel of the table (shown dotted in the right-hand side of Fig. 10.6).

In the example of Figure 10.6, the decomposition of an EMG recording is done to the SFAP level. If the recording comprises only one MUAP (as we will show in the next section), the MPS-based decomposition will locate the SFAPs contributing to this MUAP. In longer signal recordings the contributing SFAPs are recognized independent of their inclusion in the individual MUAPs. Nevertheless, as they are marked in the parameter

TABLE 10.1. Results of Iterative Parametric Search

						Original parameters								
Depth (mm)	1	3	4	2	1	3	4	2	3	2	7	4	8	1
Delay (ms)	1	16	21	51	82	92	107	139	73	47	32	84	95	91
						Parameters estimated in noise-free case								
Depth (mm)	1	3	4	2	1	2	4	2	3	3	6	6	—	1
Delay (ms)	1	17	21	49	82	94	107	139	76	49	48	81	—	90
						Parameters estimated in the case of SNR of 30 dB								
Depth (mm)	1	3	5	2	1	2	5	2	3	3	—	6	—	1
Delay (ms)	1	17	21	49	82	94	107	139	78	48	—	71	—	90

plane, they form similar fingerprints for similar events, such as MUAPs. In the search for such fingerprints, some idea of the number, shapes, and firing instants of different MUAPs is obtained. The same applies to subtractive wavelet transform described in the next section.

Despite the promising results noted in the example above, the MPS exhibits some drawbacks when applied to real EMG signals. The most important ones are summarized briefly [49]:

- Ambiguity in detection owing to variability of conduction velocities of the muscle fibers, which causes the SFAPs to emerge from different depths with correspondingly different conduction velocities that are very similar and thus indistinguishable by the MPS approach. The ambiguity can be reduced by introducing a three-dimensional parameter space, and modeling the SFAPs with conduction velocity as a variable. Still some ambiguity of the SFAP shape regarding the mutual influence of the signal source depth and its conduction velocity remains.

- The inherent peel-off technique cannot recognize the proper distribution of the observed parameters (depths in this case); it favors the SFAP waveforms tested first.

- The parameter values may be biased by inappropriate amplitude ratios taken between the real signal contents and the modeled SFAPs.

- The method is also termination-condition dependent: the decomposition stops when the next iteration degrades the match between the analyzed EMG and the obtained parametric approximation. For longer segments of signals, this condition can cause termination before an adequate approximation of the local phenomena can be reached.

Subtractive Wavelet Transform. Analytical dipole and tripole models for SFAPs are based on the distance between a muscle fiber and the measuring electrode, d, called the depth parameter [54]. A simple derivation shows that these models can be expressed in correspondence to the depth parameter as follows [42]:

$$\text{SFAP}_d(t) = \frac{1}{d}\text{SFAP}\left(\frac{t}{d}\right) \tag{45}$$

This expression shows behavior similar to that of the mother wavelets defined in Section 10.2. SFAPs can therefore act directly as nonorthogonal wavelets in the wavelet transform [41]. The continuous wavelet transform actually uses the calculation of correlation to detect the time interval when a signal fits best the applied wavelets (as matched filters). The correlation is observed at different values of the scaling factor (d in Eq. (45)), emerging as a time-scale representation. The wavelet coefficients exceeding a pre-selected threshold for every calculated scale indicate the firing instants of the corresponding SFAPs. Then the same technique is used as in the MPS decomposition, and these SFAPs peel off the analyzed signal before the algorithm proceeds with the next scale. The detection threshold must be computed referring to the autocorrelation value of SFAPs. Every instant where the peeling takes place is specially marked in the time-scale plane. This MPS-based wavelet transform has been called subtractive wavelet transform (SWT). Its computational algorithm is as follows:

```
for all possible depths (scales) do begin
        synthesise the SFAP waveform using the current scale;
        compute the zero-lag value of auto-correlation function
        for this SFAP and determine the threshold, e.g., as 90%
        of this value;
        compute the cross-correlation function between the
        analysed EMG signal and the current SFAP;
        repeat
                find the next time delay where a maximum in the
                cross-correlation function exceeds the pre-selected
                threshold;
                subtract the current SFAP from the EMG signal at
                the found delay;
                mark this point in time-scale representation;
        until any maximum > threshold
end
```

After the SWT algorithm is applied to the EMG signals, the two-parameter (actually, time-scale) plane acquires a kind of a fingerprint that characterizes the analyzed segment of the signal. The parameters thus obtained, namely the depths and time instants of contributing SFAPs, cannot be assumed to represent an exact signal decomposition. However, similar signal forms exhibit similar fingerprints because the parametric model preserves unique mapping. An example of such an analysis of the real single-differential surface EMG at 30% of maximum voluntary contraction (recorded in isometric contraction of the biceps muscle) is shown in Figure 10.7. Every circle in the time-scale parameter plane denotes the appearance of a SFAP being emitted from a muscle fiber at a certain depth under the measuring electrodes.

The inaccuracies of time-scale EMG fingerprints grow with the length of the processed signal. Nevertheless, even for short segments, such as a single MUAP, the uncertainty of the decomposed parameters remains the same for both the signal source depth and its firing instant in time. However, the latter can be recognized much more accurately with a novel approach based on time-scale phase representation [42].

Time-Scale Phase Representation. We have seen that SFAPs behave like a family of nonorthogonal wavelets whose scaling factor is related to the depths of individual muscle fibers if the conduction velocity is assumed constant and the same for all the fibers contributing to the analyzed part of the signal. We can denote the measured SFAP

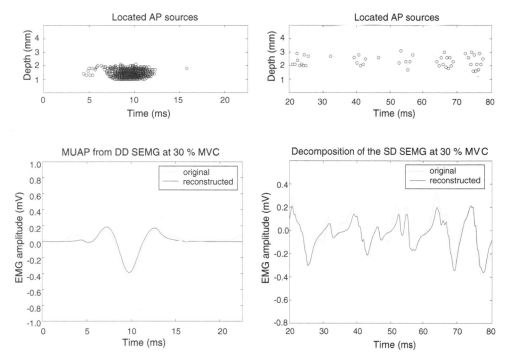

Figure 10.7. Outcomes of MPS decomposition for surface EMG at 30% of maximum voluntary contraction. (*left*) A single MUAP; (*right*) a longer segment.

waveforms by $a_d(n)$, where d stands for the depth, and the observed EMG signal by $y(n)$. The family of $a_d(n)$ may be determined a priori, because the SFAPs are assumed to be physiologically equal components with the same shape and characteristics. Their shape, as detected by the jth measuring electrode, depends on the (i, l)th source SFAP (equal for all fibers), $g_{il}(n)$, and the model filtering characteristics $f_{ij}^{(l)}(n)$ (see Section 10.2.1).

We can define a novel time-scale phase transform (TSPT) as follows:

$$\text{TSPF}(d, \tau) = \sum_{k=0}^{N-1} \left| \arg[Y(k)A_{d,\tau}^*(k)] \right| \tag{46}$$

where $Y(k)$ stands for discrete Fourier transform of $y(n)$ of length N, $A_{d,\tau}^*(k)$ for discrete Fourier transform of $a_d(n - \tau)$, and τ for the time shifts (delays of the SFAP firings). The * superscript indicates conjugation. The phase function *arg* must be computed according to

$$\arg(X) = \arctan\left(\frac{\text{Im}(X)}{\text{Re}(X)}\right) + \pi \cdot u(-\text{Re}(X)) \cdot \text{sgn}(\text{Im}(X)) \tag{47}$$

where $u(.)$ is the unit-step function and sgn stands for the sign function. The *arg* function from (47) gives values in the half-open interval $[0, 2\pi)$, which is of a crucial importance for the desired decomposition properties of TSPT. Because of the fact that $\arg[X(k)X_\tau^*(k)] = 0$ only if $\tau = 0$ and if $X_\tau(k)$ stands for discrete Fourier transform of $x(n - \tau)$, we can expect the minima in the TSPT(d, τ) plane for those spots where a corresponding SFAP

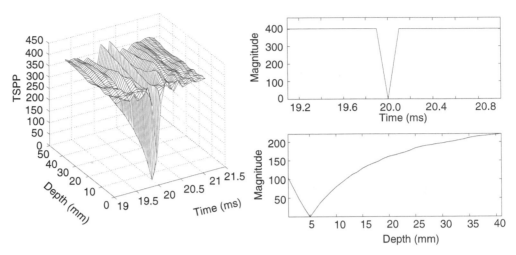

Figure 10.8. Time-scale phase analysis of a single SFAP. (*left*) Time-scale phase transform; (*upper window*) orthogonal slices of TSPT at time shift 20 ms and (*lower window*) depth $d = 5$ mm.

waveform becomes optimally aligned in time ($\tau = 0$) with its counterpart hidden in the observed EMG signal.

Figure 10.8 illustrates the power of TSPT to locate time events very sharply using only a single SFAP whose source depth and time shift are at $d = 5$ mm and $\tau = 20$ ms. Two cross sections taken at the point of minimum in TSPT (Fig. 10.8, left) confirm exactly the decomposed parameters of the modeled SFAP (Fig. 10.8, right). Additionally it is evident that the minimum is very sharp in time, while the resolution is much more vague in the depth parameter direction.

TSPT can be used in an iterative decomposition procedure similar to MPS and SWT. Therefore the obtained results may be presented in the two-dimensional parameter plane as well [56]. The algorithm is as follows:

```
generate a bank of signal building blocks (i.e., wavelets or
SFAPs);
repeat
        calculate the TSPT;
        search for a global minimum and determine the best
        matching building block;
        subtract the selected building block from the signal;
until threshold for the model error is reached;
```

The decomposition of a synthetic MUAP built of seven very similar SFAPs is shown for the noise-free case in Figure 10.9. The two-parameter plane, which was reconstructed with the help of the TSPT and the aforementioned algorithm, depicts the original and decomposed time-scale parameters by circles and pluses, respectively. All the time shifts are detected precisely, whereas the method performs worse for the depths, especially when some of the SFAPs originate from "deeper" fibers shown in Figure 10.9 (top panel). This technique is a promising tool for the estimation of the spatial dispersion of fibers of a MU and of the scatter of arrival times (due to the scatter of neuromuscular junctions and conduction velocities) along the direction of the fibers.

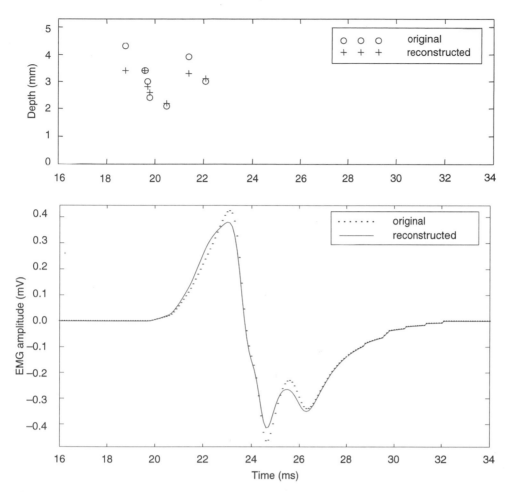

Figure 10.9. Results of SWT. Time and scale parameters (*top*); the original and reconstructed signal (*bottom*).

10.3.2 Decomposition of EMG Signal Using Higher Order Statistics

The MIMO model in Figure 10.2 can be applied to EMG signals at two different levels. The more detailed scheme depicted in Figure 10.1 assumes that transfer channels describe an SFAP and its modification on the path from the lth muscle fiber in the ith MU to the jth measuring electrode. This suits best the deterministic approach with wavelet transform because the signal building components must be known a priori, which can only be obtained by synthesizing the SFAPs based on physiological data (see Chapters 4 and 8). The second possible view of the model in Figure 10.2 sees transfer channels $h_{ij}(n)$ standing for MUAPs and their filtering on the paths to the electrodes. This modeling replaces the problem of EMG decomposition by MIMO channel identification.

The EMG decomposition may employ the MIMO system identification scheme because the excitation pulse trains, $w_i(n)$, are considered to have interspike intervals distributed as Bernoulli white noise [21,53]. Therefore blind identification based on higher order statistics can solve the problem. The most important feature of statistical identification methods is their capability to extract patterns that repeat unchanged throughout the signal at random intervals. Moreover, even if the patterns vary with a Gaussian distribution, the identification outcomes are not affected. If these facts are considered in conjunction with the stationary EMG recordings, individual MUAPs will be the signal components that repeat throughout the recordings. Even if the individual sample values vary from one appearance of the same MUAP (because of nonsynchronized sampling and possible small changes in the MUAP shape) to another according to a Gaussian distribution, these variations will be suppressed by the higher order statistical identification (see Section 10.2.2). At the same time a MUAP represents unit-pulse system response of a channel in MIMO modeling scheme, as mentioned above (see also Fig. 10.2).

A multi-electrode surface EMG measurement corresponds to a MIMO output vector $\mathbf{y}(n)$ that is assumed stationary M-variate, non-Gaussian MA process of length Q:

$$\mathbf{y}(n) = \sum_{i=0}^{Q} \mathbf{H}^T(i)\mathbf{w}(n-i) + \mathbf{v}(n) \tag{48}$$

where $\mathbf{v}(n)$ describes a possible M-variate, Gaussian, zero-mean additive noise, and the rest of notation is the same as in Eq. (4). The components of $\mathbf{H}(n)$, $h_{ij}(n)$, correspond to the ith MUAP measured by the jth electrode.

Suppose that the identification uses third-order cumulants of M-variate process $\mathbf{y}(n)$. They are calculated as $M \times M$ matrices

$$\mathbf{C}_m(\tau_1, \tau_2) = \text{cum}[\mathbf{y}(n+\tau_1)\mathbf{y}^T(n)y_m(n+\tau_2)], \qquad m = 1, \dots, M \tag{49}$$

where $y_m(n)$ denotes the mth component of the vector process $\mathbf{y}(n)$. Following the derivation in [18], the MA coefficient matrices $\mathbf{H}(k)$, $k = 1, \dots, Q$, are given by the closed-form expression

$$\mathbf{H}(k) = \mathbf{L}^{(i)T}(q, k)\mathbf{L}^{(i)}(q, 0)[\mathbf{L}^{(i)T}(q, 0)\mathbf{L}^{(i)}(q, 0)]^{-1} \tag{50}$$

where

$$\mathbf{L}^{(i)}(q, k) = [\mathbf{c}_1^{(i)}(q, k), \dots, \mathbf{c}_M^{(i)}(q, k)] \tag{51}$$

and $\mathbf{c}_1^{(i)}(q, k)$ denotes the ith column of matrix $C_m(q, k)$. Derivation (50) assumes $\mathbf{H}(0) = \mathbf{I}_{M \times K}$. This is equivalent to both temporal and spatial third-order independence among components $w_i(n)$.

Equation (50) is actually a multivariate version of the $C(q, k)$ formula from (20). Again, its importance lies in the fact that theoretically an identification of the channel characteristics in MIMO models is feasible and obtainable by higher order statistics. Although the performance of multivariate $C(q, k)$ is much too weak to succeed in decomposing the multichannel EMGs, we exemplify its capabilities by a simple 2-input 2-output system whose channel responses h_{ij}; $i, j = 1, 2$ were generated randomly as very short, four-sample sequences and are depicted with solid lines in Figure 10.10. Figure 10.10 actually depicts only four channel-response sample values in each graph; nevertheless, these samples are linearly interpolated (connected) by solid lines for easier visual interpretation. The two system inputs had been synthesized as two independent Bernoulli noises of length 16,384 samples with a 0.1 probability of firing. System identification was performed on simulated MIMO output signals, $y_1(n)$ and $y_2(n)$, according to Eq. (50) in 100 Monte Carlo simula-

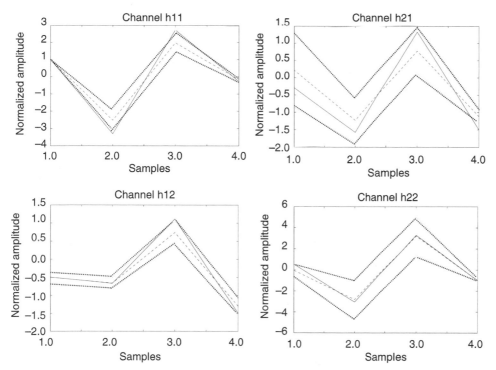

Figure 10.10. Identification of a MIMO(2, 2) system with Bernoulli input noise (p = 0.1); the channel output lengths = 16,384 samples; channel responses = h_{11}, h_{12}, h_{21}, and h_{22} generated randomly for testing purposes. The original channel responses are plotted with solid lines, the estimated responses dashed lines, and the estimation standard deviation dotted lines.

tion runs [19]. No additive noise, $v(n)$, was applied. The estimated channel responses are shown dashed, while dotted lines in Figure 10.10 depict the standard deviation of resulting estimates. The depicted identification results show that a satisfactory channel-response identification of MIMO(2, 2) model is feasible for short channel responses even with multivariate $C(q, k)$ approach. The estimated samples of all four $h_{ij}(n)$ closely correspond with the original values (dashed lines along with the solid ones), whereas also the sample standard deviation estimated from 100 simulation runs (dotted lines) exhibit acceptable levels.

A Possible Surface EMG Decomposition. Statistical system identification is often difficult because of the variance of the system output, which heavily depends on the system response characteristics [28] and additive noise. The higher the variance is, the longer the signals needed for successful identification. In decomposing the EMG signals, the estimated MUAPs will get closer to their original form with the inclusion of more MU firings, on the one hand, which decreases the inter-MUAP variance and the taking of more signal samples, on the other hand, to decrease the variance owing to additive noise. Considering that a few thousand, or even a few ten thousand, replicas of a system response are needed in order to obtain a successful identification statistically, an HOS-based EMG decomposition requires recordings of length up to 10^5 to 10^6 samples [55]. To see the figures from another point of view, for an MUAP firing frequency at about 20 Hz, a few thousand repetitions of the same MUAP would take a few minutes. If, at the same time, the signal sampling frequency is 1 kHz, every one-minute signal segment contributes 6×10^4 samples. Because of the statistical nature of HOS-based decomposition, it is important that the analyzed signal be stationary throughout the recording.

To be applied for decomposition, the HOS-based method must be general and independent of any a priori information about the estimated systems. It must handle equally all the system models (MA, AR, and ARMA), it must not depend on the knowledge about model order, and should be available for the SISO as well as the MIMO models. There are two robust approaches that satisfy all these conditions. The first is called the w-slice method [45]. The second is based on the bicepstral decomposition algorithm [38] which has been extended to polycepstral cases [35] and improved using the interpolation-based computation [50,51,52]. It has been shown [35] that the outcomes of the HOS-based identification approaches may further be improved by additional optimisation.

Finally, we will illustrate the HOS-based EMG decomposition by an example of synthetic surface EMG. A MIMO(2, 3) model was taken into account. The model output signals were generated using the EMG simulator built by the LISiN laboratory (Prof. Roberto Merletti), Torino, Italy [33]. The main simulation parameters were set as follows:

- ARMA(2, 3) model was chosen, and 2 MUAPs were generated and measured by 3 surface pickups.
- One MU was assumed to be 3 mm, the other 6 mm, deep; the former with 5 fibers and the latter with 20 fibers.
- The fibers of the first MU were aligned with the electrodes' placement, while the second one had fibers inclined by 10 degrees and shifted 10 mm in the x direction from the electrode array.
- The spread of the innervation zone was taken 0 mm for the first, and 10 mm for the second MU.
- The conduction velocity was taken as 4 m/s for all fibers.
- Measurements were double differential.
- Rectangular electrodes 5 by 1 mm were simulated.
- The interelectrode distance was taken as 10 mm.
- The electrode array was assumed placed between the innervation zone and tendons for fibers of length of 70 mm.
- A sampling frequency of 1024 Hz was used for the generated EMG signal.
- Three synthetic SEMG signals were generated in duration of 100 s.

The three synthetic recordings were included in a least-mean square (LMS) cumulant-based optimization [35] in order for MUAPs to be extracted. The basis for such decomposition is a system of nonlinear equations of the following form:

$$f_j(\tau_1, \tau_2) = \sum_{i=1}^{K} \gamma_{3,w_i} h_{ij}(n) \cdot h_{ij}(n + \tau_1) \cdot h_{ij}(n + \tau_2) - \hat{C}_{3,y_j}(\tau_1, \tau_2) \qquad (52)$$

where K stands for the number of different signal sources (i.e., the number of MUs), $h_{ij}(n)$ for the model MUAP shape as detected in a certain step (iteration) of decomposition, and the hat ($\wedge$) denotes the cumulant estimation based on the measured output $y_j(n)$.

To be solvable, the system of equations in Eq. (52) must have a number of equations that is at least $K \cdot N$, where the length of K source signals (MUAPs) equals N. To decom-

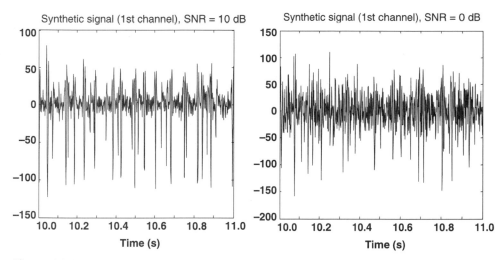

Figure 10.11. Synthetic SEMG signal (first channel) with additive Gaussian white noise: SNR = 10 dB (*left*) and 0 dB (*right*).

pose the generated synthetic SEMGs, we applied the MATLAB function called *lsqnonlin*, which minimizes the least square norm of the system using the large-scale, trust-region reflective Newton algorithm [29]. The number of linear equations we selected was 36, meaning that for each of the two expected source MUAPs, 18 samples were estimated. The cumulant estimates used in these equations were taken from 1/6 of the cumulant matrix in such a manner that $\tau_1 = 0, \ldots, (\sqrt{8KN+1} - 3)/2$ and $\tau_2 \leq \tau_1$.

The described decomposition approach exhibits satisfactory robustness, which can be seen from the results of three experiments. The first is done with the synthesized SEMGs in a noise-free condition. In the second and the third cases, the outcomes are decomposed from the same synthetic SEMGs but with additive Gaussian white noise, with SNR = 10 dB and SNR = 0 dB, respectively. Two segments of a noisy synthesized signal from channel 1 are depicted in Figure 10.11.

Decomposition of the synthetic SEMGs obtained by the cumulant-based nonlinear LMS optimization has proved to be successful even in rather noisy circumstances, as shown in Figure 10.11. In Figure 10.12 the decomposed MUAPs are depicted dashed in their original form, their estimates are shown solid at SNR = 10 dB, and dotted at SNR = 0 dB.

Finally, there is the fact that a MIMO decomposition using higher order statistical approaches returns the estimates according to the model shown in Figure 10.2. In the case of the surface EMG this implies that the measured waveforms of MUAPs, do not have the same shapes as generated by the source motor units. The filtering effect depicted in Figure 10.2 changes, by additional filters $f_{ij}^{(l)}(n)$, the original MUAP from the ith motor unit to the measured shape at the jth electrode. This filtering effect can be eliminated if another decomposition is applied by single-input, multiple-output (SIMO) modeling. The SIMO model is extracted as a submodel of MIMO in such a way that only the channels starting from one common MIMO input are respected (one horizontal layer in Fig. 10.2). For this kind of SIMO model, the outputs are known from the MIMO decomposition. Now, in general, we deal with M known channel responses, while a common excitation for all of them is unknown and looked for. A solution can be found, for example, by cepstral approaches published in [35] and [48].

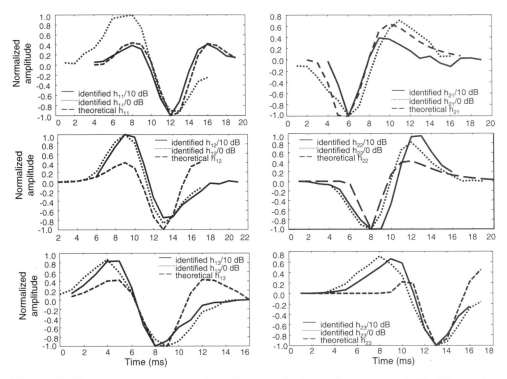

Figure 10.12. MUAPs decomposed from three synthetic SEMG channels of 102,400 samples in length. MUAPs are denoted by $h_{ij}(n)$ where indexes i and j indicate the signal source and measurement, respectively, as well as the column and row of the corresponding subfigure. The original MUAPs are depicted dashed, the decomposition results at SNR = 10 dB solid, and the results at SNR = 0 dB dotted.

10.4 APPLICATIONS TO MONITORING MYOELECTRIC MANIFESTATIONS OF MUSCLE FATIGUE

Mechanical manifestations of muscular fatigue have been defined as the "failure to maintain the required or expected force" [14] or "any exercise-reduction reduction in the capacity to generate force or power output" [46]. Chaffin [7] established the term "localized muscular fatigue" to refer to manifestations of fatigue that are localized to a single muscle or a group of synergistic muscles performing the contraction. According to the definition of myoelectric manifestations of muscle fatigue (see Chapter 9), the involved muscles continuously become fatigued from the onset of the contraction. Physiological and biochemical data can show time-dependent changes indicating the fatigue process. The rate of fatigue highly depends on the properties of the muscles being activated, the level of contraction, and if the contractions are continuous or intermittent. There are two types of fatigue depending on the site of failure: *central* or *peripheral*. They are respectively caused by alterations in neural pathways (central fatigue) or within the muscle fiber itself, including the neuromuscular junction (peripheral fatigue). While very little is actually known about central muscular fatigue, more is known about factors involved in peripheral fatigue such as increased lactate concentration and decreased pH, and changes in ionic concentrations, but there is uncertainty about their relative contributions.

Frequency analysis of EMG signals has been widely used to characterize muscle fatigue during isometric, constant-force contractions. At contraction levels sufficiently high to generate myoelectric manifestations of muscle fatigue detectable in a short time (less than one minute), the EMG signal may be considered as a band-limited stochastic process with a Gaussian distribution and zero mean [2]. It is also generally accepted that the signal can be assumed to be wide-sense stationary for epochs lasting 0.5 to 2 seconds, depending on contractile force level and the properties of the investigated muscle. The mean and median frequencies of the power spectrum are commonly used in the literature as spectral change indicators for the surface EMG signal (see Chapter 6). Recently also higher order moments have been introduced to analyze surface EMG signals [24,32].

Pure isometric contractions represent a simple and reproducible bench-test condition but are not common in daily activities. In the fields of rehabilitation medicine, sport medicine, and ergonomics, it is more "natural" for the subject to perform functional tasks similar to daily life activities or movement (i.e., dynamic contractions). However, during dynamic contractions, spectral analysis must be handled with great care because there are changes in several parameters: (1) the number of active motor units, (2) the geometric relation between the active muscle fibers and the electrode, (3) the geometric relation to the innervation zone, (4) the muscle-fiber lengths, and so on. All such factors contribute to the nonstationarity of the EMG signal. Thus Fourier and other classical signal processing methods may not be suitable for the analysis of EMG signals during dynamic conditions. Therefore time-frequency methods, which do not require stationarity of the signal, have been introduced to improve the EMG analysis.

10.4.1 Myoelectric Manifestations of Muscle Fatigue during Static Contractions

The most widely used method for estimating the spectrum of the EMG signal is the Fourier transform, partly because of the computationally effective fast Fourier transform (FFT) algorithm. However, stationary conditions are required for spectral estimation using the Fourier transform—otherwise, information about spectral changes will be lost. A common method to study a "quasi-stationary" signal is to divide the long-term signal into segments of short duration (0.5 to 2 seconds) where the wide-sense stationarity hypothesis holds; namely the first and second statistical moments of the stochastic process do not change in time [36].

Despite the popular FFT method, parametric (or autoregressive) identification methods have also been used to estimate the spectrum of the EMG signal [20,32,37] (see Chapter 6). Some of the problems associated with the Fourier transform can be avoided by using the parametric approach: (1) frequency leakage (related to pre-windowing), (2) frequency resolution (related to the stationarity problem and selection of time segments), (3) large variance of estimate, and (4) assumption of signal periodicity or signal equal to zero outside of the analysis segment. Parametric methods are specifically useful when short data segments are available due to time-varying EMG signals (nonstationary), such as during isokinetic contractions. In fact these methods extrapolate the values of the autocorrelation for lags greater than the analyzed segment length. The main disadvantage of the parametric identification methods is the assumption that a mathematical model of signal generation (AR or ARMA) may not always describe the signal faithfully.

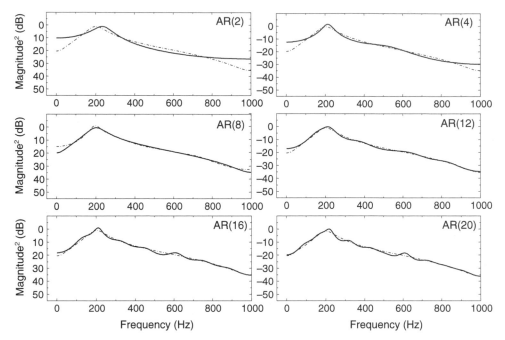

Figure 10.13. Different orders of AR spectral estimates for a synthesized EMG signal. Dotted lines show the spectral shape of the ARMA model used.

Comparison of Different Methods to Estimate PSD of a Synthesized EMG Signal. How do we compare different spectral estimation approaches for analyzing EMG signals? The major problem is that there is no way to know the actual EMG spectra. One possible method is to create synthesized signals generated from known models that mimic the "true" EMG signal. One such model is the Shwedyk model (see Chapter 6), and others have also been used. In [23] white noise was filtered by an ARMA model with spectral characteristics based on a true EMG signal measured during an isometric contraction.

Figure 10.13 shows AR spectral estimates of a synthesised EMG signal for some AR-models of order between 2 and 20. The main problem with parametric modeling methods is the difficulty to select the model order and the calculation of the parameters. Large model orders tend to model some noise, as can be seen in the figure, where orders larger then 8 start to generate peaks in the region above 400 Hz. In general, a model order of 2 to 7 is enough, but if spectral peaks such as the firing rate of dominant MUs are also of interest, a model order of at least 25 is needed [37].

In Figure 10.14a and e the same synthesized EMG signal was analyzed using FFT and WP, respectively. Some different methods (Bartlett, Welch, and wavelet shrinkage) to improve the spectral estimate are also shown. Figure 10.14c shows how averaging of two FFT periodograms reduced the variance. In Figure 10.14d averaging was based on the Welch method by dividing the signal into seven 50% overlapping segments. This reduced the variance even more. However, it is important to note that the frequency resolution was also decreased. In using the wavelet shrinkage method, the variance was reduced remarkably more without loss of frequency resolution, which coincides with the theoretical properties of the wavelet shrinkage method. Compared to Fourier, statistics based on

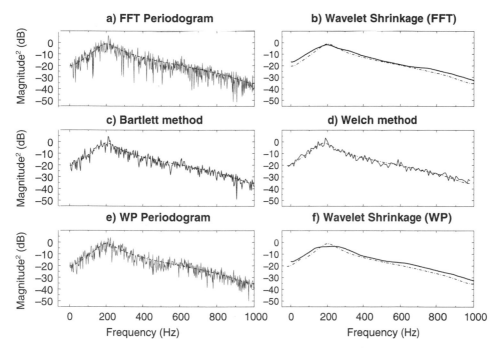

Figure 10.14. Comparison between estimated spectrum based on (*a*) FFT and (*e*) WP for the same synthesized EMG signal as in Figure 10.13. Different methods were used to reduce the variance in the spectral estimates: (*b*) Wavelet shrinkage of the FFT; (*c*) Bartlett's method by dividing the signal into two segments; (*d*) the Welch method using seven segments with 50% overlap; (*f*) wavelet shrinkage of the WP periodogram. Dotted lines show the spectral shape of the ARMA model used.

100 synthesized EMG signals [23] have shown that WP, choosing a level with the same resolution as Fourier, gives a similar performance.

Application Examples. In the comparison above shows only synthesized EMG signals used. A fair question is if similar performance can be observed with real EMG signals. Since the mean frequency of the power spectrum (MNF) is a commonly used parameter for estimating muscle fatigue, let us calculate MNF based on FFT and WP spectral estimation and compare the results.

Figure 10.15 shows one of six tests where force (1) and EMG signal (from vastus lateralis muscle) (2) were measured from healthy subjects performing sustained isometric knee extension contractions at 70% of a maximum voluntary contraction (MVC). Figure 10.15*c* to *f* shows the FFT and WP periodograms and their wavelet shrinkage enhancement (dotted lines). The level-dependent thresholds, soft shrinkage function, and Symlet (order 4) wavelet at two different time points (3 and 37 seconds) correspond to nonfatigued and fatigued muscle. MNFs for each spectrum were also calculated, and they indicate the frequency shift caused by muscle fatigue.

The following steps can be used to monitor muscle fatigue:

1. Select a segment (typically 1024 samples for a signal sampled at 2 kHz) from the EMG signal.

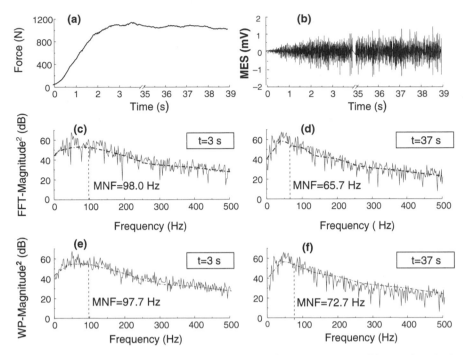

Figure 10.15. Typical static contraction during 70% of MVC. (a) Force; (b) myoelectric signal (MES) obtained from the vastus lateralis muscle (with sampling frequency 2 kHz); (c–d) PSD estimate using FFT; (e–f) PSD estimate using wavelet packets (WP). All spectral estimates are based on 1024 samples (c, e) taken at 3 seconds, and (d, f) taken at 37 seconds. Dotted lines show the wavelet shrinkage of the PSD estimates.

2. Estimate a spectrum using FFT or WP (or other method; e.g., AR) based on the selected segment.
3. Calculate the MNF from the estimated spectrum.
4. Move to the next segment and repeat from step 2 until the end of the EMG signal.

Figure 10.16 shows an example of MNF estimates based on FFT (solid lines) and WP (dotted lines) for the same subject and muscle as in Figure 10.15. Note that the results are quite similar, but the WP method gives slightly less variance.

10.4.2 Myoelectric Manifestations of Muscle Fatigue during Dynamic Contraction

The usual way to overcome the difficulties with the Fourier transform and the parametric methods, and to satisfy the stationarity requirements, is to introduce time dependency into the Fourier analysis while preserving linearity. The idea is to introduce a "local frequency" parameter (local in time) so that the "local" Fourier transform looks at the signal through a window during which the signal is approximately stationary. The first attempt to construct a function that represents both time and frequency was proposed by Gabor [15] who developed the short-time Fourier transform. Hannaford and Lehman [22] used this short-time technique to locate time-dependencies in the EMG signals. The time dependencies

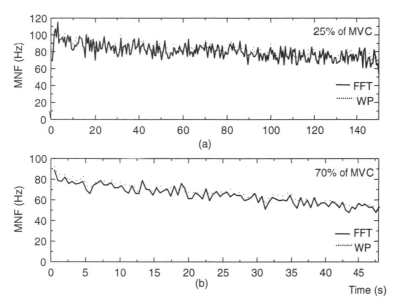

Figure 10.16. Comparison between Fourier (*solid lines*) and wavelet packets (*dotted lines*) methods to estimate the MNF variable during 25% and 70% of the maximum voluntary contraction (MVC) for quantifying muscle fatigue (during sustained isometric contractions obtained from the vastus lateralis muscle).

that are normally averaged in the Fourier analysis were picked from the typical "three-burst pattern" in the EMG signals of both wrist and head movements. Later Duchéne et al. [13] used the Wigner-Ville distribution (WVD) methods to analyze uterine electromyogram to track instantaneous burst frequency of the EMG signal. Their results suggested that the time-frequency representation of the EMG signal using the WVD method was promising because it correlated the surface electromyogram and electrical activity of the uterine muscle at a cellular level. Furthermore the results showed that the choice of the specific kernels influenced the time-frequency distribution of the EMG signals and created a confounding factor in the interpretation of the data [13].

To analyze the EMG signals from skeletal muscles during dynamic contractions, Bonato et al. [3] compared several time-frequency methods based on different types of kernels, including the smoothed WVD, the cone kernel, the reduced interference, and the Choi-Williams Distribution (CWD). Their results showed that the Choi-Williams method was the most suitable method to represent the nonstationary EMG signals during dynamic contractions. However, the use of the CWD requires careful selection of the parameters in the kernel [3,13]. Later Roy et al. [40] used this approach during cyclical dynamic exercise and observed nonlinear and more complex behavior of the instantaneous median frequency compared with static contractions. During the static contraction a more or less linear decrease of the mean or median frequency is often reported in the literature.

One of the most critical problems is the dependence of the estimates with respect to the signal-to-noise ratio. For example, when white Gaussian noise is superimposed on the surface EMG signal, the signal-to-noise ratio increases with increasing force; that is, the EMG signal increases but the noise level remains the same. Roy et al. modified an algorithm proposed by D'Alessio [9] to estimate a threshold, which was used to discriminate between the signal and the noise components. That threshold can be an upper frequency (integration limit) when calculating time-dependent mean or median frequency. Different methods to find the integration limits have been compared by Östlund and Karlsson [37]. A characterization of the algorithm can be found in [4].

Recently the wavelet transform method has been proposed in an effort to overcome the limitations of the traditional time-frequency methods. The wavelet transform acts as a "mathematical microscope" by which one can observe different parts of the signal by just "adjusting the focus." This allows the detection of short-lived time components within the EMG signals. The reasoning behind this "adapted" method is that high-frequency components such as short bursts need high time resolution (i.e., short basis functions). In contrast, low-frequency components often need a detailed frequency analysis (i.e., long basis functions). Tscharner [44] suggested that methods based on Wigner-Ville are not well suited when applied to multicomponent signals and that the wavelet method is therefore a more appropriate tool for EMG signals analysis.

There are two main trade-offs involved when STFT, CWT, and WVD (including filtered versions) are considered. Potential increases in performance for a given application must be balanced against computational complexity and storage requirements. However, the discrete version of the CWT, associated with orthonormal representation (i.e., DWT), is the most efficient. The computational complexity of FFT is $O(n \log_2(n))$, and for the fast (dyadic) wavelet transform it goes down to $O(n)$, where O represents the degree of complexity in the order of the signal of length, n. The two extreme situations, the dyadic (linear independent) and CWT (completeness), can be used to optimize the properties of the transform with respect to a given signal processing application.

Comparison of Different TFR Methods by Analyzing Synthesized EMG Signal.
To compare different TFR methods we need nonstationary synthesized EMG signals with known characteristics. Estimation accuracy and precision can be investigated and compared using computer-synthesized EMG signals with a linearly decreasing MNF. Figure 10.17 shows some time-frequency representations used in surface EMG applications—STFT, PWVD, RWED, and CWT—for one of 100 synthesized EMG signals. In each method, IMNF estimates and the "true" IMNF signals (dotted lines) are included. Statistics (i.e., bias, variance, and relative error for four spectral moments) based on all 100 synthesized EMG signals show that the CWT method performed somewhat better in all cases. In contrast to TFRs based on the Wigner-Ville transform using kernels, the wavelet method offers a technique that is almost free from "tricky" parameter selection. For more details regarding statistical results and generation of nonstationary synthesized EMG signals, see [24].

Application Examples Using the CWT.
In sports, ergonomics, and rehabilitation medicine, isokinetic exercise contractions are commonly used [6]. Isokinetic contractions are often performed in an electromechanical device (isokinetic dynamometer) in which the angular velocity can be controlled. When the subject performs maximal or submaximal contractions, the device limits the joint angular velocity to a pre-set value during the contraction cycle. These controlled movements at relatively high angular velocities are useful for clinical evaluation of work-related myalgia and upper motor neurone disorders (e.g., stroke). It is known that work, which involves highly repetitive contractions, is a risk factor for work-related trapezius myalgia. The analysis of the EMG signal can help us understand the development of fatigue during repetitive dynamic work cycles at high angular velocities. In patients with upper motor neurone diseases (e.g., stroke with hemi-

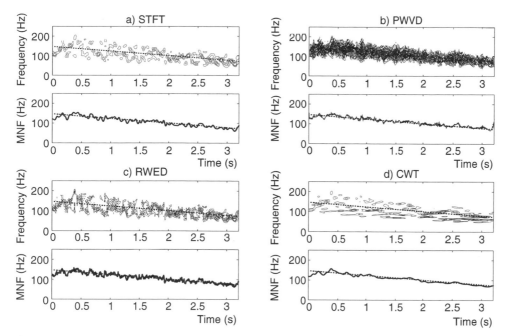

Figure 10.17. Time-frequency representations and the instantaneous mean-frequency (IMNF) estimates based on a linearly decreasing simulated myoelectric signal using (a) short-time Fourier transform (STFT), (b) pseudo–Wigner-Ville distribution (PWVD), (c) running-windowed exponential distribution (RWED), and (d) continuous wavelet transform (CWT).

plegia), the magnitude of spasticity and other disturbances increases with increasing angular velocity.

Figure 10.18 shows three-selected isokinetic contractions extracted from 100 repeated knee extensions performed with maximum effort. They were selected from the beginning, middle, and end of the sequence, respectively. For each selected contraction the raw surface EMG signal, the contour plot of its time-frequency distribution, the MNF estimates using the CWT method, the corresponding force, and the knee movement velocity are presented for one of the tested subjects.

The results suggest that the force and MNF decreased as the number of contractions increased. The hypothesis is that a decrease in the muscle-fiber conduction velocity, especially of the fast-twitch fibers (type IIB) could partly be responsible for the MNF decrease. During repetitive maximum dynamic (isokinetic) contractions, force and MNF decrease during the initial 40 to 60 contractions, followed by a stable level with no further decrease [17]. The decrease in biomechanical output is significantly related to the change in MNF, thus indicating that MNF is a fatigue indicator also during dynamic contractions. However, the correlations are not high, which cannot be expected since they both are multifaceted. The majority of subjects (17–18 out of 21 subjects), performing gradually increasing contractions up to 100% of a maximum voluntary contraction, showed a significant positive correlations between MNF (based on CWT) and force in the three muscles of quadriceps [26]. This opens the possibility that during dynamic contractions the decrease in MNF could be, in part, due to the decrease in force [25].

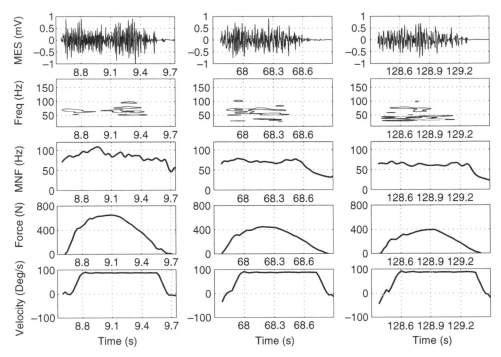

Figure 10.18. MNF estimates based on the CWT method for three contraction, selected in the beginning (*left*), the middle (*middle*), and the end (*right*) of 100 repeated dynamic contractions until exhaustion. For each selected contraction are plotted, starting at the top, the raw surface EMG signal measured from the right vastus lateralis muscle, the contour plot of its time-frequency distribution, the MNF estimates using the CWT method, the corresponding force, and the knee movement velocity.

10.5 CONCLUSIONS

The advanced signal processing methods described and exemplified in this chapter suit several SEMG analysis needs, but each method has specific applications and some known drawbacks. Let us summarize their most important features.

Time-frequency representations are well-known tools for nonstationary signals analysis but can also be used for decision or classification. These are nonparametric approaches that are directly related to the data and are independent of the model choice. The main difficulties are the selection of the TFR kernel and of the interest zones of the time-frequency plane, but the flexibility and the easiness of TFR-based methods make them a potentially useful candidate for source (MUAP) separation in complex SEMG.

Unlike the linear representations (i.e., STFT and wavelets) there are more than a dozen candidates for the bilinear (or quadratic) time-frequency representation. In the group of bilinear TFRs, which also includes the square of the STFT (the spectrogram) and wavelets (the scalogram), the Wigner-Ville distribution (WVD) has many more desirable time-frequency and mathematical properties than many other representations. For example, time and frequency shifts of the signal are preserved. The WVD satisfies frequency and time marginal properties, namely the signal's instantaneous power and the spectral energy density, respectively. However, the big problem with the WVD is cross-term interference, and this limits its use in several applications, particularly in analyzing multicomponent nonstationary signals such as EMG signals. Therefore most of the research field of time-frequency analysis has involved finding TFRs that preserve the properties of WVD but have reduced cross-term interference. For $s(t) = s1(t) + s2(t)$, the WVD is

$$WVD_s(t, \omega) = WVD_{s1}(t, \omega) + WVD_{s2}(t, \omega) + 2 \operatorname{Re}\{WVD_{s1,s2}(t, \omega)\}$$

which shows that the WVD of a two-component signal is the sum of two auto-terms that are nonnegative, but also a cross-term that oscillates with a magnitude twice as large as the auto-terms and an average equal zero.

All bilinear representations can be written in a general form introduced by Cohen called Cohen's class. The difference between the WVD and the member of Cohen's class is a typical linear (low-pass) filtering performed by the kernel function. This filtering suppresses the cross-terms and also reduces the resolution. Due to the fact that the signals are real valued, the spectrum is always symmetric, which introduces redundancy and cross-terms. Therefore sampling at twice the Nyquist rate, or converting the real signal to the so-called analytical signal, must be performed before transformation.

Arbitrary time-frequency resolution can be obtained from the library of orthonormal bases of wavelet packets. The wavelet packets (see Section 10.2.4) method involves a generalized multiresolution approach that represents the entire family of subband coded decompositions. From this family of bases, the best basis can be chosen to match the signal's nonstationarity statistics. An adaptive filtering of the signal can be performed by using an entropy criterion and choosing among parents and children in a tree.

The issue of performance and computational complexity of various time-frequency and time-scale approaches was discussed in Section 10.4.2. Sections 10.4.1 and 10.4.2 also focused on the application of these methods to SEMG. It is clear that all of them greatly outperform the classical Fourier transform in estimating the EMG spectral densities. In particular, wavelet packets and wavelet shrinkage (Sections 10.2.4 and 10.4.1) produce robust and accurate results. Also the mean frequency estimates of the power spectrum, used in muscle fatigue monitoring, are considerably improved during both static and dynamic contractions.

In the last decade higher-order statistics entered biomedical research and was mainly incorporated in feature extraction and classification solutions. The power of statistics in the area of system response identification has not been noticeably utilized. This is probably because of certain drawbacks that hinder wider implementation of HOS-based biomedical signal processing. The first is the need for long steady stationary signals; they involve very high computational complexity. Next is the problem of dealing with multi-channel models whose model order should be known a priori. Both features, long stationary recordings of up to a few minutes duration and advance knowledge of the model order, (i.e., the number of active MUs) cannot always be fulfilled but are critical to SEMG decomposition.

Nevertheless, HOS-based methods have many useful properties. The most attractive is a capability to considerably suppress Gaussian additive noise. As we showed in this chapter, they play an important role in the EMG feature extraction for further signal recognition or classification. Finally, we looked at the way SEMG decomposition can be supported by HOS (Section 10.3.2). It is evident that several MUAPs can be decomposed from fewer parallel SEMG recordings, possibly from only one-channel measurement. The MUAP superimpositions do not hinder a thorough decomposition. However, while the

MUAP shapes can be obtained directly, the innervation trains must be recognized by some other means. One way is to use the time-scale phase representations, as we showed in Section 10.3.1.

ACKNOWLEDGMENT

Most of the algorithms and results presented on parametric EMG decomposition and the application of higher-order statistics were developed by Dr. Danilo Korže, Dr. Andrej Šoštarič, and Dr. Dean Korošec in their doctoral research work at the University of Maribor, Slovenia. The implementation of nonlinear LMS decomposition of SEMG was contributed by Eric Plévin of the Ecole Centrale de Nantes, France. The use of wavelets in EMG spectral estimation applications was investigated in cooperation with Prof. Jun Yu (Umeå University, Sweden) and Prof. Metin Akay (Dartmouth College), and the work was partly supported by the European Union Regional Fund.

REFERENCES

1. Akay, M., *Detection and estimation methods for biomedical signals*, Academic Press, San Diego, 1996.

2. Basmajian, J. V., and C. J. DeLuca, *Muscles alive: Their functions revealed by electromyography*, 5th ed., Williams and Wilkins, Baltimore, 1985.

3. Bonato, P., G. Gagliati, and M. Knaflitz, "Analysis of myoelectric signals recorded during dynamic contractions," *IEEE Eng Med Biol Mag* **15**, 102–111 (1996).

4. Bonato, P., S. H. Roy, M. Knaflitz, and C. J. De Luca, "Time-frequency analysis of the surface myoelectric signal," *IEEE Trans BME* **48**, 745–753 (2001).

5. Bruce, A. G., and H. Y. Gao, "Understanding wave shrink: Variance and bias estimation," *Biometrika* **83**, 727–745 (1996).

6. Cabri, J. M. H., "Isokinetic strength aspects of human joints and muscle," *Crit Rev Biomed Eng* **19**, 231–259 (1991).

7. Chaffin, D. B., "Localized muscle fatigue—Definition and measurement," *J. Occup. Med.* **15**, 346–354 (1973).

8. Coifman, R. R., Y. Meyer, S. Quake, and M. V. Wickerhauser, "Signal processing and compression with wavelet packets," in J. S. Byrnes, ed., *Wavelets and their applications*, Kluwer, Norwell, MA, 1994, pp. 363–379.

9. D'Alessio, T., "Objective algorithm for maximum frequency estimation in Doppler spectral analysers," *Med Biol Eng Comput* **23**, 63–68 (1985).

10. Daubechies, I., *Ten lectures on wavelets*, SIAM, Philadelphia, 1992.

11. Donoho, D. L., "De-noising by soft-thresholding," *IEEE Trans Informat Theor* **41**, 613–627 (1995).

12. Donoho, D. L., and I. M. Johnstone, "Ideal spatial adaptation by wavelet shrinkage," *Biometrika* **81**, 425–455 (1994).

13. Duchéne, J., D. Devedeux, S. Mansour, and C. Marque, "Analyzing uterine EMG: Tracking instantaneous burst frequency," *IEEE Eng Med Biol Mag* **14**, 125–132 (1995).

14. Edwards, R. H. T., "Human Muscle Function and Fatigue," in R. Porter, and J. Whelan, eds., *Human muscle fatigue: Physiological mechanisms*, Pitman Medical, London, 1981.

15. Gabor, D., "Theory of communication," *J IEE* **93**, 429–457 (1946).

16. Gao, H. Y., "Choice of thresholds for wavelet shrinkage estimate of the spectrum," *J Time Series Anal* **18**, 231–251 (1997).

17. Gerdle, B., S. Karlsson, A. G. Crenshaw, J. Fridén, and I. Nilsson, "Characteristics of the shift from the fatigue phase to the endurance level (breakpoint) of peak torque during repeated dynamic maximal knee extensions are correlated to muscle morphology," *Isokinet Exer Sci* **7**, 49–60 (1998).

18. Giannakis, G. B., Y. Inouye, and J. M. Mendel, "Cumulant based identification of multichannel moving-average models," *IEEE Trans Autom Control* **34**, 783–787 (1989).

19. Gould, H., and J. Tobochnik, *An introduction to computer simulation methods*, Addison-Wesley, Reading, MA, 1996.

20. Graupe, D., and W. K. Cline, "Functional separation of EMG signals via ARMA identification methods for prothesis control purposes," *IEEE Trans Syst Man Cybern* **5**, 252–259 (1975).

21. Gygi, A., "Analyse des Nadel- und Oberflächen-Elektromyogramms mittels Statistiken höherer Ordnung," PhD thesis 10863, ETH Zurich, 1994.

22. Hannaford, B., and S. Lehman, "Short time Fourier analysis of the electromyogram: Fast movements and constant contraction," *IEEE Trans BME* **33**, 1173–1181 (1986).

23. Karlsson, S., J. Yu, and M. Akay, "Enhancement of spectral analysis of myoelectric signals during static contractions using wavelet methods," *IEEE Trans BME* **46**, 670–684 (1999).

24. Karlsson, S., J. Yu, and M. Akay, "Time-frequency analysis of myoelectric signals during dynamic contractions: A comparative study," *IEEE Trans BME* **47**, 228–238 (2000).

25. Karlsson, J. S., N. Östlund, B. Larsson, and B. Gerdle, "A method to distinguish the effects of force and muscle fatigue upon the shift of the mean frequency of the surface EMG during maximum isokinetic knee extensions," *Proc XIVth ISEK Congr Vienna*, 2002, pp. 333–334.

26. Karlsson, S., and B. Gerdle, "Mean frequency and signal amplitude of the surface EMG of the quadriceps muscles increase with increasing torque—A study using the continuous wavelet transform," *J Electromyogr Kinesiol* **11**, 131–140 (2001).

27. Kay, S. M., *Fundamentals of statistical signal processing: Estimation theory*, Prentice Hall, Englewood Cliffs, NJ, 1993.

28. Korže, D., and D. Zazula, "Variance analysis for identification methods using third-order cumulants," *Electrotechn Rev* **64**, 7–14 (1997).

29. Magrab, E. B., *An engineer's guide to MATLAB*, Prentice Hall, Upper Saddle River, NJ, 2000.

30. Mallat, S., "A theory for multiresolution signal decomposition: The wavelet representation," *IEEE Trans Pattern Anal Mach Intell* **11**, 674–693 (1989).

31. Mendel, J. M., "Tutorial on higher-order statistics (spectra) in signal processing and system theory: Theoretical results and some applications," *Proc IEEE* **79**, 278–305 (1991).

32. Merletti, R., A. Gulisashvili, and L. R. Lo Conte, "Estimation of shape characteristics of surface muscle signal spectra from time domain data," *IEEE Trans BME* **42**, 769–776 (1995).

33. Merletti, R., L. Lo Conte, E. Avignone, and P. Guglielminotti, "Modelling of surface myoelectric signals. Part I: Model implementation," *IEEE Trans BME* **46,** 810–820 (1999).

34. Moulin, P., "Wavelet thresholding techniques for power spectrum estimation," *IEEE Trans Sig Process* **42**, 3126–3136 (1994).

35. Nikias, C. L., and A. P. Petropulu, *Higher-order spectra analysis: A nonlinear signal processing framework*, Prentice Hall, Englewood Cliffs, NJ, 1993.

36. Oppenheim, A. V., and R. W. Schafer, *Discrete-time signal processing*, 2nd ed., Prentice-Hall, Upper Saddle River, NJ, 1999.

37. Östlund, N., and J. S. Karlsson, "Methods to improve mean frequency measurements of surface EMG signals during dynamic contractions," *Proc XIVth ISEK Congr, Vienna*, 2002, pp. 49–50.

38. Pan, R., and C. L. Nikias, "The complex cepstrum of higher order cumulants and nonminimum phase system identification," *IEEE Trans Acous Speech Sig Process* **36**, 186–205 (1988).

39. Papoulis, A., *Probability, random variables, and stochastic processes*, McGraw-Hill, New York, 1991.

40. Roy, S. H., P. Bonato, and M. Knaflitz, "EMG assessment of back muscle function during cyclical lifting," *J Electromyogr Kinesiol* **8**, 233–245 (1998).

41. Šoštarič, A., "Application de la transformée en ondelettes et des representations temps-échelle à la décomposition de signaux composites," PhD thesis, Univeristy of Maribor and Ecole Centrale de Nantes, 1998.

42. Šoštarič, A., D. Zazula, and C. Doncarli, "Time-scale decomposition of compound (EMG) signal," *Electrotechn Rev* **67**, 69–75 (2000).

43. Therrien, C. W., *Discrete random signals and statistical signal processing*, Prentice Hall, Englewood-Cliffs, NJ, 1992.

44. Tscharner, V., "Intensity analysis in time-frequency space of surface myoelectric signals by wavelets of specified resolution," *J Electromyogr Kinesiol* **10**, 433–445 (2000).

45. Vidal, J., and J. A. R. Fonollosa, "Impulse response recovery of linear systems through weighted cumulant slices," *IEEE Trans Sig Process* **44**, 2626–2631 (1996).

46. Vøllestad, N. K., "Measurement of human muscle fatigue," *J Neurosci Meth* **74**, 219–227 (1997).

47. Wahba, G., "Automatic smoothing of the log-periodogram," *J Am Stat Assoc* **75**, 122–132 (1980).

48. Zazula, D., "Cepstral modelling and decomposition of the multilead ECG signals," in *Systems, modelling, Control*, vol. 2, Zakopane, Poland, 1995, pp. 416–421.

49. Zazula, D., "Computer-assisted decomposition of the electromyograms," *CBMS '98, Lubbock, USA*, 1998, pp. 26–31.

50. Zazula, D., "Asymptotically exact computation of differential cepstrum using FFT approach," *Electr Lett* **34**, 842–844 (1998).

51. Zazula, D., "Interpolated differential bicepstrum and its application," *ICECS '99*, Pafos, Cyprus, 1999.

52. Zazula, D., "System identification using polycepstra with interpolation," *WCC 2000*, Beijing, China, 2000, pp. 215–222.

53. Zazula, D., "Experience with surface EMG decomposition using higher-order cumulants," *Signal processing 2001*, Poznań, Poland, 2001, pp. 19–24.

54. Zazula, D., D. Korošec, and A. Šoštarič, "Parametric decomposition of the SEMG," in H. J. Hermens, D. Stegeman, J. Blok, and B. Freriks, eds., *SENIAM*, vol. 6, 1998, pp. 68–75.

55. Zazula, D., D. Korže, A. Šoštarič, and D. Korošec, "Study of methods for decomposition of superimposed signals with application to electromyograms," in Pedotti et al., eds., *Neuroprosthetics*, Springer Verlag, Berlin, 1996.

56. Zazula, D., and A. Šoštarič, "Possible approaches to surface EMG decomposition," in H. J. Hermens, R. Merletti, H. Rix, and B. Freriks, eds., *SENIAM*, vol. 7, 1999, pp. 169–176.

11

SURFACE MECHANOMYOGRAM

C. Orizio

Department of Biomedical Sciences and Biotechnologies
Section of Human Physiology
University of Brescia, Brescia, Italy

11.1 THE MECHANOMYOGRAM (MMG): GENERAL ASPECTS DURING STIMULATED AND VOLUNTARY CONTRACTION

Even during muscular isometric contraction, when the muscle tendon unit is kept at constant length, the sliding of the actomyosin filaments determines a reduction of the length of the contractile elements of the muscle with a shortening of its long axis [68]. Since muscle can be considered as a near constant volume system [8], these changes in muscle length are paralleled by changes in the transverse axis dimension. Indeed, already at the end of the nineteenth century the French physiologist Marey designed specific myographs to record the changes in the muscle thickness during evoked contraction (Fig. 11.1). Analysis of the transverse diameter changes of muscle was considered as a reliable tool, and since the beginning of twentieth century was used to describe the muscle contraction process in several handbooks of physiology [31].

Because of dimensional changes of active muscle fibers the muscle surface oscillations are described as a mechanical event. Muscle contractions can actually be detected by several transducers such as piezoelectric contact sensors, microphones, accelerometers, and laser distance sensors. To underscore the mechanical nature of the generated electrical signal it has been recently suggested to use the term surface mechanomyogram (MMG) independently from the transducer employed. In Figure 11.2 are reported the simultane-

Electromyography: Physiology, Engineering, and Noninvasive Applications, edited by Roberto Merletti and Philip Parker.
ISBN 0-471-67580-6 Copyright © 2004 Institute for Electrical and Electronics Engineers, Inc.

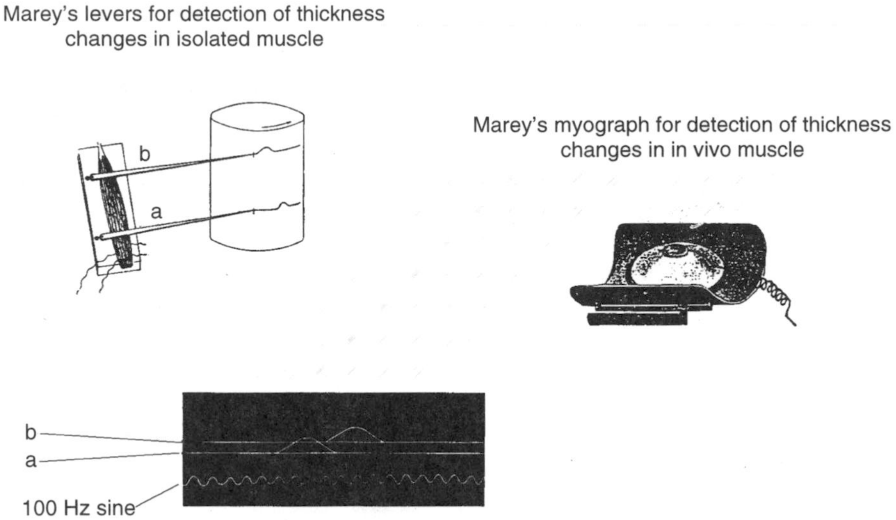

Marey's levers for detection of thickness
changes in isolated muscle

Marey's myograph for detection of thickness
changes in in vivo muscle

b

a

b
a
100 Hz sine

Figure 11.1. Marey's myographs and paper recording for the detection of the muscle thickness changes during evoked muscle activity.

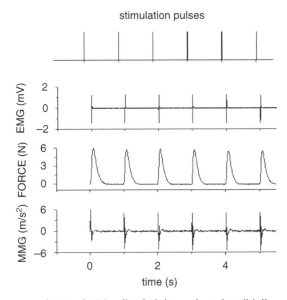

Figure 11.2. (From top to bottom): Stimuli administered at the tibialis anterior motor point, compound muscle action potential (EMG), force twitches, and MMG. The signals were recorded at the Human Physiology Laboratory of the University of Brescia (I).

ous recording of the electrical activity, the force twitches, and the accelerometer detected MMG during stimulation of the human tibialis anterior.

During voluntary contraction the active fibers of the recruited motor units generate pressure waves that determine specific muscle surface oscillations. A schematic drawing

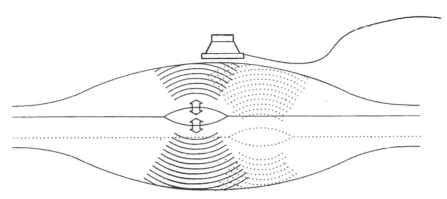

Figure 11.3. Schematic representation of the hypothesised MMG generation process.

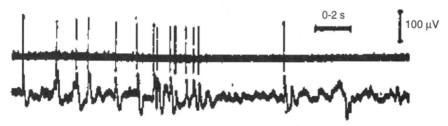

Figure 11.4. Single motor unit action potentials (*upper trace*) and corresponding MMG (*lower trace*) in orbicularis oris muscle. (Redrawn from [21] with permission)

of the generation process of the muscle surface oscillation due to the activity of the recruited fibers is reported in Figure 11.3. The mechanical events due to the activity of single motor units were first detected by piezoelectric contact sensors in the 1940s. An example is shown in Figure 11.4; the surface MMG and the motor unit action potential from the orbicularis oris are registered in the lower and upper traces respectively.

In sum, it appears that during electrical stimulation the MMG provides information about the muscle mechanics while during voluntary contraction, the MMG provides information about the number and the firing rate of the recruited motor units that combines all the contributions of the active motor units. This electromechanical conception of muscle contraction will be discussed and outlined in the sections below.

11.2 DETECTION TECHNIQUES AND SENSORS COMPARISON

The papers on MMG published in the last two years deal with signals detected mostly by accelerometers. In some cases a laser distance sensor or microphones were used. Because the MMG is related to the muscle surface movement, my description of detecting techniques will begin with the laser transducer. This tool will be then compared with others.

11.2.1 MMG Detected by Laser Distance Sensors

In two papers Orizio et al. [41,42] used an optical distance sensor (MEL M5L/20, Germany) that provides an output dc voltage (±10 V) proportional to the distance (linear

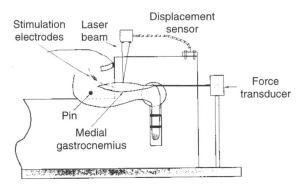

<u>Figure 11.5.</u> Schematic of experimental setup for the detection of the muscle surface displacement during electrically evoked activity of the cat medial gastrocnemius. (Redrawn from [41] with permission)

response within ±10 mm) between the laser-beam head emitter and the reflecting surface of the target. As the muscle surface moves toward or away from the laser head, the reflected laser beam takes different positions along a sensor in the head. The electrical signal provided is proportional to the surface's absolute position. The measure of the distance of the reflecting surface from the laser source is not affected by the surface's rotation if it is within ±15° and ±30° with respect to the short and long axes of the laser head, respectively. The sensitivity is 1 V/mm. The application of the MMG detection system is reported in Figure 11.5. The detected MMG, from in vivo human muscles, allows the study of force and geometrical changes during stimulated muscular contraction.

In Figure 11.6 (data from the Human Physiology Laboratory of the University of Brescia) the force and the surface displacement of the human tibialis anterior are reported during electrical activity evoked by increasing the stimulation frequency from 1 to 50 Hz. The use of the laser sensor for the detection of the surface MMG allows the muscle to change its position without the additional inertial load of a transducer and the output voltage converted into mm of displacement allows the comparison of the results from different studies. Moreover the bandwidth of the instrument is very large (0–10 kHz) and the resolution is very high (about 6 μm). The MMG results [41,42,47] compare well with those of the force signal, and the similarity of the two signals produced during electrically evoked contraction (see Fig. 11.6) suggest that the transverse dimensional changes of the muscle may be an adjunct tool to the force signal in studying muscle mechanics.

11.2.2 MMG Detected by Accelerometers

One of the first papers reporting the use of an accelerometer to detect the transverse mechanical output of a muscle was authored by Lammert, Jorgensen, and Einer-Jensen [30]. In particular, in an earlier work Jorgensen and Lammert [25] were able to detect the contribution of single motor units to surface oscillations. The later use of the accelerometer for MMG detection was due to the very low mass (less than 2 g) of the transducers available on the market. This made it possible to affix the transducer with simple double-sided adhesive tape. No additional pressure, that can interfere with the muscle surface dynamics during contraction, is required. This technique has been used to detect MMG during both voluntary [2,3] and stimulated [4,7,22,41,43,64,71] contractions. Also the

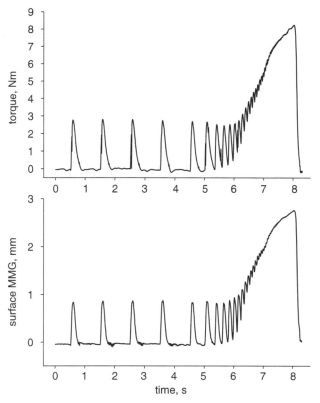

Figure 11.6. The force signal (*top panel*) and the laser detected MMG (*lower panel*) from human tibialis anterior supramaximally stimulated at the most proximal motor point. In 8 seconds the stimulation rate increased from 1 to 50 Hz, stepped 1 Hz with every stimulus increment. The similarity between the two signals dynamics suggests the possibility to use the laser detected MMG to study the muscle mechanical response.

transducer made it possible to compare the results from different studies, since the acceleration of muscle surface could be expressed in physical units (m $\ast$ s^{-2}) rather than transducer dependent units [7].

11.2.3 MMG Detected by Piezoelectric Contact Sensors

This kind of transducer has been widely used since the 1980s. In particular, most investigators use the HP-21050-A device. Its physical dimensions (mass: 40 g; diameter: 3 cm; height: 2.5 cm) limit its application to large muscles. Indeed, the possibility of damping the muscle surface dynamic by the MMG detecting system has to be taken into account. This is due not only to the physical dimensions of the piezoelectric transducer but also to the pressure as it has to be forcefully applied to allow correct mechanical coupling between the sensor and the investigated muscle. This mechanical coupling is obtained by elastic bands, stretchable adhesive nonwoven fabric, or external supports. This is to prevent movement of the transducer with respect to the skin as well as overpressure on the piezo-element during the bulging of the muscle due to contraction. For these reasons the force applied to the sensor cannot exceed 2 N. Below this value the frequency response of the sensor is flat, in the 5 to 150 Hz range. This is the range of interest for the MMG analysis [39]. Smith and Stokes [54] found that the amplitude of the MMG was dependent on the coupling pressure, and that with 1200 Pa a larger signal was recorded than that using 180 or 790 Pa for the same contraction intensity. Smith and Stokes [54] concluded that to compare the MMG amplitude the signal must be detected at the same piezo sensor coupling pressure.

11.2.4 MMG Detected by Microphones

The literature on MMG reports widespread use of the capacitive microphone to detect the surface mechanomyogram. The way to secure the sensor to the muscle is similar to that used for the piezoelectric sensor. The coupling between the muscle surface and the microphone is done by air [49], by ultrasound gel [60], or by surgical cement filling the chamber in which the sensor is placed [20]. Also in this case the pressure exerted for the mechanical coupling with the muscle influences the signal's amplitude [15].

11.3 COMPARISON BETWEEN DIFFERENT DETECTORS

The first paper describing and comparing MMG sensors was published on 1989 [15]. A comparison was made between the air-coupled condenser microphone (Tandy Electronics) and the piezoelectric contact sensor (Hewlett-Packard 21050-A). The conclusion was that the two transducers provide similar signals during evoked single twitches of the thenar muscles during supramaximal stimulation of the median nerve at the wrist. The two output signals were strongly dependent on the "pressure of the microphones on the skin." Moreover Bolton and colleagues ascertained that during the single twitch even a 1 Hz cutoff for high-pass filtering can alter the shape of the recorded signal. The duration of the evoked MMG was 800 ms with a "prominent positive pressure wave" lasting 150 ms.

From this paper it appeared that the duration of the mechanical event recorded at the tendon (force) and that recorded at the muscle surface (MMG) was about the same. The differences in the electromechanical delay (5.5 ms EMG-MMG and 9.6 ms EMG-force) were attributed to the specific role of the elastic elements in series (acting on the force signal) or in parallel (acting on the MMG) to the contractile machinery [15]. The MMG detected during single twitches by means of light mass accelerometers presented a duration of about 50 to 70 ms [4,7,45,47]. This means that the information contained in the accelerometer detected MMG is mainly related with the force generation process (see Section 11.6) during muscle-fiber activity.

Watakabe et al. [65] compared the MMG detected by a piezoelectric contact sensor or accelerometer during voluntary contraction and during sinusoidal mechanical excitation. One of their conclusions was that the piezo sensor behaves as a muscle surface displacement meter. Indeed, the spectral analysis of the double integration of the accelerometer MMG produced spectra resembling those of the piezo MMG. This conclusion should be carefully considered, as it is known that in the case of a constant muscle surface displacement, the piezoelectric crystal is unable to maintain a steady voltage when a constant force is applied [15]. This phenomenon can impair the precise description of the MMG dynamics at low rates of muscle surface displacement.

More recently Watakabe et al. [66] compared the response of an air-coupled microphone to that of an accelerometer. They concluded that the spectra of the double integral of the accelerometer MMG resemble that of the microphone detected MMG. As a conse-

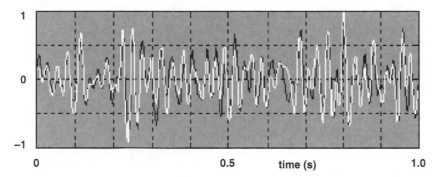

Figure 11.7. MMG detected by an accelerometer and second derivative of the simultaneously detected MMG by the laser sensor. Note that the two signals are nearly completely overlapped. This means that they are detecting the same phenomenon with the same accuracy, only the units of measurement are different because of the peculiarity of the two transducers. This figure is from an experimental session on voluntary contraction of biceps brachii carried out in the Laboratory of Human Physiology of the University of Brescia with the cooperation of the Department of Bioengineering of the Politecnico of Milan.

quence the microphone detecting device seems to act as a muscle surface displacement meter. This paper also demonstrates that the air-coupled microphone time and frequency responses are strongly influenced by the geometry of the air chamber between the muscle surface and the sensitive element. A positive feature of this detecting device seems to be its lower sensitivity to the motion artifacts with respect to the accelerometer, and so it was suggested to be the candidate to pick up MMG even during the movement of the limbs [66]. Generally speaking both microphones and piezoelectric transducers investigate circular and rather large areas isolated from the rest of the muscle by the rigid sensor perimeter. This system introduces a rigid mechanical discontinuity on the muscle surface. This should be taken into account when a decision about the transducer to be used for MMG detection has to be made.

Our laboratory, in collaboration with the Department of Bioengineering of Politecnico of Milan, is evaluating the information contained in the accelerometer (acc-MMG) and laser detected MMG (laser-MMG). The off-line analysis of the two signals detected during voluntary contraction allows the comparison between the second derivative of laser MMG and the acc-MMG as well as between the laser-MMG double integral of the acc-MMG. In Figure 11.7 the second derivative of the laser-MMG is shown together with the acc-MMG. The second derivative is filtered with a Chebicev digital filter in the 5 to 85 Hz band to damp the noise in the high-frequency range. The acc-MMG has been filtered in the same way to allow a convenient comparison. The signals were normalized and the second derivative of the laser MMG was multiplied by -1 to compensate for the $180°$ phase shift due to the mathematical procedure.

Figure 11.7 and the spectral analysis of the reported signals confirm findings that the signals monitor the same phenomenon with the same frequency content. Even the reverse procedure, a comparison between the laser-MMG and the double integral of the acc-MMG, provides the same overlap of the investigated signals. Thus the evidence is that during either voluntary or stimulated contraction the movement of the muscle surface can be detected and transduced into voltage by several devices. The accelerometer, because of its very light mass, and the laser distance sensor may be the most reliable tools for MMG detection during isometric contraction of the muscle. They do not interfere with the muscle surface dynamics, they produce signals with physical units (m/s^2 for accelerometer, mm for laser distance sensor) able to be used at different laboratories, and finally their sensitivities are independent from the method adopted for mechanically coupling the detector and the muscle. Because of its lower sensitivity to movement artifacts, the air-coupled microphone seems to be more suitable to detecting MMG from a dynamically contracting muscle.

11.4 SIMULATION

The concept that the mechanomyogram is an interferential signal encompassing the muscle surface's displacement due to the recruited motor units is supported by the evidence that the isolated contribution of each active motor unit (MU) can be retrieved in the signal [5,10,21,30,43,49,69]. Contrary to the surface EMG the summation of MU contributions is not linear nor algebraic [9]. Indeed, it has been reported that the MMG, related with a single motor unit activity, changes dramatically its properties with the additional recruitment of one MU (see Fig. 11.8 in [35]). This may be due to the long duration of the mechanical event recorded by the MMG.

To study the effect of simultaneous MUs activities on the signal, Orizio et al. [43] simulated the MMG generation stimulating at two motor points, with different frequencies, the third digit of the extensor digitorum communis of the hand (see Fig. 11.8). This can be regarded as the electrical activation of two artificial motor units (MU1 and MU2) at known frequency rates (f1 and f2) in which to study the summation pattern of the MUs' mechanical contributions to MMG. Trains of stimuli were separately administered to MU1 and MU2 for 5 seconds at different times in order to define MMG signals reference trains during 3 and 9 Hz for MU1, 8 and 20 Hz for MU2. A second stimulation pattern provided a simultaneous stimulation of MU1 and MU2 at 3 (f1) and 8 Hz (f2) and at 9 (f1) and 20 (f2) Hz. The time relationship between the two stimulation rates was known and used to align the MMGs of the reference trains detected during the separate stimulation of MU1 and MU2. After alignment the MMG from MU1 and the MMG from MU2 were point by point summated in order to generate a MMG (MMGg) to be compared with the one obtained during the simultaneous stimulation with the same frequency pairs.

In Figure 11.8 it is evident that the MMG generated (MMGg) by the summation of the MMG obtained at f1 and f2 separately administered is similar to the MMG recorded during simultaneous stimulation of the two motor points (MMGs) only for very low frequency rates. The conclusion that can be drawn is that the MMG can be considered a compound signal in which the transverse mechanical activity of the active muscle fibers is summated. However linear summation of the mechanical contribution of the recruited MUs does not occur in the whole range of the physiological motor units firing rates. The frequency domain analysis by the application of high-order spectra to MMGg and MMGs indicated that a nonlinear behavior of the artificial MU2 took place during its stimulation at 20 Hz. It was hypothesized that this could be due to the mechanical influence of the MU1 twitching at 9 Hz on the MU2 in a situation where the internal visco-elastic properties are strongly conditioned by the higher frequency of the electrical stimuli. However, it was clear that more detailed studies have to be carried out in order to find the model capable of describing the summation process of the transverse mechanical contribution of the recruited MUs to the MMG.

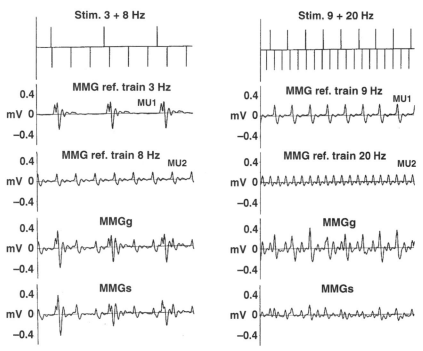

Figure 11.8. Time window of 1 second with the MMG recorded during simultaneous stimulation (MMGs) of motor unit 1 (MU1) and motor unit 2 (MU2), and the MMG generated (MMGg) by the linear summation of the two reference trains (MMG ref. train) obtained during the separate stimulation of MU1 and MU2. Note that only for low frequencies of stimulation (3 Hz (MU1) + 8 Hz (MU2)) the summation of the two motor units contribution to MMG is linear, and that MMGs and MMGg are mostly overlapped. (Redrawn from [43])

11.5 MMG VERSUS FORCE: JOINT AND ADJUNCT INFORMATION CONTENT

In Figure 11.2 it is evident that the accelerometer detected signal's MMG duration is within the on-phase of the single twitch. This suggests that MMG may reflect some aspects of the force generation process and the changes of the muscle contractile properties at fatigue. To verify this hypothesis, Orizio et al. [47] recorded and analyzed the MMG and force (F) from the human anterior tibialis before and immediately after a fatiguing stimulation as well as during 6 minutes of recovery (single twitches evoked at the end of the fatiguing stimulation and after 30, 60, 120, 240, and 360 seconds). Figure 11.9 shows the result with the amplitude of MMG highly correlated with the maximum d^2F/dt^2 during the contraction phase of the single twitches. The open circles refer to pre-fatigue data, while the solid circles refer to recovery data at six time instants.

The d^2F/dt^2 appears to reflect the intensity of the active state related to the amount of Ca^{++} released by the sarcoplasmatic reticulum [63]. This led Orizio et al. [47] to conclude that the changes of MMG at fatigue may reflect the Ca^{++} transient modifications, and hence specific aspects of the electromechanical coupling efficiency, at fatigue and during the recovery phase. A reduction of the accelerometer-detected MMG at fatigue, even if not correlated with specific parameters of the force twitch, was already reported by Barry [7].

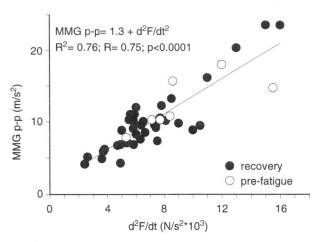

Figure 11.9. Relationship between MMG peak to peak (MMG p-p) versus the second derivative of the force increase during the single twitch (d^2F/dt^2), 49 points from 7 subjects. Each subject provided one point for the pre-fatigue condition (open circle) and 6 points related to the recovery period (6 minutes long). The relationship demonstrates that the changes in the force acceleration are mirrored by the changes in the MMG p-p. (Redrawn from [47] with permission)

A comparison of the force twitch and MMG was recently made also for the single motor unit activity [10]. In this work the relationship between the twitch parameters and the amplitude of the MMG was examined. The MMG was detected by a piezoelectric sensor with the active surface parallel to the long axis of the rat medial gastrocnemius, which was placed in a pool filled with paraffin oil. The experimental setup allowed the stimulation of single motor units. The results indicated that the amplitude of the MMG was related mainly to the velocity of the force changes during the contraction phase of the single twitch of each recruited motor unit. This velocity was low, intermediate, and high for slow-twitch MUs, fast-twitch fatigue-resistant MUs, and fast-twitch not fatigue-resistant MUs, respectively. On this basis a close relationship can be intuited between the MMG time domain properties and the functional/contractile properties of the active motor units.

Since the MMG can be regarded as the result of muscle surface deformation of active MUs [10], a combined analysis of the force (F) and the muscle transverse diameter changes (MMG) was performed during electrical stimulation of the motor nerve of cat medial gastrocnemius in isometric conditions [41]. A laser distance sensor measuring the muscle surface displacement detected the MMG. Different levels of activity of the MUs' pool were investigated by changing the number and the firing rate of the active motor units. It was found that the signal rising phase was always earlier in MMG than in F. Increasing the MUs' firing rate from 5 to 50 Hz or the number of recruited motor units (twitching at a fixed frequency of 40 Hz) shows a nonlinear force/MMG relationship, with a large displacement with low force production in the first part and the opposite in the second (see Fig. 11.10).

The authors then analyzed the different behaviors of F and MMG from low to high levels of activity in the MUs' pool. They split the series elastic component of the muscle mechanical model into two components: high and low compliance. The high compliance

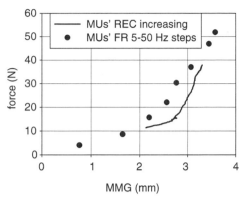

Figure 11.10. Laser-detected MMG and force during motor nerve stimulation of the cat medial gastrocnemuis. The circles show the average values, over the central 1 second time window, of the two signals during the steady phase of 8 different stimulation rates at 5, 10, 15, 20, 25, 30, 40 and 50 Hz for 10 seconds. The continous line shows the relationship between the two signals over a short period in which the number of the recruitable motor units was increased while the firing rate was kept constant. In both cases the increase in the level of motor unit pool activity induced first a large muscle geometry change (reflected by the MMG) and second a large force increment while the muscle surface displacement was mostly constant. (Redrawn from [41] with permission)

element, with a nonlinear stress–strain relationship, seemed to allow the muscle fibers to increase their transverse diameter with little tension transmitted at the tendon. When the extension limit of this nonlinear elastic component was reached, some displacement with high force increases could be observed. The different effects of specific components of the muscle mechanical model (for details, see [68]) on the MMG and the force signal were already studied by Bolton et al. [15] to determine the electromechanical delays (EMD) between the two signals. A shorter EMD for MMG than for force was observed by the authors in the influence of the series elastic element on tension at the tendon and of parallel elastic elements on MMG properties.

When the frequency response of the muscle, obtained from force and laser-detected MMG during a sinusoidal (0.4–6 Hz range) change of the number of orderly recruited MUs, was studied, a critically damped second-order system with a pure time delay was observed with two coincident poles at 1.83 Hz (with 22.6 ms delay). When the frequency response, obtained from force and MMG, was investigated, two coincident poles at 2.75 Hz (with 38 ms delay) were observed. Also in this case the differences in the second-order system properties, even if not statistically significant, could be attributed to the specific influence of the passive elements of the muscle mechanical model on each of the two signals. In particular, it was thought that one of the viscoelastic structures, such as the aponeurosis, had more direct influence on the force, because of its series placement between the contractile machinery and the force transducer, than did the MMG [42].

In conclusion, the compared analysis of the force and MMG showed that the MMG is able to monitor specific aspects of muscle mechanics and of fiber contractile properties as well as the possible influence of the different muscle mechanical components on the two mechanical outcomes (force production and muscle dimensional changes) during muscular contraction.

11.6 MMG VERSUS EMG: JOINT AND ADJUNCT INFORMATION CONTENT

During voluntary contraction the motor unit activation pattern of a specific muscle is mostly based on the interaction of two factors: the recruitment (REC) and the firing rate (FR) of the active MUs. The surface EMG analysis shows that the root mean square (RMS) increases monotonically from low to maximal voluntary contraction (MVC), with both REC and FR increments contributing to its value [9]. The frequency content of the signal is strongly influenced by the shape of the motor unit action potentials with the largest part of the power distributed beyond 40 Hz. The minor peaks below this frequency may be related to the motor units firing rate [9]. On this basis it is possible to interpret the increase of the mean or median frequency of the EMG spectrum with the increase of the effort intensity as a consequence of the increase of the overall conduction velocity of the active muscle fibers due to the recruitment of larger MUs [62]. When REC is finished and no other MUs can be recruited, only an FR increase can achieve higher contraction forces [9,29] as the mean frequency/%MVC levels off both during voluntary [13,52] and during stimulated [57] contraction. The pattern, however, is influenced by the geometrical location of the motor units.

The influence of REC and FR on MMG properties has been investigated both during stimulated and voluntary contraction. Orizio et al. [40] used the experimental setup reported in Figure 11.5 to detect MMG using a piezoelectric sensor that recorded the muscle surface displacement during the stimulation of the cat medial gastrocnemius. The stimulator was able, in a given time window, to orderly recruit the motor units, from slow to fast contracting units with a fixed FR, an increased FR at different stable levels of REC, and both REC and FR increased from a minimal to maximum value. It was found that REC increases the MMG amplitude and RMS while, beyond 5 Hz, the motor units' firing rate is always inversely related to the MMG amplitude and RMS. As a consequence the effect of REC appeared to be strongly conditioned by the motor units FR. It was hypothesized that high frequency creates a fusion-like situation where the active fibers have no time to relax and make dimensional changes, so muscle surface displacement (see Fig. 11.3) takes place from one stimulation pulse to the following pulse. The results were obtained in a well-controlled situation, so likely the decrease of the MMG RMS at high contraction intensities is due to the high global frequency rate of the active MUs. This result, together with an increase in the MMG mean frequency, was in fact observed in the biceps brachii [2,3,38,44], rectus femoris [1,53], soleus [69], and jaw elevator muscles [58] during voluntary contractions. It should be noted that here, contrary to EMG, the MMG frequency content seemed to be strongly influenced by the global motor units' firing rate. Further studies on the simulation of the MMG generation process by the different activated MUs should be devoted to this aspect. The possible relationship between the biceps brachii motor units activation pattern and the EMG and MMG parameters (RMS and mean frequency) is schematized in Figure 11.11.

It is well known that in the small muscles of the hand REC finishes much earlier than in large muscles. As a consequence the FR is the sole way to generate more force which is about 30% and about 70% to 80% MVC in small and large muscles, respectively [9,29]. To test this, preliminary experiments were carried out on the first dorsal interosseus (FDI) in our laboratory. A careful interpretation was made of the MMG RMS and MF behavior during the increase of the force output as a reflection of REC and FR relationship. In FDI, within the force range in which REC should take place, the MMG RMS was found to

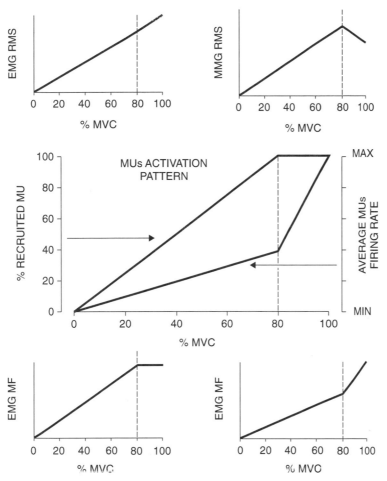

Figure 11.11. Schematic drawing of the possible correlation between the crucial changes in the biceps brachii motor unit activation patterns and their possible effects on the EMG and MMG time (RMS) and frequency (mean frequency, MF) domains parameters. The end of the motor unit recruitment (REC) determines a leveling off of the EMG MF and % MVC relationship, and the use of high levels of motor unit firing rates (FR) determines the increase and the decrease in the MMG MF and the MMG RMS and %MVC relationships, respectively.

decrease (instead of increase), and beyond this range and up to MVC when the FR increase should be the only tool to increment the force, the MMG RMS increased (instead of decreasing) and the MMG MF decreased (instead of increasing). An explanation may be that the way the motor units are distributed in biceps (slow in the core and fast more externally in the muscle) is different than that in FDI. As a result the effect of the orderly recruitment of MUs may be reduced as the sensor comes closer and the force output changes from low to high intensity. Moreover it cannot be ignored that FDI presents progressively larger synchronization of MUs as the isometric effort increases [26]. This may contribute to the increase of the MMG RMS and the reduction of MF from the level of effort corresponding to the end of REC to MVC.

In sum, MMG analysis of voluntary contractions may provide data on motor unit activation patterns the same as surface EMG. However, one has to carefully consider that the changes in the signal parameters, either in time or frequency domain, can have different meanings in different muscle groups for specific phenomena associated with changes in REC and FR.

11.7 AREA OF APPLICATION

MMG and force relationship in different muscle groups during isometric contractions are the subject of many muscle physiology studies. Related investigations of the influence of the motor unit activation pattern on MMG time and frequency domain parameters can be found in [2,3,32,40,59]. The relationship between the MMG and the mechanical variables of movement during dynamic contractions has been described by Bodor [14], Cramer et al. [16], and Cramer et al. [17]; the influence of the motor task on the MMG properties has been also studied [18,32,48,55,56]. The MMG properties as a function of the exercise intensity have been reported for both the overall [61] and local [70] metabolic rate. The influence of muscle fatigue on MMG characteristics, due to changes in the muscle contractile properties as well as in motor units activation pattern, have been considered in a vast number of papers and contributions. For reviews, see [36,59]; for more recent pubblications, see [23,67,70]. The MMG is being considered a tool to distinguish the different properties of muscles in diseases [6,24,27,28,45,46,50,51] in athletically trained muscles [19,30,33,37] and in young developing muscle fibers [34]. Comparisons among muscle electromechanical performances of these different groups are computed using a parameter called the electromechanical coupling efficiency (EMCE) [6] in a MMG/EMG activity ratio. This parameter is a useful means of investigating deviations of the muscle electromechanical coupling from a reference level [6,34,45].

There are more basic studies relating the MMG signal to the contribution of every active motor unit [5,10,11,12,21,25,43,49,69] that provide the background on the modeling of MMG. These studies provide a solid basis for the interpretating the MMG and are more focused on its applications.

REFERENCES

1. Akataki, K., K. Mita, and Y. Itoh, "Relationship between mechanomyogram and force during voluntary contractions reinvestigated using spectral decomposition," *Eur J Appl Physiol* **80**, 173–179 (1999).

2. Akataki, K., K. Mita, M. Watakabe, and K. Itoh, "Mechanomyogram and force relationship during voluntary isometric ramp contractions of the biceps brachii muscle," *Eur J Appl Physiol* **84**, 19–25 (2001).

3. Akataki, K., K. Mita, M. Watakabe, and K. Itoh, "Age related change in motor unit activation strategy in force production: a mechanomyographic investigation," *Muscle Nerve* **25**, 505–512 (2002).

4. Barry, D. T., "Vibrations and sounds from evoked muscle twitches," *Electromyogr Clin Neurophysiol* **32**, 35–40 (1992).

5. Barry, D. T., S. R. Geiringer, and R. D. Ball, "Acoustic myography: A noninvasive monitor of motor unit fatigue," *Muscle Nerve* **8**, 189–194 (1985).

6. Barry, D. T., K. E. Gordon, and G. G. Hinton, "Acoustic and surface EMG diagnosis of paediatric muscle disease," *Muscle Nerve* **13**, 286–290 (1990).

7. Barry, D. T., T. Hill, and I. Dukjin, "Muscle fatigue measured with evoked muscle vibrations," *Muscle Nerve* **15**, 303–309 (1992).

8. Baskin, R. J., and P. J. Paolini, "Volume change and pressure development in muscle during contraction," *Am J Physiol* **213**, 1025–1030 (1967).

9. Basmajian, J. V., and C. J. De Luca, *Muscle alive: Their functions revealed by electromyography*, Williams and Wilkins, Baltimore, 1985.

10. Bichler, E., "Mechanomyograms recorded during evoked contractions of single motor units in the rat medial gastrocnemius muscle," *Eur J Appl Physiol* **83**, 310–319 (2000).

11. Bichler, E., and J. Celichowsky, "Mechanomyographic signals generated during unfused tetani of single motor units in the rat medial gastrocnemius muscle." *Eur J Appl Physiol* **85**, 513–520 (2001).

12. Bichler, E., and J. Celichowsky, "Changes in the properties of mechanomyographic signals and in the tension during the fatigue test of the rat medial gastrocnemius muscle motor units," *J Electromyogr Kinesiol* **11**, 387–394 (2001).

13. Bilodeau, M., M. Cincera, S. Gervais, A. B. Arsenault, D. Gravel, Y. Lepage, and P. Mac Kinley, "Changes in the electromyographic spectrum power distribution caused by a progressive increase in the force level," *Eur J Appl Physiol* **71**, 113–123 (1995).

14. Bodor, M., "Mechanomyographic and electromyographic muscle responses are related to power," *Muscle Nerve* **22**, 649–650 (1999).

15. Bolton, C. F., A. Parkes, T. R. Thompson, M. R. Clark, and C. J. Sterne, "Recording sound from human skeletal muscle: Technical and physiological aspects," *Muscle Nerve* **12**, 126–134 (1989).

16. Cramer, J. T., T. J. Housh, G. O. Johonson, K. T. Ebersole, S. R. Perry, and A. J. Bull, "Mechanomyographic amplitude and mean power output during maximal, concentric, isokinetic muscle actions," *Muscle nerve* **23**, 1826–1831 (2000).

17. Cramer, J. T., T. J. Housh, T. K. Evetovich, G. O. Johonson, K. T. Ebersole, S. R. Perry, and A. J. Bull, "The relationships among peak torque, mean power output, mechanomyograpy and electromyography in men and women during maximal eccentric, isokinetic muscle actions," *Eur J Appl Physiol* **86**, 226–232 (2002).

18. Ebersole, K. T., T. J. Housh, J. P. Weir, G. O. Johnson, T. K. Evetovich, and D. B. Smith, "The effects of leg angular velocity on mean power frequency and amplitude of the mechanomyographic signal," *Electromyogr Clin Neurophysiol* **40**, 49–55 (2000).

19. Evetovich, T. K., T. J. Housh, J. P. Weir, D. J. Housh, G. O. Johnson, K. T. Ebersole, and D. B. Smith, "The effect of leg extension training on the mean power frequency of the mechanomyographic signal," *Muscle Nerve* **23**, 973–975 (2000).

20. Goldenberg, M. S., H. J. Yack, F. J. Cerny, and H. W. Burton, "Acoustic myography as an indicator of force during sustained contractions of a small hand muscle," *J Appl Physiol* **70**, 87–91 (1991).

21. Gordon, G., and H. S. Holbourn, "The sounds from single motor units in a contracting muscle," *J Physiol* **107**, 456–464 (1948).

22. Herzog, W., Y. T. Zhang, M. A. Vaz, A. C. S. Guimaraes, and C. Janssen, "Assessment of muscular fatigue using vibromyograpy," *Muscle Nerve* **17**, 1156–1161 (1994).

23. Housh, T. J., S. R. Perry, A. J. Bull, G. O. Johnson, K. T. Ebersole, D. J. Housh, and H. A. de Vries, "Mechanomyographic and electromyographic responses during submaximal cycle ergometry," *Eur J Appl Physiol* **83**, 381–387 (2000).

24. Hufshmidt, A., P. Shubnell, and I. Schwaller, "Assessment of denervation by recording of muscle sound following direct stimulation," *Electromyogr Clin Neurophysiol* **27**, 301–304 (1987).

25. Jorgensen, F., and O. Lammert, "Accelerometermyography (AMG) II: Contribution of the motor unit," in P. V. Komi, ed., *International series on biomechanics*, Vol 1a, University Park Press, Baltimore, 1976.

26. Kamen, G., and A. Roy, "Motor unit synchronization in young and elderly adults," *Eur J Appl Physiol* **81**, 403–410 (2000).

27. Keidel, M., Th. Flick, and W. S. Tirsch, "Abnormal vibromyogram spectra in motor disorder," *Electroencephalogr Clin Neurophysiol* **66**, S122 (1987).

28. Keidel, M., G. Mayer-Kress, W. S. Tirsch, Th. Flick, and S. J. Poppl, "Characterisation of chaotic dynamics in human vibratory output," *Electroencephalogr Clin Neurophysiol* **70**, 59–60 (1988).

29. Kukulka, C. G., and H. P. Clamann, "Comparison of the recruitment and discharge properties of motor units in human brachial biceps and adductor pollicis during isometric contractions," *Brain Res* **219**, 45–55 (1981).

30. Lammert, O., F. Jorgensen, and N. Einer-Jensen, "Accelerometermyography (AMG). I: Method for measuring mechanical vibrations from isometrically contracted muscles," in *Biomechanics V-A*, University Park Press, Baltimore, pp. 152–156.

31. Luciani, L., *Fisiologia dell'uomo*, Vol. 3, Società Editrice Libraria, Milan, 1923.

32. Madeleine, P., P. Bajaj, K. Sogaardand, and L. Arendt-Nielsen, "Mechanomyography and electromyography force relationships during concentric, isometric and eccentric contractions," *J Electromyogr Kinesiol* **11**, 113–121 (2001).

33. Marchetti, M., F. Felici, M. Bernardi, O. Minasi, and L. di Filippo, "Can evoked phonomyography be used to recognise fast and slow muscle in man?" *Int J Sports Med* **13**, 65–68 (1992).

34. Nonaka, H., K. Mita, K. Akataki, M. Watakabe, and K. Yabe, "Mechanomyographic investigation of muscle contractile properties in preadolescent boys," *Electromyogr Clin Neurophysiol* **40**, 287–293 (2000).

35. Orizio, C., "Muscle sound: bases for the introduction of a mechanomyographic signal in muscle studies," *Crit Rev Biomed Eng* **21**, 201–243 (1993).

36. Orizio, C., "Muscle fatigue monitored by the force, surface mechanomyogram and EMG," in B. M. Nigg, B. R. MacIntosh, and J. Mester, eds., *Mechanics and biology of movement*," Human Kinetics Publishers. Champaign, IL, 2000.

37. Orizio, C., and A. Veicsteinas, "Soundmyogram analysis during sustained maximal voluntary contraction in sprinters and long distance runners," *Int J Sports Med* **13**, 594–599 (1992).

38. Orizio, C., R. Perini, and A. Veicsteinas, "Muscular sound and force relationship during isometric contraction in man," *Eur J Appl Physiol* **58**, 528–533 (1989).

39. Orizio, C., R. Perini, B. Diemont, and A. Veicsteinas, "Muscle sound and electromyogram spectrum analysis during exhausting contractions in man," *Eur J Appl Physiol* **65**, 1–7 (1992).

40. Orizio, C., M. Solomonow, R. Baratta, and A. Veicsteinas, "Influence of motor units recruitment and firing rate on the soundmyogram and EMG characteristics in cat gastrocnemius," *J Electryogr Kinesiol* **2**, 232–241 (1993).

41. Orizio, C., R. Baratta, B. H. Zhou, M. Solomonow, and A. Veicsteinas, "Force and surface mechanomyogram relationship in cat gastrocnemius," *J Electromyogr Kinesiol* **9**, 131–140 (1999).

42. Orizio, C., R. Baratta, B. H. Zhou, M. Solomonow, and A. Veicsteinas, "Force and surface mechanomyogram frequency responses in cat gastrocnemius," *J Biomech* **33**, 427–433 (2000).

43. Orizio, C., D. Liberati, C. Locatelli, D. De Grandis, and A. Veicsteinas, "Surface mechanomyogram reflects muscle fibres twitches summation," *J Biomech* **29**, 475–481 (1996).

44. Orizio, C., R. Perini, B. Diemont, M. Maranzana Figini, and A. Veicsteinas, "Spectral analysis of muscular sound during isometric contraction of biceps brachii," *J Appl Physiol* **68**, 508–512 (1990).

45. Orizio, C., F. Esposito, I. Paganotti, L. Marino, B. Rossi, and A. Veicsteinas, " Electrically elicited surface mechanomyogram in myotonic dystrophy," *Ital J Neurol Sci* **18**, 185–190 (1997).

46. Orizio, C., F. Esposito, V. Sansone, G. Parrinello, G. Meola, and A. Veicsteinas, "Muscle surface mechanical and electrical activities in myotonic dystrophy," *Electrmyogr Clin Neurophysiol* **37**, 231–239 (1997).

47. Orizio, C., B. Diemont, F. Esposito, E. Alfonsi, G. Parrinello, A. Moglia, and A. Veicsteinas, "Surface mechanomyogram reflects the changes in the mechanical properties of muscle at fatigue," *Eur J Appl Physiol* **80**, 276–284 (1999).

48. Perry, S. R., T. J. Housh, G. O. Johnson, K. T. Ebersole, and A. J. Bull, "Mechanomyographic responses to continuous, constant power output cycle ergometry," *Electromyogr Clin Neurophysiol* **41**, 137–144 (2001).

49. Petitjean, M., and B. Maton, "Phonomyogram from single motor units during voluntary isometric contraction," *Eur J Appl Physiol* **71**, 215–222 (1995).

50. Ratighan, B. A., K. Mylrea, E. Lonsdale, and L. S. Stern, "Investigation of sounds produced by healthy and diseased human muscular contraction," *IEEE Trans BME* **33**, 967–971 (1986).

51. Rodriquez, A. A., J. C. Agre, T. M. Franke, E. R. Swiggum, and J. T. Curt, "Acoustic myography during isometric fatigue in postpolio and control subjects," *Muscle Nerve* **19**, 384–387 (1996).

52. Sanchez, J., M. Solomonow, R. Baratta, and R. D'Ambrosia, "Control strategies of the elbow antagonist muscle pair during two types of increasing isometric contraction," *J Electromyogr Kinesiol* **3**, 33–40 (1993).

53. Shinoara, M., M. Kouzaki, T. Yoshihisa, and T. Fukunaga, "Mechanomyogram from the different heads of the quadriceps muscle during incremental knee extension," *Eur J Appl Physiol* **78**, 289–295 (1998).

54. Smith, T. G., and M. J. Stokes, "Technical aspects of acoustic myography (AMG) of human skeletal muscle; contact pressure and force/AMG relationships," *J Neurosci Meth* **47**, 85–92 (1993).

55. Smith, D. B., T. J. Housh, G. O. Johnson, T. K. Evetovich, and K. T. Ebersole, "Mechanomyographic responses to maximal eccentric isokinetic muscle actions," *J Appl Physiol* **82**, 1003–1007 (1997).

56. Smith, D. B., T. J. Housh, G. O. Johnson, T. K. Evetovich, K. T. Ebersole, and S. R. Perry, "Mechanomyographic and electromyographic responses to eccentric and concentric isokinetic muscle actions of the biceps brachii," *Muscle Nerve* **21**, 1438–1444 (1998).

57. Solomonow, M., C. Baten, J. Smit, R. V. Baratta, H. Hermens, R. D'Ambrosia, and H. Shoji, "Electromyogram power spectra Frequencies associated with motor unit recruitment strategies," *J Appl Physiol* **68**, 1177–1185 (1990).

58. Stiles, R., and D. Pham, in *Proc. IEEE-EMBS 13th An Int Conf*, IEEE, New York, 1991, pp. 946–947.

59. Stokes, M. J., and M. Blythe, *Muscle sounds in physiology, sports science and clinical investigation: Applications and history of mechanomyography*. Medintel Monographs, Oxford, 2001.

60. Stokes, M. J., and P. A. Dalton, "Acoustic myography for investigating human skeletal muscle fatigue," *J Appl Physiol* **71**, 1422–1426 (1991).

61. Stout, J. R., T. J. Housh, G. O. Johnson, T. K. Evetovich, and D. B. Smith, "Mechanomyography and oxigen consumption during incremental cycle ergometry," *Eur J Appl Physiol* **76**, 363–367 (1997).

62. Stulen, F. B., and C. J. De Luca, "Frequency parameters of the myoelectric signal as a measure of muscle conduction velocity," *IEEE Trans BME* **28**, 515–523 (1981).

63. Takamori, M., L. Gutman, and S. R. Shane, "Contractile properties of human skeletal muscle," *Arch Neurol* **25**, 535–546 (1971).

64. Vaz, M. A., W. Herzog, Y. T. Zhang, T. R. Leonard, and H. Nguyen, "The effect of muscle length on electrically elicited muscle vibrations in the in-situ cat soleus muscle," *J Electromyogr Kinesiol* **7**, 113–121 (1997).

65. Watakabe, M., Y. Itoh, K. Mita, and K. Akataki, "Technical aspects of mechanomyography recording with piezoelectric contact sensor," *Med Biol Eng Comput* **36**, 557–561 (1998).

66. Watakabe, M., K. Mita, K. Akataki, and K. Itoh, "Mechanical behaviour of condenser microphone in mechanomyography," *Med Biol Eng Comput* **39**, 195–201 (2001).

67. Weir, J. P., K. M. Ayers, J. F. Lacefield, and K. L. Walsh, "Mechanomyographic and electromyographic responses during fatigue in humans: Influence of muscle length," *Eur J Appl Physiol* **84**, 19–25 (2000).

68. Winter, D. A., *Biomechanics and motor control of human movement*, Wiley, New York, 1990.

69. Yoshitake, Y., and Moritani, T., "The muscle sound properties of different muscle fiber types during voluntary and electrically induced contractions." *J Electromyogr Kinesiol* **9**, 209–217 (1999).

70. Yoshitake, Y., H. Ue, M. Miyazaki, and T. Moritani, "Assessment of lower-back muscle fatigue using electromyography, mechanomyography and near-infrared spectroscopy," *Eur J Appl Physiol* **83**, 174–179 (2001).

71. Zhang, Y. T., C. B. Frank, R. M. Rangayyan, and G. D. Bell, "A Comparative study of simultaneous vibromyography and electromyography with active human quadriceps," *IEEE Trans Biom Eng* **39**, 1045–1052 (1992).

12

SURFACE EMG APPLICATIONS IN NEUROLOGY

M. J. Zwarts and D. F. Stegeman

Department of Clinical Neurophysiology
Institute of Neurology
University Medical Center, Nijmegen
Interuniversity Institute of Fundamental
and Clinical Human Movement Sciences (IFKB), The Netherlands

J. G. van Dijk

Department of Neurology and Clinical Neurophysiology
Leiden University Medical Centre
Leiden, The Netherlands

12.1 INTRODUCTION

The nervous system consists of a central and peripheral part. The central nervous system (CNS) consists of the brain (enclosed in the skull) and the spinal cord within the vertebral column. Grossly, the brain consists of two hemispheres, several deep-lying nuclei, the brainstem, and cerebellum. The spinal cord serves as the connection between the body and the brain (via the tracts) and contains local "smart terminals" translating descending commands from the higher levels into control signals of the individual motor units.

The peripheral nervous system (PNS) consists of the axons, namely the fibers of individual neurons, that run together grouped in nerves that in turn may form a network or peripheral plexus, the neuromuscular junction (connecting axon and muscle membrane), and finally the common effector of the central nervous system: the muscle.

Electromyography: Physiology, Engineering, and Noninvasive Applications, edited by Roberto Merletti and Philip Parker.
ISBN 0-471-67580-6   Copyright © 2004 Institute for Electrical and Electronics Engineers, Inc.

Neurology is the medical specialization concerned with all aspects of diagnosis, assessment, and treatment of diseases of the nervous system. Since the nervous system is very close to all aspects of the patient's life that are fundamental for self-consciousness and human life, the consequences and burden of these diseases can be devastating. The variety of diseases, age range of the patients, and course of symptoms is very large. Well known disorders are stroke, multiple sclerosis, Parkinsons's disease, and epilepsy.

The nervous system consists of different parts, as outlined above, but for the purpose of this chapter the effector or motor parts are the most important. Almost all aspects of our daily life involve some kind of motor activity. Think of laughing, eating, conversation, walking, playing sports, or making music. The ingenious machinery necessary for the precise and coordinated actions of hundreds of muscles usually goes unnoticed until some defect or problem occurs. The patient can complain about stiffness, lack of coordination, difficulty in walking or running, weakness or some kind of involuntary movements such as a tremor. The symptoms can evolve insidiously, intermittently, or occur suddenly, stroke-like.

After taking a thorough history the neurologist performs a physical exam that includes testing the muscles, the coordination, reflexes, and so on. Depending on the problem, further investigations are ordered. These can include imaging techniques (e.g., CT or MRI scan), blood tests, or some kind of neurophysiological (also called electrodiagnostic) test. Possible tests include electroencephalography (measuring brain electricity), nerve conduction tests and electromyography. The latter two are the subject of this chapter.

Analysis, assessment, and sometimes diagnosis of neurological disease is possible with the aid of recording from the output of the nervous system in terms of muscle activation by surface EMG. It should be emphasized that an important part of electromyography is performed with needle electrodes enabling the measurement of spontaneous single muscle-fiber depolarizations and intramuscular MU action potentials. However, this essential part of EMG is not the subject of this chapter (see Chapter 2). Here we confine ourselves to surface EMG applications. Thus the following sections will give a short introduction to clinical and experimental SEMG applications and provide insight into some recent developments.

12.2 CENTRAL NERVOUS SYSTEM DISORDERS AND SEMG

Diseases of the CNS are often accompanied by changes in motor output. This can be either a loss or diminished output resulting in paresis (weakness) or paralysis (total loss of the ability to contract a muscle at will). Interruption of central connections usually also results in a loss of inhibition of the local, spinal circuitry giving rise to enhanced reflexes and muscle tone (spasticity). Alternatively, the CNS output can be heightened on account of dysregulation of neural systems or hyperexcitability of groups of neurons. An example of the latter is epilepsy in which the patient can show involuntary, often rhythmic jerks of the limbs (convulsions). Involuntary movements such as tics, tremor, rigidity, chorea (dance-like movements), and dystonia (resulting from altered tone) often arise from diseases in the basal ganglia, a major conglomerate of nuclei deep in the brain hemispheres.

Essentially two approaches using SEMG are available for evaluating these patients. The first is by polymyography (multi-channel SEMG applied to many muscles). In this case the natural (or pathological) patterns of muscle activation by the CNS are studied either in rest conditions or during some kind of activity such as walking and alternating hand movements. These measurements are often combined with other assesments of move-

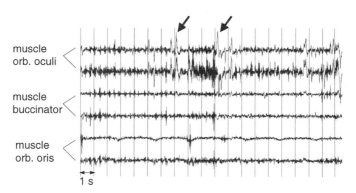

muscle orb. oculi

muscle buccinator

muscle orb. oris

1 s

Figure 12.1. Spontaneous SEMG recording of three facial muscles (in pairs of right and left, respectively) showing intense involuntary contraction in the orbicularis oculi muscles with erratic bursts of increased activity. The arrows denote movement artefacts due to eye blinks. The vertical divisions indicate a time scale of 1 second.

ment using sensors such as accelerometers, foot to floor contacts, during walking, and video recording. The other approach is based on eliciting some kind of reflex—usually by electrically stimulating a peripheral nerve and measuring the response of the CNS with SEMG over an appropriate muscle. In this way abnormal motor control and the spread of the abnormalities over different muscle groups can be quantified and objective measures for treatment intervention and evaluation are at hand. A precise analysis of the relative timing of activation patterns by the central nervous system using SEMG has been a major tool in numerous rehabilitation and motor control studies. In this context the SEMG is only used as a marker of the way in which the CNS controls muscles during different tasks, such as walking or running. In the same way analysis of movement disorders such as tremor, dystonia, or myoclonus can be performed with SEMG [10]. The high temporal resolution with which the SEMG can be measured using many channels on different muscles simultaneously makes SEMG a superior tool in the analysis of complex movements and movement disorders. In such an approach the muscle is regarded as a single entity and a common effector of the CNS. Placing the electrodes in a bipolar configuration with an interelectrode distance of 2 to 4 cm, usually suffices to detect the activity of the muscle under investigation without excessive crosstalk.

Figure 12.1 shows the recording from a patient with dystonia (involuntary, continuous movements) of the face with recording of three muscles of the face: the orbicularis oculi (top), buccinator (middle), and orbicularis oris (bottom). Note the large movement artifact due to eye blinks, only visible in the top channel (arrows). There is continuous activity in the orbicularis oculi with an erratic pattern of bursts of EMG activity that is not present in the other muscles in the lower half of the face.

By stimulating a nerve twig of the trigeminal nerve (innervating the face skin) at the orbital rim, it is possible to measure reflex activity in the muscles around the eye (blink reflex). The EMG activity accompanied with these blinks can be measured with SEMG. Stimulating twice within a short interval results in diminishing of the reflex response (habituation). In certain central nervous disorders (blepharospasm) this habituation is lost as an expression of increased central excitability. This can be shown by evoking 2 blink reflexes within a short interval. As evident from Figure 12.2, in the patient the second stimulus results in similar response amplitude indicating a loss of habituation. This is an example of using surface EMG to quantify abnormal reflex activity of the CNS.

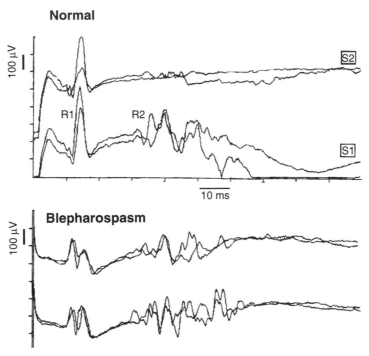

Figure 12.2. Electrical evoked blink reflex activity measured with SEMG around the eye (orbicularis oculi). The stimulus is applied at the start of the traces. Two recordings are superimposed to judge the reproducibility. S1 is the conditioning respons while S2 (0.5 seconds later) is the test response showing an almost complete habituation of the R2 part of the reflex in normals (*top panel*) and a loss of inhibition in the blepharospasm patient (*bottom panel*). Calibration: Vertical: 100 μV/division; horizontal: 10 ms/division.

Figure 12.3 shows an example of the recording of an abnormal, involuntary myoclonic response evoked by a tap to the shoulder in a patient with a corticobasal degeneration (a focal loss of cortical gray matter in the motor areas of the brain resulting in disinhibition). Note the highly synchronized SEMG activity in the different muscles and the longer and higher activity in the hand (bottom traces) as compared with the proximal arm muscles (top traces).

12.3 COMPOUND MUSCLE ACTION POTENTIAL AND MOTOR NERVE CONDUCTION

Electrical stimulation of motor nerves causes the corresponding set of muscles to contract; when electrodes are positioned over these muscles a compound muscle action potential (CMAP) can be recorded. Although this recording is wholly generated by a muscle, it is used to infer properties of the nerve rather than of the muscle. Readers more familiar with other uses of surface electromyography may wish to contrast the two approaches (Fig. 12.4). In non-stimulated EMG, motor units fire in response to central nervous system (CNS) commands. During a weak contraction, only a few motor units are active, and these

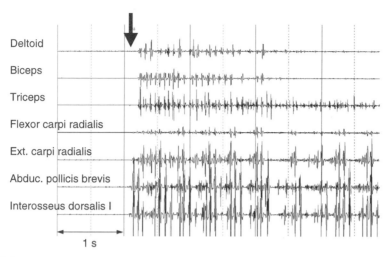

Figure 12.3. Polymyography of many muscles of one arm in a patient with corticobasal degen-eration and myoclonus induced by a tap (*arrow*). Note the highly synchronized SEMG bursts over all muscles with a higher amplitude and longer duration in the hand muscles as compared with proximal muscles. The solid vertical lines denote 1 second intervals.

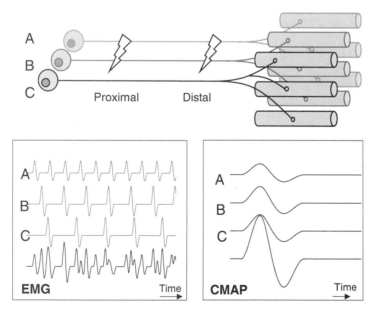

Figure 12.4. Surface EMG and nerve stimulation. (*Top panel*) Three motor units labeled A, B, and C, with nerve cells, axons, and their associated muscle fibers. A proximal and a distal site along the nerve is indicated by lightning symbols. When the motor units are driven by the central nervous system (*lower left panel*), each motor unit fires at a particular frequency and generates a particular waveform. Their summed activity forms the surface EMG. When the nerve is stimu-lated, all motor units fire in synchrony. This is shown (*lower right panel*) for the same MUAPs but at another time scale. The motor units again generate their particular waveforms, and the summed activity is now apparent as the compound muscle action potential, or M-wave.

fire at fairly low frequencies (5–10 Hz). Because of the limited activity, single motor unit action potentials (MUAPs) can be recognized, enabling various factors to be studied, such as the number of muscle fibers in the motor unit (its size) and its spatial characteristics. To increase the force of contraction, the CNS increases the firing frequency of motor units that are already active and also recruits new motor units that were previously silent. In the recording, MUAPs will gradually overlap and interfere with one another, so individual motor units can no longer be identified as such. However, the resulting recruitment pattern now opens venues to study matters such as movement patterns, the nature of CNS activation and fatigue.

The two elements that most distinguish surface EMG in motor nerve conduction (MNC) testing from other uses are the massive nature and the synchrony of the discharge. Nerves are stimulated at supramaximal levels for MNC purposes; that is, the current is increased and stimuli repeated until the CMAP no longer grows in amplitude. The investigator is then certain that all excitable motor axons are stimulated. As all axons are stimulated at the same time, the corresponding motor unit potentials are also formed almost, but not completely synchronously (why this is so is discussed below). To understand the limitations and uses of motor nerve conduction (MNC) testing in more detail, aspects of CMAP generation will be discussed. Readers interested in a more clinical approach are referred to sources [7,23].

12.4 CMAP GENERATION

12.4.1 CMAP as a Giant MUAP

The electrical stimulus given to the nerve ensures that all motor axons are excited synchronously near the point of stimulation. From there, impulses travel to the muscle where they give rise to MUAPs (Fig. 12.5). Due to larger or smaller differences in conduction velocity between motor units, there will be differences in timing of the onset of firing of the various MUAPs. We will start the explanation by assuming small differences in onset. This is, for instance, the case in normal nerves, and when the stimuli are delivered close to the muscle (Fig. 12.5). The MUAPs will be formed largely synchronous, causing the resulting CMAP to resemble a giant MUAP [13].

A good example of the CMAP/MUAP similarity in normal nerves is formed by the effect of the site of the recording electrode on the waveform of the potential. In the common monopolar type of recording one electrode is placed over the synaptic zone, which is usually situated in the middle of the fiber, and the other is placed at a fairly large and hopefully electrically neutral distance. This is also the method commonly used in motor nerve testing, where it is referred to as the "belly-tendon" montage. In MNC, the first electrode is often called the "active electrode," but that term should be reserved for electrodes with attached electronics; here the term "belly electrode" will be used. With this arrangement the MUAP waveform will be biphasic, starting with a negative phase with a high amplitude. When, however, the belly electrode is placed over either half of the muscle fiber, well away from the synapse, the potential will be triphasic: started with a positive phase, followed by a negative phase, and ended by a positive phase (Fig. 12.6). Similar features can also be seen in many CMAPs [2,19]: the amplitude of the negative phase will be highest over the synaptic zone, while an initial positive phase signifies that the belly electrode does not lie directly over the synapses of a large number of muscle

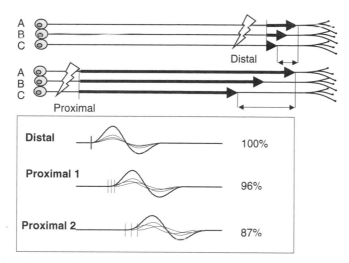

Figure 12.5. Synchrony of MUAP summation. The two upper panels represent three motor units. After distal stimulation, impulses travel towards the muscle and give rise to MUAPs (they also travel in the other direction, but that is not relevant here). Impulses in fast fibers travel farther than in slow fibers in the same interval; this is depicted by the varying lengths of the arrows. In other words, impulses in slow fibers take longer to arrive at the muscle than impulses traveling in fast fibers. The difference in arrival time becomes more pronounced as the nerve is stimulated further away from the muscle. (Compare the upper two panels to see this effect.) The lower panel shows the same effect for three instead of two stimulation points. The three MUAPs are shown in thin lines, whereas their sum, the CMAP, is shown with a bold line. The onset of each MUAP is shown as a thin vertical line. For distal stimulation the onsets are virtually identical, and the three MUAPs are formed with almost complete synchrony. For proximal stimulation, the degree of synchrony decreases. Because of this the CMAP waveform changes, in that amplitude decreases slightly and its duration increases. The amplitude is shown as a percentage of that of the distal CMAP. These changes in waveform are more pronounced as the stimulation site lies farther away from the muscle.

fibers. For instance, moving the belly electrode over the thenar muscles by just 10 mm caused CMAP amplitude to drop from 9.7 to about 8.0 mV, or by 18% [20].

A recording mode that is often used to study MUAPs (but less frequently the CMAPs) is the bipolar lead, often in the form of a chain of electrodes in which the potential differences between successive pairs are recorded (A–B, B–C, C–D, etc.). When such a chain, or linear array, is placed parallel to the fiber direction over a muscle fiber or motor unit, the potential can be seen to change polarity over the synapse, and the potential starts later as the recording pair is situated further away from the synapse. With this lead, one can visualize the site of the synapse, and can see the potential traveling over the muscle fiber(s) at a certain speed. When CMAPs are recorded with similar arrays or grids of electrodes, the same features can be detected in the most commonly tested muscle sets of the hands and feet [19]. This is not really surprising: as long as most motor units under the array have a similar configuration, with parallel muscle fibers and synapses close together. In this case the CMAP may readily be seen as one giant MUAP. Unfortunately, this is often not the case, which brings us to the subject of muscle cartography.

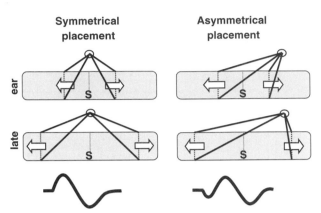

Figure 12.6. CMAP waveform basics. In this highly schematic view, two phases of the depolarization of a muscle fiber are shown as two electrical fronts traveling away from the synapse (*S*). When the electrode is placed directly over the synapse (*left upper panel*), the electrode "sees" the two fronts traveling away from the electrode, meaning they present the same polarity to the electrode. This view is maintained throughout the entire depolarization phase, causing the negative phase of the potential. If the electrode is placed asymmetrically over the fiber, one front will initially be seen to travel toward the electrode (*upper right panel*). This causes an initial positive wave. During depolarization the front will pass underneath the electrode, presenting its other side and causing a negative wave. The final positive waves of the CMAPs are caused by repolarization fronts (not shown) that resemble the depolarization fronts but have the reverse polarity.

12.4.2 Muscle Cartography

Only few muscles are organized in such a way that their fibers are parallel, with all synapses situated closely together in one location along the muscle. The biceps brachii muscle comes close, but there are frequently several separate motor points and zones of innervation even in that muscle. In other muscles the direction and three-dimensional arrangement of muscle fibers is much more complicated [3,4,5]. Fibers may run at an angle to a tendon in a feather-shaped pattern, called "pennate" for that very reason. In other muscles, such as the adductor pollicis muscle, the fibers are attached to central tendons resembling the fingers of a hand (Fig. 12.7). Such differences in muscle architecture no doubt arose in evolution to accommodate varying demands regarding both strength and the degree of shortening of a muscle; they affect CMAP waveforms considerably.

It is simply physically impossible to position an electrode over a muscle and have it sit atop the synapses of all fibers. Accordingly, some motor units will register as a triphasic, positive–negative–positive, waveform, and others will show up as biphasic waveform. How much each contributes to the recorded CMAP depends on the size of that motor unit, on how far it is away from the electrodes, and on properties of volume conduction. The more complex the architecture is, the less can the CMAP be seen as one giant MUAP. Note that the resemblance to a MUAP is affected by *spatial* matters, even when all MUAPs fire in synchrony (*temporal* influences are discussed below). The CMAP of the thenar (ball of the thumb) muscle innervated by the median nerve resembles a MUAP fairly well [13], probably because the fibers of the muscle closest to the electrode (the abductor pollicis brevis muscle) lie in a neat parallel bundle.

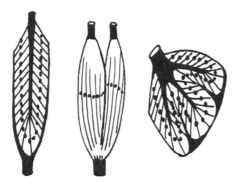

Figure 12.7. Muscle architecture. Tendons are represented by black structures while muscle fiber directions are indicated by parallel lines. Synapses (*black dots*) generally lie in the middle of the muscle fibers. It can be seen that synapses may be distributed in fairly complex three-dimensional patterns. The simplest pattern would be a disk in the middle of a cylindrical muscle; only then would it be feasible to locate a recording (monopolar) electrode in such a way that it directly overlies all synapses. In the more complex patterns this is physically impossible.

In effect, changing the position of the muscle will alter the lengths as well as the three-dimensional position of the muscle fibers relative to one another, causing a different summation pattern and a different CMAP. Altering the site of the electrode also provides another summation, and therefore a different CMAP. Because there are systematic differences in architecture between muscle groups, there are also systematic differences in CMAP waveforms between them [19]. For instance, CMAPs of the hypothenar eminence (muscles in the hand below the little finger) often exhibit split negative and positive peaks, probably due to the contribution of more than one muscle to the CMAP [23]. The extensor digitorum muscle on the dorsum of the foot consists of several small muscle bellies with different orientations.

The more complex the spatial anatomical arrangement, the stronger will the waveform dependency on the recording site. For reasons discussed later, CMAP amplitude is important for clinical purposes. CMAP waveform and amplitude prove to be highly dependent on recording site for the extensor digitorum muscle: drawing an amplitude map (Fig. 12.8) allows the spatial distribution of amplitude to be studied. The area where amplitudes are at least 80% of the maximum amplitude was only 1.7 cm^2 over several subjects. If this area would be circular, moving the electrode just 7 mm away from the maximum site would cause amplitude to drop more than 20% [19]. From this small distance it may easily be understood that identifying the site with the highest amplitude need not be successful at every try. Other muscle sets have other characteristics: when foot sole muscles are studied, the dependency of amplitude on recording site is much less critical. Here the area where amplitudes are at least 80% of the maximal amplitude measures 18.4 cm^2, providing much more freedom in placing the electrode [19].

It is not likely that the exact same electrode site can be found again in a repeated study. That is to say, the measured amplitude in repeated tests can differ merely as a result of a different recording site [18,20,22]. This is a pity, as amplitude may decrease in a whole range of diseases of the nerve as well as of the muscle itself, so it would be very useful to be able to detect small decreases over time. Unfortunately, amplitude reproducibility is rather poor for the reasons outlined here. Still, knowledge of physiology can be put to

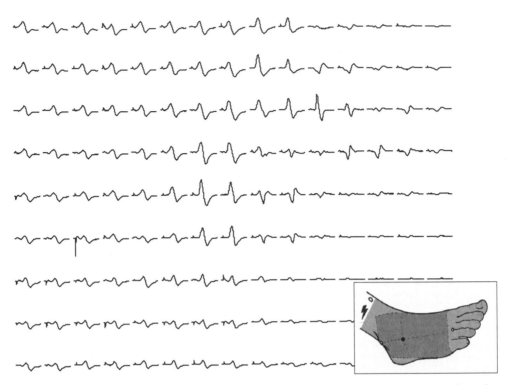

Figure 12.8. CMAP map. The inset shows the outline of a 9 × 15 cm measurement grid on the foot. A CMAP was recorded for each grid site with a small pin electrode; all were referred to the same site on a toe. The various CMAPs are shown according to the grid orientation in the inset. It can be seen that CMAP shape and amplitude depend very strongly on the measurement site.

good use: amplitude reproducibility is worse for muscle groups in which slight electrode shifts cause a large change in waveform [18,19]. This is simply because of slight differences in recording site. For example, the peroneal and median nerves show worse reproducibility than the tibial nerve. The effects of site dependency could be limited by using large recording electrodes. Shifting a 50 mm diameter electrode over 10 mm will alter the electrode's "view" of the muscle only slightly. Shifting a 5 mm electrode over the same 5 mm will, however, alter its view profoundly. In other words, effects of shifting the site can be minimized by using large electrodes [18,19,23]. Obviously this can only work if the views from different locations indeed differ. Since the foot sole muscles CMAPs do not differ much between adjacent sites, there is little to gain from using large electrodes. Accordingly, large electrodes improved reproducibility less for the tibial nerve than for other nerves [18].

12.5 CLINICAL APPLICATIONS

12.5.1 Amplitude: What Does It Stand For?

CMAP amplitude can be used to infer the number of functional axons (motor units), functional muscle fibers or functional synapses. It is easy to see that CMAP amplitude will grow as more motor units take part, until all MUAPs do so. As all axons are stimulated in MNC studies, CMAP amplitude is commonly used as a measure of the number of functioning motor units. This is, however, not entirely correct for a number of reasons. First, there are the measurement problems due to effects of posture, muscle architecture, electrode site and size, as discussed above. These problems, coupled with interindividual variability, are serious enough to limit amplitude reproducibility. Of course, the number of functional motor units is extremely important from a clinical point of view: it reflects muscle strength or weakness, the most important aspect of muscle function. Second, muscle strength directly depends on the number of muscle fibers, and it cannot be assumed that motor units always have a fixed numbers of muscle fibers. The number of muscle fibers in a motor unit (its "size") does in fact change in several disorders.

The best example is a disease called amyotrophic lateral sclerosis (ALS), in which the motoneurones in the spinal cord die one by one over many years. After the death of a cell, its muscle fibers are denervated. This causes surrounding axons of healthy neurones to send out "sprouts" that reinnervate these muscle fibers. If the rate of motoneurones death is low, surviving cells can keep up, and the number of functional muscle fibers remains normal. Strength and CMAP amplitude will also remain normal. In fact there are estimates that muscle weakness in ALS only becomes apparent when half the original number of motoneurones has died. The remaining motoneurones should then innervate twice as many muscle fibers as normal. However, an increasingly smaller number of neurones then have to innervate an increasingly larger number of muscle fibers. This process fails progressively, so amplitudes drop and weakness ensues. In disorders in which motoneurones or their axons die off at a faster rate, reinnervation fails more quickly, so CMAP amplitude is then a better indicator of the number of motor units.

Faced with this limitation, several techniques of "motor unit counting" have been developed, but all are fairly cumbersome, and their reproducibility is rather limited. Most rely on dividing CMAP amplitude by the mean amplitude of a number of separately obtained individual motor units. CMAP amplitude is also used for the diagnosis of disorders of neuromuscular transmission [17], such as myasthenia gravis (MG) and the Lambert-Eaton myasthenic syndrome (LEMS). In these diseases both axons and muscle fibers are normal, but impulse transmission at the synapse is at risk of failing. Normally, each nerve impulse is with certainty converted into a contraction of the muscle fibers. In these diseases, transmission may fail when the synapses are stimulated at a frequency above a critical value. To observe the effect, a nerve (usually the ulnar one) is stimulated at 3 to 5 Hz and the accompanying CMAPs recorded (Fig. 12.9). If the amplitude drops below a certain level during such a train of stimuli, a "decrement" is said to be present, signifying that an increasing number of synapses failed to transmit the nerve impulses. The reverse is also possible: synapses may start out by being "blocked" in the resting state, and will only start responding normally after a number of stimuli. This is typically the case in LEMS, where trains of stimuli at over 10 Hz may cause CMAP amplitude to rise from very low values of about 0.5 mV to normal values of up to 10 mV. In this case the increase is known as an "increment". Not surprisingly, just as for individual CMAPs recordings [21], testing of the neuromuscular transmission is influenced by the recording site.

12.5.2 Deriving Conduction Properties from Two CMAPs

Nerves may be stimulated at any accessible site. Obviously the more proximal the site is, the longer it will take for the muscle to respond. By measuring the latencies of two sites

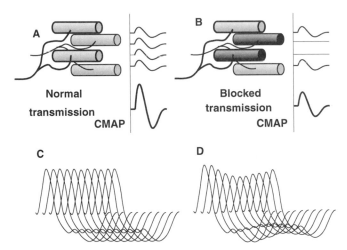

Figure 12.9. Repetitive stimulation. Panel *A* shows a number of muscle fibers with their inner-vating axon branches. When neuromuscular transmission is normal, nerve impulses are trans-mitted to all corresponding muscle fibers. Their action potentials are summed to form MUAPs, and together these form the CMAP. When impulse transmission fails in some synapses (panel *B*), the corresponding fibers (shown in a dark gray) do not form potentials, ultimately resulting in a CMAP with an abnormally low amplitude. Panel *C* shows a typical clinical representation of a repetitive nerve stimulation test: 10 CMAPs were obtained with stimulation at 3 Hz. The CMAPs are overlaid in a staggered fashion to visualize the results; normally amplitude differs by only a few percent from one CMAP to the next. In a decrement, however (panel *D*) the consecutive stimuli cause an increasing number of synapses to fail, and amplitude to drop. Due to physio-logical mechanisms transmission may then start to work again, causing a resurge of amplitude.

and the distance between them, the conduction velocity is calculated (Fig. 12.10). The latencies are measured from the first comparable deflection of the baselines, as the fastest nerve fibers in both cases cause these. The velocity reflected by these fibers is estimated to be about 5% of all fibers. If this velocity is too low, there is certainly something wrong with conduction: mild slowing is compatible with axonal damage, while severe slowing points to demyelination. This is important because establishing the pathophysiology of the damage narrows the range of possible causes. Note that slowing in itself is not associated with any loss of function. Suppose a motor unit is driven with a frequency of 15 Hz, meaning it fires every 67 ms; will it matter if each impulse arrives 20 ms late? No, func-tion will only be lost if impulses fail to arrive at all.

But what of the remaining 95% of the axons? The exact time of onset of firing of their motor units is lost in the activity of faster motor units, and cannot be measured (Fig. 12.10). Although some attempts have been made to measure the distribution of conduction veloc-ities in a nerve, these require numerous unpleasant stimuli ("collision neurography") [7]. Still, comparing proximal and distal CMAP amplitudes can help. If the intervals between arrivals of various motor units increase, MUAPs will be temporally dispersed, and the CMAP, as a summation of their activity, will be subject to phase cancellation. The effect of this is that CMAP amplitude will be lower and duration longer (Fig. 12.10). These con-comitant changes therefore signify the presence of temporal dispersion. The degree of tem-poral dispersion depends on a number of factors. One physiological factor is distance: the larger the distance between two sites of nerve simulation, the larger will be the degree of

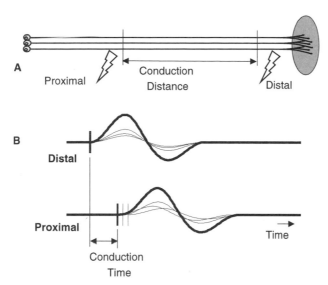

Figure 12.10. Nerve conduction testing. Panel *A* shows a schematic nerve (see Fig. 12.5) with two stimulation sites. The distance between these sites is measured in mm. The resulting CMAPs are shown in panel *B*. The earliest activity in both is formed by the MUAPs belonging to motor units with fastest axons. The onset of activity is measured, and the difference is the conduction time in ms. Conduction velocity is calculated by dividing distance by time. Note that the onset of activity of the slower motor units cannot be measured directly, and so their conduction velocity cannot either.

temporal dispersion. Demyelinating diseases can cause profound slowing in some fibers while leaving other fibers undamaged. As a result there will be abnormal temporal dispersion, apparent as abnormally low amplitude decreases and duration increases over a nerve segment (Fig. 12.11).

Finally, demyelinating disorders may also cause impulse transmission to fail completely in a portion of the nerve, while leaving surrounding segments wholly functional. This phenomenon, known as blocking, becomes apparent as a large drop in amplitude over a nerve segment (Fig. 12.11).

12.6 PATHOLOGICAL FATIGUE

Fatigue refers to the well-known state and feelings of exhaustion after a strenuous exercise (see Chapter 9), although it can also have a broader meaning, such as a feeling of fatigue following a day of mental work. In this chapter the physiological definition of local muscle fatigue is used: the decline of maximal force (or capacity for maximal force) during voluntary exercise. This decline in force or force-generating capacity can have its origin at many levels of the neural axis from motor cortex, spinal cord to neuromuscular junction, muscle membrane, and metabolism. It is often difficult to decide which changes are a result of adaptation of the system and which are due to limiting factors in the force-generating capacity. Simultaneous registration of surface EMG and force provides useful information concerning several mechanisms responsible for fatigue. The amplitude of the SEMG signal (often expressed by the integrated EMG over a specified time, that is the area under the full-wave

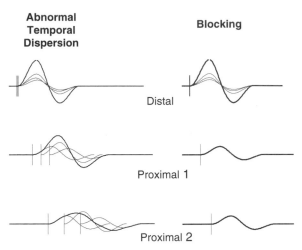

Figure 12.11. Abnormal conduction. The same model as in Figures 12.5 and 12.10 is used to illustrate abnormal transmission. When some fibers are slowed more than others, the onsets of MUAPs become less synchronous (abnormal temporal dispersion). The difference in firing becomes more pronounced as the distance from the stimulation point to the muscle increases. Because of this dispersion, the CMAP becomes longer in duration and lower in amplitude. When impulse conduction halts altogether in some fibers, blocking occurs. When a distal site is stimulated, all fibers (here all three) may still contribute with their MUAPs to the CMAP. When stimulation is performed more proximally, some impulses will not reach the muscle at all. In the example, only one of the three MUAPs can be made to fire from proximal sites; the other two are blocked somewhere between the distal site and the proximal 1 site. CMAP amplitude drops sharply as a result.

rectified signal within a time interval: IEMG) is a measure of the voluntary drive to the muscle. During submaximal contractions the IEMG usually increases considerably due to recruitment of extra motor units and increase in firing frequency. Both are mechanisms to cope with the declining force output during a sustained contraction of the motor units. During maximal contractions the amplitude usually declines, concomitant with a change of the muscle-fiber action potentials. Changes in the frequency spectrum and muscle-fiber conduction velocity during fatigue are well known and mainly due to changes in the metabolic environment of the muscle (and in the muscle membrane) during fatigue [14,23]. These changes include accumulation of lactate and lowering of pH, increase of potassium concentration in the T tubuli system, and depletion of substrates (phosphocreatinin and ATP). Most of the shift in the frequency spectrum to the lower frequencies during fatigue is due to the slowing of the conduction along the sarcolemma, although central changes in motor unit firing (especially synchronization) also play a role [24]. (See Chapter 9 and [6] for the myoelectric manifestations during local muscle fatigue.)

A diminished ability for sustained voluntary contractions accompanies many neurological disorders. If the problem lies in the central nervous system, it is logical to assume a lack in central drive of the muscle during the course of the contraction. In this case, changes in the muscle itself are limited and the lack of drive indeed *protects* the muscle from getting exhausted. Consequently changes in the SEMG (in terms of decline in muscle-fiber conduction and lowering of frequency content) will also be very limited, thereby showing that the problem resides in the central nervous system. Alternatively, the

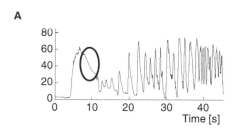

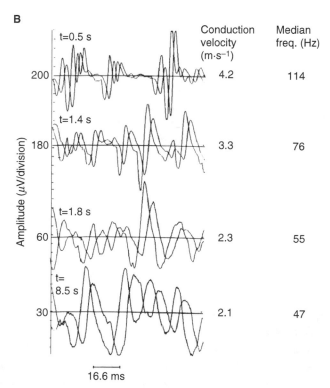

Figure 12.12. (*A*) Isometric force measurement of the elbow flexion of a patient with abnormal fatigue: myotonia congenita. Note the erratic force profile during an attempt to hold the force at 80% of the maxiaml voluntary contraction. (*B*) The SEMG during the phase of paresis (encircled in panel *A*). Surface EMG results of a myotonia congenita patient measured with bipolar recordings and a linear array. The distance between the two displaced curves reflects the average muscle-fiber conduction velocity (MFCV). Concomitant with a strong decline in MFCV, the frequency content of the signals shifts to left (lower frequencies).

muscle could be prone to abnormal fatigue due to an intrinsic disorder, for example, a metabolic disease or a disorder in the muscle membrane or contractile mechanism.

The use of SEMG in the assessment of abnormal fatigue in neurological disorders is, as yet, mainly scientific. For example, the changes in force and SEMG of myotonia congenita (MC), an inherited disease of the muscle, are characterized by myotonia (spontaneous discharges of muscle fibers resulting in cramps) and a peculiar transient paresis following rest [25]. Figure 12.12*A* shows the changes in force during a maximal voluntary contraction. In healthy subjects, such an instruction results in a slow and steady decline over time (approximately 50% in 1 minute). The patients show a strong and erratic force decline after several seconds. The SEMG measured in the phase of force decline indicated by a circle shows concomitantly a sharp decline in muscle-fiber conduction velocity (measured with an array; see Chapter 7), EMG amplitude, and frequency content (Fig. 12.12*B*). The sudden and fast change of all SEMG variables point to the muscle membrane as the cause of the contraction failure. Indeed, an abnormal chloride ion channel conductivity in the muscle membrane is responsible for both the myotonic discharges and the temporary loss in force following an initial normal contraction [12].

Abnormal behavior of MFCV during fatigue tests in McArdle's disease and type I fiber dominance was shown by Linssen et al. [15,16]. The absence of decline of MFCV in the McArdle patients underscores the importance of lactate accumulation as a cause of the decline in MFCV during fatigue. In a patient with carnitine deficiency an unexpected increase in median power frequency of SEMG of the vastus lateralis was evident [1]. The abnormal rise in SEMG amplitude during prolonged, low-level contractions in this metabolic myopathy indicates the recruitment of additional motor units in an effort to maintain force. Due to the metabolic compromise in these diseases, the energy supply, which is necessary to continue the contractions, fails. The recruitment of extra motor units is an adaptation strategy of the CNS in order to maintain the desired force and is evident from the abnormal rise in SEMG amplitude [1].

12.7 NEW AVENUES: HIGH-DENSITY MULTICHANNEL RECORDING

In principle, there is no hindrance in increasing the number of electrodes over the muscle, the number of channels, or arranging complex montages from different electrodes for each channel, as done in linear array recording [8,9,24]. This enables the measurement of sarcolemmal propagation by comparing the time delay of the consecutive signals, provided they are aligned with the muscle-fiber direction (Fig. 12.13). Here we encounter the first spatial aspect of the MUP: its origination at the endplate zone, propagation along the sarcolemma and extinction at the tendon.

In a bipolar recording, the end plate zone is characterized by a low amplitude (the two electrodes above the place of origination of the AP "see" almost the same potential). The propagation of the MUP is seen from the later arrival of the potentials at the consecutive electrode pairs. The lines connecting the peaks (Fig. 12.13) indicate a constant propagation along the muscle fibers. Note that the polarity of the signals is reversed with respect to the two directions of MUP propagation. This results in a phase reversal at the end plate zone, facilitating its recognition [8].

A logical extension of the electrodes is the configuration of a two-dimensional grid of electrodes with an arbitrary number of electrodes and interelectrode distances (Fig. 12.14). Such a configuration enables the localization and size estimation of MUs as well as the determination of the position of the endplate zone. With a two-dimensional grid it is also possible to display the SEMG activity in an amplitude map (Fig. 12.15). So we recommended for all SEMG measurements, especially for multichannel SEMG as discussed here, to record and store the signals of the individual electrodes in the array or grid referenced to a remote electrode (in a monopolar fashion). We found this approach to enable a versatile and purpose-dependent (re-) selection afterward, both with respect to the desired montages (bipolar, Laplacian, etc.) and the way in which the EMG activity is displayed (column, row, map, etc.).

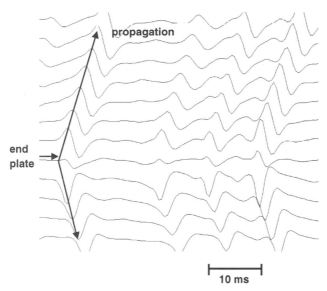

Figure 12.13. Example of a linear array recording over the biceps showing the end plate position and propagation of MUP's. Horizontal calibration: 10 ms.

Figure 12.14. The 126 channel surface EMG grid with gold-coated electrodes mounted on a flexprint base.

In most SEMG studies, it is best to measure the exerted force simultaneously with the SEMG of the muscle under investigation. With the use of direct feedback of the force on a display, the subject undergoing the study can be instructed to contract at the desired force level (usually expressed as a percentage of the maximal force). This way SEMG measurements can be done at different force levels or during a certain period of time enabling the measurement of fatigue. Usually these measurements are restricted to isometric conditions.

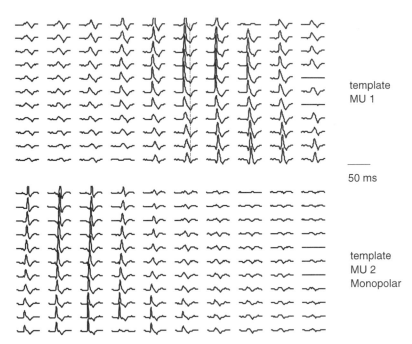

template
MU 1

50 ms

template
MU 2
Monopolar

Figure 12.15. 3-D surface MUP templates recorded from two MUs. The signals are represented in monopolar montage. The two templates were constructed from a single recording from the vastus lateralis. Note the different amplitude profiles, especially in medio-lateral (left–right) direction. The propagating surface MUP component and the nonpropagating final component (*dashed line*) can be recognized in proximo-distal (*vertical*) direction.

The motor control within one muscle is a complicated process, and unraveling it requires the recognition and measurement of single motor units. The extraction of MUPs during voluntary contraction using a standard concentric needle electrode is known to be difficult and fraught with problems. At low force levels it is possible to follow the firing pattern of a few MUs during isometric contraction. The recognition of MUs using SEMG is possible using the MU's unique configuration of the spatial potential distribution over the skin (its fingerprint or signature). The 126-channel SEMG grid allows specific topographic configuration of each MU to be recognized: its end plate zone, direction of propagation, extent and rate of decline in lateral direction, are all variables contributing to its specific topographical distribution of activity (Fig. 12.15). The study of Kleine et al., showed how these characteristics could be used to unravel MU firing patterns [11].

In a low-level voluntary contraction with several MUs active, a first extraction was done by using a peak detection at a user-defined level. Such a peak detection should be done repeatedly in several channels, since it is not known a priori where the maxima of different MUs lie. The resultant MUP topograms were clustered on the basis of their similarity by using a clustering algorithm and visual review. Next, averaging the SEMG signals of all electrodes and treating the peaks as a time-lock trigger can be used to form templates. Each resulting template thus consisted of 126 MUPs (Fig 12.15). Note that the main difference between the topograms is not in the temporal but in the spatial domain. The fingerprints can be used to resolve all firings of the MU, including superimpositions.

Up to five simultaneously active MUs can be detected in this way. The measured MU's are the consequence of single neuron firings, implying that with such a noninvasive technique the behavior at a single cell level is accomplished. This way changes in neuronal excitability of Parkinson's disease patients can be recorded [11]. The advantage of the multichannel SEMG approach, compared with needle EMG measurements, is that besides a list of all firing moments, also basic information concerning MU properties becomes available.

12.8 CONCLUSION

SEMG has many different applications in the diagnosis of neurological disorders. It can be used to monitor and measure the motor output of the CNS and measure reflex activity. For the peripheral nervous system it allows the valuable measurement to be made of the conduction velocity of motor nerves and the amplitude of the compound muscle action potential in terms of the number of motor axons. Changes in SEMG signal during normal and pathological fatigue can be followed reflecting different aspects of the metabolic state of the muscle and the changing motor command during exercise. New developments are the application of high-density multichannel grids that enable noninvasive measurement of motor unit characteristics such as firing behavior, position, and sarcolemmal properties of motor units.

REFERENCES

1. Braakhekke, J. P., D. F. Stegeman, and E. M. G. Joosten, "Increase in median power frequency of the myoelectric signal in pathological fatigue," *Electroenceph Clin Neurophysiol* **73**, 151–156 (1989).

2. Bromberg, M. B., and T. Spiegelberg, "The influence of active electrode placement on CMAP amplitude," *Electroenceph Clin Neurophysiol* **105**, 385–389 (1997).

3. Christensen, E., "Topography of terminal motor innervation in striated muscles from stillborn infants," in H. D. Bouman, and A. L. Woolf, eds., *The Utrecht symposium on the innervation of muscle*, Williams and Wilkins, Baltimore, 1960, pp. 17–30.

4. Coërs, C., and A. L. Woolf, "Normal histology of the intramuscular nerves and nerve endings," in C. Coërs, and A. L. Woolf, eds., *The innervation of muscle*. Blackwell, Oxford, 1959, pp. 12–22.

5. Del Toro, D. R., and T. A. Park, "Abductor hallucis false motor points: electrophysiologic mapping and cadaveric dissection," *Muscle Nerve* **19**, 1138–1143 (1996).

6. DeLuca, C. J., "Myoelectrical manifestations of localized muscular fatigue in humans," *CRC Crit Rev Biomed Eng* **11**, 251–279 (1984).

7. Dumitru, D., A. Amato, and M. J. Zwarts, "Nerve conduction studies," in D. Dumitru, A. M. Amato, and M. J. Zwarts, eds., *Electrodiagnostic medicine*, 2nd ed., Hanley and Belfus, Philadelphia, 2002, pp. 159–223.

8. Falla, D., P. Dall'Alba, A. Rainoldi, R. Merletti, and G. Jull, "Location of innervation zones of sternocleidomastoid and scalene muscles—A basis for clinical and research electromyography applications," *Clin Neurophysiol* **113**, 57–63 (2002).

9. Farina, D., W. Muhammad, E. Fortunato, O. Meste, R. Merletti, and H. Rix, "Estimation of single motor unit conduction velocity from surface electromyogram signals detected with linear electrode arrays," *Med Biol Eng Comput* **39**, 225–236 (2001).

10. Fasshauer, K., "Elektromyographische untersuchungen bei extrapyramidal-motorischen erkrankungen," *Fortschr Med* **99**, 740–744 (1981).

11. Kleine, B. U., P. Praamstra, D. F. Stegeman, and M. J. Zwarts, "Impaired motor cortical inhibition in Parkinson's disease: Motor unit responses to transcranial magnetic stimulation," *Exp Brain Res* **138**, 477–483 (2001).

12. Koch, M. C., K. Steinmeyer, C. Lorenz, et al., "The skeletal muscle chloride channel in dominant and recessive human myotonia," *Science* **257**, 797–800 (1992).

13. Lateva, Z. C., K. C. McGill, and C. G. Burgar, "Anatomical and electrophysiological determinants of the human thenar compound muscle action potential," *Muscle Nerve* **19**, 1457–1468 (1996).

14. Lindström, L., R. Magnusson, and I. Peterson, "Muscular fatigue and action potential conduction velocity changes studied with frequency analysis of EMG signals," *Electromyography* **4**, 341–356 (1970).

15. Linssen, W. H. J. P., M. Jacobs, D. F. Stegeman, E. M. G. Joosten, and J. Moleman, "Muscle fatigue in McArdle's disease. Muscle fiber conduction velocity and surface EMG frequency spectrum during ischaemic exercise," *Brain* **113**, 1779–1793 (1990).

16. Linssen, W. H. J. P, D. F. Stegeman, E. M. G. Joosten, H. J. H. Merks, H. J. Laakter, R. A. Binkhorst, and S. L. H. Notermans, "Force and fatigue of human type I muscle fibers. A surface EMG study in patients with congenital myopathy and type I fiber predominance," *Brain* **114**, 2123–2132 (1991).

17. Oh, S. J., *Electromyography: Neuromuscular transmission studies*, Williams and Wilkins, Baltimore, 1988.

18. Tjon-A-Tsien, A., H. H. P. J. Lemkes, A. Kamp van der-Huyts, and J. G. Van Dijk, "Large electrodes improve nerve conduction repeatability in controls as well as in patients with diabetic neuropathy," *Muscle Nerve* **19**, 689–695 (1996).

19. Van Dijk, J. G., I. van Benten, C. G. S. Kramer, and D. F. Stegeman, "CMAP amplitude cartography of muscles of the median, ulnar, peroneal and tibial nerves," *Muscle Nerve* **22**, 378–389 (1999).

20. Van Dijk, J. G., A. M. L. Tjon-A-Tsien, and W. van der Kamp, "CMAP variability as a function of electrode size and site," *Muscle Nerve* **18**, 68–73 (1995).

21. Van Dijk, J. G., B. J. van der Hoeven, and J. H. van der Hoeven, "Repetitive nerve stimulation: Effects of recording site and the nature of 'pseudofacilitation,'" *Clin Neurophysiol* **111**, 1411–1419 (2000).

22. Van Dijk, J. G., W. van der Kamp, B. J. van Hilten, and P. van Someren, "Influence of recording site on CMAP amplitude and on its variation over a length of nerve," *Muscle Nerve* **17**, 1286–1292 (1994).

23. Van Dijk, J. G., "'Rules of conduct': Some practical guidelines for testing motor nerve conduction," *Arch Physiol Biochem* **108**, 229–247 (2000).

24. Zwarts, M. J., T. W. van Weerden, and H. T. M. Haenen, "Relationship between average muscle fiber conduction velocity and EMG power spectra during isometric contraction, recovery and applied ischemia," *Eur J Appl Physiol* **56**, 212–216 (1987).

25. Zwarts, M. J., and T. W. van Weerden, "Transient paresis in myotonic syndromes: A surface EMG study," *Brain* **112**, 665–680 (1989).

26. Zwarts, M. J., and D. F. Stegeman, "Multichannel surface EMG: scientific basis and clinical utility," *Muscle Nerve* **28**, 1–17 (2003).

APPLICATIONS IN ERGONOMICS

G. M. Hägg and B. Melin

Department for Work and Health
National Institute for Working Life, Stockholm, Sweden

R. Kadefors

Industry and Human Resources
National Institute for Working Life, Göteborg, Sweden

13.1 HISTORIC PERSPECTIVE

Electromyography was, in the early decades of the twentieth century, a matter mostly for the clinical neurophysiologists and neurologists. The chief aims were widening the diagnostic repertoire with respect to differentiation between myopathy and neuropathy in patients with suspected neuromuscular disorders. However, the potential of electromyography as a tool in studies of the normal human muscle was realized by several early investigators. The observation that an increased muscle force is reflected in the EMG as an increase in activity level was applied in functional anatomy, for instance, in studies of the shoulder [32], postural muscles [37], and masticatory muscles [8]. In these investigations electromyography made it possible, for the first time, to elucidate the interplay between agonists and antagonists, and the role of different muscles in synergistic action. The contribution of electromyography in this field of studies was significant and advanced our understanding of motor action and control in normal human beings.

The approximately linear relationship between isometric muscular force and the so-called integrated EMG (its rectified and averaged value) was clarified further [33,53]. Findings from these studies provided the basis for subsequent quantitative research into muscle

Electromyography: Physiology, Engineering, and Noninvasive Applications, edited by Roberto Merletti and Philip Parker.
ISBN 0-471-67580-6   Copyright © 2004 Institute for Electrical and Electronics Engineers, Inc.

action, primarily in isometric–isotonic loading situations. At the same time some of the complexities of electromyographic methodologies were revealed, for instance, the dependence of EMG variables on anatomical properties of muscle, crosstalk, electrode location, and effects of muscle length. Nevertheless, surface electromyography developed to become a standard method in biomechanics, particularly following publication of the book, *Muscles Alive—Their Functions Revealed by Electromyography* by J. V. Basmajian [4], and the formation of the International Society of Electrophysiology and Kinesiology, also recognizing ergonomics as a viable area of application of EMG methodologies.

As early as 1912 it was observed already, by Piper [82], that fatigue caused "slowing" of the myoelectric signal. Research into changes in the myoelectric signal in relation to fatigue and endurance, aiming at an advanced understanding of the basic physiological metabolic processes, identified EMG signal augmentation and an increased duration of action-potentials [12,54]. Statistical techniques employing spectral analysis of interference EMG pattern were developed and applied in fatigue studies [9,40,44]. The results from these investigations provided a noninvasive methodology for application in work physiology studies at the local muscular level.

The potential of electromyographic techniques in studies of the effects of exposures at work was realized gradually. From the point of view of ergonomics, both the results from research relating to the muscle force dimension and to the muscle fatigue dimension were found useful, although most studies focused on biomechanical rather than physiological effects. Already in the beginning of the 1950s, when Lundervold [59,60] published studies on muscular activation in typewriting, using surface electromyography as a tool. Here the focus was not primarily on static loading but on the dynamic action of the prime movers in the forearm. Similar studies on other occupational groups followed, for instance, in the use of hand tools [95], and in car driving [36]. Spectral changes as an indicator of localized muscle fatigue were studied in work on the car assembly line [79], using telemetry for signal acquisition, and in welding workshops [41].

In the early years, instrumentation deficiencies presented a problem to investigators, for instance, concerning sensitivity, reliability, and capacity. However, through the development of high-performance amplifiers and powerful digital instrumentation for ambulatory monitoring, these problems were overcome gradually. It is at this time possible to store 12 channels of surface EMG for 8 hours on a solid-state memory card. Using surface electromyography as a method for ergonomic studies outside the laboratory is nowadays technically feasible not only for researchers but for a wide range of practitioners as well. What still remains an obstacle is the signal analysis and interpretation of the results. In this chapter we set out to describe the theoretical basis for using SEMG in ergonomic studies, and to give examples of areas of application relevant to issues encountered in work science and occupational health.

13.2 BASIC WORKLOAD CONCEPTS IN ERGONOMICS

In ergonomics, physical load represents one of the most important aspects to be covered in the design of work and workplaces. Even though new technologies have helped alleviating many problems associated with, for example, heavy materials handling, they have tended at the same time to introduce new types of adverse load situations. In many sectors of the labor market this has entailed shorter work cycles, more monotonous job tasks, and more localized muscular loading. The problem focus has shifted from circulatory strain to localized load on certain parts of the musculoskeletal system. This is also an area where surface electromyography has contributed substantially and where electromyographic methods represent useful techniques for the practitioner.

There is a wide spectrum of shown or suspected work-related musculoskeletal disorders (WMSDs), affecting various anatomical structures in different parts of the body. Authoritative critical reviews [6,49,78,84] have discussed ailments that include disorders in the hand–arm system (e.g., carpal tunnel syndrome, epicondylitis, tenosynovitis), in the shoulder and neck (rotator cuff tendinitis, shoulder myalgia, thoracic outlet syndrome, rhizopaty), and in the postural system (thoracic and low back pain, patella bursitis). Even though the specific contribution of exposures at work to the development of musculoskeletal ailments is under debate with respect to some specific disorders, it is obvious that physical load represents a major concern with respect to health and well-being in working life.

In taking into account a spectrum of possibly work-related disorders, it is unlikely, however, that a single causality mechanism can be identified. There is rather a number of factors that need to be taken into account. Putz-Anderson [84] has suggested a generic formula implying that the combination of high exerted force, awkward posture, high repetitivity, and lack of recovery provides a high risk for work-related musculoskeletal disorders. Other researchers have further emphasized the relation between the background factors by developing conceptual models for their interplay [42,48,93].

What is the contribution of surface electromyography in this context? Here it should be realized that it is often not possible to evaluate the strain on the vulnerable body tissues by observation techniques alone. Although external forces may be measured, there remains always an uncertainty concerning how reaction forces are distributed among the body tissues. Electromyography may be instrumental in this regard by providing insight in the degree of individual muscular involvement in a certain task, thereby making possible to arrive at evaluation of the strain of associated tissues (tendons, ligaments), known to be common sites of, for instance, chronic inflammations.

Because of the important role ascribed to muscular recovery and rest, the specific information that may be provided by surface EMG is of high interest. Work–rest patterns at the muscular level may differ considerably from what can be expected in sheer task analysis, for instance, due to the need for joint stabilization or effects from mental stress.

13.3 BASIC SURFACE EMG SIGNAL PROCESSING

The raw EMG signal provides without further signal processing limited information to an ergonomist. Visual inspection of the raw EMG on an oscilloscope screen or on a fast chart recorder may however give crude estimates of muscle activation over time expressed as, for example, no–low–medium–high activation. While this type of crude estimate provided valuable contributions in functional muscle studies in the past, with modern signal processing techniques we can extract much more specific information from a raw EMG signal.

EMG data processing applications in ergonomics can usually be separated into two main steps: the initial step that converts the raw EMG signal into some relevant parameter while preserving the variation over time, and a second step where further data reduction takes place by applying different types of statistical models and algorithms. In this section the first step is the focus while the second step is addressed later in this chapter.

For the initial analysis one may consider amplitude or spectral parameters, or both. Amplitude measures are mainly related to forces, torques, and muscle activation, while spectral variables mainly relate to different aspects of fatigue. However, a combination of these two variable classes can in some cases be fruitful as will be described below in Section 13.6.

Amplitude variables mostly demand a rectification of the raw EMG signal, either linearly, by forming the absolute values of the raw samples, or quadratically, by calculating the squares of the raw samples. Either way these rectified data must undergo some kind of (moving or weighted) averaging over a suitable time window. In the linear case the outcome of this process is denoted electrical activity (EA), average rectified value (ARV), or, somewhat inappropriately, integrated EMG (IEMG). In the quadratic case the root is taken of every average reading and the outcome is denoted root mean square (RMS), which is a well-known measure for an electronics engineer. An alternative to conventional mean values in the linear case is the median as advocated by Nieminen and Hämeenoja in 1995 [77] which is claimed to improve the relation between dynamic properties and precision.

The choice of the averaging time window duration must be a compromise between demands on precision and dynamic response. The most commonly adopted duration is 100 ms, but shorter as well as considerably longer windows may be appropriate for rapidly or slowly changing contractions, respectively. The choice between the RMS and EA/ARV/IEMG parameters is of little importance from a biologic/ergonomic point of view and is more related to practical circumstances. RMS and EA/ARV/IEMG show a monotonic, sometimes linear, relationship to the force developed in the muscle. More about this is found in Section 13.4. Additional information concerning specific signal processing techniques, such as "whitening," wavelet and time-frequency representations, may be found in Chapters 6 and 10.

Another class of parameters is derived from consecutive Fourier power spectra calculated on time epochs of raw data with a duration of usually 0.25 to 1 second, depending on the degree of signal stationarity (see Chapters 6 and 10). From each such spectrum further data reduction yields one numerical parameter reading for each epoch. The two parameters mostly used are the median frequency (MDF, [11]) and the mean (or centroid) power frequency (MNF, or MPF as it was denoted by the authors first suggesting it [47]). They are calculated as the median or mean frequency of the power density spectral function (usually referred to as power spectrum). The application and interpretation of these parameters are further described in Section 13.6 (as well as in Chapters 6, 9, and 10).

Still another surface EMG parameter that has been applied in ergonomics is the number of zero crossings (ZC) of the raw EMG per time unit [21]. This parameter has been shown to have properties resembling MDF and MNF [22]. (More details on these basic parameters are found in Chapter 6.)

13.4 LOAD ESTIMATION AND SEMG NORMALIZATION AND CALIBRATION

Surface EMG provides a unique tool for on-line assessment of muscular activation and internal loads on muscles, tendons, and other tissues. Two different approaches can be discerned here, a "physiological" and a "biomechanical." Depending on research hypothesis, we may choose one or the other.

Choosing the physiological approach we are only interested in the degree of muscular activation, which implies that we use ARV or RMS expressed in μV and then apply a normalization coefficient. It should be emphasized that absolute values recorded in μV from different individuals should never be compared without some kind of normalization. Raw EMG amplitude depends on a large number of individual factors of no interest in this context but relevant as confounding factors. Two normalization alternatives are available. The first alternative is to relate all measurements to the electrical activity at maximal voluntary contraction (MVC). This is denoted maximal voluntary electrical activation (MVE), and the muscular activity during work is expressed in terms of %MVE. Before the real measurements the test subject is instructed to perform a few MVCs during which the MVE level is measured. It should be noted that given a certain electrode position, different MVEs (and MVC values) can be obtained at various muscle lengths and postures [67].

However, there are situations where a reliable and safe MVC is hard to obtain, such as upper trapezius muscle measurements [67]. In such cases normalization can be performed using a standardized reference contraction, and the muscle activation during work is expressed as a percentage of the reference voluntary electrical activity (%RVE). For the upper trapezius muscle a suggested reference contraction is to hold the arms horizontally in the frontal plane [67].

When we want to express the load on muscles or other tissues in biomechanical terms, we have to make a transformation of the electrical activity into forces or torques. The relation between muscle force and electrical activity is mostly a progressively increasing function. This calibration function has to be established, and it can be performed in different ways. The most elaborate way is to register the force or torque and EMG during an isometric contraction in which the subject is increasing the muscle exertion slowly up to MVC. By regression analysis the best-fit regression curve is established as the calibration curve [34]. The load variable in this case usually is %MVC but can also be expressed in absolute mechanical units. A somewhat simpler method is to use a number of voluntary reference contractions (RVCs) at known loads and establish a piecewise linear calibration curve between these points [34].

A problem with these biomechanical calibrations is that several new error sources are introduced. First of all the calibration curve has to be established at isometric conditions and slowly varying forces while the real measurements usually take place during varying muscle length and with more dynamic activation, which alters the force-EMG relationship substantially. A second error source is the fatigue effect, which implies that gradually higher EMG levels are generated at the same mechanical load level. Unfortunately, too many of these error sources have been overlooked or underestimated in several investigations [64].

In sum, the MVE/RVE procedures are simpler to apply and introduce less error sources compared to MVC/RVC procedures. On the other hand, load data expressed in mechanical units demand a MVC/RVC calibration. Amplitude increase due to fatigue, which at least to some extent is caused by additional recruitment of motor units, can be interpreted with an MVE/RVE approach as increased muscular activity, whereas it should be treated as an error source in the MVC/RVC case, altering the force-EMG relationship [61]. An attempt to eliminate the fatigue influence on the EMG amplitude has been made by introducing the ACT variable [20]. The ACT variable is built up according to a recursive algorithm by multiplying the previous activity sample by a constant, 0.9875 (at 500 Hz), and adding the root of the absolute difference between the present and previous raw-EMG sample. However, the general applicability of this variable remains to be proved.

13.5 AMPLITUDE DATA REDUCTION OVER TIME

When the EMG raw data are transformed into a suitable amplitude parameter, time plots estimating the muscle load are obtained, one plot for each muscle. In situations where an instantaneous load estimate is desired, no further data reduction is needed. However, the aim is normally to characterize the load profile over a longer period of time in a specific task or job. Such recordings may extend over several hours and even over a whole working day, using modern data logger technology [2,83]. In this case, of course, further data reduction is inevitable.

The simplest approach for data reduction is to calculate a time average of the total recording [88]. This measure gives a crude estimate of the mean exposure level, but we know today that more detailed information regarding exposure dynamics is needed to give an adequate description of the load in terms of risks for musculoskeletal disorders [103].

The APDF method (amplitude probability distribution function) introduced by Jonsson and colleagues is the traditional solution for such a data reduction [35]. The APDF is calculated based on a recording over time. This function is integrated to yield a cumulative probability function. The amplitude can be normalized/calibrated according to any of the approaches described above (RMS or EA, %RVE, %MVE, or %MVC). The 10% percentile of the function is defined as *the static load level*, the median (50% percentile) as *the median load level*, and the 90% percentile as *the peak load level* (see Figure 13.1). This method has been established as a standard for describing occupational loads [35]. However, a drawback with APDF is that the repeatability aspects on the load are not reflected in these three standard parameters. Another is that these parameters are not statistically independent [67].

A further development of the APDF method is the EVA method (exposure variation analysis) [65]. It is as the APDF based on the relative occurrence (% of total time) of different amplitudes recorded over time. Contrary to the APDF method the occurrence figures are not accumulated. The main new feature of EVA is the introduction of a second independent time axis, in addition to amplitude, that describes the distribution of durations or stays within each amplitude class. The outcome of an EVA analysis is a three-dimensional plot with percent of total time on the vertical z-axis (see Fig. 13.2). The amplitude axis (facing the reader in the figure) is divided into exponentially growing classes [65]. The unit on this axis can be any amplitude measure described above. The time axis is the horizontal axis to the left divided into exponentially growing time classes. High bars close to the $t = 0$ plane (farther wall in Fig. 13.2) indicate higher repetitivity, while bars closer to the viewer indicate activity that remains within the same amplitude class for longer times. This method also provides good possibilities to analyze "EMG gaps" (see below). The lowest amplitude class can be set according to a gap definition (e.g., 0–0.5 %MVC), and the gap duration distribution is shown by the bars within this class [27]. The EVA method is aimed at the analysis of exposure measurements, in general, but so far it has mainly been applied to EMG amplitude data [3,31,66]. Even if it may give the most relevant exposure description so far, too little is known about what should be considered "bad" or "good" EVA profiles. Other EMG data reduction parameters may in the future prove to be more relevant.

13.6 ELECTROMYOGRAPHIC SIGNAL ALTERATIONS INDICATING MUSCLE FATIGUE IN ERGONOMICS

Fatigue is a fundamental concept in ergonomics. It is a basic principle that fatigue should be avoided. However, fatigue can be defined in many different ways involving a large number of independent physiological as well as psychological phenomena. The concept

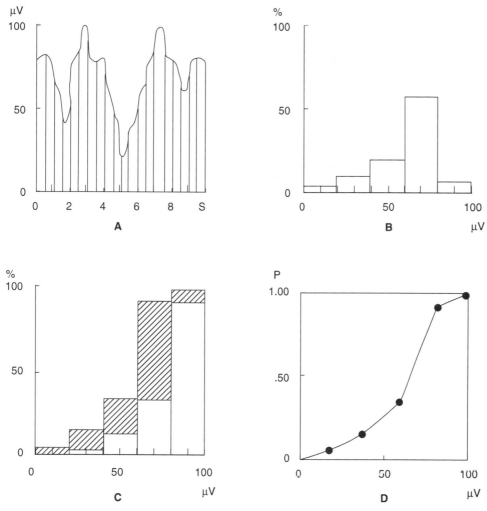

Figure 13.1. Amplitude probability distribution function calculation according to Jonsson [35]. Plot A shows samples of an amplitude variable (RMS or EA) over time, B the relative occurrence of different amplitude classes (20 μV in this case), C the cumulative procedure, and D the final APDF plot. The x-axis unit in B–D can as well be %MVC, %MVE, or any other amplitude measure.

of localized fatigue has been introduced in relation to the application of EMG in ergonomics [10]. A basic definition of fatigue in muscle physiology is that a subject can no longer maintain a required force. This concept differs from the concept of myoelectric manifestations of muscle fatigue as proposed in many contributions by C. J. De Luca, R. Merletti, and many others (and discussed in depth in Chapter 9). Studies of recovery, where localized fatigue indicated by EMG has been compared to MVC capacity, show diverging results. After fatiguing contractions Kroon and Naeije showed fairly good resemblance between the time course of MVC and EMG fatigue variables [45,46], while Byström et al. found that EMG fatigue variables recovered much faster than the force-generating capacity [7]. Different muscles (biceps and triceps) have also been reported to show different response patterns during sustained contractions over long times [38]. Hence the user

% of total time

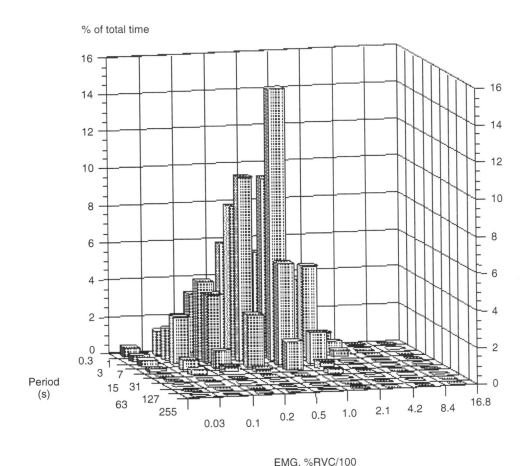

EMG, %RVC/100

Figure 13.2. Exposure variation analysis (EVA) according to Mathiassen and Winkel [65]. Example of upper trapezius activity during office paper handling (modified from Hägg and Åström [27]). The EVA plot shows that the EMG amplitude most of the time is below reference contraction (1.0 on *x* axis) while the dominating duration of contractions within all amplitude classes is 1 to 3 seconds (*y* axis). However, there are some contributions in the range 3 to 7 seconds and even some minor ones in the range 7 to 15 seconds.

of electromyographic fatigue indices should be well aware of the differences between EMG responses at fatigue and other fatigue definitions prevailing in other areas of application.

During a fatiguing contraction the amplitude (EA, RMS, or ARV) gradually increases while the EMG spectrum is compressed toward lower frequencies. The frequency shift is mainly related to motor unit action potential conduction velocity (MUAPCV) decrease and possibly motor unit firing synchronization [24] while the amplitude increase to a large extent seems to be a consequence of gradual additional recruitment of new MUs [1,76]. Hence the amplitude and spectrum alterations have their origins in partly different physiological phenomena.

The most widely applied EMG fatigue indexes are the spectral parameters MDF and MNF. The normal procedure is to normalize all readings by an initial well-rested reading,

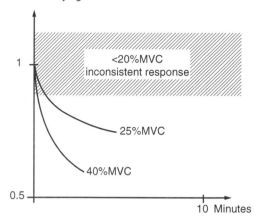

Typical average response of a normalized
EMG fatigue index (median frequency,
mean power frequency or zero-crossing rate)
at varying static loads

Figure 13.3. Typical electromyographic fatigue index (MNF, MDF, or ZC) development over time at different load levels.

yielding plots over time starting from unity [52]. These are often referred as "fatigue plots" (see Chapter 9), and they allow comparison of percent variations of different physical variables such as muscle-fiber conduction velocity, amplitude, and spectral variables. During static isometric contractions the normalized indexes demonstrate an exponential decrease with a time constant inversely related to the load level and final index readings at exhaustion decreasing with increasing load level [52] (see Fig. 13.3). Hence the endurance time, which is inversely related to load level [74,85], can be fairly well predicted from the initial slope of MDF or MNF. However, a disadvantage in using these indexes is that the consistent exponential decreasing behavior over time tends to be unreliable at load levels below approximately 20 %MVC where increasing indexes over time sometimes have been reported [18,28]. Linear regression is more suitable in these cases. The conventional interpretation of lower index readings indicating a "higher" state of fatigue is hard to apply under these circumstances.

These fatigue indexes have also been applied during occupational work, simulated in the laboratory as well as in real environment. A major obstacle here is that instant absolute readings of these indexes vary with instant load level, mainly in the range 0% to 30 %MVC [15,16,17,75]. This specially creates problems during vocational recordings where large variations of instant load level occur frequently. Two solutions have been suggested and applied to overcome these problems.

The first one is to apply short, nonfatiguing, test contractions at a known load level during short intermissions of ordinary work during which EMG is recorded. This solution was originally suggested by Kadefors [39] and applied in practice by Hägg and Suurküla [29,30,91]. The latter investigators used, as the fatigue index, the zero-crossing rate (ZC), which showed properties close to MDF and MNF [22].

A second solution, advocated by Luttmann and colleagues, involves analyses of amplitude and spectral alterations jointly from continuously recorded vocational EMG. The method is called JASA (joint analysis of EMG spectrum and amplitude) [61,62,63]. In previous investigations of vocational work reported from this group, increasing trends in the amplitude only have been studied as fatigue indications [50,51]. A prerequisite here was that the biomechanical load remain constant over the study period. This is often hard to control in an occupational setting. With the JASA approach the criteria for fatigue are more rigorous, defining fatiguing work as periods with both amplitude increase (EA) and MNF decrease.

13.7 SEMG BIOFEEDBACK IN ERGONOMICS

Biofeedback of EMG has been widely applied in various fields for decades [5]. Basics of the technique and clinical applications are described in Chapter 17. Here some special aspects on the applications in ergonomics will be given.

The hardware usually consists of a miniaturized EMG amplifier to be worn in a pocket with electrodes attached to a suitable muscle and with feedback of the EMG amplitude to the wearer as a sound signal in an earphone or some kind if visual display. Most apparatuses are equipped with a threshold setting that represents a safe limit that should not be exceeded.

In ergonomics the technique has been applied mainly in the Scandinavian countries for individual work technique training [81]. Personal communication with occupational physiotherapists reveals that the technique also can be used as an effective tool for group demonstrations of inappropriate work techniques with a demonstration subject.

The threshold option, available in all equipment, indicates that the underlying philosophy is based on the notion that you can identify a safe upper extremity muscle load limit in accordance with, for example, the limits suggested by Jonsson [35]. More modern hypotheses like the Cinderella model [23] suggest that muscular activity at any load level exerted over a too long time without a break is harmful. In this perspective the amplitude threshold concept is not functional. An alternative design of an EMG biofeedback device where the activation duration is focused has been suggested [25]. However, this approach remains to be practically evaluated.

13.8 SURFACE EMG AND MUSCULOSKELETAL DISORDERS

It is clear from the previous paragraphs that the amplitude, the frequency, and the time dimension parameters of the surface EMG may contain important information for practical ergonomics work and help reveal situations where an elevated risk exists of contracting musculoskeletal disorders. Since the SEMG amplitude dimension reflects force output, it is possible to study individual muscular involvement in work tasks (rather than absolute force), and thereby to evaluate the ergonomic design of tools and workplaces. It should be realized, however, that for the practitioner, external force measurements or even estimations thereof may in many cases suffice, and it is often possible to rely on subjective assessment of perceived effort, pain, or discomfort. Moreover, on the basis of external force measurements and observations or measurement of joint angles, mathematical models (indeed often validated by means of SEMG) may provide predictions of the internal force distribution among muscles. This means that SEMG amplitude measurements are not so commonly used when it comes to ergonomic evaluation in relation to specific diseases, although there are situations where they would be of very high value, for instance, in workplace evaluation of shoulder muscle load (rotator cuff tendinitis) and forearm muscle load (epicondylitis, compartment syndrome). The same observation may be made with respect to fatigue assessment, where SEMG methods are replaced by subjective fatigue assessment due to the additional complexity of evaluating the mean frequency shift of the SEMG power spectrum. Nevertheless, SEMG has an important role as a reference method also in these fields of study.

It is from the ergonomic practitioner's point of view, rather from amplitude variables in the time domain, that the unique contributions of SEMG are to be found. The temporal muscle activity patterns are considered to be the relevant dimension of study with respect to risk in occupational groups to acquire chronic muscular pain, and there is no alternative methodology available to study work–rest patterns at the individual muscle level.

In particular, chronic ailments of the myalgia type have been found to be prevalent in many occupational groups, where the exposure patterns are characterized by repetitive or sustained work, often at low voluntary effort. These ailments are most often localized to the shoulder–neck region. Veiersted and co-workers [96], in a prospective study of women employed in a chocolate factory, demonstrated that occurrence of so-called gaps, namely short incidents of absence of SEMG activity in the upper trapezius muscle during work, reduced the risk to contract trapezius myalgia. Although the specific mechanisms behind the pain development are not fully understood, a prevailing view is that sustained muscular activation may lead to physiological overuse. Hägg [23] formulated the so-called Cinderella theory, which hypothesizes that there are low-threshold motor units that are always recruited as soon as the muscle is activated, and stay active until total muscular relaxation. These motor units would face energy crisis at the membrane level, which may lead to degenerative processes, necrosis, and pain.

This theory has been subject to considerable physiological research, where motor unit recruitment and firing patterns have been studied under a variety of conditions, and the links between sustained muscular activation and pain development scrutinized (for a review, see Sjøgaard et al. [89]). These studies incorporate the dimension of psychosocial stress, since it has been shown that the same low-threshold motor units tend to be recruited in voluntary activation and psychological stress-induced muscular tension. There are still important areas of scientific debate, for instance, with respect to the role of possible protective mechanisms [100], and lack of evidence for cell damage in myalgic muscle [26]. Nevertheless, the bulk of the research carried out has entailed much support to the Cinderella hypothesis.

Practical implications of this research are plentiful. The temporal patterns and their relation to development of muscular pain of the myalgia type have been studied in particular with respect to computer operators, and have formed the basis for the EU project "PROCID" and the "Recommendations for Healthier Computer Work" [86]. This also means that SEMG may be used in ergonomic intervention research to identify adverse working situations, which could be followed up by a combination of technical and organizational measures. SEMG monitoring devices and methodologies are on their way to becoming important aids in ergonomic fieldwork.

13.9 PSYCHOLOGICAL EFFECTS ON EMG

Disease patterns have shifted dramatically since the beginning of the twentieth century. Morbidity and mortality rates have dropped for infectious diseases but have risen sharply for cardiovascular, immune, and musculoskeletal related disorders. This shift calls for a new approach to study illness and health. Historically illness has been viewed as a biomedical phenomenon caused by germs, injuries, or by some other internal malfunction. Today several serious illnesses, for example, cancer, hypertension, myocardial infarction, stroke, and diabetes, and different pain syndromes are known to be, at least partly, caused or enhanced by behavioral and psychosocial factors. In regard to musculoskeletal disorders the possible role of an interaction between a more traditional biomechanical approach and psychosocial factors has received an increasing scientific and clinical interest [72].

13.9.1 Definitions of Stress

Stress has been defined as the individual's cognitive assessment of the balance between situational demands and his/her coping resources as reflected in mood, activity level, health-related behavior, and psychobiological reactions [14]. Most psychobiological definitions of stress share the long-held notion of homeostasis, typically involving the maintenance of a constant interior milieu in the face of external and internal perturbations. In healthy organisms deviations from normal may occur in response to environmental demand, but then are quickly corrected in order to maintain a constant state.

The term "allostasis" was introduced in an effort to take into account the more "dynamic" conditions under which physiological systems typically operate [90]. Allostasis describes the operating range of healthy systems and their ability to increase or decrease vital functions to new levels in response to changing demands. Allostasis acknowledges that physiological variables are not maintained at a steady state, but that regulation is achieved through change. Extending this concept over the dimension of time, McEwen and Stellar [70] describe the state of allostatic load as potentially damaging effects of repeated strain and elevated activity levels on organs and tissues. Excessive allostatic load is the "hidden price" that the body pays for anticipatory and compensatory control of stressful events. Thus these new conceptualizations extend the notion of homeostasis to account for the dynamic quality of physiological regulation. McEwen [69] has proposed a model called the allostatic load model, which predicts under what conditions physiological stress responses are adaptive and when they may cause health problems.

13.9.2 Psychological and Physical Stress and the Total Workload on the Organism

Throughout our lives we are exposed to both psychological and physiological demands. Sometimes both types of demands interact such as when heavy physical demands in repetitive lifting of items contribute to increased psychological strain. In the working life many individuals are highly exposed to both psychological and physiological demands because of their work. Examples of work conditions associated with both types of stress are traditional assembly line work and work at checkout counters at supermarkets [56,57,73]. These jobs are characterized by conditions such as monotonous and repetitive tasks, machine- (or customer-) paced work, lack of influence and control over the content and conditions at work, physical restriction to the workplace, and low occupational status and income. Due to traditional sex roles, women might especially suffer from a combination of high demands from both their paid work and their unpaid duties performed at home such as household chores and childcare. This means that the total load on the organisms can be very high and last over very long periods of time.

13.9.3 Psychological Stress and Musculoskeletal Disorders

Today a large body of evidence shows that physical conditions alone cannot explain the development of musculoskeletal pain in the modern work environment. For example, despite considerable ergonomic improvements of the work environment during the last decades and attempts to eliminate as much heavy lifting and pushing as possible, the incidence of musculoskeletal disorders (MSD) has remained high and still constitutes a major health problem in the industrialized part of the world. MSD are common not only in physically demanding jobs but also in light physical work, such as data entry at video display, where only a very small fraction of the worker's physical capacity is used.

Both short- and long-lasting demands on the organism leads to several biological changes within the body. Within a biopsychosocial frame, these physiological changes are viewed as a part of a stress reaction. In psychophysiological studies related to work organizational factors, the neuroendocrine and the cardiovascular responses triggered by, for example, heavy workload, monotonous work, and low autonomy are the reactions most investigated. More recently the biopsychosocial stress approach has become highly important also in the functioning of the immune and the musculoskeletal systems. This means that new parameters have been added to the field of stress research. In regard to the musculoskeletal system, traditional stress measures referring to the cardiovascular system and the neuroendocrine system have been added with measures such as surface and/or intramuscular EMG (see [72] among other references).

13.9.4 Two Neuroendocrine Systems Sensitive to Psychological Stress

Two neuroendocrine systems are particularly sensitive to psychological stress: the sympathetic-adrenal medullary (SAM) system influencing blood pressure, heart rate, and the secretion of catecholamines, and the hypothalamic pituitary-adrenal cortical (HPA) system affecting the secretion of corticosteroids. Blood pressure, heart rate, catecholamines, and cortisol have been assessed in laboratory settings as well as in natural stressful situations [55]. These studies indicate that the SAM system mainly reflects the intensity of psychological stress and arousal, whereas the HPA system is more sensitive to the affective aspects of a situation.

Physiological measures may thus serve as objective indicators of work-related stress. However, psychophysiological responses to psychological stress are also assumed to link psychosocial stressors to increased health risks [68]. Psychological stress induced by a perceived threat to control, activates the SAM defense reaction: energy is mobilized as blood glucose and fat from the liver and blood is directed to the muscles. But elevated catecholamines also enhance blood clotting, and the risk of arterial obstruction.

Catecholamines and cortisol can be measured in several ways. A fairly constant fraction of epinephrine, norepinephrine, and cortisol in the blood is excreted into urine. Thus assessment can be made in both blood plasma and in urine. Cortisol can also be measured in saliva and correlations with plasma cortisol are high [43].

13.9.5 Is It Justified to Include EMG in the Field of Stress?

The answer to this question must be a clear yes. Consider the following example. Imagine yourself in a seated position, with movement under control and with your trapezius muscle connected to an EMG recorder. You are asked by an experiment leader to start counting from number 7000 and continuously add 10, during two minutes, as correctly as possible. Your trapezius will show almost no sign of activity, and neither will this simple task produce any cardiovascular or neuroendocrine responses. However, if the same procedure is kept but this time the experiment leader ask you to count from 7006 and continuously take away seven, an increase in neuroendocrine responses, for instance, and in the EMG activity of the trapezius muscle will be seen. In short, a psychological demand has triggered the activity of the trapezius muscle. This indicates that increased EMG activity can be regarded as a part of, or a parallel process to, a general stress reaction. Some empirical studies regarding mental stress and EMG activity are presented below.

13.9.6 Mental Stress Increases EMG Activity

Several experimental studies have shown that psychological demands such as mental stress or cognitive factors, even in the absence of physical demands, can increase muscle tension [58,71,92,97]. In order to investigate the effects of mental demands and physical load (separately and combined) on muscular tension, Lundberg et al. [58] studied 62 women in a laboratory experiment. Measures were obtained on perceived mental stress, physiological stress responses (catecholamines, cortisol, blood pressure, and heart rate), and muscular tension, as reflected in EMG activity of the trapezius muscle. Each subject was individually exposed to one stress session (60 min), and in order to control for the diurnal variation in physiological variables, a corresponding baseline session (60 min) on a separate day. During the stress period, the subjects were individually exposed to psychological stressors such as mental arithmetic and the Stroop color-word test (CWT), and a standardized physical load, namely a test contraction (subjects kept their arms straight and elevated to 45 degrees of abduction in the scapular plane, halfway between the frontal and sagittal planes). One of the stressors, the CWT, was also combined with the standardized physical load.

The results show statistically significant positive correlations between changes in EMG activity and corresponding measures in cardiovascular parameters, which is in accordance with findings by Helin et al. [19], for example, and thus indicate that increased EMG activity might be part of a general stress response. Lundberg et al. [58] also found that the combination of mental stress and physical load increased the surface electromyographic (SEMG) activity of the trapezius muscle more than when mental stress and physical load were administered separately (see Fig. 1 of [58]). This indicated a possible synergistic effect, which may be of importance in jobs characterized by low to moderate physical demands and a negative psychosocial milieu. However, there are also studies that that do not find any effects of psychological demands on EMG [13]. In this study the psychological stressor used might not have been stressful enough since no effects were reflected in the excretion of catecholamines. Experimental data regarding this issue are still inconsistent (e.g., [99,102]), but that may be due partly to methodological factors in the experiments carried out so far. For a more detailed review of these issues, see Waersted [98].

13.9.7 Is the Trapezius Muscle Special in Its Response to Psychological Stress?

Too little research has been performed to answer this question. However, McNulty et al. [71] used intramuscular EMG and showed that the trapezius muscle was active under an arithmetic test performance, whereas no EMG reactions were observed in surrounding

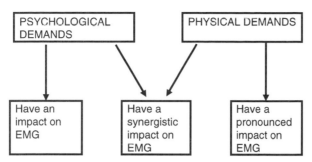

Figure 13.4. How psychological and physical demands, separately and in combination, effect the EMG pattern.

muscles. Furthermore Palmerud et al. [80] showed that the trapezius muscle reacted within an EMG based biofeedback procedure, whereas no similar pattern was recorded from other muscles. Waersted and colleagues demonstrated that the trapezius and facial muscles are more sensitive to stress than other skeletal muscles [101].

13.9.8 Psychological Factors and Possible Links to Musculoskeletal Tension

It has been proposed that psychosocial factors in and outside the workplace may induce psychological stress that increases the risk for musculoskeletal disorders via elevated muscle tension as reflected in EMG activity. The muscle tension itself may then constitute the link from psychological stress to musculoskeletal disorders, for example, via low-threshold motor units that are damaged due to long lasting activation described in the Cinderella hypothesis earlier in this chapter. However, psychological stress also causes breathing pattern changes, and a possible link between psychological stress, hyperventilation, and muscle tension is proposed by Schleifer and Ley [87]. Their view is that stress-induced hyperventilation decreases the peak CO_2 levels and increases the blood pH level. This contributes to elevated muscular tension and a suppression of parasympathetic activity. The sympathetic dominance may amplify the responses to catecholamines.

Psychology may also contribute in a different way to muscle tension. Stress levels may not only be elevated during work but also off the job. From a health perspective, the importance of being able to unwind or deactivate after work has been emphasized in earlier studies [14,94]. The speed at which a person unwinds after, for example, a day at work, reflects the "total load" on the organism. A quick return to physiological baseline indicates that the energy resources are demobilized when no longer needed. Monotonous and repetitive work seems to slow down the speed of unwinding [72]. Role conflicts between demands from paid work and unpaid duties (childcare, household chores) and work overload further contribute to reducing this process. Assuming that muscular tension is part of a general work-related stress reaction, psychosocial demands, during and after work, are relevant for activating physiological mechanisms that contribute to upper limb disorders. However, muscle tension in relation to a disturbed unwinding process after exposure to psychological stress has still to be further investigated.

13.9.9 Conclusions

Several experimental studies have confirmed that psychological stress or cognitive factors, even in the absence of physical demands, can increase muscle tension as reflected in surface or intramuscular EMG. The combination of mental stress and physical load seems to increase the EMG activity of the trapezius muscle more than what mental stress and physical load do when administrated separately. Only few studies have included traditional stress measures when investigating EMG activity, but in one study there were found significant positive correlations between blood pressure and SEMG activity and between mean norepinephrine excretion (urine reactivity level) and SEMG activity, indicating that the increased muscle tension (SEMG activity) might be part of a general stress response. However, experimental data regarding this issue are still inconsistent, but that may be due to methodological factors in the experiments carried out so far.

Assuming that muscular tension, as reflected in EMG, is part of a general work-related stress reaction, psychosocial demands, during and after work, are relevant for activating physiological mechanisms that contribute to sustained muscle tension. However, muscle tension in relation to a disturbed unwinding process after exposure to psychological stress has still to be further investigated.

REFERENCES

1. Arendt-Nielsen, L., R. K. Mills, and A. Forster "Changes in muscle fiber conduction velocity, mean power frequency, and mean EMG voltage during prolonged submaximal contractions," *Muscle Nerve* **12**, 493–497 (1989).
2. Asterland, P., G. Å. Hansson, and M. Kellerman "New data logger system for workload measurements—Based on PCMCIA memory cards," 25th Int Congr Occup Health, Stockholm, 1996, p. 273.
3. Bao, S., S. E. Mathiassen, and J. Winkel, "Ergonomic effects of a management-based rationalization in assembly work—A case study," *Appl Ergon* **27**, 89–99 (1996).
4. Basmajian, J. V., *Muscles alive: Their functions revealed by electromyography*, Williams and Wilkins, Baltimore, 1967.
5. Basmajian, J. V., *Biofeedback: Principles and practice for clinicians*, Williams and Wilkins, Baltimore, 1983.
6. Buckle, P., and J. Devereux, "Work related neck and upper limb musculoskeletal disorders: European Agency for Safety and Health at Work," Office for Official Publications of the European Communities, Luxembourg, 1999.
7. Byström, S., S. E. Mathiassen, and C. Fransson-Hall, "Physiological effects of micropauses in isometric handgrip exercise," *Eur J Appl Physiol* **63**, 405–411 (1991).
8. Carlsöö, S., "Nervous coordination and mechanical function of the mandibular elevators: An electromyographic study of the activity, and an anatomic analysis of the mechanics of the muscles," *Acta odont Scand* **10** (suppl), 1–132 (1952).
9. Chaffin, D. B., "Electromyography—A method of measuring local muscle fatigue," *J Meth Time Measurem* **14**, 29–36 (1969).
10. Chaffin, D., "Localized muscle fatigue-definition and measurement," *J Occ Med* **15**, 346–354 (1973).
11. De Luca, C. J., M. A. Sabbahi, and S. H. Roy, "Median frequency of the myoelectric signal," *Eur J Appl Physiol* **55**, 457–464 (1986).

12. de Vries, H. A., "Method for evaluation of muscle fatigue and endurance from electromyographic fatigue curves," *Am J Phys Med* **47**, 125–135 (1968).

13. Ekberg, K., B. Björkqvist, P. Malm, B. Bjerre-Kkiely, M. Karlssson, and O. Axelson, "Case-control study of risk factors for disease in neck and shoulder are ••," *Occup Environ Med* **51**, 262–266 (1994).

14. Frankenhaeuser, M., U. Lundberg, M. Fredrikson, B. Melin, M. Tuomisto, A. L. Myrsten, M. Hedman, B. Bergman-Losman, and L. Wallin, "Stress on and off the job as related to sex and occupational status in white-collar workers," *J Org Beh* **10**, 321–346 (1989).

15. Gerdle, B., N. E. Eriksson, and L. Brundin "The behaviour of the mean power frequency of the surface electromyogram in biceps brachii with increasing force during fatigue: With special regard to the electrode distance," *Electromyogr Clin Neurophysiol* **30**, 483–489 (1990).

16. Hagberg, C., and M. Hagberg "Surface EMG amplitude and frequency dependence on exerted force for the upper trapezius: A comparison between right and left sides," *Eur J Appl Physiol* **58**, 641–645 (1989).

17. Hagberg, M., and B. Ericson "Myoelectric power spectrum dependence on muscular contraction level of elbow flexors," *Eur J Appl Physiol* **42**, 147–156 (1982).

18. Hansson, G. Å., U. Strömberg, B. Larsson, K. Ohlsson, I. Balogh, and U. Moritz, "Electromyographic fatigue in neck/shoulder muscles and endurance in women," *Ergonomics* **35**, 1341–1352 (1992).

19. Helin, P., K. Kuoppasalmi, J. Laakso, and M. Härkönen, "Human urinary biogenic amines and some physiological responses during situation stress," *Int J Psychophysiol* **6**, 125–132 (1988).

20. Hermans, V., A. J. Spaepen, and M. Wouters, "Relation between differences in electromyographic adaptions during static contractions and the muscle function," *J Electromyogr Kinesiol* **9**, 253–261 (1999).

21. Hägg, G., "Electromyographic fatigue analysis based on the number of zero crossings," *Pflügers Arch* **391**, 78–80 (1981).

22. Hägg, G. M., "Comparison of different estimators of electromyographic spectral shifts during work when applied on short test contractions," *Med Biol Eng Comput* **29**, 511–516 (1991).

23. Hägg, G. M., "Static work load and occupational myalgia—A new explanation model," in P. Anderson, D. Hobart, and T. Danoff, eds., *Electromyographical kinesiology*, Elsevier Science Amsterdam, 1991.

24. Hägg, G. M., "Interpretation of EMG spectral alterations and alteration indexes at sustained contraction," *J Appl Physiol* **73**, 1211–1217 (1992).

25. Hägg, G. M., "Gap-training—A new approach and device for prevention of shoulder/neck myalgia by EMG feedback," in P. Seppälä, T. Luopajärvi, C. H. Nygård, and M. Mattila, eds., *13th Proc Triennial Congress of the International Ergonomics Association*, Tampere, 1997, pp. 283–285.

26. Hägg, G. M., "Muscle fibre abnormalities related to occupational load—A review," *Eur J Appl Physiol* **83**, 159–165 (2000).

27. Hägg, G. M., and A. Åström, "Load pattern and pressure pain threshold in the upper trapezius muscle and psychosocial factors in medical secretaries with and without shoulder/neck disorders," *Int Arch Occup Environ Health* **69**, 423–432 (1997).

28. Hägg, G. M., and R. Ojok, "Isotonic and isoelectric tests for the upper trapezius muscle," *Eur J Appl Physiol* **75**, 263–267 (1997).

29. Hägg, G. M., and J. Suurküla, "Zero crossing rate of electromyograms during occupational work and endurance test as predictors for work related myalgia in the shoulder/neck region," *Eur J Appl Physiol* **62**, 436–444 (1991).

30. Hägg, G. M., J. Suurküla, and M. Liew, "A worksite method for shoulder muscle fatigue measurements using EMG, test contractions and zero crossing technique," *Ergonomics* **30**, 1541–1551 (1987).

31. Hägg, G. M., J. Öster, and S. Byström, "Forearm muscular load and wrist angle among automobile assembly line workers in relation to symptoms," *Appl Erg* **28**, 41–47 (1997).

32. Inman, V. T., J. B. Saunders, and L. C. Abbott, "Observations on the function of the shoulder joint," *J Jt Bone Surg* **26**, 1–30 (1944).

33. Inman, V. T., H. J. Ralston, J. B. Saunders, B. Feinstein, and E. W. Wright, "Relation of human electromyogram to muscular tension," *Electroenceph Clin Neurophysiol* **4**, 187–194 (1952).

34. Jonsson, B., "Quantitative electromyographic evaluation of muscular load during work," *Scand J Rehab Med* **6** (suppl), 69–74 (1978).

35. Jonsson, B., "Measurement and evaluation of local muscular strain on the shoulder during constrained work," *J Human Ergol* **11**, 73–88 (1982).

36. Jonsson, S., and B. Jonsson, "Function of the muscles of the upper limb in car driving," *Ergonomics* **18**, 375–388 (1976).

37. Joseph, J., and A. Nightingale, "Electromyography of muscles of posture: Thigh muscles in males," *J Physiol (Lond)*, **117**, 484–491 (1952).

38. Jørgensen, K., N. Fallentin, C. Krogh-Lund, and B. Jensen "Electromyography and fatigue during prolonged, low-level static contractions," *Eur J Appl Physiol* **57**, 316–321 (1988).

39. Kadefors, R., "Application of electromyography in ergonomics: New vistas," *Scand J Rehab Med* **10**, 127–133 (1978).

40. Kadefors, R., E. Kaiser, and I. Petersén, "Dynamic spectrum analysis with special reference to muscle fatigue," *Electromyography* **8**, 38–74 (1968).

41. Kadefors, R., I. Petersén, and P. Herberts, "Muscular reaction to welding work: An electromyographic investigation." *Ergonomics* **19**, 543–558 (1976).

42. Kadefors, R., Å Kilbom, and L. Sperling, "An ergonomic model for workplace assessment," *Proc. IEA'94*, Vol. 5, International Ergonomics Association, Toronto, Canada, 1994, pp. 210–212.

43. Kirschbaum, C., and D. H. Hellhammer, "Salivary cortisol in psychobiological research: An overview," *Neuropsychobiology* **22**, 150–169 (1989).

44. Kogi, K., and T. Hakamada, "Slowing of surface electromyogram and muscle strength in muscle fatigue," *Rep Inst Sci Lab* **60**, 27–41 (1962).

45. Kroon, G. W., and M. Naeije, "Recovery following exhaustive dynamic exercise in the human biceps muscle," *Eur J Appl Physiol* **58**, 228–232 (1988).

46. Kroon, G. W., and M. Naeije, "Recovery of the human biceps electromyogram after heavy eccentric, concentric or isometric exercise," *Eur J Appl Physiol* **63**, 444–448 (1991).

47. Kwatny, E., D. H. Thomas, and H. G. Kwatny, "An application of signal processing techniques to the study of myoelectric signals," *IEEE Trans BME* **17**, 303–312 (1970).

48. Kumar, S., "A conceptual model of overexertion, safety, and risk of injury in occupational settings," *Human Factors* **36**, 197–209 (1994).

49. Kuorinka, I., and L. Forcier, *Work related musculoskeletal disorders (WMSDs)*, Taylor and Francis, London, 1995.

50. Laurig, W., "Methodological and physiological aspects of electromyographic investigations," in P. Komi, ed., *5th International Congress on Biomechanics*, 1976, pp. 219–230.

51. Laurig, W., A. Luttmann, and M. Jäger, "Evaluation of strain in shop-floor situations by means of electromyographic investigations," in S. S. Asfour, ed., *Trends in ergonomics*, Elsevier Science, Amsterdam, 1987, pp. 685–692.

52. Lindström, L., R. Kadefors, and I. Petersén, "An electromyographic index for localized muscle fatigue," *J Appl Physiol* **43**, 750–754 (1977).

53. Lippold, O. C. J., "The relation between integrated action potentials in a human muscle and its isometric tension," *J Physiol (Lond)* **117**, 492–499 (1952).

54. Lippold, O. C. J., J. W. T. Redfearn, and J. Vucô, "The electromyography of fatigue," *Ergonomics* **3**, 121–131 (1960).

55. Lundberg, U., U. Hansson, Eneroth., M. Frankenhaeuser, and K. Hagenfeldt, "Anti-androgen treatment of hirusute women: A study of stress responses," *J Psychosom Obstet Gynaecol* **3**, 79–92 (1984).

56. Lundberg, U., M. Granqvist, T. Hansson, M. Magnusson, and L. Wallin, "Psychological and physiological stress responses during repetitive work at an assembly line," *Work Stress* **3**, 143–153 (1989).

57. Lundberg, U., I. Elfsberg Dohns, B. Melin, L. Sandsjö, G. Palmerud, R. Kadefors, M. Ekström, and D. Parr, "Psychophysiological stress responses, muscle tension and neck and shoulder pain among supermarket cashiers," *J Occup Health Psychol* **4**, 1–11 (1999).

58. Lundberg, U., R. Kadefors, B. Melin, G. Palmerud, P. Hassmén, M. Engström, and I. Elfsberg Dohns, "Psychophysiological stress and EMG activity of the trapezius muscle," *Int J Beh Med* **1**, 354–370 (1994).

59. Lundervold, A. J. S, "Electromyographic investigation of position and manner of working in typewriting," *Acta Physiol Scand* **84** (Suppl), •• (1951).

60. Lundervold, A. J. S, "Electromyographic investigations during sedentary work, especially typewriting," *Brit J Phys Med* **14**, 32–36 (1951).

61. Luttmann, A., M. Jäger, and W. Laurig, "Electromyographical indication of muscular fatigue in occupational field studies," *Int J Ind Erg* **25**, 645–660 (2000).

62. Luttmann, A., M. Jäger, J. Sökeland, and W. Laurig, "Electromyographical study on surgeons in urology. Part II: Determination of muscular fatigue," *Ergonomics* **39**, 298–313 (1996).

63. Luttmann, A., J. Sökeland, and W. Laurig, "Electromyographical study on surgeons in urology. Part I: Influence of the operating technique on muscular strain," *Ergonomics* **39**, 285–297 (1996).

64. Mathiassen, S., "The statistical confidence of load estimates based on ramp calibration of upper trapezius EMG," *J Electromyogr Kinesiol* **6**, 59–65 (1996).

65. Mathiassen, S. E., and J. Winkel, "Quantifying variation in physical load using exposure-vs-time data," *Ergonomics* **34**, 1455–1468 (1991).

66. Mathiassen, S. E., and J. Winkel, "Physiological comparison of three interventions in light assembly work: Reduced work pace, increased break allowance and shortened working days," *Int Arch Occup Environ Health* **68**, 94–108 (1996).

67. Mathiassen, S. E., J. Winkel, and G. M. Hägg, "Normalization of surface EMG amplitude from upper trapezius muscle in ergonomic studies—A review," *J Electromyogr Kinesiol* **5**, 197–226 (1995).

68. McCabe, P. M., N. Schneiderman, T. M. Field, and J. S. Skyler, "Stress, coping and disease," Lawrence Erlbaum, Hillsdale, NJ.

69. McEwen, B. S., "Stress, adaptation and disease: Allostasis and allostatic load," *N Eng J Med* **840**, 33–44 (1998).

70. McEwen, B. S., and E. Stellar, "Stress and the individual," *Arch Intern Med* **153**, 2093–2101 (1993).

71. McNulty, W. H., R. N. Gevirtz, D. R. Hubbard, and G. M. Berkoff, "Needle electromyographic evaluation of trigger point response to a psychological stressor," *Psychophysiology* **31**, 313–316 (1994).

72. Melin, B., and U. Lundberg, "A biopsychosocial approach to work-stress and musculoskeletal disorder," *J Psychophysiol* **11**, 238–247 (1997).

73. Melin, B., U. Lundberg, J. Söderlund, and M. Granqvist, "Psychophysiological stress reactions of male and female assembly workers: A comparison between two different forms of work organizations," *J Org Beh* **20**, 47–61 (1999).

74. Merletti, R., and S. Roy, "Myoelectric and mechanical manifestations of muscle fatigue in voluntary contractions," *J Occup Sport Phys Ther* **24**, 342–353 (1996).

75. Moritani, T., and M. Muro "Motor unit activity and surface electromyogram power spectrum during increasing force of contraction," *Eur J Appl Physiol* **56**, 260–265 (1987).

76. Moritani, T., M. Muro, and A. Nagata "Intramuscular and surface electromyogram changes during muscle fatigue," *Eur J Appl Physiol* **60**, 1179–1185 (1986).

77. Nieminen, H., and S. Hämeenoja, "Quantification of the static load component in muscle work using nonlinear filtering of surface EMG," *Ergonomics* **38**, 1172–1183 (1995).

78. NIOSH, *Musculoskeletal disorders and workplace factors: A critical review of epidemiologic evidence for work-related musculoskeletal disorders of the neck, upper extremity, and low back,* U. S. Department of Health and Human Services, National Institute for Occupational Health and Safety, Cincinnati, OH, 1997.

79. Örtengren, R., B. G. J. Andersson, H. Broman, R. Magnusson, and I. Petersén, "Vocational Electromyography: studies of localized muscle fatigue at the assembly line," *Ergonomics* **18**, 157–174 (1975)

80. Palmerud, G., R. Kadefors, H. Sporrong, U. Järvholm, P. Herberts, C. Hogfors, and B. Peterson, "Voluntary redistribution of muscle activity in human shoulder muscles," *Ergonomics* **38**, 806–815 (1995).

81. Parenmark, G., B. Engvall, and A. K. Malmkvist, "Ergonomic on-the-job training of assembly workers," *Appl Erg* **19**, 143–146 (1988).

82. Piper, H., *Elektrophysiologie Menschlicher Muskeln*, Springer-Verlag, Berlin, 1912.

83. Pozzo, M., and R. Ferrabone, "A portable multichannel EMG acquisition system," *Proc Int Soc for Electrophysiology and Kinesiology*, Vienna, 2002, pp. 264–265.

84. Putz-Anderson, V., *A textbook of cumulative trauma disorders*, Taylor and Francis, London, 1988.

85. Rohmert, W., "Die Beziehung zwischen Kraft und Ausdauer bei Statischer Muskelarbeit," in *Schriftenreihe Arbeitsmedizin, Sozialmedizin, Arbeitshygiene,* Gentner Verlag, Stuttgart, 1968.

86. Sandsjö, L., D. Zennaro, and R. Kadefors, and the PROCID Group, PROCID recommendations for healthier computer work. *Proc HCI Int 2001, 9th Int Conf Human-Computer Interaction*, New Orleans, LA, 2001, pp. 295–297.

87. Schleifer, L. M., and R. Ley, "End-tidal PCO2 as an index of psychophysiological activity during VDT data-entry work and relaxation," *Ergonomics* **37**, 245–254 (1994).

88. Silverstein, B., L. J. Fine, and T. J. Armstrong, "Hand wrist cumulative trauma disorders in industry," *Br J Ind Med* **43**, 779–784 (1986).

89. Sjøgaard, G., U. Lundberg, and R. Kadefors, "The role of muscle activity and mental load in the development of pain and degenerative processes at the muscle level during computer work," *Appl Physiol* **83**, 99–105 (2000).

90. Sterling, P., and J. Eyer, "Allostasis: A new paradigm to explain arousal pathology," in S. Fisher, and J. Reason, eds., *Handbook of life stress*, Wiley, New York, 1988.

91. Suurküla, J., and G. M. Hägg, "Relations between shoulder/neck disorders and EMG zero crossing shifts in female assembly workers using the test contraction method," *Ergonomics* **30**, 1553–1564 (1987).

92. Svebak, S., R. Anjia, and S. I. Kårstad, "Task-induced electromyographic activation in fibromyalgia subjects and controls," *Scand J Rheumatol* **22**, 124–130 (1993).

93. Tanaka, S., and J. D. McGlothlin, "A conceptual model for prevention of work-related carpal tunnel syndrome (CTS)," *Int J Industr Erg* **11**, 181–193 (1993).

94. Theorell, T., R. A. Karasek, and P. Eneroth, "Job strain variation in relation to plasma testosterone fluctuation in working men—A longiudinal study," *J Intern Med* **227**, 31–36 (1990).

95. Tichauer, E. R., "Electromyographic kinesiology in the analysis of work situations and hand tools," *Electromyography* **1** (suppl), 197–211 (1968).

96. Veiersted, B., R. H. Westgaard, and P. Andersen, "Electromyographical evaluation of muscular work pattern as a predictor of trapezius myalgia," *Scand J Work Environ Health* **19**, 284–290 (1993).

97. Wærsted, M., "Attention-related muscle activity—A contributor to stustained occupational muscle load," PhD thesis, National Institute of Occupational Health, Oslo, Norway, 1997.

98. Wærsted, M., "Human muscle activity related to muscular load," *Eur J Appl Physiol* **83**, 151–158 (2000).

99. Westgaard, R. H., and R. Björklund, "Generation of muscle tension additional to postural muscle load," *Ergonomics* **39**, 911–923 (1987).

100. Westgaard, R., and C. J. De Luca, "Motor control of low-threshold motor units in the human trapezius muscle," *J Neurophysiol* **85**, 1777–1781 (2001).

101. Wærsted, M., and R. H. Westgaard, "Attention-related muscle activity in different body regions during VDU work with minimal physical activity," *Ergonomics* **39**, 661–676 (1996).

102. Wærsted, M., R. Björklund, and M. Westgaard, "Shoulder muscle tension induced by two VDU-based tasks of different complexity," *Ergonomics* **34**, 137–150 (1991).

103. Winkel, J., and S. E. Mathiassen, "Assessment of physical work load in epidemiologic studies: Concepts, issues and operational conciderations," *Ergonomics* **37**, 979–988 (1994).

14

APPLICATIONS IN EXERCISE PHYSIOLOGY

F. Felici

Exercise Physiology Laboratory
Department of Human Movement and Sport Sciences
University Institute of Motor Sciences (IUSM)
Faculty of Motor Sciences, Roma, Italy

14.1 INTRODUCTION

Motor performance is the most qualifying activity of human beings. It is the product of the coordinated action of all body systems and organs. The complexity of voluntary movements has always challenged—and still challenges—researchers. It was immediately evident from the more recent history of the studies on movement that an interdisciplinary approach was needed, now commonly known by the term bioengineering. The many interests involved had produced a variety of competences and applications, as are well documented in the contents of this book. In this context, it is very difficult, if not impossible, to limit the description of human movement to one particular aspect. Despite this objective limitation, the discussion in this chapter will deal with a very confined topic: the use of surface electromyography (sEMG) in the study of various motor performances.

The extraction of information from the electrical signal generated by the activated muscle has always been regarded as an easy way to gain access to hardly accessible activity of motor control centers. This can be achieved invasively, by wires or needles inserted directly into the muscle(s), or noninvasively, by recording electrodes placed over the skin surface overlying the investigated muscles (sEMG). The use of this latter modality is preferable in healthy voluntary sedentary subjects and in athletes, despite its limitations and drawbacks. To mention just few of them, single-channel sEMG signals provide an average information on the activity of many concurrently active motor units (MU), the

Electromyography: Physiology, Engineering, and Noninvasive Applications, edited by Roberto Merletti and Philip Parker.
ISBN 0-471-67580-6   Copyright © 2004 Institute for Electrical and Electronics Engineers, Inc.

reproducibility of the results is often difficult, and standard recording procedures are still confined to few laboratories, therefore limiting comparisons among results obtained by clinical researchers. Finally, as Merletti and Rix [34] recently pointed out, the information that can be extracted from sEMG through complex algorithms is often difficult to interpret. However, there is no doubt that sEMG studies have provided relevant contributions to the understanding of human movement as is witnessed by the enormous number of papers in which sEMG data are discussed (300 papers listed in the review from Clarys and Cabri [11]).

A list of the principal areas of research follows: description of normal muscle function during selected movements and postures, investigation of muscle activity in complex sports, validation of classical anatomical studies, coordination and synchronization studies, fatigue and force/EMG relationship studies, and so on. Thus in the next sections, after a short comment on sEMG technical aspects that should be taken care of to improve the quality of sEMG recordings during dynamic exercise, I will expand on some examples of sEMG applications in exercise physiology limiting to the description of results obtained from studies where only strictly noninvasive techniques (or of very limited invasivity) were applied.

14.2 A FEW "TIPS AND TRICKS"

Issues related to signal detection and conditioning during isometric exercise are discussed in detail in Chapter 5. However, when trying to collect reliable sEMG data during non-isometric exercise (NIE), the reliability of collected data depends on a number of further aspects. Having in mind the final remark of a recent paper from Clarys [10], "Twenty years ago the first warnings were given about use and misuse of sEMG in physical education and the misuse still exists. These warnings stay valid today and apply to all movement studies, in both sport and occupational context," it would be worth to recall a number of warnings.

A major source of artifacts (unwanted signals) is represented, in NIE, by the movements of the electrodes and cables. These movements generate signals, often called motion artifacts, that are easily recognizable being ample and slow. In some cases, however, they may assume the aspect of "spikes," usually provoked by brisk impacts with the ground as during hopping or running and walking at high speed. In the first instance, motion artifacts can be greatly reduced, or even eliminated, by carefully fixing all cables over the subject. Unwanted spikes can be eliminated off line, keeping in mind that some "good" sEMG segments will be lost too.

Another important issue is represented by the maintenance of a good contact between electrodes and skin throughout the whole experiment duration. To prevent sudden detachments or change of contact quality of the electrodes during movement, it is convenient to prepare the subject in fully extended position of the joints that the cables have to cross over (typically the ankle, knee, and elbow joints) leaving an extra length of cable to prevent "pulling" on the electrodes.

During NIE, the problem of sweating is very often underestimated. Sweating causes modification of electrode–skin impedance and reduces contact stability. Many different tapes are commercially available to prevent electrodes detachment due to sweating. To limit the rate of sweating, care must be taken in keeping the laboratory environment in a proper condition of relative humidity and temperature.

Provided that we collected reliable sEMG data, we may pose the problem of the repeatability of measurement. Without going into much detail, it is clear that repeatability concerns a number of aspects common to static and dynamic contractions: intra- and inter-subject variability; intra- and inter-experiments variability. Most problems arise from proper localization of the site of application of the electrodes on the skin with respect to the innervation zone (IZ), electrodes repositioning when measurements are performed in different experimental sessions, and from cumulative fatigue due to exercise repetition in the same experimental session. The optimal place where to locate the electrodes is between the IZ and the tendon most distal from it. However, this region may be not large enough in all muscles to accommodate the electrodes.

It must be considered that the IZ moves in accordance with the level of contraction and the complexity and speed of movement. In many sports, characterized by complex skills as well as in many occupational settings, muscles shorten and surface electrodes can explore, at the end of movement, a completely different muscle region, if not the IZ or a tendon [16]. Clarys [10] refers of a simple method to place the electrodes in such circumstances: over the visual midpoint of the fully "contracted" muscle. The issue of relative movements between electrodes and muscle during dynamic exercise is still not satisfactorily addressed. An approach that can be valuable to exploit could be the use of electrode arrays and "smart" spatial filtering (as described in Chapters 5 and 7).

Within the general framework of exercise physiology, it is possible to make a gross distinction between two main lines of sEMG application to the study of human movement and postures: (1) the nonisometric exercise and (2) the isometric exercise. As we already have outlined, in the first case sEMG recording provides information about muscle activation patterns (i.e., timing of relative activation), amplitude of myoelectric signal, and, in short, a rough indication about muscle coordination during movement. Isometric contraction represents an attempt to reduce the confounding factors and variables to be controlled and, generally speaking, has a very poor relationship with the real life muscle action. This case represents a sort of "simplified" version of movement (no kinematic changes though with some initial muscle geometry changes), which allow the experimenter to overcome conceptual and practical problems related to the analysis of sEMG data collected during variable muscle length exercises. This is not obtained at low cost: the price to be paid to the absence (for practical purposes) of movement is the lack of representativeness of the ordinary neural muscle control strategy.

Muscle geometry deserves a further comment. Muscle geometry is different for different muscles, and this has many implications. First of all, it is quite well established that at least two conditions are needed to have stable and repeatable sEMG recording during movement: electrodes must be arranged with their joining axes running parallel to the muscle fibres and the latter can be modeled as acting along a line. This last assumption may be considered valid only in the case of unipennate muscles. In most instances neither one of these conditions is met. It is not true that that during isometric contractions muscle does not move: its geometry changes in various ways, the effects of this change being neglected by many researchers.

The general aim of the so-called kinesiological EMG [29] is the analysis of the function and coordination of muscles in different movements and postures in healthy subjects as well as in the disabled, in sedentary as well as in athletes, under laboratory conditions as well as in the field. In this context a further important aspect is raised by dynamic exercise. In other words, while measuring something in a laboratory environment, it is customary to devise a simplified version, a model, of the phenomenon under study. The ideal situation, however, should be represented by experiments in which the test exercise is the "actual" exercise, namely the experimental test exactly reproduces what happens outside the lab.

An alternative approach is represented by laboratory simulation of the exercise under investigation, as close to the real exercise as possible. This has been successfully exploited for very common exercises, such as walking, race walking, running, cycling, and rowing among many others. In these cases laboratory simulation can be conveniently adopted. As far as sEMG data recording devices are concerned, field measurements, and possibly laboratory testing, are made by the use of portable instruments. This can be achieved by adopting different solutions, such as teletransmission of data, storage of data aboard the portable sEMG recording unit, and off-line data exchange with a remote data analysis unit.

14.3 TIME AND FREQUENCY DOMAIN ANALYSIS OF sEMG: WHAT ARE WE LOOKING FOR?

The aim of this section is not to provide a mathematical description of the algorithms used to describe the sEMG, which are discussed in detail in other chapters of this book. Rather, a brief commentary on the various application of sEMG analysis to the study of exercise will be presented.

Common sEMG time domain data processing techniques are represented by the so-called integrated EMG (IEMG), linear envelope, mean rectified EMG, along with time synchronization tools and data normalization procedures. If the aim is the determination of the timing of muscle(s) activation, the fact that contraction is isometric or nonisometric is not so relevant. However, it is important to record from the muscle of interest (which is not as obvious as we can think!) and ensure that signal is not contaminated by a nearby muscle's electrical activity. This phenomenon is known as crosstalk. Recommendations for correct electrode placement over the intended muscle have been recently provided by SENIAM concerted action [27]. Provided that no crosstalk is present, then all of the above-mentioned techniques of time domain signal processing are fairly convenient. Co-activation of other muscles participating in the motor action should also be monitored. In NIE this has an evident relevance in terms of accuracy of muscle intervention timing and does not imply particular consideration other than the warnings just recalled. In isometric contractions co-activation, apart from being a matter of study per se, might represent a major source of errors.

The linear envelope is a representation of the rectified and low-pass filtered signal [10]. Linear envelope can be more or less smoothed using low-pass filters with different properties; this is convenient especially when, as in cyclical motor tasks such as walking, running, and so forth, a grand average of many sEMG cycles of a given muscle is wanted (see Fig. 14.1). In this cases additional smoothing is introduced by the averaging process, and on-line low-pass filtering may be less severe.

A final remark concerns sEMG amplitude normalisation procedures. To date there is no agreement on the procedure to be adopted (see for reference the recent paper from Burden and Bartlett [4]. This is of critical importance in order to compare results from different experiments or even to compare results obtained from different tests performed on the same subject. In gait analysis it is customary to represent sEMG data collected from NIE as % of cycle (e.g., see Fig. 14.1). Obviously this implies recording of some events or variables versus time duration of the cycle, such as stride phases, pedaling phases, and ground reaction forces. Data normalization is important also when dealing with IE, although it is, to some extent, less relevant than in NIE.

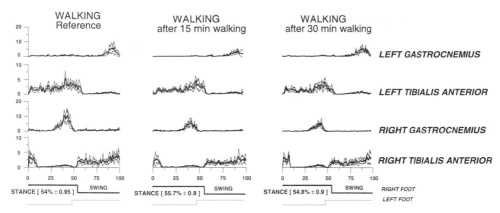

Figure 14.1. Averaged rectified and smoothed sEMG signals obtained from leg muscles during walking at 5 km h^{-1} for 30 minutes. Ten consecutive steps were used for average computation. Thin lines are the standard deviation. Time is expressed as percentage of stride cycle. Percentage duration of the stance phase is also indicated for the right foot.

The amount of scientific work done on the frequency domain analysis of sEMG is enormous. It can be said that all of the theoretical and practical implications have been faced at least once. However, once again, no general agreement exists, to the point that each research group seems to have its own "frequency recipe." Only in recent years a European concerted action attempted to correct this situation [27].

It is not a new idea that a time series, such as the sEMG, can be described in terms of its harmonic components. However, application of the most commonly used algorithms for the analysis of the sEMG in the frequency domain will mask time varying parameters, if present. Appropriate time-frequency methods are required in this case (see Chapter 10). It is evident that during dynamic contractions, muscles activate in burst-like fashion, making the sEMG signal highly nonstationary. Burst duration also plays a critical role, being extremely variable and spanning from few milliseconds of ballistic (very swift, brisk) contraction to several hundred milliseconds in slower contractions. It is clear that passing, for example, from normal walking to race walking and running will imply a reduction of burst duration as the speed of progression increases. This obviously complicates enormously the quantitative analytical approach to the sEMG and limits its clinical diffusion outside the research laboratories.

Also the relative movements intervening between recording electrodes and the explored muscle region are of importance for frequency domain analysis (what is generating what?), and this, again, questions the frequency domain transformation of sEMG data recorded during dynamic contractions. According to Merletti and Lo Conte [35], during isometric contractions the sEMG signal can be considered quasi-stationary over short time intervals (0.5–2 s), but this condition is not met during NIE. Movements of muscle under the electrodes, introduces a number of confounding factors. Many different alternative analytical approaches have been proposed (e.g., the Choi-Williams transformation), but their application to kinesiological studies is still quite limited [2].

14.4 APPLICATION OF sEMG TO THE STUDY OF EXERCISE

14.4.1 Walking versus Race Walking and Running

Human deambulation, in its various forms, is used in this context as a paradigmatic example of application of sEMG to the study of dynamic exercise. Many other forms of exercise have been explored with sEMG. A complete review of all these studies would expand well over the limits of this chapter (see Chapter 15).

Walking involves a series of coordinated movements of the body segments, implying an interplay of muscular forces and external forces (inertial, gravitational, and reaction forces) in order to achieve locomotion of the body [6]. The importance of having a complete and precise description of human walking is evident: this knowledge provided significant contributions in various fields: from rehabilitation to exercise. The results of the biomechanical data analysis, including sEMG data in the time domain, of normal walking on flat terrain can be summarized as follows: (1) there is no antagonistic muscular activity during the whole stride (except for the heel strike phase), (2) biarticular muscles are active only if this action is consistent with the moments required at the joints on which they act, and (3) muscles are activated on lengthening (i.e., eccentric action eliciting reflex responses). However, as already pointed out, gait analysis data, to be used at their best, should be organized according to some standards. As pointed out by Davis et al. [12], to facilitate the systematic interpretation of sEMG gait data, stride-to-stride variability needs to be assessed before any particular stride is considered representative of subject's performance. For sEMG data from multiple (possibly consecutive) strides are needed. Averaging multiple data will provide linear envelopes or ensemble averages of sEMG data (see Chapter 15). These can then be used to identify gait deviation or changes intervening because of fatigue (see Fig. 14.1), a change in speed of progression (see Fig. 14.2), or walking style (see Fig. 14.3). It must be stressed once again that sEMG data alone are not enough, in the majority of cases, to obtain a complete and meaningful picture. Pertinent

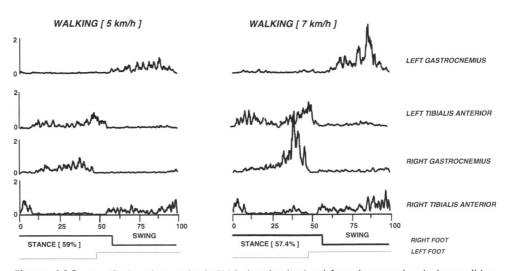

Figure 14.2. Rectified and smoothed sEMG signals obtained from leg muscles during walking at 5 and 7 km h⁻¹. A representative stride is reported. The increase of sEMG amplitude, passing from low to high speed of progression, is evident.

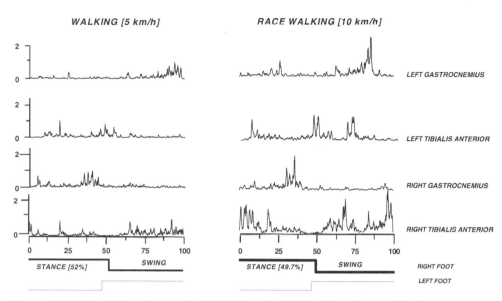

Figure 14.3. Rectified and smoothed sEMG signals obtained from leg muscles during walking at 5 km h^{-1} and race walking at 10 km h^{-1}. A representative stride is reported. It is evident a reduction of the stance phase during race walking along with a reduction of double support phase.

temporal parameters that should be included are walking velocity, cadence, stride time, step and stride length and duration, and double support and single support intervals. From the point of view of both dynamics and kinematics, race walking can be regarded as an intermediate and somewhat artificial form of walking between normal gait and running.

14.4.2 Gait Analysis Results

In previous studies [33] it was shown that there is no evident discontinuity in passing from ambulation to race walking in terms of either kinematics or energetic. Nevertheless, the knowledge of the sEMG profile during race walking may not be as obvious considering the particular rules imposed to athletes. Namely the swinging leg must be kept straight until its heel strike, and the double support phase should be maintained. As shown in Figure 14.1, sEMG can be used to evaluate the response of selected muscles to fatigue; the figure shows EMG of the tibialis anterior, before and after 15 and 30 minutes of walking at 7 kmh^{-1} on a motor driven treadmill in a normal (nonathlete) subject. A reduction in sEMG amplitude of selected muscles, particularly of the right gastrocnemious is evident after 15 minutes of exercise. On the other hand, since the treadmill speed is not changed, the absolute stride duration remains unchanged.

Different results are obtained when passing from low to high speed of walking (Fig. 14.2) in a sedentary subject (same subject as in Fig. 14.1). The pattern of muscle activation in passing from 5 to 7 kmh^{-1} in control subjects presented the expected increase in amplitude, the peak of the right gastrocnemious sEMG activity was delayed and the percentage of the single stance duration decreased by 2%. During race walking (Fig. 14.3), in addition to the sEMG amplitude increase, an earlier activation of the gastrocnemius muscle is evident on both sides. The major sEMG peak of this muscle was anticipated by almost 20% with respect to walking, while during walking the timing of this muscle was comparable in athletes and normal subjects.

The same methodological approach can be used for the description of lower limb selected muscles activation pattern during running [22]. While running, the sEMG pattern of gastrocnemius and tibialis anterior muscles are quite different from those observed during walking. The maximum activity of gastrocnemius is observed around the very last part of the stance phase, in the range between 90% to 95% of this phase, while Tibialis anterior is active mostly at the end of the swing phase. Apart from the mere description of the activation phases of single muscles during a given motor task, this comparison provides evidence of different motor control strategies in the three locomotion modalities studied.

Electromyography provided interesting results also for the characterization of muscle adaptation to different running specialization. Sadoyama and co-workers [39], studying the vastus lateralis muscle in long-distance runners and sprinters, found a positive correlation between muscle-fiber action potential conduction velocity (CV) and the percentage of FT fiber areas (see Fig 9.10). This means that muscles with a higher percentage of FT fiber area had higher CVs with respect to muscles with a lower percentage of FT fiber area. From this work emerges that power athletes (sprinters) have higher CVs than resistance athletes (long-distance runners). Most important this work provided convincing evidence of the feasibility of sEMG as a noninvasive tool for the characterization of muscle. An important and still open question arises from this and similar studies: Namely can training modify the amount of a specific type of muscle fiber in the target muscle or muscle group? Experimental evidence [30] seems to indicate that in monozygous twins there is an essentially identical representation of muscle-fiber types and that sEMG modification after training was correlated with muscle-fiber type composition. The authors thus concluded that there is a significant influence of genetic factors in determining muscle-fiber type composition in humans that is not significantly modified by training. It must be stressed, however, that sEMG ability in sensing specific muscle adaptation to training is still under study.

14.5 STRENGTH AND POWER TRAINING

Physiological mechanisms of adaptation to strength and power training include a variety of aspects, from morphological to functional and have been recently reviewed by Kraemer, Fleck, and Evans [31]. Electromyography has been largely used to study the adaptations of the neuromuscular system to heavy resistance training and to examine the adaptational time course for these changes. Many theoretical paradigms, summarized in [31], have been proposed to interpret the various adaptations of the neuromuscular system to training exercises. It is evident that although sEMG investigates only and indirectly the lowest part, namely the α-motoneuron level, of the complex motor command system, sEMG studies have provided the most direct evidence of neural adaptation to training [40]. However, considering the many factors that affect sEMG signal and their interactions [14], inferences on neural motor control modifications due to a specific training protocol can be made only with extreme care. This is reinforced by the fact that strength/power training protocols can be differently configured and can be described in terms of five (acute) program variables [44]: (1) type of exercise, (2) order of the exercise, (3) resistance or intensity adopted, (4) number of repetitions per set, and (5) rest period duration between sets. Because of the many possible configurations of training protocols, muscle adaptations are thought to be exercise-specific. The specificity of the physiological muscle adaptation induced would then be dependent on the specific modality of neuromuscular activation elicited by the given exercise [23]. It must be kept in mind that adaptation starts with the first exercise session (in untrained, usually students, subjects) and probably follows a time course that is specific to the individual and the type of exercise protocol adopted.

It is generally thought that the aim of force training is the improvement of the MVC level. However, in many sport activities, trainers focus their programs on the improvement of muscle power, which is dependent on the interplay of both muscle-shortening velocity and force. Generally speaking, pushing the training protocol toward the extreme values of these two variables, results in detrimental effects on the athletic performance. In other words, when the desired effect of training is the improvement of muscle power, training programs for athletes are designed in order to have a muscle that is neither too "fast" nor too "strong." In obedience to the "size principle" [26] MUs activation seems to be controlled by a relationship between MUs twitch forces and their individual recruitment thresholds and firing rates [14], which result in a continuum of force in the agonist muscle. The key point for the voluntary achievement of a true MVC is thus that all the motor units are recruited and all are firing at their maximum firing rates. When sedentary, or non-accustomed, subjects undergo strength training, it is speculated that force gain achieved during the early phases of training protocol (within the first 4 weeks of training, i.e., without apparent muscle hypertrophy) is due to an improvement in subject's ability in recruiting high-threshold MUs [37]. Although some concerns about this conclusion still exists [5], favorable evidence emerges from studies on healthy young [1] and older subjects [32].

Experienced weight-lifting athletes represent a good model for the study of neuromuscular adaptation occurring when a subject is seriously trained to strength (as opposed to laboratory training protocols). Some research partially attributed the enhancement in strength of weight-trained subjects to an increase in MUs recruitment and firing rate of motor units [24,25]. Furthermore structural adaptations of muscle fibers can imply a change in the conduction velocity properties of the fiber membrane [24,25]. A greater MU synchronization in weight-lifting athletes with respect to untrained persons has been noted since the work of Milner-Brown et al. [36]. Besides the enhanced MU synchronization, common simultaneous fluctuations in MU firing rate have been described [42], which are much more pronounced in strength trained than in controls and minimal in skilled trained subjects. These results have been confirmed recently [17]; in the same study greater myoelectric manifestations of muscle fatigue—steeper decay of the MDF—in weight lifters with respect to controls is also reported. Once again this should indicate a greater and more precocious recruitment of FT fibers in well-trained power athletes.

Felici and co-workers [17] also applied a new analytical tool to the sEMG, namely a nonlinear approach called recurrence quantification analysis (RQA; described in Chapter 6). One of the variables obtained by RQA analysis, the percentage of determinism (%DET; see Chapter 6), is effective in detecting subtle changes of the underlying dynamics in sEMG signal [19] attributable to ongoing changes in muscle activation [43]. In particular, during continuous heavy isometric contractions, the increase in %DET has been taken as an index of myoelectric fatigue [43]. In weight lifters, even in the cases where evident muscular burst activity was not observed, a %DET increase was noted; conversely, in the controls this %DET modification was not observed [17]. In accordance with Sale [40] a prevalent MUs synchronization seems to be a special feature of people trained for brief, maximal efforts. This author speculated about the usefulness of any sort of synchronization in improving the force output; in his opinion, it does not seem convincing that

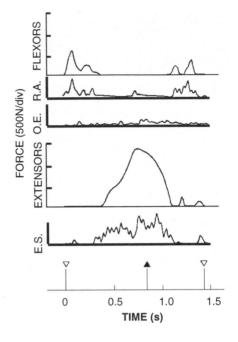

Figure 14.4. Filtered sEMG recordings of the erector spinae (ES), obliquus externus (OE), and rectus abdominis (RA) muscles and predicted extensor and flexor equivalent muscle forces during one cycle of half-squat exercise. (*Empty triangle*) Beginning of the yielding phase; (*filled triangle*) beginning of the lifting phase. (Redrawn, with permission, from [7])

synchronization or common drive can increase the force. Conversely, he suggested that the grouping of firings could be effective in increasing the rate of force development during brief maximal contractions. A weight can be lifted only if it is maximally accelerated when body lever arms are in the most efficient position as at the start of lifting. Results provided by the RQA [17] can be interpreted as a peculiar neuromuscular adaptation in these athletes, who seem to have developed a particular ability in activating their muscles as swiftly and completely as possible. This interpretation has received a recent support from a work from Farina et al. [15] on simulated and real sEMG signals, which showed that %DET senses both motor unit CV and synchronization.

It should be clear, at this point, that among exercises adopted to train athletes, weight-training programs have a large diffusion. However, considering the high potential for injuries associated with this modality of training, a little digression is needed. Epidemiological studies [3,8,28] show that musculoskeletal injuries are most frequently due to overload and, in particular, that the lumbar spine is one of the most frequent sites of injury (for further details on muscle damage, see below). A biomechanical approach [7] that encompasses measurements of mechanical and sEMG variables can be extremely useful in determining the magnitude of the compressive load acting on the L3–L4 segment during half-squat exercises. An example of calculated equivalent extensor and flexor muscle forces (on the basis of a muscle model) is presented in Figure 14.4. The filtered and rectified sEMG recordings are also shown from erector spinae (ES), obliquus externus (OE), and rectus abdominis (RA) muscles. In this particular case only data computed when the muscles were active are reported, assuming a time delay between mechanical and electrical activity of about 80 ms. During half-squat exercise with weights ranging from 80% to 160% of body weight, the trunk extensor muscle contraction forces were predicted to be between 30% and 50% of their MVC. The compressive load acting on the L3–L4 segment varied from approximately 6 to 10 times body weight. In this context, epidemiological studies on muscle pain/damage induced by muscle overload, are still the most reliable source of information [8]. However, the possibility of predicting the magnitude of spinal loading and muscle involvement permits, through a comparative approach, identification of training exercises and the way of performing them with minimal risks for the integrity of the spine.

14.6 MUSCLE DAMAGE STUDIED BY MEANS OF sEMG

It is now generally accepted that the delayed onset of muscle soreness (DOMS) is associated with a damage of muscle provoked by muscle overload. Skeletal muscles can be damaged in many ways by intense and active exercise, especially in nonaccustomed subjects. Generally speaking, muscle overload is the most frequent cause of damage; however, symptoms may appear acutely—after an apparent trauma—or some time after exercise, possibly without a clear link with a specific event responsible for the damage. Any muscle, regardless of its training status, can be damaged by overload. However, greater loads are required to damage trained muscle than untrained muscle. Many causes are thought to be involved in causing a DOMS pattern, such as tissue compression caused by muscle swelling and edema, local muscle pH modifications, and vasoactive substances.

In order to study muscle damage in a systematic way, a number of experimental conditions have been devised. From laboratory data it was found that the most effective way to induce muscle damage along with a DOMS syndrome was represented by the eccentric contraction (EC), rather than isometric and concentric exercises. EC is a particular kind of muscle contraction that occurs when a contracting muscle is lengthened and stretched by external forces. In many instances of ordinary life, our muscles are forced to lengthen while they are actively contracting. This is the case for the quadriceps shock absorption function during walking and running and during stair descent. By reproducing this condition in their lab, Friden and Lieber [21] showed that muscle damage is mostly dependent on the speed of extension, rather than the absolute value of force output during EC.

As a consequence of EC-induced muscle damage, functional muscle impairment occurs as an immediate marked decrease in muscle force exertion capability. This is followed, at some distance from EC exercise, by muscle pain/sore sensation, reduction in joint range of motion, increased muscle stiffness, and plasmatic release of enzymes typical of muscle damage. All these features are typical of the so-called DOMS.

Muscle damage and recovery have been studied using, among other means, magnetic resonance [20], ultrasonography [9,38], and surface electromyography [13,41]. With the aim of being as noninvasive as possible, these techniques seem to be the most effective in detecting the early changes occurring within a damaged muscle. Magnetic resonance shows an increase in the cross-sectional area of the elbow flexor muscles and a marked increase of T2 relaxation time [20,38]. On the other hand, with ultrasonography—a less expensive and time-consuming technique—it has been shown that muscle edema induced by EC can be identified by an increase in the echo intensity within the injured muscle, and an increase in muscle belly thickness [9,38].

Recently [18], the use of the sEMG from the damaged muscle to detect the first signs of EC-induced damage was proposed. The rationale was the evidence, provided by bioptic studies, of a prevalent involvement of type IIB muscle fibers. It is generally accepted that these fibers introduce specific components into the sEMG signal, characterized by high-amplitude and high-frequency content. Thus the hypothesis was that this selective damage could produce changes in selected parameters of the sEMG signal. The parameter chosen

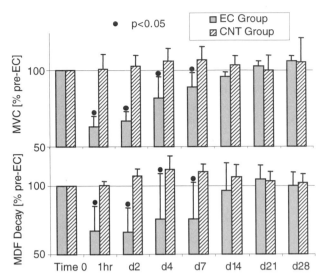

Figure 14.5. (*Upper panel*) Left biceps MVC, and elbow angle variation over time in the EC group. Mean values ± standard deviations are shown. (*Lower panel*) Left biceps MDF Decay. Mean values ± standard deviation vs time in the EC and CNT group are shown. All data are represented as a percentage of variation with respect to the pre-EC values. EC: Muscle damage (ECcentric) group; CNT: Control group. All data are represented as a percentage of variation with respect to the pre-EC values. Significant variations with respect to pre EC values are indicated with the symbol • ($p < 0.05$).

to describe sEMG modifications was the median frequency (MDF) of the power density spectrum of sEMG. After the damage an early sEMG modification was present, consisting of a shift of the MDF of sEMG power spectra towards the lower frequencies at 80% and 50% MVC. This modification, evident from one hour after EC, was still present at the end of the observation period (one week), and was assumed to be an indirect early marker of EC-induced muscle damage of type IIB fibers.

In a more recent work [41] muscle recovery from damage was studied by means of muscle force production capacity, sEMG analysis, ultrasonography, and blood enzymes. Damaged muscle status was monitored for a period of four weeks after the damage, and as can be seen from Figure 14.5 (upper panel), MVC decreased immediately after EC performance and was the last parameter to recover (between the second and third week after exercise). The sEMG was analyzed in time and frequency domains. The MDF immediately after EC (1 hour after) behaved as described above for the MVC (Fig. 14.5, lower panel). The MDF recovered somewhat earlier than MVC (between the first and second week after the damage). It must be stressed that apart from the availability of more sophisticated techniques, there is an uncontroversial parameter that can be used, in the field, to monitor for the presence of muscle damage and for the recovery status, namely the maximal muscle force output (isometric). In fact, while there is no consensus on the behavior of the many parameters studied, all authors agree on the time course of MVC modification after EC exercise. This might be particularly useful for monitoring, in an easy and nonexpensive way, muscle functional status in athletes after EC and will help avoid overloading-induced muscle traumas. The example provided by this particular application confirmed the effectiveness of sEMG analysis, coupled with other tools such as ultrasonography and dynamometry, in revealing the early signs of muscle damage and in monitoring muscle recovery from damage.

REFERENCES

1. Bernardi, M., M. Solomonow, and R. V. Baratta, "Motor unit recruitment strategy of agonist muscle pair during linearly increasing contraction," *Electromyogr Clin Neurophysiol* **37**, 3–12 (1997).

2. Bonato, P., S. Cheng Ming-Shun, J. Gonzalez-Cueto, A. Leardini, J. O'Connor, and S. H. Roy, "Assessment of dynamic conditions can provide information about compensatory muscle function in ACL patients," *IEEE Eng Med Biol* **20**, 133–143 (2001).

3. Brady, T. A., R. C. Bernard, and L. M. Bodnar, "Weight training related injuries in the high-school athlete," *Am J Sports Med* **10**, 1–5 (1982).

4. Burden, A., and R. Bartlett, "Normalisation of EMG amplitude: An evaluation and comparison of old and new methods," *Med Eng Phys* **21**, 247–257 (1999).

5. Cannon, R. J., and E. Cafarelli, "Neuromuscular adaptations to training," *J Appl Physiol* **63**, 2396–2497 (1987).

6. Cappozzo, A., F. Figura, M. Marchetti, and A. Pedotti, "The interplay of muscular and external forces in human ambulation," *J Biomech* **9**, 35–43 (1976).

7. Cappozzo, A., F. Felici, F. Figura, and F. Gazzani, "Lumbar spine loading during half-squat exercises" *Med Sci Sports Exer* **17**, 613–620 (1985).

8. Castellano, V., and M. Marchetti, "Exercise induced myalgia: An Epidemiological study. (Mialgie da sport. Uno studio epidemiologico)." *Erre Come Riabilitazione* **48**, 32–43 (2001).

9. Chleboun, G. S., J. N. Howell, R. R. Conaster, and J. J. Giesey, "Relationship between muscle swelling and stiffness after eccentric exercise," *Med Sci Sports Exer* **30**, 529–533 (1998).

10. Clarys, J. P., "Electromyography in sports movement and occupational settings: An updateof its limits and possibilities," *Ergonomics* **43**, 1750–1762 (2000).

11. Clarys, J. P., and J. Cabri, "Electromyography and the study of sports movement: A review," *J Sports Sci* **11**, 379–448 (1993).

12. Davis, R. B., S. Ounpuu, and P. A. DeLuca, "Gait data: Reporting, archiving and sharing," in A. Cappozzo, P. Allard, A. Lundberg, and C. L. Vaughan, eds., *Three dimensional analysis of human locomotion*, Wiley, New York, 1997, pp. 389–406.

13. Day, S. H., A. E. Donnelly, S. J. Brown, and R. B. Child, "Electromyogram activity and mean power frequency in exercise-damaged human muscle," *Muscle Nerve* **21**, 961–963 (1998).

14. De Luca, C. J., "The use of surface electromyography in Biomechanics," *J Appl Biomech* **13**, 135–163 (1997).

15. Farina, D., L. Fattorini, F. Felici, and G. Filligoi, "Nonlinear surface EMG analysis to detect changes of motor unit conduction velocity and synchronization," *J Appl Physiol* **93**, 1753–1763 (2002).

16. Farina, D., M. Merletti, M. Nazzaro, and I. Caruso, "Effect of joint angle on EMG variables in leg and thigh muscles," *IEEE Eng Med Biol* **20**, 62–71 (2001).

17. Felici, F., A. Rosponi, P. Sbriccoli, G. C. Filligoi, L. Fattorini, and M. Marchetti, "Linear and non-linear analysis of surface electromyograms in weightlifters," *Eur J Appl Physiol* **84**, 337–342 (2001).

18. Felici, F., L. Colace, and P. Sbriccoli, "Surface EMG modifications after eccentric exercise," *J Electromyogr Kinesiol* **7**, 193–202 (1997).

19. Filligoi, G. C., and F. Felici, "Detection of hidden rhythms in surface EMG signals with a non-linear time-series tool," *Med Eng Phys* **21**, 439–448 (1999).

20. Foley, J. M., R. C. Jayaraman, B. M. Prior, J. M. Pivarnik, and R. A. Meyer, "MR measurements of muscle damage and adaptation after eccentric exercise," *J Appl Physiol* **87**, 2311–2318 (1999).

21. Fridén, J., and R. L. Lieber, "Structural and mechanical basis of exercise-induced muscle injury," *Med Sci Sports Exer* **24**, 521–530 (1992).

22. Guidetti, L., G. Rivellini, and F. Figura, "EMG patterns during running: intra- and inter-individual variability," *J Electrom Kinesiol* **6**, 37–48 (1996).

23. Hakkinen, K., "Neuromuscular adaptation during strenght training, detraining and immobilization," *Crit Rev Phys Rehabil Med* **6**, 161–198 (1994).

24. Hakkinen, K., M. Alen, and P. V. Komi, "Changes in isometric force-relaxation-time, electromyographic and muscle fibre characteristics of human skeletal muscle during strength training and detraining," *Acta Physiol Scand* **125**, 573–585 (1985).

25. Hakkinen, K., A. Pakarinen, M. Alen, H. Kauhanen, and P. V. Komi, "Neuromuscular and hormonal adaptations in athletes to strength training in two years," *J Appl Physiol* **65**, 2406–2412 (1988).

26. Henneman, E., G. Somjen, and D. O. Carpenter, "Functional significance of cell size in spinal motoneurons," *J Neurophysiol* **28**, 560–580 (1965).

27. Hermens, H., et al., eds., *European recommendations for surface electromyography: Results of the SENIAM project*, Roessingh Research and Development, Enschede, Netherlands, 1999.

28. James, S. L., B. T. Bates, and L. R. Osterning, "Injuries to runners," *Am J Sports Med* **6**, 40–50 (1978).

29. Jonsson, B., "Electromyographic kinesiology: Aims and fields of use," in J. Desmedt, ed., *New development in EMG and clinical neurophysiology*, Karger: Basel 1973, pp. 498–501.

30. Komi, P. V., J. H. Viitasalo, M. Havu, A. Thorstensson, B. Sjodin, and J. Karlsson, "Skeletal muscle fibres and muscle enzyme activities in monozygous and dizygous twins of both sexes," *Acta Physiol Scand* **100**, 385–392 (1977).

31. Kraemer, W. J., S. J. Fleck, and W. J. Evans, "Strength and power training: physiological mechanisms of adaptation," in J. O. Holloszy, ed., *Exercise and sport sciences reviews*, Vol. 24, Williams and Wilkins, Baltimore, 1996, pp. 363–397.

32. Macaluso, A., G. De Vito, F. Felici, and M. Nimmo, "Electromyogram changes during sustained contraction after resistance training in women in their 3rd and 8th decades," *Eur J Appl Physiol* **82**, 418–424 (2000).

33. Marchetti, M., A. Cappozzo, F. Figura, and F. Felici, "Race walking versus ambulation and running," in H. Matsui, and K. Kobayashi, eds., *Biomechanics VIII-B, international series on biomechanics*, Vol. 4B, Human Kinetic Publishers, Champaign, IL, 1982, pp. 669–675.

34. Merletti, R., and H. Rix, in H. Hermens, R. Merletti, H. Rix, and B. Freriks eds., *SENIAM— Deliverable 7: State of the art on signal processing methods for SEMG*. Biomedical and Health Research Program, Deliverable of the SENIAM project. 1999.

35. Merletti, R., and L. Lo Conte, "Surface EMG signal processing during isometric contractions," *J Electromyogr Kinesiol* **7**, 241–250 (1997).

36. Milner-Brown, H. S., R. B. Stein, and R. G. Lee, "Synchronization of human motor units: possible roles of exercise and supraspinal reflexes," *Electroencephalogr Clin Neurophysiol* **38**: 245–254 (1975).

37. Moritani, T., and H. A. deVries, "Neural factors versus hypertrophy in the time course of muscle strength gain," *Am J Phys Med* **58**, 115–130 (1979).

38. Nosaka, K., and P. M. Clarkson, "Changes in indicators of inflammation after eccentric exercise of the elbow flexors," *Med Sci Sports Exer* **28**, 953–961 (1996).

39. Sadoyama, T., T. Masuda, H. Miyata, and S. Katsuta, "Fibre conduction velocity and fibre composition in human vastus lateralis," *Eur J Appl Physiol* **57**, 767–771 (1988).

40. Sale, D. G., "Neural adaptation to resistance training." *Med Sci Sports Exer* **20**, S135–S145 (1988).

41. Sbriccoli, P., F. Felici, A. Rosponi, A. Aliotta, V. Castellano, C. Mazzà, M. Bernardi, and M. Marchetti, "Exercise induced muscle damage and recovery assessed by means of linear and non-linear sEMG analysis and ultrasonography," *J Electromyogr Kinesiol* **11**, 73–83 (2001).

42. Semmler, J. G., and M. A. Nordstrom, "Motor unit discharge and force tremor in skill- and strength-trained individuals," *Exp Brain Res* **119**, 27–38 (1998).

43. Webber, C. L. Jr, M. A. Schmidt, and J. M. Walsh, "Influence of isometric loading on biceps EMG dynamics as assessed by linear and non linear tools," *J Appl Physiol* **78**, 814–822 (1995).

44. Weineck, J., "Optimales training." in *Leistungsphysiologische Trainingslehre unter besonderer Berücksichtigung des Kinder—und Jugendtrainings*. 10. Aufl. First Italian Edition, Calzetti-Mariucci ed. Ponte S. Giovanni (Perugia) (2001).

15

APPLICATIONS IN MOVEMENT AND GAIT ANALYSIS

C. Frigo

Department of Bioengineering
and Laboratory of Biomedical Technologies
Polytechnic of Milano, Italy

R. Shiavi

Departments of Biomedical Engineering
and Electrical Engineering, Vanderbilt University, Nashville, TN

15.1 RELEVANCE OF ELECTROMYOGRAPHY IN KINESIOLOGY

The study of human movement and motor control takes a great advantage from the use of the electromyogram, in particular, the surface electromyogram (SEMG). Its usage was precipitated by the discovery that under certain conditions a linear relationship could be demonstrated between the integrated myoelectric signals measured either intramuscularly or on surface and some biomechanical variables. The first set of conditions tested were for isometric contractions [45]. Later a relationship was found to hold even for variable force or dynamic voluntary contractions of limited intensity [9] provided that intramuscular signals from multiple recording sites are summed. In current practice, except for the cases in which the activity of deep muscles is to be studied, the SEMG is the preferred technique to complement movement analysis.

Depending on the objective of the study, the procedures for data collection and processing are different. The information obtained can also be different, as well as the reliability of some parameters. However, it is essential to state that in many movement analysis applications the relevance of the SEMG does not reside in a single signal/muscle EMG

Electromyography: Physiology, Engineering, and Noninvasive Applications, edited by Roberto Merletti and Philip Parker.
ISBN 0-471-67580-6   Copyright © 2004 Institute for Electrical and Electronics Engineers, Inc.

analysis, but in the comparison among signals and between signals and other quantities of mechanical nature. A good description of movement measurement and analysis can be found in [69]. There are many applications of SEMG in movement analysis. The following ones will be mentioned in this chapter:

1. Study of motor control strategies, that is the activation of agonist, synergistic and antagonist muscles
2. Investigation of mechanics of muscle contraction
3. Gait analysis
4. Identification of pathophysiologic factors
5. Work load assessment in occupational biomechanics
6. Biofeedback in motor rehabilitation

Some of these issues (1, 5, and 6) are also discussed in other chapters.

15.2 TYPICAL ACQUISITION SETTINGS

In applications of kinesiology we are more interested in comparing the activity of different muscles than in analyzing the details of the wave shape of the signal. The SEMG technique is preferred to the needle and wire methodology for the following reasons: (1) the needle technique is inapplicable because during movement laceration of muscle tissue and pain can occur, (2) the use of fine, flexible wires (10–15 μ) inserted by hypodermic needles and held in place by small hooks at the end of the wires [2] is an invasive technique requiring medical personnel, and (3) with wire electrodes the volume of muscle from which signals are recorded is relatively small (few cubic millimeters). On the contrary, SEMG is readily obtainable even by nonmedical personnel, it is noninvasive, it provides information from a large mass of muscle tissue (though the superficial fibers contribute more than deep fibers) and thus more directly correlated to the mechanical outcome, and so it is more suitable to provide insights into the neural motor control mechanism.

Typically low-frequency artefacts due to large changes of impedance in the electrode–skin coupling and to skin–muscle movement are in the range from 0 to 15 Hz. Thus high-pass filters with a cutoff frequency of 15 to 20 Hz are applied to SEMG, even if a considerable amount (5–10%) of signal power is rejected in this way. The problem has been recently addressed by a new approach in which wavelet decomposition is exploited to reject motion artifacts during walking [13]. The maximum frequency still carrying information in a SEMG signal does not usually exceed 300 Hz. As the number of channels to be acquired increases (eight or more is very common), the maximum sampling frequency (see Chapter 5) might be limited by the technical characteristics of the acquisition board. A sampling frequency of 1 kHz per channel would be desirable, but in some cases half of that is still acceptable, providing that proper anti-aliasing filters are implemented before data sampling (see Chapter 5). The practice of rectifying and low-pass filtering the signal by analog circuitry before acquisition, widely used in the past, is no longer recommended because the signal can be considerably delayed and distorted by the filter, and after such operations the recognition of motion artifact becomes problematic.

The typical electrode configuration is differential with a single reference electrode positioned sufficiently apart from all the pairs of electrodes. The recording surface is in the range 0.5 to $1 \, cm^2$, and a conductive gel is used to reduce skin impedance. Very efficient skin interfaces can be obtained with electrodes that are covered by a conductive adhesive gel (pre-gelled electrodes). They are easy to apply, can absorb sweat, can be repositioned without special skin preparation, and can stay on the site for several hours. The amplifiers usually have two stages. The first stage is an operational amplifier (j-fet or c-mos technology) in proximity to the electrodes. This stage can have unity gain (in this case they are called buffers) but definitely has a very high input impedance (usually 10^{12}–10^{15} ohms or more). This specification minimizes the artifacts produced by electrode–skin impedance fluctuations. The second stage, usually contained in a portable device, provides the additional gain necessary to obtain a signal with a magnitude in the range of ± 5 or $\pm 10 \, V$, which are the typical input ranges of A/D converters (see Chapter 5). Recommendations about all these aspects are reported in the technical deliverables of the H. Hermens et al., 1999 European project [26].

In most movement analysis applications a variety of kinematic and kinetic quantities, such as limb and joint movements, ground reaction forces, and contact time of foot switches are also measured [47]. Whenever SEMGs are also collected, the signals must be synchronized with some of the mechanical events. If a single acquisition system is used to collect all quantities, as typically occurs in multifactorial gait analysis, synchronization is intrinsically guaranteed. Even in this case, however, a time event has to be identified on the recorded signals in order to recognize the beginning of a period of interest (e.g., the initiation of a stride). The vertical component of the ground reaction force can be used for this purpose. In many cases a second time marker is defined at the end of the useful period (the stride), to allow for time normalization of all the recorded and computed quantities. Data from a second force platform, if available, or a foot switch signal can be used. In case the different variables are collected by different acquisition systems, at least one synchronization signal must be in common with all of them. The alternative of using an external trigger to start acquisition by different instruments in synchronization with an external device or event has, in some cases, the drawback that a different delay can occur at each of the systems.

Procedures for signal processing are usually relatively simple. If cyclic movements are analyzed, the time shift of different SEMGs can be obtained from the cross-correlation between two signals [43]. The initiation and termination of SEMG bursts in many cases are of interest. These time events are not easy to define, either because the background noise activity can be considerable or because a functionally relevant SEMG burst is preceded by a slowly rising SEMG phase. If this condition occurs and the event of interest is not an abrupt increase of activity but a slowly increasing activity, more sophisticated systems should be used.

Techniques to detect initiation of SEMG are either based on the recognition of MUAPs within the signal or on amplitude threshold detection [4]. Recent approaches refer to the use of a double threshold, one for the signal amplitude, and a second one for the duration of the above threshold time [5]. This prevents considering isolated above threshold spikes as SEMG initiation instants.

Until recently signal processing in the frequency domain for dynamic SEMG has not attracted much interest, except for specific applications, such as analysis of pathological tremor [34]. This is probably related to the fact that, under dynamic conditions, the SEMG is nonstationary and there are artifact components that come directly from movement. This requires special techniques of spectrum estimation [12,63]. Time-frequency techniques seem more appropriate for dynamic SEMG analysis (see Chapters 6, 7, and 10).

15.3 STUDY OF MOTOR CONTROL STRATEGIES

Examples concerning the use of SEMG in studying motor control strategies are so numerous that whatever summary can be developed will be incomplete. Let us refer to a few relevant cases only. In 1976 and following years L. M. Nashner made several studies using a movable platform to produce postural perturbations. The concept was that, depending on the amplitude of the horizontal perturbation and on the initial posture, the subject was able to switch from one compensatory reaction in which the ankle plantarflexors were the dominant controllers to one in which the hip muscles were activated. The surface EMG signals were analysed together with joint kinematics [50,51,52]. Four SEMG signals were measured from tibialis anterior (TA) and the medial portion of the gastrocnemius muscle (GM) of both legs. Reportedly the raw EMG signals were band-pass filtered with roll-off points at 10 Hz and 2 kHz. These signals were then full-wave rectified and low-pass filtered. By using a filter time constant similar to that of the ankle muscle dynamics (100 ms), an electrical measure was obtained that bore an approximately linear relationship to the active contractile force under isometric or nearly isometric conditions. As inter-subject differences and electrode placement can both affect the amplitude of the EMG, the integrated EMG was calibrated by comparing it to brief voluntary changes in the level of the ankle torque at the beginning of each session. Thus the integrated EMG could be used as a measure of the functional stretch reflex gain in arbitrary units so long as responses to stimuli of the same size were compared. An interesting data representation was adopted to facilitate analysis. The agonist and antagonist muscle signals were reported with a different polarity [32]. The activation delays were detected by visual inspection of the record. Despite the fairly simple procedure, the information was very clear, and was confirmed by many subsequent studies. This procedure thus became quite common.

Another example is the study of axial movements. These are relatively large flexion-extension movements of the trunk in the sagittal plane [14]. The signals were collected from eight channels with differential amplification, band-pass filtering (16–250 Hz), and a 500 Hz sampling frequency. Joint kinematics, center of pressure, center of mass displacement, and displacement of the main anatomical landmarks were analyzed. The results, which appeared in a number of subsequent papers, demonstrated that many motor task conditions could affect the synergy employed by the subjects, and also that different training activities could modulate the sequencing and timing [15,57].

The initiation of gait is another example of motor coordination study. Lepers [42] made several studies on this topic and compared EMG signals with ground reaction forces and kinematic and kinetic variables. It appeared that distal muscles were among the first to be involved, and this occurred consistently with the mechanical requirements. A distal-proximal synergy was also recognised in a number of different, forward-oriented, motor tasks [14]. It appeared that a motor strategy was used that was interestingly modulated by the initial conditions (bending) and velocity of execution. A disruption of such a synergy and lack of modulation could explain the difficulty to initiate gait typically encountered by Parkinsonian patients. In fact the smooth coordination between tibialis anterior and triceps surae activity seems to be lost in these patients [15].

15.4 INVESTIGATION ON THE MECHANICAL EFFECT OF MUSCLE CONTRACTION

The idea that the electrical activity of muscles is tightly related to the mechanism of contraction, and thus to the force generated by the fibers, has induced many scientists to investigate the relationship between the SEMG and the torque generated at a given joint [7,49,68]. When studying the results of these investigations, one must be very aware of the assumptions that have been made for developing the biomechanical equations and for interpreting the results: (1) that the muscle from which the signal was detected was the only muscle that produced a torque at the joint, and (2) that the lever arm of the muscle did not change during the movement [31].

The relationship between the integrated SEMG and muscle mechanical work was also investigated [8] and was found to be linear for isolated flexion movements of the elbow. The relationship was found to be quite general and consistent and appeared to originate from a property of the contractile component of the muscle. Actually it implies proportionality between the mechanical work of the muscle and the two physiological mechanisms essentially involved in the grading of the muscular contraction: the increase in the number of recruited motor units and the increase of their rate of firing. However, this relationship does not always hold, and it may be different in different muscles and must be considered with caution.

The main technical difficulty in these studies is to ascertain that no other muscle, agonistic or antagonistic, with single-joint or double-joint arrangement is contributing to the externally measured force generated. This is quite impossible in physiological conditions. Another point to consider is that the portion of muscle that contributes to the signal is not necessarily the one that contributes to the force. In fact several large muscles have separate compartments that are proved to be recruited differentially in a task-dependent mode. Several studies performed to ascertain the functional differentiation in skeletal muscles have proved that even slightly different postural or motor conditions can produce different electrical signals with (apparently) the same mechanical output [6,11,53,54,64,66]. Furthermore electrical crosstalk between adjacent muscles might affect the measurements. In some instances a contribution from antagonistic muscles has been demonstrated in adverse conditions (small limbs, thick subcutaneous tissue), even if in most cases crosstalk from neighboring muscles is estimated to be limited to 5–15% [20,41,65].

Muscle kinematics must also be considered thoroughly. It is well known that the same level of activation (the central command expressed as number of motor units and the frequency of activation of each) can produce dramatically different forces when the muscle is allowed to shorten (concentric contraction) relatively to when the muscle is stretched (eccentric contraction). Dependence on shortening velocity in humans has been investigated by Wilkie [69]. In isometric contraction the force depends on the muscle length. The physiologic cross-sectional area, which depends not only on muscle mass and length but to a great extent on the fibers pinnation angle, is to be considered when comparing myoelectric signals from muscles having different structure [44].

Caution has to be taken when comparing SEMG between different subjects and between normal and pathologic cases. Electrode location with respect to the innervation zone of the muscle is a very important factor [26]. Frequently a lower than normal SEMG signal in a given muscle is interpreted as weakness, and a higher than normal signal as stronger than normal. In many cases the opposite holds true. In fact all depends on how the mechanical output is correlated to the electrical activity. If the forces produced by two subjects are the same, a higher than normal activity means weakness, since more motor units have to be recruited to produce the same effect. For the same reason it may happen that a weak signal is generated by a strong muscle that can produce the required force with a relatively low motor unit recruitment or whose motor units are deeper than those of another muscle. This is an intriguing aspect of muscle involvement interpretation, because in pathologic cases the same motor task can be performed with strategies different from normal, so that different parts of a muscle or of the locomotor system are involved, velocities and timing can be different, passive structures like ligaments, connective tissue, fibrotic components, and subcutaneous tissue can have an important effect. A careful analysis of all the kinematic and kinetic aspects together with SEMG is fundamental to understanding the role of the muscles in a given motor task. In such situations modeling is a fundamental tool, and many approaches to predict muscle force have been developed based on optimization criteria [19,22,56], hypotheses of motor control [59], SEMG signal processing [46], and EMG-driven models [27,28,29,30]. However, existing models associating EMG and mechanical variables are not yet sufficiently advanced as to predict the force in individual muscles with acceptable accuracy. This is particularly so in pathologic cases.

In sum, it seems appropriate to say that the SEMG represents muscle "involvement" in a general sense rather than muscle force.

15.5 GAIT ANALYSIS

SEMG signals have been used in gait analysis even before the advent of modern optoelectronic systems for kinematic investigations. Definition of the time sequencing of muscle activation in the normal population was achieved in several studies. Jacqueline Perry contributed to understanding muscle involvement in normal and pathologic cases and a compilation of her work can be found in [58]. Pedotti [55] was among the first to establish a relationship between muscle activity and joint moments in functional activity, and noticed the considerable consistency of these data even in presence of individual variability.

To help reinforce the concept of muscle synergy David Winter [21,70,72] analyzed SEMG by systematically applying time normalization and averaging to data recorded from normal subjects at different walking velocities. The profile represented the typical patterns and possible variations in activity. To minimize the variation in estimating these profiles, Richard Shiavi and co-workers [62] concentrated their study on the variability of the signal recorded during gait, and defined the parameters and number of strides to be averaged in order to obtain a good representation of the muscle activation patterns. Accordingly, at least 6 to 10 strides are needed to form a representative profile, and an envelope filter with a minimum cutoff frequency of approximately 9 Hz is necessary.

Extensive analyses of SEMG patterns have been carried out by many authors in several pathologies, and particularly in hemiplegia [40]. To help understand particular features of the SEMG signal, the analysis of muscle kinematics is fundamental [23,24,25]. David Winter [71] elaborated some models and a particular kind of representation in which the lengthening, shortening, and isometric contractions were displayed in different colours.

Equipment exists for automatic SEMG processing and extraction of useful parameters [37]. The European Recommendations on surface electromyography [26] and other recent reports [47] outlined the importance of carefully placing the electrodes between the innervation zone and the tendon and not, as often done, on "the belly of the muscle." However we must be aware that in dynamic conditions the size of the subject (children) and the kind of movement performed can make this recommendation difficult to observe.

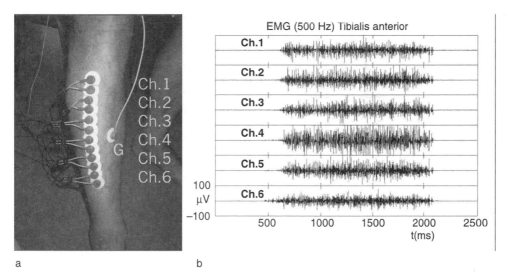

EMG (500 Hz) Tibialis anterior

Figure 15.1. (*a*) Location of six bipolar recording electrodes on the tibialis anterior muscle for testing; (*b*) raw signals recorded from the six positions reported in (*a*) during isometric maximum voluntary contraction.

Figure 15.1 reports signals recorded with the same gain and filter parameters from different locations along the course of the tibialis anterior muscle in a normal subject during a isometric, maximum voluntary contraction (see the electrodes positions on the left side of the figure). Channel 4, whose recording position was the most complying to the recommendation above, indeed produced the higher signal amplitude. However, the other recording sites were not appreciably different, with the exception of channel 6, which was almost over the tibialis anterior tendon, and probably was detecting activity form other, neighbouring, muscles. Figure 15.2 shows the signals recorded from the same electrodes location during walking at natural speed. Qualitative differences can be appreciated and can be explained by considering that the different recording areas were not aligned with the muscle fibers. However, the signals were not dramatically different, and it seems that all of them contain useful information. The most distal recording site has, of course, the higher possibility to occasionally go over the tendon when muscle belly shortens during the movement analyzed.

15.6 IDENTIFICATION OF PATHOPHYSIOLOGIC FACTORS

The major potential of a multifactorial movement analysis is the possibility it offers to understand the role of the different factors in a given pathology. In fact interest is not focused on a single signal or biomechanical quantity in a clinical context but rather on a set of correlations, or a set of derived quantities that cannot be directly estimated by visual inspection or manual analysis. An example of this approach is presented in a paper by Crenna and colleagues [18]. In that paper four main pathophysiologic factors were identified in cerebral palsied children. They were paresis, spasticity, distonia (co-contraction), and nonneural factors (muscle contractures, fibrosis). They were defined in terms of the relation between the SEMG signals and the velocity of change in muscular length, as determined by joint kinematics, or joint moments as determined by dynamic analysis. These

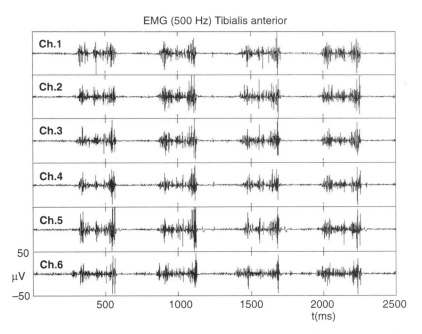

Figure 15.2. Raw signals recorded from the six positions reported in Figure 15.1a during walking at natural speed.

factors were scored in a scale from 0 (not present) to 4 (maximally present). A pathophysiologic profile can then be drawn to synthetically represent the relative importance of the four pathophysiologic factors in the patient's performance. Such a procedure can help the clinician to make decisions about the therapy. For example, where the relevant component is spasticity, he can look into details such as which muscles present clonus or hyperactivity during stretching, and consequently decide to treat them locally by botulin toxin or prescribe surgery. In these cases SEMG information is very important and should be considered jointly with the other biomechanical variables.

15.7 WORKLOAD ASSESSMENT IN OCCUPATIONAL BIOMECHANICS

Another important field of application for SEMG in movement analysis is ergonomics. This application is presented in greater detail in Chapter 13. In this application the interest is to understand the degree of muscular involvement in a motor task, and what could be the detrimental effect to the worker. Many of the work-related musculoskeletal diseases affect the joint cartilage, tendons, and ligaments, and occasionally the muscles. Muscle contractures and inflammations are quite common and also can be hindering factors and a possible cause of soft tissue damage.

There are many ways of using the SEMG signal in occupational biomechanics. One objective is to identify muscles that are activated for a long period of time in a given occupational task. The technique proposed by Bengt Jonsson [33] can be suitable to monitor and characterize the involvement of a muscle. It basically consists of analyzing the amplitude histogram of a SEMG. In this approach a calibration scale is defined in order to convert the SEMG amplitude into force units. This requires a series of measures of force and SEMG amplitude, during voluntary, static contractions. This procedure is disputable. However, the important aspect of the proposed approach is that a simple algorithm can be implemented to obtain an amplitude histogram of a signal. It can be estimated on line and allows for long-time acquisitions without a need for large storage memory. The technique can be profitably used for comparing different workplace arrangements in given subject provided that the muscles analyzed are the same.

The SEMG signal can also provide insight into the interplay between active and passive structures. For example, when a joint is approaching the limit of his range of motion, the external moment can be progressively supported by ligaments and fibrous tissue of the joint capsula. A reduced SEMG in that case does not mean the joint is unloaded but simply that the muscles are reducing their role. As the ligaments have usually a smaller lever arm than muscle tendons, this situation can represent a risk for overloading the internal structures. This is easy to see when analyzing forward trunk bending. As the trunk flexes, the spinal ligaments become more important to support the flexion moment produced by the gravity force, and the SEMG activity of the erector spinae muscles decreases. Models to predict the intervertebral disk load based on SEMG must take this aspect into consideration.

15.8 BIOFEEDBACK

In motor rehabilitation EMG biofeedback has a role and an interesting perspective. The possibility of enhancing voluntary control of muscles by providing auditory or visual feedback has been demonstrated [3], and applications are of current interest [67]. They are more thoroughly dealt with in Chapter 17 of this book.

15.9 THE LINEAR ENVELOPE

The use of the linear envelope (LE) in representing electromyographic (EMG) patterns, and the average LE for representing a muscle's activation profile has become common practice. The LE model of nonstationary EMG is a signal modulating a random carrier. The signal, or the relevant information, is the envelope of the EMG that is retrieved by linear envelope detection methods [36,60] (see also Chapter 6 of this book for further discussion on amplitude estimator).

Initially it was hypothesized that the LE has a relationship to the force generated by the muscle. While this is in part true, the relationship is very complex because many confounding factors are present and are still being investigated. Various procedures are utilized to form the LE and to reduce the variability or increase the repeatability of the final estimate of the profiles [35,61,73]. All of these procedures implement some type of time domain filtering and ensemble averaging, and depend on measuring a sufficient number of strides.

There are many sources of pattern variability. One is inherent in the measurement process and another is the natural adaptability of the control process. The variability in the amplitude of the recorded EMG among individuals occurs because of the variations in electrode location on the muscles, electrode size, differences in equipment across studies, skin resistance across subjects, and so on [75]. The adaptability is demonstrated by studies showing that there are changes in EMG patterns from stride to stride within the same individual, differences in patterns among individuals walking at the same speed, and

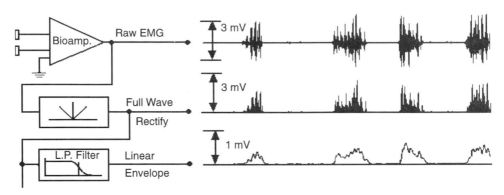

Figure 15.3. Formation of the linear envelope of the electromyogram (ISEK Standards).

changes in patterns as walking speed increases [1,38,61]. These aspects are reviewed below.

15.9.1 Construction of the Linear Envelope

The linear envelope is formed from the SEMG through the operations of rectification and envelope detection. These operations are depicted in Figure 15.3 (ISEK Standards). The linear envelopes are generated through software by taking the absolute value of the EMG and passing it through a low-pass filter. The characteristics of the filter determine the smoothness and duration of the bursts of activity. One implementation consists of a third-order Paynter finite-time integrator. The time constant of the integrator can vary from 25 to 200 ms, depending on the desired smoothness [60]. As the time constant is increased, or as the cutoff frequency of the low-pass filter is reduced, the phases of activity become smoother in amplitude and longer in duration [35]. In order not to overestimate the phase of activity, the minimum cutoff frequency is 9 Hz [62]. Analog filters are not recommended because they introduce a delay, which can be avoided when digital filters are adopted.

15.9.2 EMG Profiles

The EMG profiles are calculated in a three-step process by ensemble averaging the LEs for a number of strides. Figure 15.4 shows the LE of a rectus femoris muscle for three consecutive strides.

Because the stride duration varies from stride to stride, it is necessary to measure the time of heel strike and toe-off of each foot accurately. First, the foot-contact sensors are used to determine the beginning and end of each stride. The second step is to ensure that each LE has an equal number of points; 256 is a sufficient number. This is accomplished through a process called interpolation. Finally all of the interpolated LEs are averaged together in an ensemble manner. The ensemble variance is calculated during the last step. The relevant formulas are

$$M(i) = \frac{1}{K}\sum_{j=1}^{K} LE(j, i), \quad S(i)^2 = \frac{1}{K}\sum_{j=1}^{K} (LE(j, i) - M(i))^2$$

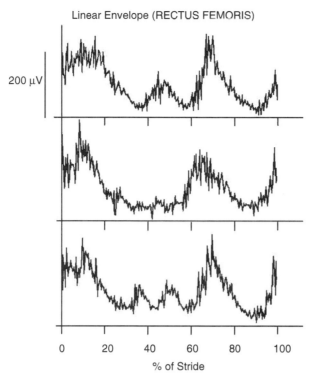

Figure 15.4. Three consecutive strides of the linear envelope from a rectus femoris muscle.

where i is the time index, j is the envelope index, K is the number of LEs, $M(i)$ is the ensemble mean at time index i, and $S(i)^2$ is the ensemble variance at time index i. The profiles (ensemble mean and standard deviation) for several muscles are shown in Figure 15.5.

15.9.3 Repeatability

A study of EMG profiles often requires the comparison of results found among various investigators and laboratories. The reliability of these comparisons depends on the repeatability of the measured SEMG and resultant profile. The SEMG varies from stride to stride; that is why the profile is calculated. However, other causes of variation are the walking speed, and the number of strides [61,62,74]. Also there are day-to-day variations that can cause some differences in profiles produced by the same individual. These differences are increased when measurements are performed in different laboratories [39]. It has been shown that 6 to 10 strides are needed to produce a representative profile [62]. A measure to quantify the repeatability is the variance-to-signal ratio (*V/S*) defined as

$$\frac{V}{S} = \frac{\sum_{i=0}^{255} S(i)^2}{\sum_{i=0}^{255} M(i)^2}$$

A good example of the usage of this measure is shown in Figure 15.6. The *V/S* is plotted for several muscles from adults who were walking at various walking speeds. Notice that

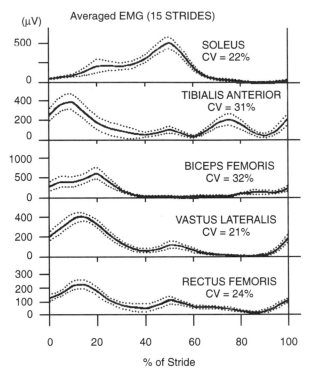

Figure 15.5. Profiles of EMG from a normal subject. The ensemble average is the solid line; the dotted lines are plus and minus one standard deviation.

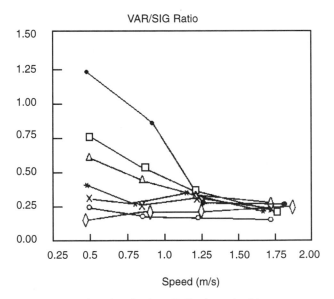

Figure 15.6. The variance-to-signal ratio (VAR/SIG) plotted with respect to average walking speed. Muscles symbols: ○, tibialis anterior; ◇, soleus; X, gastrocnemius; △, rectus femoris; □, medial hamstring; •, vastus lateralis; *, gluteus medius.

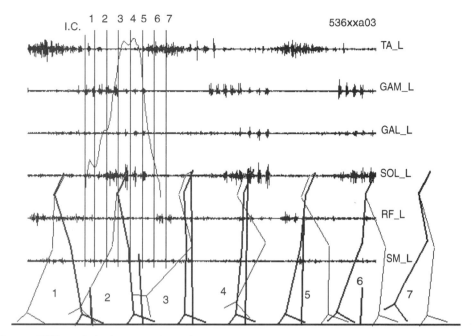

Figure 15.7. Multifactorial analysis of a hemiplegic subject with evident clonic activity of the triceps surae muscle. EMG signals from top to bottom refer to tibialis anterior, gastrocnemius medial head, gastrocnemius lateral head, soleus, rectus femoris, and semimembranosus. The vertical lines indicate the times at which the stick diagrams correspond (N = 1 to 7). The vertical component of the ground reaction force for one stride are overlaid.

at slow speeds the muscles in the thigh are highly variable with respect to those in the lower leg. As the walking speed increases, all patterns become much more repeatable.

15.10 INFORMATION ENHANCEMENT THROUGH MULTIFACTORIAL ANALYSIS

Functional diagnosis is intended to provide information about the mechanisms that affect human performance and about their physiopathology. Because of the evolution of the therapeutic and surgical methodologies that are now able to offer a number of different specific approaches, functional diagnosis has become more and more important for clinical decision making [10,48]. The intent of this section is to provide an in-depth presentation of the usage of this multifactorial analysis for assessing pathologic locomotion. The analyses utilize not only measured variables but also variables derived from the measurements.

15.10.1 Measured Variables

Locomotion is one of the more severely affected functions in hemiplegia. Characterization of the main features of muscle contraction can reveal interesting aspects of the motor disorders [18]. SEMG signals are analyzed together with biomechanical data in Figures 15.7 through 15.11.

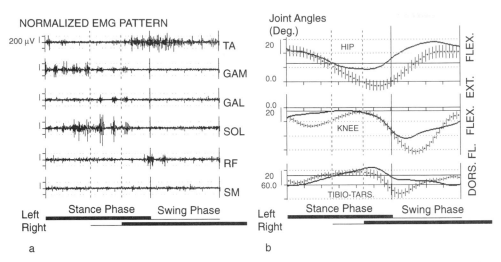

Figure 15.8. EMG pattern (*a*) and Joint angles (*b*) referred to the temporal phases of the stride. Same hemiplegic subject as in Figure 15.6. Pathological joint angles (*solid line*) are superimposed over a control pattern (average +/– SD, *N* = 25, normal healthy subjects).

Figure 15.7 shows an example of a hemiplegic subject characterized, from the neuromuscular control point of view, by a relevant clonic activity of all the three components of the triceps surae muscle. Interestingly there seems to be synchronization of the bursts among gastrocnemius medialis, lateralis, and soleus, but the relative contributions appear to be modulated in different ways. The gastrocnemius medialis starts its activity at initial contact and immediately produces a fragmented activity composed of five bursts of similar strength and by three bursts of smaller amplitude. At the lateral site of gastrocnemius only the three final bursts appear, while at the soleus muscle the same three bursts are preceded by a slowly rising phase of EMG activity. The tibialis anterior activity seems to be dominated by a reciprocal activation mechanism. Its major activity appears well modulated and corresponds to the triceps surae silent period. However, its de-activation phase appears fragmented into three or four bursts well intermingled with the gastrocnemius medialis initial activity. In Figure 15.8a the relation between EMG activity and phases of the stride are more evident (all the signals have been time windowed corresponding to the stride time). In Figure 15.8*b* the joint angles patterns have been reported on the same time scale in order to depict the main features of this pathologic gait.

It appears from Figure 15.8*b* that the knee joint does not undergo yielding flexion during early stance phase, and the large swing phase flexion is interrupted at 70% of the stride. The tibio-tarsal joint does not have the normal plantarflexion after initial contact, and progressively moves toward dorsiflexion. At late stance, dorsiflexion is higher than normal, and subsequent plantarflexion in swing phase is deficient. Note that in these phases the triceps surae activity is terminated, and tibialis anterior is strongly activated. As to the hip joint, limited extension in the second half of the stance phase and enhanced flexion in the swing phase are the most evident pathological features. The first of them could be a consequence of contracture of the hip flexors, as commonly observed in these patients. The second one can be a compensation to allow foot clearance even in presence of limited knee flexion and ankle dorsiflexion.

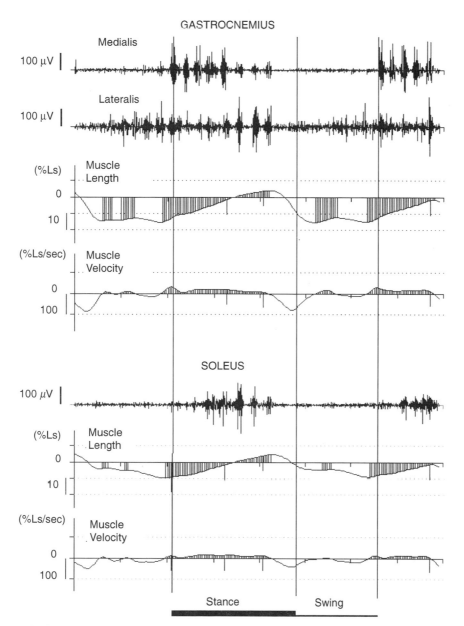

Figure 15.9. Combined analysis of EMG patterns and muscle kinematics (data refer to the same hemiplegic subject as in the previous figures). (*From top to bottom*) EMG signal from gastrocnemius medialis and lateralis, muscle length changes with respect to muscle length computed in standing upright (the model is the same for both muscles), and velocity of changing muscle length (first derivative of the previous curve); EMG from soleus, muscle length, and velocity of the same muscle.

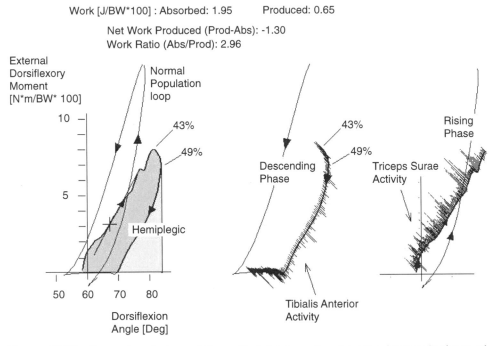

Figure 15.10. Moment-angle loop at the ankle joint (same hemiplegic subject as in the previous figures). (*Left*) The moment-angle loop of the patient (*thick line*) is compared to that obtained from an average of normal subjects (*thin line*). Note the clockwise course and the steps in the rising phase. (*Middle and right*) The descending and the rising phases are reported separately. EMG signals have been superimposed on the curves by means of segments inclined as to represent the mechanical action of the corresponding muscles. (triceps surae activity is represented as the sum of the activities of gastrocnemius medialis, lateralis, and soleus.)

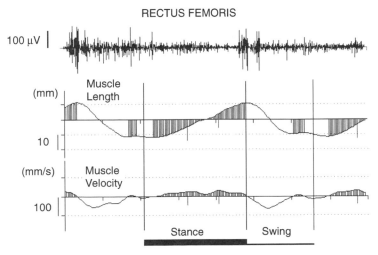

Figure 15.11. EMG signal from rectus femoris and time course of muscle length and velocity of changes of muscle length. Data refer to the same hemiplegic subject as in the previous figures.

15.10.2 Measured and Derived Variables

Figure 15.9 allows a more detailed analysis of relationships between EMG activity in gastrocnemius and soleus and their kinematics. In these figures a clear correspondence of muscle activity with the phases of positive lengthening velocity (dashed areas) is evident during stance phase. During the swing phase the reciprocal inhibition from tibialis anterior seems to prevent triceps surae to contract (however, the velocity of muscle lengthening in this phase was lower than in stance phase).

Note that in gastrocnemii EMGs a clear-cut change in the amplitude of the clonic bursts occurs when muscle length becomes greater than the standing upright value. During that period the gastrocnemius lateralis bursts become dominant. Similar changes occur at the soleus muscle. The mechanical or neurological phenomenon that triggers a different behaviour at standing upright muscle length in this subject is not clear. However, starting from this observation, a more focused clinical/functional test can be prescribed.

The mechanical outcome of the muscular activity described above can be analyzed through a moment-angle loop of the ankle joint. As depicted in Figure 15.10, the rising part of the loop presents some phases characterized by relatively high slope and subsequent slack toward dorsiflexion. The superimposed EMG activity shows that these slacking phases correspond to enhanced activity of the triceps surae as a whole (in this graphs the signals of the three muscles are summed), and thus the subsequent steeper phase can be interpreted as the result of this synchronized activity.

The descending part of the loop is considerably displaced to the right (dorsiflexion) with respect to the normal loop. This means that the plantarflexion component, instead of further increasing at maximum dorsiflexion, decreases relatively fast. Actually the activity of the triceps surae muscle is very reduced during this period, and some bursts of tibialis anterior activity appear (their mechanical effect is toward dorsiflexion and toward reducing the plantarflexor moment). As a result of this morphology the net work (represented by the area within the cycle) is negative (absorbed), and the ratio between the work absorbed (rising phase) and produced (descending phase) is well above unity. In normal subjects this ratio is between 0.5 and 0.7. This implies that considerable energy is being expended.

Looking at the problem of abnormal knee extension in the stance phase and limited flexion in the swing phase, the activity of rectus femoris can be analyzed in relation to muscle kinematics as well. Data reported in Figure 15.11 show that a constant activation, yet of medium amplitude, is present for almost all the stance phase. In the same time period rectus femoris muscle length is increasing due to two different causes: hip extension in the first instance, knee flexion in the second instance (see joint angle excursions in Fig. 15.8b). As a result, however, the velocity of muscle length change is rather constant all along the stance phase. Surprisingly, when muscle length exceeds the standing length, the rectus femoris activity seems to slightly decrease, while two bursts of relevant amplitude appear just in correspondence to the maximum length and immediately later, when the velocity becomes negative. Even if a reinforcement of the EMG activity through a stretch reflex mechanism can be hypothesized, it seems, however, that in this case a phase-dependent modulation of the gain of the reflex loop is present, and it seems not to be qualitatively different from that of normal subjects.

REFERENCES

1. Arsenault, A. B., D. Winter, R. Marteniuk, and K. Hayes, "How many strides are required for the analysis of electromyographic data in gait?" *Scand J Rehab Med* **18**, 133–135 (1986).

2. Basmajian, J. V., "Biofeedback in rehabilitation: A review of principles and practices," *Arch Phys Med Rehabil* **62**, 469–475 (1981).

3. Basmajan, J. V., and G. Stecko, "A new bipolar electrode for electromyoigraphy," *J Appl Physiol* **17**, 849 (1962).

4. Bekey, G. A., C. W. Chang, J. Perry, and M. M. Hoffer, "Pattern recognition of multiple EMG signals applied to the description of human gait," *Proc IEEE* **65**, 674–681 (1985).

5. Bonato, P., T. D'Alessio, and M. Knafkitz, "A statistical method for the measurement of muscle activation intervals from surface myoelectric signal during gait," *IEEE Trans Biomed Eng* **45**, 287–299 (1998).

6. Brown, J. M. M., C. Solomon, and M. E. Paton, "Further evidence of functional differentiation within biceps brachii," *Electromyogr Clin Neurophysiol* **33**, 301–309 (1993).

7. Buisset, S., "EMG and muscle force in normal motor activities," in E. Desmedt, ed., *New developments in electromyography and clinical neurophysiology*, vol. 1, Karger, Basel, 1973, pp. 547–583.

8. Buisset, S., and F. Goubel, "Integrated electromyographical activity and muscle work," *J Appl Physiol* **35**, 695–702 (1973).

9. Buisset, S., and B. Maton, "Quantitative relationship between surface EMG and intramuscular electromyographic activity in voluntary movement," *Am J Phys Med* **51**, 116–121 (1972).

10. Camoriano, R., A. Cama, M. Gremmo, A. Andaloro, F. Albertini, and C. Frigo, "Movement analysis and clinical application in children with Spina Bifida," *Eur J Pediat Surg* **5**, 40–41 (1995).

11. Campbell, K. M., N. L. Biggs, P. L. Blanton, and R. P. Lehr, "Electromyographic investigation of the relative activity among four components of the triceps surae," *Am J Phys Med* **52**, 30–41 (1971).

12. Capponi, M., T. D'Alessio, and M. Laurenti, "Sequential estimation of power spectra for non stationary myoelectric signals," *Electromyogr Clin Neurophysiol* **28**, 427–432 (1988).

13. Conforto, S., T. D'Alessio, and S. Pignatelli, "Optimal rejection of movement artefacts from myoelectric signals by means of a wavelet filtering procedure," *J Electromyogr Kinesiol* **9**, 47–57 (1999).

14. Crenna, P., and C. Frigo, "A motor programme for the initiation of forward-oriented movements in humans," *J Physiol* **437**, 635–653 (1991).

15. Crenna, P., C. Frigo, P. Giovannini, and I. Piccolo, "The initiation of gait in Parkinson's disease," in A. Berardelli, R. Benecke, M. Manfredi, and C. D. Marsden, eds., *Motor Disturbances II*, Academic Press, New York, 1990, pp. 161–173.

16. Crenna, P., C. Frigo, J. Massion, and A. Pedotti, "Forward and backward axial synergies in man," *Exp Brain Res* **65**, 538–548 (1987).

17. Crenna, P., C. Frigo, J. Massion, A. Pedotti, and A. Deat, "Forward and backward axial movements: two modes of central control," in V. S. Gurfinkel, M. E. Ioffe, J. Massion, and J. P. Roll, eds., *Stance and motion, facts and concepts*, Plenum Press, New York, 1988, pp. 195–201.

18. Crenna, P., M. Inverno, C. Frigo, R. Palmieri, and E. Fedrizzi, "Pathophysiological Profile of Gait in Children with Cerebral Palsy," in H. Forssberg, and H. Hirschfeld, eds., *Movements disorders in children*, vol. 36, Med Sports Sci, Basel, 1992, pp. 186–198.

19. Crowninshield, R. D., and R. A. Brand, "A physiologically based criterion of muscle force prediction in locomotion," *J Biomech* **14**, 793–801 (1981).

20. DeLuca, C., and R. Merletti, "Surface myoelectric crosstalk among muscles of the leg," *EEG Clin Neurophysiol* **69**, 568–575 (1988).

21. Dubo, H. J. C., M. Peat, D. A. Winter, A. O. Quanbury, D. A. Hobson, T. Steinke, and G. Reimer, "Electromyographic temporal analysis of gait: Normal human locomotion," *Arch Phys Med Rehab* **57**, 415–420 (1976).

22. Dul, J., M. A. Townsend, R. Shiavi, and G. E. Johnson, "Muscular synergism. I: On criteria for load sharing between synergistic muscles," *J Biomech* **17**, 663–673 (1984).

23. Frigo, C., and A. Pedotti, "Determination of muscle length during locomotion," in E. Asmussen, and K. Jorgensen, eds., *Biomechanics VI-A*, University Park Press, Baltimore 1978, pp. 355–360.

24. Frigo, C., J. Nielsen, and P. Crenna, "Modelling the triceps surae muscle-tendon complex for the estimation of length changes during walking," *J Electromyogr and Kinesiol* **6**, 191–203 (1996).

25. Frigo, C., A. Pedotti, and G. C. Santambrogio, "A correlation between muscle length and EMG activity during running," in *Science in Athletics*, Academic Publish Del Mar, 1979, pp. 61–70.

26. Hermens, H., B. Freriks, R. Merletti, D. Stegeman, J. Blok, G. Rau, C. Disselhorst-Klug, and G. Hagg, "European recommendations for surface electromyography," Results of the H. Hermens et al., 1999 project, Roessingh Research and Development, Enschede, Netherlands, 1999.

27. Hof, A. L., and J. W. Van Den Berg, "EMG to force processing. I: An electrical analogue of the Hill muscle model," *J Biomech* **14**, 747–758 (1981).

28. Hof, A. L., and J. W. Van Den Berg, "EMG to force processing. II: Estimation of parameters of the Hill muscle model for the human triceps surae by means of the calfergometer," *J Biomech* **14**, 759–770 (1981).

29. Hof, A. L., and J. W. Van Den Berg, "EMG to force processing. III: Estimation of the model parameters for the human triceps surae muscle and assessment of the accuracy by means of a torque plate," *J Biomech* **14**, 771–785 (1981).

30. Hof, A. L., and J. W. Van Den Berg, "EMG to force processing. IV: Eccentric-concentric contractions on a spring-flywheel set up," *J Biomech* **14**, 787–792 (1981).

31. Hof, A., C. Pronk, and J. van Best, "Comparison between EMG to force processing and kinetic analysis of the calf muscle moment in walking and stepping," *J Biomech* **20**, 167–178 (1987).

32. Horak, F. B., and L. M. Nashner, "Central programming of postural movements: Adaptation to altered support-surface configurations," *J Neurophysiol* **55**, 1369–1381 (1986).

33. Jonsson, B., "Measurement and evaluation of local muscular strain in the shoulder during constrained work," *J Hum Ergol* **11**, 73–88 (1982).

34. Journée, H. L., "Demodulation of amplitude modulated noise: A mathematical evaluation of the demodulator for pathological tremor EMG's," *IEEE Trans BME* **30**, 304–308 (1983).

35. Kadaba, M., M. Wooten, J. Gainey, and G. Cochran, "Repeatability of phasic muscle activity: Performance of surface and intramuscular wire electrodes in gait analysis," *J Orth Res* **3**, 350–359 (1985).

36. Kadefors, R., "Myoelectric signal processing as an estimation problem," in J. Desmedt, ed., *New developments in electromyography and clinical neurophysiology*, Karger, Basel, 1973.

37. Karlsson, S., B. E. Erlandson, and B. Gerdle, "A personal computer-based system for real-time analysis of surface EMG signals during static and dynamic contractions," *J Electromyogr Kinesiol* **4**, 170–180 (1994).

38. Kleissen, R., "Effects of electromyographic processing methods on computer-averaged surface electromyographic profiles for the gluteus medius muscle," *Phys Ther* **70**, 716–722 (1990).

39. Kleissen, R., M. Litjens, C. Baten, J. Buurke, J. Harlaar, A. Hof, and G. Zilvold, "Consistency of surface electromyographic patterns obtained during gait from three laboratories using standardised measurement technique," *Gait & Posture* **6**, 200–209 (1997).

40. Knutsson, E., and C. Richards, "Different types of disturbed motor control in gait of hemiparetic patients," *Brain* **102**, 405–430 (1979).

41. Koh, T., and M. Grabiner, "Cross talk in surface EMG of human hamstring muscles," *J Orthop Res* **10**, 701–709 (1992).

42. Lepers, R., and Y. Brenière, "The role of anticipatory postural adjustments and gravity in gait initiation," *Exp Br Res* **107**, 118–124 (1995).

43. Li, L., and G. E. Caldwell, "Coefficient of cross correlation and the time domain correspondence," *J Electromyogr Kinesiol* **9**, 385–389 (1999).

44. Lieber, R. L., *Skeletal muscle: Structure and function, implications for rehabilitation and sport medicine,* Williams and Wilkins, Baltimore, 1992.

45. Lippold, O. C. J., "The relation between integrated action potentials in a human muscle and its isometric tension," *J Physiol* **117**, 492–499 (1952).

46. Liu, M. M., W. Herzog, and H. H. C. M. Savelberg, "Dynamic muscle force predictions from EMG: an artificial neural network approach," *J Electromyogr Kinesiol* **9**, 391–400 (1999).

47. Medved, V., *Measurement of human locomotion,* Lewis Publishers, New York, 2000.

48. Monforte, S., F. Motta, R. Cardini, and C. Frigo, "Surgical treatment of Knee Dysfunction in Diplegic Child The role of Gait Analysis," in A. Pedotti and P. Rabischong, eds., *Book of abstracts of the 3rd European conference on engineering and medicine,* Edizioni Pro Juventute Don Carlo Gnocchi, Milano, 1995, p. 157.

49. Moritani, T., and H. A. deVries, "Reexamination of the relationship between the surface integrated electromyogram (IEMG) and force of isometric contraction," *Am J Phys Med* **57**, 263–277 (1978).

50. Nashner, L. M., "Adapting reflexes controlling the human posture," *Exp Br Res* **26**, 59–72 (1976).

51. Nashner, L. M., and P. J. Cordo, "Relation of automatic postural responses and reaction-time voluntary movements of human leg muscles," *Exp Br Res* **43**, 395–405 (1981).

52. Nashner, L. M., M. Woollacott, and G. Tuma, "Organization of rapid responses to postural and locomotor-like perturbations of standing man," *Exp Br Res* **36**, 463–476 (1979).

53. Paré, E. B., J. R. Stern, and J. M. Schwartz, "Functional differentiation within the tensor fasciae latae," *J Bone Joint Surg* **63**, 1457–1471 (1981).

54. Paton, M. E., and J. M. M. Brown, "An electromyographic analysis of functional differentiation in human pectoralis major muscle," *J Electr Kinesiol* **4**, 161–169 (1994).

55. Pedotti, A., "A study of motor coordination and neuromuscular activities in human locomotion," *Biol Cybern* **26**, 53–62 (1977).

56. Pedotti, A., V. V. Krishnan, and L. Stark, "Optimization of muscle-force sequencing in human locomotion," *Math Biosc* **38**, 57–76 (1978).

57. Pedotti, A., P. Crenna, A. Deat, C. Frigo, and J. Massion, "Postural synergies in axial movements: short and long-term adaptation," *Exp Brain Res* **74**, 3–10 (1989).

58. Perry, J., "Gait analysis, normal and pathological function," Slack Inc., Thorofare, NJ, (1992).

59. Pierrynowski, M. R., and J. B. Morrison, "Estimating the muscle forces generated in the human lower extremity when walking: A physiological solution," *Math Biosc* **75**, 43–68 (1985).

60. Shiavi, R., and N. Green, "Ensemble averaging of locomotor electromyographic patterns using interpolation," *Med Biol Eng Comput* **21**, 573–578 (1983).

61. Shiavi, R., H. Bugle, and T. Limbird, "Electromyographic gait assessment. Part 1: Adult EMG profiles and walking speed," *J Rehab Res Dev* **24**, 13–23 (1987).

62. Shiavi, R., C. Frigo, and A. Pedotti, "Electromyographic signals during gait: criteria for envelope filtering and number of strides," *Med Biol Eng Comput* **36**, 171–178 (1998).

63. Shwedyk, E., R. Balashubramanian, and R. N. Scott, "A nonstationary model for the electromyogram," *IEEE Trans BME* **24**, 417–424 (1977).

64. Soderberg, G. L., and W. F. Dostal "Electromyographic study of three parts of the gluteus medius muscle during functional activities," *Phys Ther* **58**, 691–696 (1978).

65. Solomonow, M., R. Baratta, M. Bernardi, B. Zhou, Y. Lu, M. Zhu, and S. Acierno, "Surface and wire EMG crosstalk in neighbouring muscles," *J Electromyogr Kinesiol* **4**, 131–142 (1994).

66. Ter Haar Romeny, B. M., J. J. Denier van der Gon, and C. C. A. M Gielen, "Changes in recruitment order of motor units in the human biceps muscle," *Exp Neurol* **78**, 360–368 (1982).

67. Tesio, L., R. Gatti, M. Monzani, and F. P. Franchignoni, "EMG-feedback from two muscles in postural reactions: a new pocket device for the patient-therapist pair," *J Electromyogr Kinesiol* **6**, 277–279 (1996).

68. Vredenbregt, J., and G. Rau, "Surface electromyography in relation to force, muscle length and endurance," in E. Desmedt, ed, *New developments in electromyography and clinical neurophysiology*, Vol. 1, Karger, Basel, 1973, pp. 547–583.

69. Wilkie, D. R., "The relation between force and velocity in human muscles," *J Physiol London*, **110**, 249–280 (1950).

70. Winter, D. A., *Biomechanics and motor control of human movement*, 2nd Ed., Wiley, New York, 1979.

71. Winter, D. A., and S. H. Scott, "Technique for interpretation of electromyography for concentric and eccentric contractions in gait," *J Electromyogr Kinesio* **4**, 263–269 (1991).

72. Winter, D. A., and H. J. Yack, "EMG profiles during normal human walking: stride-to-stride and inter-subject variability," *Electroencephalogr Clin Neurophysiol* **67**, 401–411 (1987).

73. Yang, J., and D. Winter, "Electromyographic amplitude normalizing methods: Improving their sensitivity as diagnostic tools in gait analysis," *Arch Phys Med Rehabil* **65**, 517–521 (1984).

74. Yang, J., and D. Winter, "Surface EMG profiles during different walking cadences in humans," *EEG Clin Neurophysiol* **60**, 485–491 (1985).

16

APPLICATIONS IN REHABILITATION MEDICINE AND RELATED FIELDS

A. Rainoldi

Laboratory for Engineering of the Neuromuscular System
Department of Electronics
Politecnico di Torino, Italy
University of Tor Vergata, Roma, Italy

R. Casale

Department of Clinical Neurophysiology
Rehabilitation Institute of Montescano
IRCCS, S. Maugeri Foundation, Montescano, Pavia, Italy

P. Hodges

Department of Physiotherapy
The University of Queensland
Brisbane, QLD, Australia

G. Jull

Cervical Spine and Whiplash Research Unit
Department of Physiotherapy
The University of Queensland, Brisbane, QLD, Australia

16.1 INTRODUCTION

The aim of this chapter is to provide an overview of literature in frontier fields where surface EMG may have a potentially significant role, and outline a few existing applica-

Electromyography: Physiology, Engineering, and Noninvasive Applications, edited by Roberto Merletti and Philip Parker.
ISBN 0-471-67580-6   Copyright © 2004 Institute for Electrical and Electronics Engineers, Inc.

tions. The list of fields is far from exhaustive and focuses on low back pain, pelvic floor pathologies, age-related effects, high altitude, and microgravity-induced muscle changes. Routine clinical applications of EMG based techniques are still quite limited in each of these fields. Nevertheless, it seems clear from the many studies reported in the literature that EMG analysis could have been profitably added in some investigations, and should be included in future studies to take advantage of the information that such techniques make available to the investigator.

While some fields, such as back pain and pelvic floor pathologies, have an immediate application in rehabilitation, others may appear somewhat exotic and limited to narrow research areas. This is indeed not the case. High-altitude and microgravity environments provide interesting experimental paradigms whose many fallouts in rehabilitation can be outstanding. The reader is invited to identify those of her/his interest and consider the contribution that surface EMG analysis may bring to them.

16.2 ELECTROMYOGRAPHY AS A TOOL IN BACK AND NECK PAIN (P. Hodges, G. Jull)

Back and neck pain problems are multifactorial, and there is considerable debate regarding the source of pain, and the mechanisms for development of symptoms. Despite this controversy many EMG applications are used in the assessment and management of patients with these conditions, with varying levels of physiological and clinical evidence to support the techniques. In general, EMG has been used in these populations to assess and train changes in muscle (e.g., fatiguability), and changes in the recruitment and coordination of muscle activity (e.g., reflex responses). This section will present an overview of the rationale and procedures in which EMG is used in this field.

16.2.1 Electromyography as a Tool to Investigate Motor Control of the Spine

Two principal rationales for the investigation of motor control of the trunk muscles have been presented in the literature. The first hypothesizes that appropriate activation of the trunk muscles is required to control and protect the spinal structures from injury and re-injury, while the second argues that hyperactivity of spinal muscles may be a source of pain.

The proposal that appropriate coordination of the trunk muscles is required to control and protect the spinal structures is based on the premise that the spine is unstable and relies on the contribution of muscle to maintain control in terms of its orientation and the relationship between individual segments [112]. While the muscles must have appropriate strength and endurance to generate and sustain appropriate tension in order to satisfy the demands of control, it is also necessary for the contraction of the muscles to be appropriately controlled. If contraction of a muscle is initiated too late, too slowly, or too little, then its contribution to spinal control may be compromised [112]. Hypothetically such a change in recruitment strategy may lead to compromised protection of the spinal structures. While biomechanical studies may investigate the potential contribution of muscles to control mechanisms, it is through EMG investigations that it is possible to understand the strategies used by the central nervous system to control the motion and stability of the spine. Furthermore it allows investigation of changes in recruitment strategies when people have spinal pain. Recent data argue that the deep muscles that encapsulate the lumbar spine (e.g., transversus abdominis, multifidus) act as a muscular sleeve to provide, particularly, intervertebral control [60,105] while the superficial muscles (erector spinae, oblique abdominal muscles, rectus abdominis) predominantly control the orientation of the spine [11]. Although the deep muscles are largely inaccessible to surface EMG recordings, important information can be derived from surface electrodes.

In terms of muscle as a source of pain, it has been argued that people with chronic low back pain (LBP) have elevated EMG activity, compared to nonpatients, during static and dynamic tasks [41]. This was considered to support the "muscle tension model" of chronic pain in which pain is thought to be due to factors such as the accumulation of metabolites in the active muscle. Although this model has been popularized in biofeedback training, there is conflicting evidence of increased activity, decreased activity, and no change in activity. Although there is support in the literature for the use of biofeedback training to reduce the activity of muscles such as the erector spinae in LBP, the mechanism of effect has been shown to be more related to cognitive changes rather than changes in muscle tension [62].

EMG Evaluation of Motor Control of the Trunk Muscles.

Motor control of the trunk is complicated by the large degree of redundancy in the muscle system as the trunk is surrounded by a large number of muscles and individual muscle fascicles that may contribute to spinal control to some degree. Furthermore a wide variety of strategies are available for the central nervous system to overcome the demands placed on it by internal and external forces. There is a large body of literature that has investigated the normal and abnormal control of the trunk muscles. Due to the multilayered nature of the trunk muscles these investigations have involved both surface and intramuscular EMG recordings. Although the focus here will be on recordings made with surface electrodes, several important findings from intramuscular recordings and how this relates to surface recordings will be discussed.

A variety of parameters of EMG activity are used to evaluate the temporal and spatial modalities of muscle activation in order to make judgments of the motor control of the trunk muscles. These parameters include the time of onset of EMG activity [60], time to peak activity [146], offset of activity [118], and peak amplitude [24]. In many cases measurements can be made from single trials or with data averaged over several trials when factors such as variation in background activity and response characteristics may obscure the findings in individual trials. Analysis of trunk muscle recruitment during repetitive movements can also involve analysis in the frequency domain, not only for the investigation of muscle fatigue (see Chapter 9 and Section 16.2.3 of this chapter) but also to evaluate the temporal and spatial relationship between phasic changes in EMG variables [59,89].

By these methods it has been possible to identify strategies used by the central nervous system to control and move the spine and changes that occur in this control in LBP. Two main strategies have been used to investigate the motor control of the trunk and changes in this control in LBP: (1) investigation of the response to perturbations and (2) evaluation of activity during functional movements.

PERTURBATION OF THE TRUNK TO INVESTIGATE TRUNK CONTROL. The reactive forces from limb movement are imposed on the body and present a controlled and predictable challenge to spinal stability. Numerous studies, with surface EMG electrodes over muscles such as the lumbar erector spinae and oblique abdominal muscles, have identified that the onset of EMG activity of these muscles occurs in advance of the muscle responsible for limb movement [60]. These responses are direction specific with the onset

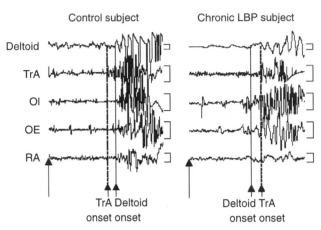

Figure 16.1. Raw EMG data recorded from transversus abdominis (TrA), obliqus internus and externus abdominis and rectus abdominis with intramuscular and surface electrodes. The onset of EMG activity of TrA is delayed in patients with low-back pain. The large arrow at the bottom left of each panel indicates the time of the movement stimulus. EMG calibration: 100 μV. (Adapted from [57] with permission)

of EMG occurring earlier and with greater amplitude (measured as the root mean square EMG for the initial 100 ms of the EMG burst) when the muscles oppose the direction of force on the trunk (see Fig. 16.1). In contrast, deep muscles such as transversus abdominis and the deep fibers of multifidus, which are less easily accessible and require intramuscular recordings, are active in a nondirection specific manner. When people have LBP, these responses are modified; however, the most consistent change is that of delayed EMG onset of the deep muscles [61].

Spatial and temporal parameters of EMG have been investigated in response to a variety of external perturbations. In its simplest form, several groups have investigated the latency to EMG onset of the paraspinal muscles to a tap to the muscle [30]. These simple reflex responses can also be investigated in terms of amplitude of EMG activity during specific time epochs that relate to short and longer latency components of the response. Zedka and colleagues [151] found an increase in the EMG amplitude of the longer latency response following experimental cutaneous pain, but not with deep muscle pain evoked by intramuscular injection of hypertonic saline (see Fig. 16.2). A variety of parameters have been measured that include the onset of EMG activity [58], the time of peak EMG [146], and the amplitude of EMG for specific epochs after the application of the perturbation [55]. In general, the normal control of the trunk muscles has been shown to involve EMG onset and amplitude that is dependent on the direction of forces, with earlier activation and greater EMG amplitude when the activity of the muscle directly opposes the forces applied to the trunk. When responses are assessed in LBP, a common finding has been that the short latency components of the reponses are unaffected (EMG onset, and amplitude for the initial 20–50 ms), while the longer latency components of responses or the early components when there is some element of predictability (e.g., visual information) are modified.

TRUNK MOVEMENT STUDIES TO INVESTIGATE TRUNK CONTROL. EMG activity of the trunk extensors at the extreme of lumbar spine flexion has been investigated extensively in the literature. Normally EMG silence of the extensor muscles occurs in this position

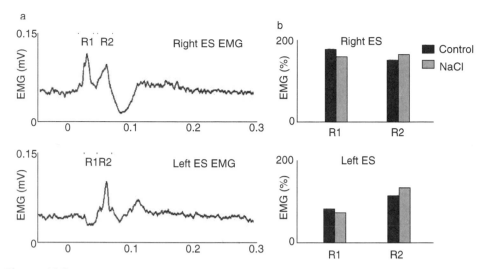

Figure 16.2. (*a*) EMG averages from multiple trials of erector spinae activity evoked by a tap to the erector spinae muscle. Short and long latency responses can be identified (R1 and R2) that are initiated by separate mechanisms. (*b*) No change in amplitude of the R1 or R2 responses can be identified after injection of hypertonic saline into the erector spinae muscles to induce pain (Redrawn from [151] with permission).

[42] as the support for the spine is provided by passive connective tissue spinal elements. This compensation effect is thought to be mediated by ligament reflexes and has been supported by animal studies [131]. However, in chronic LBP the ES muscles remain active [42,109] at the end of range of trunk flexion. This has been argued to be primarily a protective reaction [109].

There has been considerable debate regarding the appropriateness of symmetry or asymmetry of trunk muscle EMG during static and dynamic tasks. It has been argued variously that activity recorded with electrodes placed over the erector spinae muscles may be asymmetrical in tasks where it should be symmetrical (e.g., trunk flexion) and symmetrical in tasks where it should be asymmetrical (e.g., trunk rotation). However, other studies have failed to find data consistent with these proposals [109].

An area that has received a great deal of attention is the estimation of spinal loading from EMG amplitude data [31,96]. While detailed description of the methods for estimation of spinal load from EMG data are beyond the scope of this chapter, it is important to consider the main findings that have evolved from this research. The basic tenant of this research has been that people with chronic LBP experience spinal loading that differs from an asymptomatic population and that increased loading leads to accelerated degeneration of spinal structures.

EMG in Rehabilitation for LBP. EMG has a variety of applications for rehabilitation of a patient with LBP. These applications include techniques for assessment and re-assessment, and biofeedback for retraining motor control of the trunk muscles. In terms of biofeedback the aim of rehabilitation of motor control of the trunk is to retrain the components of the activation strategy that may be affected by pain. Clinically this is limited to surface EMG and is combined with other clinical techniques to gain a complete understanding of the status of motor control.

The role of EMG in assessment of the LBP patient is diverse. For example, in occupational health spinal loading can be evaluated using models that involve EMG data in combination with kinematic and force data [31,96]. Other applications include evaluation of the pattern of activation during a specific assessment task. For instance, it has been argued that the coordination of rectus abdominis and the oblique abdominal muscles during a specific abdominal maneuver is a useful outcome measure to assess for change in trunk muscles control with trunk muscle retraining [111,120]. Although many specific assessment tasks have been presented in the literature, further development is required to establish their validity.

The use of EMG in rehabilitation of motor control of the trunk lies predominantly as a technique to provide augmented feedback of muscle contraction. In motor learning several strategies are used to retrain normal movement control. These strategies include "segmentation" in which a functional movement is divided into specific component tasks that are retrained and then integrated into a complete motor task [94], and "augmented feedback" in which the awareness of a specific element of the task is heightened. EMG is used as an important tool to provide augmented feedback of muscle contraction. The use of these techniques can range from EMG biofeedback to train a patient to reduce "overactivity" of specific muscles, such as the erector spinae muscles, or to re-train activation during functional movements.

From the studies outlined above, a component of the motor control of the spine that is often deficient in LBP is the recruitment of the deep muscles. Thus one application of motor relearning principles is to focus on retraining the coordination of these muscles, namely retraining the early activation, tonic activity, and co-activation [120]. A common strategy to facilitate this change is to train the deep muscles (particularly TrA and multifidus) to function independently of the superficial muscles, then to incorporate this activation into functional tasks [120]. Although these muscles are deeply situated, EMG biofeedback can be used clinically to assess the activation of the superficial muscles. Electrodes placed over the OE (inferior to the angle of the ninth rib [108]) and over RA (at an angle of 2 degrees to the vertical 2 cm lateral to the midline and inferior to the umbilicus [108]) provide important clinical information of over-reliance of patients on these muscles. Some authors argue that electrodes placed at an angle of 6 degrees from the horizontal, below the anterior superior iliac spine, may provide an indication of TrA activity. However, in this region there is a substantial mass of OI overlying TrA, making it difficult to assess the degree to which activity of TrA contributes to the EMG signal [111]. Other biofeedback tools such as measurement of abdominal wall motion and ultrasound imaging [120] provide additional techniques to fine-tune the skilled performance of the training tasks. Training in this manner involves instructing the patient to perform low load, tonic contractions of the deep muscles, with minimal compensation by the superficial muscles. Once mastered the patient learns to integrate the activation of the trunk muscle system into functional positions and tasks [111,120].

There is experimental evidence that the reeducation of motor control of the trunk is associated with good clinical outcome, particularly when it is accompanied by EMG biofeedback of muscle activity, within the limitations outlined above. The first study investigated patients with chronic LBP associated with radiologic evidence of spondylosis/spondylolisthesis and showed a reduction in pain and disability following re-education of trunk muscle control [111]. In a second study people with their first episode of acute LBP had a reduction of recurrence of pain [56].

16.2.2 Application to Neck Pain

The flexible cervical column must function to allow appropriate head movements in three dimensions in space, to maintain mechanical stability of the head–neck system at any given orientation and to distribute load from the weight of the head as well as intrinsic and extrinsic loads of upper limb. Furthermore the neck muscle system is intimately related with reflex systems concerned with stabilization of the head and the eyes, vestibular function, and proprioceptive systems that serve not only local needs in the neck but also needs for postural orientation and stability of the whole body [72]. In an analogous situation to the lumbar spine, there is evidence to suggest that the deep muscles of the cervical region, the longus capitus and colli ventrally and the multifidus, semispinalis cervicus, and splenius capitus dorsally act as a muscle sleeve to support the cervical spinal segments in functional movements [97]. The ventral muscles also have the important function of supporting the cervical curve against the compressive weight of the head and the contraction of the larger posterior muscles, which create a bending force in the sagittal plane producing a greater lordosis [98,147].

Most of the research concerned with the patterns of neck muscle activity during orientating and stabilizing activities has been performed on animal or biomechanical models. Due to the difficulty in making EMG recordings from the deep neck muscles, EMG experiments in humans have been largely limited to superficial muscles. Clinically there has been particular interest in the ventral muscles, the cervical flexors. There is some indirect evidence derived from clinical muscle testing that the deep cervical flexors in particular, lose their endurance capacity in patients with neck pain and headache [139]. This would impact on the muscles' capacity to support the cervical curve and segments, which could contribute to the patient's pain. To better understand the differences in movement organizational strategies of the neck flexors between asymptomatic persons and those with neck pain, a novel clinical testing method was developed that might better direct the neck rehabilitation process. The test aimed to detect impaired muscle strategies in the superficial and deep neck flexor muscles. Surface EMG was used to monitor activity in the superficial flexors and, due to the inaccessibility of the deep flexor muscles to surface EMG measurement, an indirect measure of neck displacement was developed to gain some quantification of performance of the deep neck flexors.

The clinical test was based on an analysis of the craniocervical flexion action, the anatomical action of the deep longus capitus and longus colli muscles. Two presumptions were made in the development of the test. If the discreet movement was performed under low-load conditions, in supported supine lying, it could be expected that the longus capitus and longus colli muscles would perform the action and little work would be required from the sternocleidomastoid (SCM) and anterior scalene muscles. The SCM attaches to the mastoid process and effectively in this position is a craniocervical extensor, and the scalenes have no attachments to the cranium. The second presumption was based on the hypothesis of [22] who suggested that excessive activity in the superficial muscles might be a measurable compensation for poor passive or active segmental support (by the deep muscles). Indirect quantification of the deep flexor muscle contraction was gained by inserting an air-filled pressure sensor (Stabilizer, Chattanooga Pacific) between the testing surface and the back of the neck. A contraction of the longus colli causes a subtle flattening of the cervical lordosis [97], which could be registered as discreet increases in pressure in a progressively staged cranio-cervical action. Concomitantly, surface EMG electrodes were positioned over the SCM, using specific electrode location guidelines [36] to measure activity in the superficial flexors.

Results of a preliminary study on subjects with persistent whiplash-associated disorders reveal that the clinical test can elucidate differences in muscle strategies between these patients and an asymptomatic group in performances of the progressive test [65]. The patients with neck pain were unable to achieve the controlled pressure levels with the staged craniocervical test (suggesting a poorer capacity in the deep neck flexors) as achieved by the asymptomatic subjects. This was associated with higher measured activity in the superficial neck flexors, a probable compensation strategy to assist the neck pain subjects to perform the nominated task. With this knowledge of impairment in the neck flexor synergy, a specific rehabilitation strategy was developed which incorporated specific retraining of the motor strategy in craniocervical flexion. Patients are trained to perform craniocervical flexion using the deep neck flexors without undue compensatory activity of the superficial flexors, recorded with surface EMG electrodes, with the aim of improving the functional supporting capacity of these muscles. The effectiveness of this specific rehabilitation program derived from clinical research into muscle organizational strategies, has been tested in a randomized controlled clinical trial for the management of people with neck pain and headache and been shown to be efficacious [66].

16.2.3 Analysis in the Frequency Domain

As described in Chapter 9, myoelectric manifestations of muscle fatigue are a multifactorial outcome that, in general, are described by an increase of the signal amplitude and a decrease (with different rates of change) of both mean power frequency (MNF) and muscle-fiber conduction velocity (CV). One of the first studies to report a change in the EMG power spectrum of lower back muscles was done by Chapman et al. [21]. EMG signal was recorded from the fourth lumbar vertebral level during slight trunk flexion. The low-frequency components of the EMG power spectrum increased (i.e., the MNF decreased) throughout the contraction due to fatigue. Moreover the EMG signal amplitude initially increased due to fatigue and then decreased, suggesting that subjects altered their posture to alleviate muscle pain, or modified the active motor unit pool.

Roy, DeLuca, and co-workers developed a complete system for EMG acquisition and processing called Back Analysis System (BAS) [27,125,126]. It was designed to measure isometric torques providing visual feedback during trunk extension efforts. In their work of 1989, EMG signals recorded from chronic low-back pain patients and from control subjects revealed (1) high reliability estimates for repeated trials, (2) initial values and rate of change of MNF in the patient group statistically lower with respect to the control group, and (3) a proper classification of LBP and control subjects based on such parameters.

In a recent work [124] Roy et al. described a new approach to estimating frequency compression of the surface EMG signal during cyclical dynamic exercise. Signals from paraspinal muscles were recorded during repetitive lifting and processed using time-frequency representation providing the instantaneous median frequency (see Chapters 9 and 10 for further details about the use of this technique).

Mannion and colleagues [95] demonstrated a relationship between the median frequency and muscle-fiber type distribution. In their study the median frequency of the myoelectric signal from erector spinae was analyzed during isometric back extension contractions. The median-frequency time course was recorded from the thoracic and lumbar erector spine, and the data obtained were compared with biopsy samples taken from the same sites where the sEMG was recorded. The results demonstrated that the smaller is the proportion of slow-twitch fibers, the greater is the slope of median frequency, that is, the faster is the fatigue process.

In their recent work, Lariviere et al. [81] found statistically different MNF estimates obtained from two contraction modes (ramps vs. step) at variable force levels in eight back muscles. Thus, to reduce the effect of confounding factors, if median or mean frequency are used to monitor myoelectric manifestations of fatigue or to design a noninvasive tool for fiber type estimation, isometric contractions, although not physiological, seem to be preferred.

16.3 EMG OF THE PELVIC FLOOR: A NEW CHALLENGE IN NEUROLOGICAL REHABILITATION (R. Casale)

16.3.1 Introduction

Although micturition, genitosexual, and defecation disturbances are part of the vast majority of serious neurological pathologies, it is only in the last few decades that the diagnostic capacity of neurophysiological instruments and techniques has been broadened and extended to these traditionally poorly studied areas. This interest has, at least in part, been increased by the possibility of applying rehabilitation techniques to such areas and the consequent need that the specialist in neurological rehabilitation has objective reference physiopathological data in order to be able to evaluate changes induced by the rehabilitation treatments.

Although one of the main reasons for the use of invasive techniques certainly derives from the difficulty or impossibility of reaching some deep-seated muscle areas, the considerable discomfort caused by needle electrodes restricts routine use of electroneurophysiological investigations, particularly, if the examinations must be repeated.

As far as we know, the first paper on the use of surface electrodes to evaluate the electromyographic patterns of pelvic floor muscles dates back to 1976 [73]. A few years later, Vereecken and co-workers applied spectral analysis of the surface electromyographic signal to the perineal muscles [138]. Although these papers touched on a new and important topic, they were not followed by other contributions from the same groups.

A new burst of interest in this type of investigation arose from the improvement and validation of methods for recording electromyographic signals from surface electrodes; a set of recommendations was published by SENIAM (Surface Electromyography for Noninvasive Assessment of Muscle) European project for the clinical use of sEMG in the main striated muscles of the trunk and limbs. Recently the specific European project On Asymmetry in Sphincters (OASIS) was approved to investigate this issue.

These events boosted awareness of the possibility of studying the pelvic floor by evaluating the information contents of muscle signals recorded by intraluminal surface electrodes. However, the technical problems, such as having electrodes tailored for the specific anatomical site, the methodology for carrying out the recordings, the parameters to be measured, and their interpretation are still under investigation.

This section is organized on the following points:

1. An outline of the anatomy of the pelvic floor
2. Short synopsis of routine neurophysiological tests of the pelvic floor
3. Surface electromyography (sEMG) of the pelvic floor

16.3.2 Anatomy of the Pelvic Floor

The superficial muscle layer of the pelvic floor is made up of three bilateral muscles (the ischiocavernosus muscle, the bulbocavernosus muscle, and the transverse superficial muscle of the perineum), one unpaired muscle (the external anal sphincter muscle), and the superficial fascia of the perineum.

The bulbocavernosus muscle in the male is applied to the bulb of the corpus carvernosus of the urethra, united to its contralateral muscle by a midline fibrous raphe, which inserts posteriorly into the perineal body. Likewise the ischiocavernosus muscle is another bilateral muscle with a concavity sheathing the corresponding corpus cavernososus of the penis. Both muscles contribute to penile erection. In women, on the other hand, the ischiocavernosus muscle is thinner and related to the corpus cavernosus of the clitoris, while the two bulbocavernosus muscles are clearly distinct and separate. Each is related to the bulb of vestibule of the corresponding side and extends to the clitoris forming an oval-shaped ring around the vagina with its medial part forming the contractile muscle of the vestibule.

The external anal sphincter muscle is an elliptical annular muscle arranged around the anus. It has two halves that are inserted posteriorly into the anococcygeal ligament and the lower extremity of the coccyx and anteriorly into the perineal body. This region is innervated by the anterior roots of S2, S3, and S4, which form the pudendal plexus, and the anterior roots of S5 and CO1, which form the coccygeal plexus.

16.3.3 Physiopathology of the Pelvic Floor

The above-mentioned muscles of the pelvic floor and perineum are subjected to continuous work and act synergistically as a single functional unit during the performance of their numerous functions [127]:

- Ensuring pelvic statics, counterbalancing increased intra-abdominal pressures such as those to which the abdomen is subjected during pregnancy, labor, some sport activities, and prolonged periods of standing.
- Permitting faecal and urinary continence.
- Ensuring sexual activity and reproduction.

Dysfunction of the pelvic floor thus gives rise to some important and disabling conditions such as urinary incontinence or retention, faecal incontinence or constipation, and impotence. At the base of these dysfunctions there may be several and sometimes overlapping pathological conditions as somatic or autonomic peripheral nerve lesions, changes in the central nervous system, and behavioral-psychiatric disorders.

16.3.4 Routine Evaluation of the Pelvic Floor

Neurophysiological evaluation of the pelvic floor and perineum provides information on both the somatic and visceral nervous components. Furthermore it can identify whether a lesion is neurogenic and, if so, its site, extent, and completeness. Only a few and selected bibliographical references have been cited; among others, we wish to herein cite the work of Wertsch [142] as a compendium of several articles of interest to clinical electromyographers.

The neurophysiological EMG techniques routinely used are as follows:

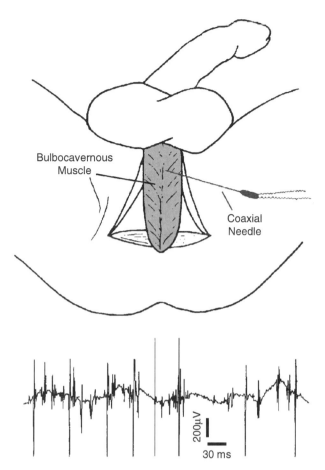

Figure 16.3. (*Top*) EMG investigation of bulbocavernous muscle through a concentric needle electrode (redrawn from [26] with permission). (*Bottom*) EMG trace of a voluntary activation.

1. Needle EMG of the perineal muscles (striated urethral sphincter and external anal sphincter, bulbocavernosus, ischiocavernosus, and transverse muscle of perineum) allows analysis and description of the characteristics of the motor units (morphology, amplitude, frequency and pattern of voluntary activation), and identification of any current neurogenic-type changes (fibrillation, slow denervation potentials, pseudomyotonic discharges at rest and weakening of the trace with temporal summation from voluntary contractions), or chronic changes (presence of long-lasting polyphasic potentials; see Fig. 16.3).

2. Surface electromyography (sEMG) is used in the diagnosis of pelvic floor disorders. While innervation deficits may be observed when needle EMG is used, surface EMG can be much more useful, for instance, in assessing coordination between bladder contraction and sphincter relaxation [129]. Several papers in the last decades have compared needle (nEMG) and surface (sEMG) electromyography, assessed reproducibility and the relationship between sEMG and manometric measurements.

Some conflicting reports deserve a more detailed analysis of data collected by needle and surface methods. In 1980 King and Teague remarked on recording artifacts associated with the surface (catheter) electrode used. In this paper cystometry (i.e., pressure and flow) was recorded during simultaneous surface and needle EMG of the pelvic floor showing considerable unreliability of the surface EMG data [74]. More recently Preece et al. [117] showed that the variability of the EMG signal between individuals appeared to be less marked when surface electrodes were used. The same authors also demonstrated that data from surface EMG recordings showed a higher degree of repeatability than those obtained by needle EMG, when collected over a period of time. This last observation is particularly important given the clinicians' need to compare data from serial muscle testing during rehabilitation of the pelvic floor. Lopez et al. demonstrated, in the diagnosis of paradoxical anal sphincter reaction, a good correlation between signals from surface electrodes applied to the perineal skin and concentric needle electrodes. The authors also suggested that noninvasive electromyography recordings of the external anal sphincter reaction should be preferred in the diagnosis of paradoxical anal sphincter [90]. sEMG appeared to be preferable also in evaluating motor conduction time to the anal sphincter by means of magnetic stimulation above the motor cortex [64].

Pinho et al. found intra-anal EMG to be reproducible when evaluated by two independent observers and may prove a reliable noninvasive method of assessing sphincter function. Significantly they also found a correlation between intra-anal EMG values and anal manometry at rest and during a maximum squeeze effort [114].

Indeed, the current improved technology applied to sEMG electrodes, the use of systems for artifacts rejection, and the use of double differential recording technique allow a quality of sEMG signals much superior to that of the 1980s. Differences between literature reports from the 1980s and more recent data are therefore probably due to technical as well as methodological factors. Today the tremendous improvements in the fields of medical technology and bioengineering make available in the research field new types of electrodes for surface EMG designed for specific studies of this particular anatomical area such as arrays of electrodes [101] or even disposable silver chloride surface electrodes mounted on a trochlear-shaped sponge for pelvic floor/external anal sphincter and intravaginal electromyography [91,92].

Cylindrical probes, with 12 mm diameter, carrying 16 silver bars (see Fig. 16.4) to record MUAPs circumferentially along the muscle fibers were used by Merletti et al. [101] to detect EMG from the external anal sphincter muscle near the anal orifice and 1 and 2 cm proximal in healthy volunteers. Signals were recorded differentially between adjacent electrodes during rest and during moderate and maximal voluntary contractions in each of the three locations, and were evaluated to detect the left and right innervation zones as well as individual MUAPs. As indicated in Figure 16.5, single MUAPs can be reliably observed during maximum and moderate contractions, and occasionally in relaxed conditions. In different subjects the innervation zone(s) appeared to be located ventrally (as in the case depicted in Fig. 16.5), dorsally, or laterally on both the left and right side. Similar observations were made about the urethral sphincter using a probe with 5 mm diameter and eight equally spaced electrode contacts. Individual MUAPs detected from the urethral sphyncter of a female subject are depicted in Figure 16.6. This type of recording allows the location of innervation zones and the identification of indexes of asymmetry in muscle anatomy and activation. This information is useful in planning episiotomy during delivery and biofeedback techniques during rehabilitation training.

A possible further advantage of using surface EMG in the pelvis district derives from the anatomical and physiological observation that the external anal and urethral sphinc-

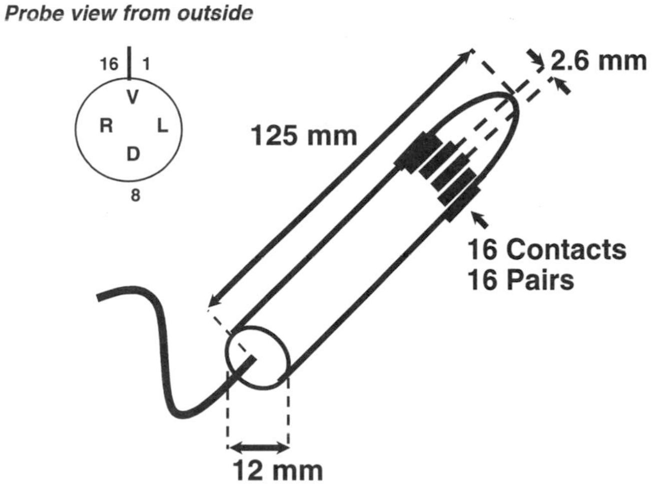

Probe view from outside

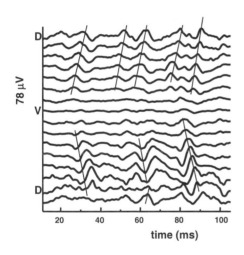

Figure 16.4. Drawing of the cilyndrical array of 16 contacts used for detecting single motor unit action potentials (MUAPs) of the striated anal sphyncter.

Figure 16.5. Example of single differential recording obtained from the array of 16 electrode pairs. Individual MUAPs showing innervation zones at the ventral side and propagation towards the dorsal side can be identified. In other subjects different locations of the innervation zones could be identified. The lines highlight individual MUAPs.

ters, the bulbocavernosus, the puborectalis, and the puboccygeus muscles have the same innervation from the pudendal nerve. This anatomical organization implies that the pelvic floor muscles behave as a single muscle unit: they contract or relax "en masse," which helps to explain some of the physiological phenomena that occur during evacuation [127].

Paradigmatic in this context is a recent study in which surface EMG showed a sort of "clinical supremacy" over needle EMG in the evaluation of the functional status of the pelvic floor. Zermann and co-workers used EMG to study the pelvic floor. They observed that fine motor changes of pelvic floor and function were the main findings in patients with incontinence after radical prostatectomy. In their study needle EMG showed no significant changes in the pre-postoperative comparison, whereas a change of myoelectric manifestations of fatigue during pelvic floor contraction (decreased mean and median frequency) was found using surface EMG polygraphy [152].

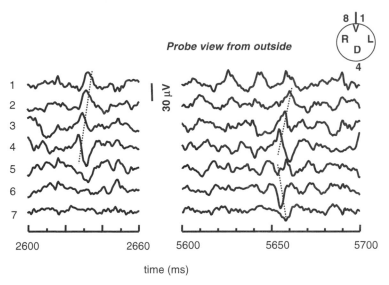

Figure 16.6. Examples of recordings of individual MUAPs obtained using an 8 contact cylindrical array with 5 mm diameter inserted in the urethra of a female subject. Innervation zones in the dorsal portion of the urethral sphincter can be detected as well as propagating patterns on the left, on the right, or on both sides of the muscle.

In the pelvic floor rehabilitation sEMG has been used in the context of biofeedback therapies. In adults sEMG has been also used in specific rehabilitation set as a measure of intravaginal and intra-abdominal activity in assessing biofeedback-assisted Kegel exercises [148], and as electromyography-assisted pelvic floor muscle rehabilitation for dysesthetic vulvodynia [113]. In recent years sEMG has been tested [141] and successfully used also in children with voiding problems [100] as suggested in an early paper by Kiesswetter [73].

Gee et al. [47] very recently found a correlation between nEMG and sEMG together with a significant positive correlation between mean fiber density on needle EMG and maximum turns rate on sEMG. They found that sEMG was able to discriminate between normal and patients with denervated muscle on the basis of the rate of neuromuscular jitter. Jitter is the variability in time between activation of the motor nerve and generation of the muscle-fiber action potential, and reflects the normality of nerve–muscle transmission. Normal jitter varies among muscles and among muscle fibers within individual muscles but is generally in the range of 10 to 50 μs. When neuromuscular transmission is sufficiently abnormal that nerve activation produces no muscle action potential, blocking is seen. Increased jitter, blocking, or both, may occur in a variety of conditions, including primary disorders of neuromuscular transmission.

In their work the authors related increased jitter in denervated patients to the reduced rectified mean surface signal observed, suggesting that sEMG may measure progressive denervation and re-innervation [47].

The possibility to longitudinally compare EMG patterns, including MU activation, the ability to evaluate the functional status, namely the activation patterns of pelvic floor muscles, makes sEMG a very useful tool in the pelvic floor rehabilitation. Indeed, there is a growing interest in pelvic floor sEMG in a rehabilitation contest not only in urology or in colon proctology field but, for instance, also in the more complex set of a comprehensive neurologic rehabilitation as in patients with multiple sclerosis [136].

It is clear from the foregoing that surface EMG is a potentially powerful tool for assessing pelvic floor function and dysfunction, outcome of surgery, and effects of rehabilitation. However, in order for this technique to become more useful, it is essential that the importance of the choice of electrodes, methodological procedures, and data processing be appreciated, and that this choice be standardized so that data from different laboratories can be compared. (The state of the art concerning this information on methodological procedures and equipment can be found in Chapter 5.)

16.4 AGE-RELATED EFFECTS ON EMG ASSESSMENT OF MUSCLE PHYSIOLOGY[1] (A. Rainoldi)

The modifications that occur in aging skeletal muscles have important consequences in several fields, including physiology, geriatrics, and rehabilitation medicine. Impaired motor performance of an older individual is often characterized by slowing of movements, a decrease in muscle strength, and a loss of fine coordination [134]. Such alterations could be due to several factors as (1) genetic factors, (2) overuse or lack of activity, (3) hormonal changes, (4) blood flow changes, and (5) anatomical and physiological changes in the central and peripheral neuromuscular system. Loss of neurons and consequent denervation and reinnervation processes, and changes in central nervous system control strategies, lead to weakness, back pain, altered work ability, and increased likelihood of accidents and falls.

16.4.1 Muscle Strength

In the elderly, muscle weakness may not be a sign of neuromuscular disease or age but rather a result of inactivity. If this is true, strength training can be used as a preventive, therapeutic, and rehabilitative procedure to enhance physical performance and improve quality of life.

Muscle strength appears to be relatively well maintained up through 50 years of age. A 15% loss in muscle strength per decade occurs between the ages of 50 and 70 years of age [83,84]. The loss of maximum isometric force during aging has been documented by a number of authors [9,25,43,103,128,137,149], and this varies among different muscle groups, being higher in the lower limb with respect to the upper limb. McDonagh et al. [99] have reported a difference in maximal isometric force of about 40% in the triceps surae muscle and a difference of only about 20% for the elbow flexors between the age of 26 years and the age of 71 years.

The decline in muscle strength may have important functional consequences: in the lower extremities it may be associated with gait disorders, falls, and hip fractures. A reduction in upper body strength increases the risk of accidents in activities that require lifting, pushing, or pulling maneuvers (e.g., housekeeping, cooking, and eating).

The muscle strength decrease with aging can be attributed to the reduced use of the lower extremities with respect to the upper extremities [6], to the loss of muscle mass (decrease in skeletal muscle-fiber number and size) [122], to some alteration of the muscle's capacity to generate force (recruitment of skeletal muscle fibers, force/cross-

[1] The authors are grateful to Dr. Dario Farina of Politecnico di Torino, Italy, for his help in preparing this section.

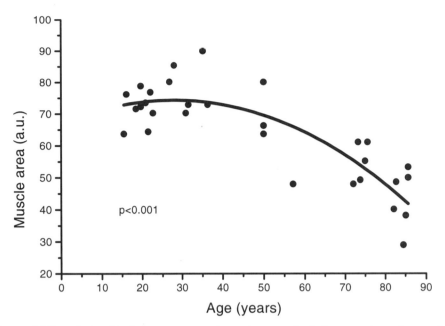

Figure 16.7. Relationship between age and muscle area of whole vastus lateralis muscle cross sections (in a.u.). (Redrawn from [122] with permission)

bridge), and to a combination of these two mechanisms, thus making older muscle intrinsically weaker [134] (see Fig. 16.7).

Lexell and co-workers have demonstrated that men, from 24 to 80 years, lose approximately 40% of total skeletal muscle mass [87]. Total muscle cross-sectional area peaks at the age of 24 years. From 24 to 50 years of age, 10% of the total muscle cross-sectional area is lost. Thereafter muscle atrophy is accelerated so that between 50 and 80 years of age, an additional 30% of total muscle cross-sectional area is lost. Older persons who remain physically active, regardless of whether they trained in youth, have only moderate losses in skeletal muscle mass [15,43,76,140], but exactly how much of the decrease in muscle is due to a consequence of aging or a reduction in physical activity or specific types of activity is currently unknown.

16.4.2 Fiber Type Composition

The main physiological changes underlying the loss of muscle strength seem to be related to the loss of muscle fibers observed in human subjects between ages 52 and 77 years, as well as to the decreased fiber cross-sectional area. It is known that progressive muscular weakness is associated with a shrinking of muscle mass and muscle cross section. Moreover the loss of muscle mass has been found to be greater in muscles that contain predominantly type II fibers.

Discrepancies among bioptic studies in regard to changes in fiber composition with aging and with respect to different muscles are reported in literature. However, most of the authors agree that type II fibers show a preferential atrophy with respect to type I fibers [19,49,50,52,82,83,88,132,149] (see Fig. 16.8).

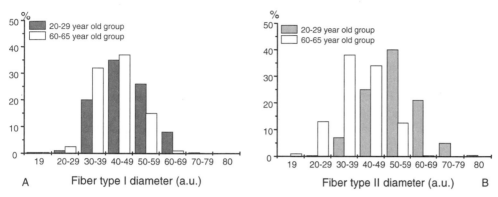

Figure 16.8. Percentage distribution of muscle-fiber diameters of fiber type I (A) and II (B) in vastus lateralis muscle from 55 subjects divided in the 20–29 and 60–69 years old groups. Arbitrary units are used for muscle-fiber diameter. (Redrawn from [82] with permission)

Motor unit (MU) remodeling (denervation and reinnervation of the muscle) seems also to be altered with aging such that type II fibers are reinnervated by collateral sprouting axons from fibers of the slow MUs [1,16,49,82,87]. It has been suggested that type II fibers innervated by slow MU axons become (or mimic) slow fibers [80]. The differences in force decline between different muscle groups seem to be reflected in the extent of the age-related changes in fiber size and in the proportion of the fiber types in different muscles [4]. For example, the age-related atrophy has been shown to be more evident in the human vastus lateralis muscle than in the biceps brachii muscle. No reduction in the mean cross-sectional area of type I fibers was observed. These findings may account, in part, for the slowing of contraction times with aging. The result of these changes is a shift toward a more uniform and slow muscle fiber constituency with few larger MUs [20,103]. As type II fibers become smaller or disappear, the statistical distribution of fiber diameters becomes progressively narrower [54,82].

In their relevant review, Enoka and Fuglevand [34] describe the very important issue of the innervation number estimate. In the case of the first dorsal interosseus muscle, for example, although it comprised of equal number of type I and II fibers, the skewed distribution of innervation number resulted in about 84% of the motor unit being type I and only about 16% being type II fibers. The clinical significance of such an anatomical design provide some insight into the adaptation that occurs when fiber type proportions change due to an intervention or a disease. Because the physiological remodeling occurring with age mainly involves type II fibers, only a few large motor units are involved in the process.

16.4.3 Myoelectrical Manifestation of Muscle Fatigue

The results reported in the previous section lead to the hypothesis that elderly subjects should show a paradoxically higher resistance to fatigue and longer endurance time. Results have been reported both in cases of voluntary and electrically elicited contractions, but comparison among the studies in the literature is difficult because of different electrode locations and adopted measures of fatigue. However, little research has been carried out with the aim of studying different fatigability between young and elderly subjects. With different protocols, some authors have reported higher fatigability in elderly subjects than younger during electrically elicited contractions [25,85], others lower fatigability [106], still others have not found any age-related change in muscle fatigability in this contraction modality [75,103].

A few studies suggest that aging has little effect on muscle endurance during voluntary contractions [84,115]. Since muscle endurance appears to be related to the rate of decrease of CV and spectral variables of the surface EMG signal [104], these findings are in contrast with results reported by Merletti et al. [102,103], who clearly found a lower level of myoelectric manifestations of muscle fatigue in the tibialis anterior muscle and in biceps brachii muscle in a group of elderly subjects, with respect to a control group of young subjects, during voluntary contractions.

Differences in ischemic conditions of the muscle between elderly and young subjects during the contraction could also introduce a further confounding factor. Hara et al. [51] recently did not find differences in CV decrement during voluntary contractions of the abductor digiti minimi in young and elderly subjects and explained the finding also on the basis of different muscle ischemic conditions. Blood flow is in fact an additional factor that can partly explain the differences reported on muscle fatigability [153]. Since the muscles of young subjects exert higher forces, they are subject to more severe ischemia than elderly subjects at high relative force levels. On the other hand, elderly persons have less capillary bed, making their muscles also susceptible to be in an ischemic condition.

Moreover, in case of voluntary contractions, other aspects related to the central control have to be taken into account, such as changes in the efficiency of activation of MUs, specifically changes in MU firing rates, MU rotation, and MU synchronization. A number of studies on intramuscular EMG signal have indicated a decrease in maximal discharge rate and an increase of its variability in the elderly [35,67,130]. Among others, Kamen et al. [68] observed a decrease of about 40% in the maximal MU firing rate with age, which might contribute to the lower manifestations of myoelectrical fatigue observed in elderly during voluntary contractions. When electrically elicited at tetanic rates, MUs of old and young subjects demonstrate similar rate of fatigue, as shown in the works of Merletti et al. [102,103], indicating that the different firing rate might contribute to the observed differences of myoelectric fatigue between young and elderly.

16.5 SURFACE EMG AND HYPOBARIC HYPOXIA (A. Rainoldi)

The effects of hypobaric hypoxia on skeletal muscle are issues of great interest for many reasons. In the field of basic research, variations of muscular fatigue patterns could be of help in clarifying and distinguishing among central and peripheral effects. Under a clinical point of view, the effects of hypoxia allow a better understanding of pathologies such as chronic hearth failure (CHF) or others in which lack of oxygen could be related to localized manifestation of muscle fatigue.

In the following, hypoxia will be considered as a consequence of high altitude (HA), a peculiar experimental environment that offers to the researcher a reversible pathological situation useful to compare HA effects with the normoxia conditions at sea level (SL). Several hypotheses have been considered to explain both physiological and biochemical changes that occur at HA [28,29,37,70,107,143,144] as well as to explain the physiological changes that take place during acclimatization [14,44,48,69,93,110] and that maybe reflected by surface EMG. The main ones are summarized below.

16.5.1 Physiological Modification Induced by Hypoxia

No agreement exists about changes induced by exposure to moderate altitude [14,33,44,45,48,69,70,107,110]. As barometric pressure falls and the oxygen availability decreases, voluntary exercise capacity and maximal aerobic power decrease, cardiac output and muscular blood flow are reduced [14,144], hyperventilation and respiratory alkalosis are induced [143]. During an acute hypoxic insult an increase in blood lactate concentration is usually observed; with acclimatization the increase in blood lactate concentration is blunted. One explanation [48] of such a reduced lactate release, the so-called lactate paradox, is that with acclimatization a reduction in maximal transport of glucose into the cell (glycolytic flux) occurs.

16.5.2 Modification of Mechanical Muscle Response Induced by Hypoxia

Except for the work of Burse et al. [17], who found a significant increase in maximum isometric handgrip strength in HA, total agreement exists in literature that both isokinetic and isometric strengths showed no significant differences at HA with respect to sea level (SL) [33,44,45,46,110,150]. Ferretti et al. [40] found that maximal muscular power decreased; however, because of the concomitant decrease in cross-sectional area (CSA) of the muscle, the muscle power, per unit cross-sectional area, was unchanged. These authors concluded that the decrease in muscle power after prolonged high-altitude exposure depends only on a net loss of muscle mass.

16.5.3 Modification of Fiber Type Induced by Hypoxia

Modifications in muscular fibers were found by several authors due to exposure over a period of a month to an altitude greater than 5000 m; such findings became more evident if extreme altitude were reached (greater than 7000 m). Boutellier et al. [14] found that, although the number of capillaries to the number of fibers ratio was unchanged, the mean fiber diameter was reduced, while the mitochondrial-to-fiber-volume ratio increased. Similar results about the reduction of muscle CSA and about the reduction in volume of muscle mitochondria and the consequent reduction of muscle oxidative capacity were found in high altitude natives [28].

In the Operation Everest II (OEII) [93], where extreme simulated conditions were obtained, all subjects displayed a decrease in total muscle volume and in subcutaneous fat volume. Needle biopsies showed that fiber type distribution was unchanged while the muscle area was reduced. This finding was substantiated by significant decreases in cross sectional fiber area of the type I fibers and in type II fiber area (although the latter was not significant, possibly due to the small number of subjects).

Results of studies about manifestations of muscle fatigue under hypobaric hypoxia conditions are contradictory and difficult to compare: some show increase of fatigue while others show no effects. Different methods were used and not always both mechanical and myoelectrical manifestations of muscle fatigue were recorded simultaneously.

16.5.4 Modification of Muscle Fatigue Induced by Hypoxia

Several authors have monitored the endurance time and/or the progressive decline of MVC torque in different exercises: during static muscle performance of the quadriceps [33], in isometric handgrip strength [17], in isometric strength of the adductor pollicis muscle [44], and in voluntary ankle dorsiflexion [46]. Others have considered indicators of myoelectric manifestations of muscle fatigue such as the integrated electromyogram activity (iEMG) [45,69,70,110] or the median and mean (centroid) frequency of the power spectrum density distribution [39,69,110] of the surface EMG signal recorded during isometric constant force voluntary contractions.

Almost all researchers found that fatigue, however estimated, was greater in HA than at SL, while consensus is lacking in explaining the causes of such behavior. Eiken et al. [33] observed the peak of torque (MVC) declining more rapidly in hypoxia (both in static and in dynamic exercise), and they interpreted these results in terms of differences in rate of intramuscular H^+ accumulation. In their works Fulco et al. [44,45] observed a shorter endurance in HA during dynamic contractions and the slope of MVC decline that for hypobaric hypoxia was almost twice that observed for normoxia although there was no difference in intercept. To explain the significance of such an increase of fatigue, several different interpretations were examined: cardiopulmonary effects (an attenuated alveolar-pulmonary capillary oxygen gradient), energetic factors (a diminished muscle capillary-to-mithochondrial oxygen gradient), and force-velocity function (the muscle exhaustion is more closely linked with impaired shortening velocity than with decreased force generating capacity). These authors' [44,45] conclusion was that the accelerated fatigue in HA should be linked with a metabolic state of fatigued muscle; muscle contraction under ischemic or hypoxic conditions in fact is linked with increased release of metabolites such as H^+, K^+ or $H_2PO_4^-$.

Kayser et al. [69] postulated the hypothesis that at HA fatigue during heavy exercise originates centrally. This contention appeared to be supported by the observation that at HA, after exhaustion is reached, 100% O_2 breathing allowed the subjects to resume exercising for another 3 minutes at a work rate close to the maximum sustained at SL. The increase in EMG amplitude and decrease in frequencies that follow the exercise resuming, were considered as EMG signs of peripheral fatigue and as the evidence of a larger recruitment of unfatigued muscle fibers and/or recruitment of previously inactive fibers. The issue of involved muscle portion was also considered in the work of these authors. In agreement with Bigland-Ritchie and Vollestad [12], the authors hypothesized that the respiratory system, which is stressed maximally during exhaustive exercise at HA, may inhibit, via a negative feedback, the maximal activation of the locomotory muscles. Hence they concluded that heavy exercise involving large muscle groups may be limited by a reduced central drive and that, by contrast, at HA small muscle groups appear to preserve their SL capacity for maximum work.

16.5.5 The Role of Acclimatization

Scientific expeditions whose results are reported here have had, in general, a duration of 13 to 40 days. During such a period all the HA measures were repeated no more than three times, at the same altitude [14,44,48,69,110] or during the ascent [93]. Hence no reports could be found about a study of possible trends in variables describing hypobaric hypoxia effects.

Only two studies explored the effects of the acclimatization and they reported opposite results. Fulco et al. [44] found shorter exhaustion time in the first day at HA, while at the thirteenth day the exhaustion time returned at values comparable with SL. During the first day at SL after altitude exposure, the exhaustion time was longer than the pre-exposure SL time. Not the same was found by Boutellier et al. [14], who attributed the apparent lack of such beneficial effects of acclimatization to the consequence of the interaction of a number of opposite changes induced by hypoxia.

16.6 MICROGRAVITY EFFECTS ON NEUROMUSCULAR SYSTEM (A. Rainoldi)

The negative effects that spaceflight and reloading upon return to Earth have on skeletal muscle must be better understood to ensure performance and prevent injury. Most of the primary changes actually represent simple deconditioning without pathology. These changes can be considered as appropriate adaptations for efficient functioning in a low-workload microgravity environment.

Microgravity has proved to be an excellent tool for noninvasively perturbing the synthesis of muscle proteins in the search for molecular signals and gene regulatory factors influencing differentiation, growth, maintenance, and atrophy of muscle. All the knowledge arising from studies in this field is likely to assist human rehabilitation on earth, especially individuals in the more vulnerable aging population and persons participating in strenuous sports [121].

16.6.1 Postflight Effects on Humans

Muscles that are composed predominantly of slow-twitch fibers tend to atrophy more than those consisting primarily of fast-twitch fibers, and in some cases, the slow-twitch muscles even acquire the characteristics of fast-twitch muscles. Since the calf muscles are an important antigravity muscle group, whereas the biceps of the arm are not, investigation of both should allow comparison of the effects of weightlessness on weight- and non-weight-bearing muscle groups [18].

Antigravity slow twitch muscles generally show the greatest deterioration following spaceflight and lower (hind) limb unloading. These fibers have low-myosin ATPase activity and slow shortening contractions, and they are specialized for oxidative metabolism that provides fatigue resistance. In contrast, muscles such as tibialis anterior and extensor digitorum longus, which contain mostly fast-twitch fibers, lack a weight-bearing function, are rapidly contracting, and are enriched with enzymes for anaerobic glycolysis. Available data from muscle biopsies obtained from flight crews before and after missions suggest that adaptations in the size, metabolic properties, and vascularization of muscle fibers can occur rapidly in the space environment [5,32,123].

Neglect of the lower limbs during spaceflight correlates with the loss of proprioception in weightlessness and reliance on visual feedback of position. The well-known Romberg test was used to identify changes in EMG latency and amplitudes that might explain the instabilities observed postflight [71]. Electromyographic activity from the tibialis anterior, the gastrocnemius and soleus muscles showed decrements in standing stability with eyes closed for several days postflight.

16.6.2 Postflight Effects on Animals

Almost all animal experiments were carried out on rats. Modifications observed in humans are qualitatively similar to those observed in animals after actual or simulated spaceflight conditions for short periods [133,135]. Chronic electromyographic recordings of soleus muscles of hindlimb adult male rats show a decrease in contractile activity of more than 60%, estimated by differences in the total number of "turns" [2].

The transversus muscle, which is presumed to be the most important expiratory muscle and to work against internal pressures that are not likely to change much between 1-g and weightlessness, did not change in size. However, both the rectus abdominis (a spinal flexor) and the external oblique (a rotator of the torso), which resist gravity in the 1-g environment, showed significant signs of atrophy after extended exposure to microgravity [38]. Significant reductions in muscle-fiber size were observed in the gastrocnemius (mainly composed of type I fibers) and estensor digitorum longus (EDL, mainly composed by fiber type IIb) muscles after spaceflight [79]. Significant MHC isoform transformations also resulted during this brief period of microgravity exposure with a significant decrease in MHC IId isoform in the EDL muscle.

Caiozzo et al. [18] found that after 14 days of microgravity the contractile properties of the soleus, an antigravity muscle, were reduced and that the MHC protein and mRNA isoform content of the soleus and of the vastus intermedius exhibited significant elevations in the fast type IId/x. On the contrary, plantaris and tibialis anterior muscles increased the fast type IIb. These results suggest that spaceflight of even short duration produces important changes in the contractile properties of antigravity skeletal muscle.

In their work, Baldwin et al. [7] suggested that myofibril protein degradation is an early event in the muscle atrophy response to 0-g. Analysis of myosin isoforms in the vastus intermedius (VI) and vastus lateralis muscles indicated that slow myosin was the primary isoform lost in the degradation of total myosin. No evidence of loss of the fast isomyosins was apparent for either muscle following spaceflight. Myofibril ATPase activity of the VI was increased in the flight group compared with controls, which is consistent with the observation of preferential slow myosin degradation and with the hypothesized shift toward glicolitic activity. Findings in the same direction were referred by the work of Staron et al. [133] and Allen et al. [3].

One work on adult Rhesus was carried out by Recktenwald et al. [119], who studied soleus, medial gastrocnemius, vastus lateralis, and tibialis anterior muscles during quadrupedal stepping before and after 14 days of either spaceflight or flight simulation. Alterations in mean EMG amplitudes reflect differential adaptations in motor unit recruitment patterns of flexors and extensors, as well as fast and slow motor pools. Their conclusions are that modulations in the nervous system controlling locomotion cannot merely be attributed to restriction of movement but appear to be the result of changes in the interpretation of load-related proprioceptive feedback to the nervous system. Moreover Leterme and Falempin [86] demonstrated that an immediate adaptation occurred in the motoneuronal recruitment of rat hindlimb muscles at the onset of changes in gravity level. The EMG signal recorded by means of implanted electrodes shows a reorganization of the recruitment patterns among the muscles biased toward an increased recruitment of fast versus slow motor units and flexor versus extensor muscles.

16.6.3 Models of Microgravity Effects

Because of the relative few data from spaceflight available to the researchers and of the importance in finding suitable countermeasures, several terrestrial models were developed to increase knowledge about microgravity effects. The models include (1) bed rest [23,78,116,145], (2) dry water immersion [77], (3) crutches [10,53], and (4) hindlimb suspension [2,13,121] for which the induced effects are in agreement with real microgravity effects; hence they could be considered as "good" models.

16.6.4 Microgravity Effect, Duration, and Countermeasures

Twenty-two days of flight have demonstrated atrophic and dystrophic developments in the soleus muscle of rats accompanied by metabolic changes in the muscle tissue [63]; unloading of the muscle caused slow (I) to fast oxidative (IIa) fiber and fast intermediate (IId/x) to fast glycolytic (IIb) fiber transformations for the whole duration of the spaceflight [133]. The changes were reversible but did not disappear entirely 27 days postflight. A recent NASA panel on countermeasures concluded that many questions still have to be investigated so that researchers can identify effective countermeasure protocols [8]. Up to now surface EMG has received limited attention as a means to monitor either neuromuscular deterioration or countermeasure effectiveness. Its ability to extract individual MU recruitment and firing strategy will probably be exploited in the future to monitor changes in microgravity environments.

REFERENCES

1. Albert, M. L., *Clinical neurology of aging*, Oxford University Press, Oxford, 1984.
2. Alford, E. K., R. R. Roy, J. A. Hodgson, and V. R. Edgerton, "Electromyography of rat soleus, medial gastrocnemius, and tibialis anterior during hind limb suspension," *Exp Neurol* **96** (3), 635–649 (1987).
3. Allen, D. L., W. Yasui, T. Tanaka, Y. Ohira, S. Nagaoka, C. Sekiguchi, W. E. Hinds, R. R. Roy, and V. R. Edgerton, "Myonuclear number and myosin heavy chain expression in rat soleus single muscle fibers after spaceflight," *J Appl Physiol* **81**, 145–151 (1996).
4. Aniansson, A., M. Hedberg, G. B. Henning, et al., "Muscle morphology, enzymatic activity, and muscle strength in elderly men: A follow-up study," *Muscle Nerve* **9**, 585–591 (1986).
5. Antonutto, G., C. Capelli, M. Girardis, P. Zamparo, and P. E. di Prampero, "Effects of microgravity on maximal power of lower limbs during very short efforts in humans," *J Appl Physiol* **86**, 85–92 (1999).
6. Asmussen, E., "Aging and exercise," in S. M. Horvath, and K. M. Yousef, eds., *Environmental physiology: Aging, heat and altitude*, Elsevier, New York, 1980, 419–428.
7. Baldwin, K. M., R. E. Herrick, E. Ilyina-Kakueva, and V. S. Oganov, "Effects of zero gravity on myofibril content and isomyosin distribution in rodent skeletal muscle," *FASEB J* **4**, 79–83 (1990).
8. Baldwin, K. M., T. P. White, S. B. Arnaud, V. R. Edgerton, W. J. Kraemer, R. Kram, D. Raab-Cullen, and C. M. Snow, "Musculoskeletal adaptations to weightlessness and development of effective countermeasures," *Med Sci Sports Exer* **28**, 1247–1253 (1996).
9. Bemben, M., B. Massey, D. Bemben, J. Misner, and R. Boileau, "Isometric muscle force production as function of age in healthy 20–74 year old men," *Med Sci Sports Exer* **11**, 1302–1310 (1991).
10. Berg, H. E., G. A. Dudley, T. Haggmark, H. Ohlsen, and P. A. Tesch, "Effects of lower limb unloading on skeletal muscle mass and function in humans," *J Appl Physiol* **70**, 1882–1885 (1991).
11. Bergmark, A., "Stability of the lumbar spine: A study in mechanical engineering," *Acta Orthoped Scand* **60**, 1–54 (1989).

12. Bigland-Ritchie, B., and N. K. Vollestad, "Hypoxia and fatigue: how are they related?" in J. R. Sutton, C. S. Huston, and G. Coates, eds, *Hypoxia: The tolerable limits*, Benchmark Press, Indianapolis, 1988.

13. Blewett, C., and G. C. Elder, "Quantitative EMG analysis in soleus and plantaris during hindlimb suspension and recovery," *J Appl Physiol* **74**, 2057–2066 (1993).

14. Boutellier, U., H. Howald, P. E. di Prampero, D. Giezendanner, and P. Cerretelli, "Human muscle adaptations to chronic hypoxia," *Prog Clin Biol Res* **136**, 273–285 (1983).

15. Brown, A., N. McCartney, and D. Sale, "Positive adaptation to weight lifting training in the elderly," *J Appl Physiol* **69**, 1725–1733 (1990).

16. Brown, W., M. J. Strong, and R. Snow, "Method for estimating number of motor units in biceps and brachialis muscles and losses of motor units with aging," *Muscle Nerve* **11**, 423–430 (1988).

17. Burse, R. L., A. Cymerman, and A. J. Young, "Respiratory response and muscle function during isometric handgrip exercise at high altitude," *Aviat Space Environ Med* **58**, 39–46 (1987).

18. Caiozzo, V. J., F. Haddad, M. J. Baker, R. E. Herrick, N. Prietto, and K. M. Baldwin, "Microgravity-induced transformations of myosin isoforms and contractile properties of skeletal muscle," *J Appl Physiol* **81**, 123–132 (1996).

19. Campbell, M. J., A. J. McComas, and F. Petito, "Physiological changes in aging muscles," *J Neurol Neurosurg Psych* **36**, 174–182 (1973).

20. Capodaglio, P., and M. Narici, "Muscle atrophy: Disuse and disease," *FSM series on Advances in occupational medicine and rehabilitation*, 1998. FSM, Pavia, Italy, ISBN 88-7963-078-4.

21. Chapman, A. E., and J. D. Troup, "Prolonged activity of lumbar erectores spinae: An electromyographic and dynamometric study of the effect of training," *An Phys Med* **10**, 262–269 (1970).

22. Cholewicki, J., M. M. Panjabi, and A. Khachatryan, "Stabilizing function of trunk flexor-extensor muscles around a neutral spine posture," *Spine* **22**, 2207–2212 (1997).

23. Convertino, V. A., D. F. Doerr, K. L. Mathes, S. L. Stein, and P. Buchanan, "Changes in volume, muscle compartment, and compliance of the lower extremities in man following 30 days of exposure to simulated microgravity," *Aviat Space Environ Med* **60**, 653–658 (1989).

24. Cresswell, A. G., "Responses of intra-abdominal pressure and abdominal muscle activity during dynamic trunk loading in man," *Eur J Appl Physiol* **66**, 315–320 (1993).

25. Davies, C., D. Thomas, and M. White, "Mechanical properties of young and elderly human muscles," *Acta Med Scand* **711** (suppl), 219–226 (1985).

26. DeLisa J. A., H. J. Lee, K. S. Lai, N. Spielholz, and K. Mackenzie, *Manual of nerve conduction velocity and clinical neurophysiology*, Raven Press, New York, 1994.

27. De Luca, C. J., "Use of the surface EMG signal for performance evaluation of back muscles," *Muscle Nerve* **16**, 210–216 (1993).

28. Desplanches, D., H. Hoppeler, L. Tüscher, M. H. Mayet, H. Spielvogel, G. Ferretti, B. Kayser, M. Leuenberger, A. Grünenfelder, and R. Favier, "Muscle tissue adaptations of high-altitude natives to training in chronic hypoxia or acute normoxia," *J Appl Physiol* **81**, 1946–1951 (1996).

29. Di Prampero, P. E., U. Boutellier, and P. Pietsch, "Oxygen deficit and stores at onset of muscular exercise in humans," *J Appl Physiol* **55**, 146–153 (1983).

30. Dimitrijevic, M. R., M. R. Gregoric, A. M. Sherwood, and W. A. Spencer, "Reflex responses of paraspinal muscles to tapping," *J Neurol Neurosurg Psychiatry* **43**, 1112–1118 (1980).

31. Dolan, P., I. Kingma, M. P. De Looze, J. H. van Dieen, H. M. Toussaint, C. T. Baten, and M. A. Adams, "An EMG technique for measuring spinal loading during asymmetric lifting," *Clin Biomech (Bristol, Avon)* **16**, 17–24 (2001).

32. Edgerton, V. R., M. Y. Zhou, Y. Ohira, H. Klitgaard, B. Jiang, G. Bell, B. Harris, B. Saltin, P. D. Gollnick, and R. R. Roy, "Human fiber size and enzymatic properties after 5 and 11 days of spaceflight," *J Appl Physiol* **78**, 1733–1739 (1995).

33. Eiken, O., and P. A. Tesch, "Effects of hyperoxia and hypoxia on dynamic and sustained static performance of the human quadriceps muscle," *Acta Physiol Scand* **122**, 629–633 (1984).

34. Enoka, R. M., and A. J. Fuglevand, "Motor unit physiology: Some unresolved issues," *Muscle Nerve* **24**, 4–17 (2001).

35. Erim, Z., M. F. Beg, D. T. Burke, and C. J. De Luca, "Effects of aging on motor unit control properties," *J Neurophysiol* **82**, 2081–2091 (1999).

36. Falla, D., P. Dall'Alba, A. Rainoldi, R. Merletti, and G. Jull, "Location of innervation zones of sternocleidomastoid and scalene muscles—A basis for clinical and research electromyography applications," *Clin Neurophysiol* **113**, 57–63 (2002).

37. Favier, R., H. Spielvogel, D. Desplanches, G. Ferretti, B. Kayser, A. Grünenfelder, M. Leuenberger, L. Tüscher, E. Caceres, and H. Hoppeler, "Training in hypoxia vs. training in normoxia in high-altitude natives," *J Appl Physiol* **78**, 2286–2293 (1995).

38. Fejtek, M., and R. Wassersug, "Effects of spaceflight and cage design on abdominal muscles of male rodents," *J Exp Zool* **289**, 330–334 (2001).

39. Felici, F., A. Rosponi, P. Sbriccoli, M. Scarcia, I. Bazzucchi, and M. Iannattone, "Effect of human exposure to altitude on muscle endurance during isometric contractions," *Eur J Appl Physiol* **85**, 507–512 (2001).

40. Ferretti, G., H. Hauser, and P. E. di Prampero, "Maximal muscular power before and after exposure to chronic hypoxia," *Int J Sports Med* **11**, 31–34 (1990).

41. Flor, H., and D. C. Turk, "Psychophysiology of chronic pain: Do chronic pain patients exhibit symptom-specific psychophysiological responses?" *Psychol Bull* **105**, 215–259 (1989).

42. Floyd, W. F., and P. H. S. Silver, "The function of the erectores spinae muscles in certain movements and postures in man," *J Physiol (Lond)* **129**, 184–203 (1955).

43. Frontera, W., et al., "A cross sectional study of muscle strength and mass in 45 to 78 old men and women," *J Appl Physiol* **71**, 644–650 (1991).

44. Fulco, C. S., A. Cymerman, S. R. Muza, P. B. Rock, K. B. Pandolf, and S. F. Lewis, "Adductor pollicis muscle fatigue during acute and chronic altitude exposure and return to sea level," *J Appl Physiol* **77**, 179–183 (1994).

45. Fulco, C. S., S. F. Lewis, P. N. Frykman, R. Boushel, S. Smith, E. A. Harman, A. Cymerman, and K. B. Pandolf, "Muscle fatigue and exhaustion during dynamic leg exercise in normoxia and hypobaric hypoxia," *J Appl Physiol* **81**, 1891–1900 (1996).

46. Garner, S. H., J. R. Sutton, R. L. Burse, A. J. McComas, A. Cymerman, and C. S. Houston, "Operation Everest II: neuromuscular performance under conditions of extreme simulated altitude," *J Appl Physiol* **68**, 1167–1172 (1990).

47. Gee, A. S., R. S. Jones, and P. Durdey, "On-line quantitative analysis of surface electromyography of the pelvic floor in patients with faecal incontinence," *Br J Surg* **87**, 814–818 (2000).

48. Grassi, B., G. Ferretti, B. Kayser, M. Marzorati, A. Colombini, C. Marconi, and P. Cerretelli, "Maximal rate of blood lactate accumulation during exercise at altitude in humans," *J Appl Physiol* **79**, 331–339 (1995).

49. Grimby, G., and B. Saltin, "The aging muscle," *Clin Physiol* **3**, 209–218 (1983).

50. Gutmann, E., and V. Hanzlikova, "Fast and slow motor units in aging," *Geront* **22**, 280–300 (1976).

51. Hara, Y., T. W. Findley, A. Sugimoto, and K. Hanayama, "Muscle fiber conduction velocity (MFCV) after fatigue in elderly subjects," *Electromyogr Clin Neurophysiol* **38**, 427–435 (1998).

52. Harridge, S. D., and A. Young, *Principles and practice of geriatric medicine*, J. Wiley and Sons, Pathy ed., London, UK, 1997.

53. Hather, B. M., G. R. Adams, P. A. Tesch, and G. A. Dudley, "Skeletal muscle responses to lower limb suspension in humans," *J Appl Physiol* **72**, 1493–1498 (1992).

54. Henriksson-Larsen, K., J. Friden, and M. Whetling, "Distribution of fiber sizes in human skeletal muscle. An enzyme histochemical study in muscle tibialis anterior," *Acta Physiol Scand* **123**, 171–177 (1985).

55. Henry, S. M., J. Fung, and F. B. Horak, "EMG responses to maintain stance during multidirectional surface translations," *J Neurophysiol* **80**, 1939–1950 (1998).

56. Hides, J. A., C. A. Richardson, and G. A. Jull, "Multifidus muscle recovery is not automatic after resolution of acute, first-episode low back pain," *Spine* **21**, 2763–2769 (1996).

57. Hodges, P. W., "Changes in motor planning of feedforward postural responses of the trunk muscles in low back pain," *Exp Brain Res* **141**, 261–266 (2201).

58. Hodges, P. W., A. G. Cresswell, and A. Thorstensson, "Perturbed upper limb movements cause short-latency postural responses in trunk muscles," *Exp Brain Res* **138**, 243–250 (2001).

59. Hodges, P. W., and S. C. Gandevia, "Activation of the human diaphragm during a repetitive postural task," *J Physiol* **522**, 165–175 (2000).

60. Hodges, P. W., and C. A. Richardson, "Feedforward contraction of transversus abdominis is not influenced by the direction of arm movement," *Exp Brain Res* **114**, 362–370 (1997).

61. Hodges, P. W., and C. A. Richardson, "Inefficient muscular stabilisation of the lumbar spine associated with low back pain: A motor control evaluation of transversus abdominis," *Spine* **21**, 2640–2650 (1996).

62. Holroyd, K. A., D. B. Penzien, K. G. Hursey, D. L. Tobin, L. Rogers, J. E. Holm, P. J. Marcille, J. R. Hall, and A. G. Chila, "Change mechanisms in EMG biofeedback training: Cognitive changes underlying improvements in tension headache," *J Consult Clin Psychol* **52**, 1039–1053 (1984).

63. Ilyina-Kakueva, E. I., V. V. Portugalov, and N. P. Krivenkova, "Space flight effects on the skeletal muscles of rats," *Aviat Space Environ Med* **47**, 700–703 (1976).

64. Jost, W. H., K. W. Ecker, and K. Schimrigk, "Surface versus needle electrodes in determination of motor conduction time to the external anal sphincter," *Int J Colorectal Dis* **9**, 197–199 (1994).

65. Jull, G. A., "Deep cervical neck flexor dysfunction in whiplash," *J Musculoskel Pain* **8**, 143–154 (2000).

66. Jull, G. A., P. Trott, H. Potter, G. Zito, K. Niere, D. Shirley, J. Emberson, I. Marschner, and C. Richarson, "A randomized controlled trial of exercise and manipulative therapy for cervicogenic headache," *Spine* **27**, 1835–1843 (2002).

67. Kamen, G., and C. J. De Luca, "Unusual motor firing behavior in older adults," *Brain Res* **482**, 136–140 (1989).

68. Kamen, G., S. V. Sison, C. C. Du, and C. Patten, "Motor unit discharge behavior in older adults during maximal-effort contractions," *J Appl Physiol* **79**, 1908–1913 (1995).

69. Kayser, B., M. Narici, T. Binzoni, B. Grassi, and P. Cerretelli, "Fatigue and exhaustion in chronic hypobaric hypoxia: Influence of exercising muscle mass," *J Appl Physiol* **76**, 634–640 (1994).

70. Kayser, B., M. Narici, and F. Cibella, "Fatigue and performance at high altitude" in J. R. Sutton, C. S. Houston, and G. Coates, eds., *Hypoxia and molecular medicine*, Burlington, VT, USA, Queen City Printers, 1993.

71. Kenyon, R. V., and L. R. Young, "M.I.T./Canadian vestibular experiments on the Spacelab-1 mission: 5. Postural responses following exposure to weightlessness," *Exp Brain Res* **64**, 335–346 (1986).

72. Keshner, E. A., "Controlling stability of a complex movement system," *Phys Ther* **70**, 844–854 (1990).

73. Kiesswetter, H., "EMG-patterns of pelvic floor muscles with surface electrodes," *Urol Int* **31**, 60–69 (1976).

74. King, D. G., and C. T. Teague, "Choice of electrode in electromyography of the external urethral and anal sphincters," *J Urol* **124**, 75–77 (1980).

75. Klein, C., D. A. Cunningham, D. H. Peterson, and A. W. Taylor, "Fatigue and recovery of contractile properties of young and elderly men," *Eur J Appl Physiol* **57**, 684–690 (1988).

76. Klitgaard, H., M. Mantoni, S. Schiaffino, S. Ausoni, L. Gorza, C. Laurent-Winter, P. Schnohr, and B. Saltin, "Function, morphology and protein expression of aging skeletal muscle: A cross-sectional study of elderly men with different training background," *Acta Physiol Scand* **140**, 41–54 (1990).

77. Koryak, Y., "Electromyographic study of the contractile and electrical properties of the human triceps surae muscle in a simulated microgravity environment," *J Physiol* **1**, 287–295 (1998).

78. Koryak, Y., "The effects of long-term simulated microgravity on neuromuscular performance in men and women," *Eur J Appl Physiol Occup Physiol* **79**, 168–175 (1999).

79. Kraemer, W. J., R. S. Staron, S. E. Gordon, J. S. Volek, L. P. Koziris, N. D. Duncan, B. C. Nindl, A. L. Gomez, J. O. Marx, A. C. Fry, and J. D. Murray, "The effects of 10 days of spaceflight on the shuttle Endeavor on predominantly fast-twitch muscles in the rat," *Histochem Cell Biol* **114**, 349–355 (2000).

80. Kugelberg, E., "Adaptive transformation of rat soleus motor units during growth," *J Neurol Sci* **27**, 269–289 (1976).

81. Lariviere, C., A. B. Arsenault, D. Gravel, D. Gagnon, and P. Loisel, "Effect of step and ramp static contractions on the median frequency of electromyograms of back muscles in humans," *Eur J Appl Physiol* **85**, 552–559 (2001).

82. Larsson, L., "Morphological and functional characteristics of the aging skeletal muscle in man, A cross-sectional study," *Acta Physiol Scand* **457** (suppl), 1–36 (1978).

83. Larsson, L., G. Grimby, and J. Karlsson, "Muscle strength and speed of movement in relation to age and muscle morphology," *J Appl Physiol* **46**, 451–456 (1979).

84. Larsson, L., and J. Karlsson, "Isometric and dynamic endurance as a function of age and skeletal muscle characteristics," *Acta Physiol Scand* **104**, 129–136 (1978).

85. Lennmarken, C. T., T. Bergman, J. Larsson, and L. E. Larsson, "Skeletal muscle function in man: force, relaxation rate, endurance and contraction time-dependence on sex and age," *Clin Physiol* **5**, 243–255 (1985).

86. Leterme, D., and M. Falempin, "EMG activity of three rat hindlimb muscles during microgravity and hypergravity phase of parabolic flight," *Aviat Space Environ Med* **69**, 1065–1070 (1998).

87. Lexell, J., D. Downham, and M. Sjostrom, "Distribution of different fibre types in human skeletal muscle: fiber type arrangement of m. vastus lateralis from three groups of healthy men between 15 and 83 years," *J Neurol Sci* **72**, 211–222 (1986).

88. Lexell, J., K. Henriksson-Larsen, B. Winblad, and M. Sjostrom, "Distribution of different fiber types in human skeletal muscles: effects of aging studied in whole muscle cross sections," *Muscle Nerve* **6**, 588–595 (1989).

89. Li, L., and G. E. Caldwell, "Coefficient of cross correlation and the time domain correspondence," *J Electromyogr Kinesiol* **9**, 385–389 (1999).

90. Lopez, A., B. Y. Nilsson, A. Mellgren, J. Zetterstrom, and B. Holmstrom, "Electromyography of the external anal sphincter: comparison between needle and surface electrodes," *Dis Colon Rectum* **42**, 482–485 (1999).

91. Lose, G., J. T. Andersen, and J. K. Kristensen, "Disposable vaginal surface electrode for urethral sphincter electromyography," *Br J Urol* **59**, 408–413 (1987).

92. Lose, G., J. K. Kristensen, and J. T. Andersen, "A disposable anal plug electrode for pelvic floor/anal sphincter electromyography," *J Urol* **137**, 249–252 (1987).
93. MacDougall, J. D., H. J. Green, J. R. Sutton, G. Coates, A. Cymerman, P. Young, and C. S. Houston, "Operation Everest II: structural adaptations in skeletal muscle in response to extreme simulated altitude," *Acta Physiol Scand* **142**, 421–427 (1991).
94. Magill, R. A., *Motor learning: Concepts and applications*, McGraw-Hill, New York, 2001.
95. Mannion, A. F., G. A. Dumas, J. M. Stevenson, and R. G. Cooper, "The influence of muscle fiber size and type distribution on electromyographic measures of back muscle fatigability," *Spine* **23**, 576–584 (1998).
96. Marras, W. S., K. G. Davis, S. A. Ferguson, B. R. Lucas, and P. Gupta, "Spine loading characteristics of patients with low back pain compared with asymptomatic individuals," *Spine* **26**, 2566–2574 (2001).
97. Mayoux-Benhamou, M. A., M. Revel, and C. Vallee, "Selective electromyography of dorsal neck muscles in humans," *Exp Brain Res* **113**, 353–360 (1997).
98. Mayoux-Benhamou, M. A., M. Revel, C. Vallee, R. Roudier, J. P. Barbet, and F. Bargy, "Longus colli has a postural function on cervical curvature," *Surg Radiol Anatomy* **16**, 367–371 (1994).
99. McDonagh, M. N., M. J. White, and C. M. Davies, "Different effects of ageing on the mechanical properties of human arm and leg muscles," *Geront* **30**, 49–54 (1984).
100. McKenna, P. H., C. D. Herndon, S. Connery, and F. A. Ferrer, "Pelvic floor muscle retraining for pediatric voiding dysfunction using interactive computer games," *J Urol* **162**, 1056–1062 (1999).
101. Merletti, R., P. Enck, M. Gazzoni, and H. Hinninghofen, "Surface EMG recording of single motor unit action potentials from the external anal sphincter," *XIVth Congr Int Society of Electrophysiology and Kinesiology*, Vienna, June 22–25, 2002.
102. Merletti, R., D. Farina, M. Gazzoni, and M. P. Schieroni, "Effect of age on muscle functions investigated with surface electromyography," *Muscle Nerve* **25**, 65–76, Jan 2002.
103. Merletti, R., L. Lo Conte, C. Cisari, and M. V. Actis, "Age related changes in surface myoelectric signals," *Scand J Rehab Med* **24**, 25–36 (1992).
104. Merletti, R., and S. Roy, "Myoelectric and mechanical manifestations of muscle fatigue in voluntary contractions," *J Orthop Sports Phys Ther* **24**, 342–353 (1996).
105. Moseley, G. L., P. W. Hodges, and S. C. Gandevia, "Deep and superficial fibers of lumbar multifidus are differentially active during voluntary arm movements," *Spine* **27**, 27–36 (2002).
106. Narici, M. V., M. Bordini, and P. Cerretelli, "Effect of aging on human abductor pollicis muscle function," *J Appl Physiol* **71**, 1277–1281 (1991).
107. Narici, M. V., and B. Kayser, "Hypertrophic response of human skeletal muscle to strength training in hypoxia and normoxia," *Eur J Appl Physiol* **70**, 213–219 (1995).
108. Ng, J. K., V. Kippers, and C. A. Richardson, "Muscle fibre orientation of abdominal muscles and suggested surface EMG electrode positions," *Electromyogr Clin Neurophysiol* **38**, 51–58 (1998).
109. Nouwen, A., P. F. VanAkkerveeken, and J. M. Versloot, "Patterns of muscular activity during movement in patients with chronic low back pain," *Spine* **12**, 777–782 (1987).
110. Orizio, C., F. Esposito, and A. Veicsteinas, "Effect of acclimatization to high altitude (5,050 m) on motor unit activation pattern and muscle performance," *J Appl Physiol* **77**, 2840–2844 (1994).
111. O'Sullivan, P. B., L. T. Twomey, and G. T. Allison, "Evaluation of specific stabilizing exercise in the treatment of chronic low back pain with radiologic diagnosis of spondylolysis or spondylolisthesis," *Spine* **22**, 2959–2967 (1997).

112. Panjabi, M. M., "The stabilizing system of the spine. Part I: Function, dysfunction, adaptation, and enhancement," *J Spinal Dis* **5**, 383–389 (1992).

113. Pfister, C., J. N. Dacher, S. Gaucher, A. Liard-Zmuda, P. Grise, and P. Mitrofanoff, "The usefulness of a minimal urodynamic evaluation and pelvic floor biofeedback in children with chronic voiding dysfunction," *BJU Int* **84**, 1054–1057 (1999).

114. Pinho, M., K. Hosie, K. Bielecki, and M. R. Keighley, "Assessment of noninvasive intra-anal electromyography to evaluate sphincter function," *Dis Colon Rectum* **34**, 69–71 (1991).

115. Porter, M. M., A. A. Vandervooert, and J. Lexell, "Aging of human muscle: Structure, function and adaptability," *Scand J Med Sci Sports* **5**, 129–142 (1995).

116. Portero, P., C. Vanhoutte, and F. Goubel, "Surface electromyogram power spectrum changes in human leg muscles following 4 weeks of simulated microgravity," *Eur J Appl Physiol Occup Physiol* **73**, 340–345 (1996).

117. Preece, A. W., H. S. Wimalaratna, J. L. Green, E. Churchill, and H. M. Morgan, "Non-invasive quantitative EMG," *Electromyogr Clin Neurophysiol* **34**, 81–86 (1994).

118. Radebold, A., J. Cholewicki, M. M. Panjabi, and T. C. Patel, "Muscle response pattern to sudden trunk loading in healthy individuals and in patients with chronic low back pain," *Spine* **25**, 947–954 (2000).

119. Recktenwald, M. R., J. A. Hodgson, R. R. Roy, S. Riazanski, G. E. McCall, I. Kozlovskaya, D. A. Washburn, J. W. Fanton, and V. R. Edgerton, "Effects of spaceflight on rhesus quadrupedal locomotion after return to 1G," *J Neurophysiol* **81**, 2451–2463 (1999).

120. Richardson, C. A., G. A. Jull, P. W. Hodges, and J. A. Hides, *Therapeutic exercise for spinal segmental stabilisation in low back pain: Scientific basis and clinical approach*, Churchill Livingstone, Edinburgh. 1999.

121. Riley, D. A., G. R. Slocum, J. L. Bain, F. R. Sedlak, T. E. Sowa, and J. W. Mellender, "Rat hindlimb unloading: Soleus histochemistry, ultrastructure, and electromyography," *J Appl Physiol* **69**, 58–66 (1990).

122. Rogers, M. A., and W. J. Evans, "Changes in skeletal muscle with aging: effects of exercise training," *Exer Sport Sci Rev* **21**, 65–102 (1993).

123. Roll, J. P., K. Popov, V. Gurfinkel, M. Lipshits, C. Andre-Deshays, J. C. Gilhodes, and C. Quoniam, "Sensorimotor and perceptual function of muscle proprioception in microgravity," *J Vestib Res* **3**, 259–273 (1993).

124. Roy, S. H., P. Bonato, and M. Knaflitz, "EMG assessment of back muscle function during cyclical lifting," *J Electromyogr Kinesiol* **8**, 233–245 (1998).

125. Roy, S. H., C. J. De Luca, and D. A. Casavant, "Lumbar muscle fatigue and chronic lower back pain," *Spine* **14**, 992–1001 (1989).

126. Roy, S. H., C. J. De Luca, M. Emley, L. I. E. Oddsson, R. J. C. Buijs, J. Levins, D. S. Newcombe, and J. F. Jabre, "Classification of back muscle impairment based on the surface electromyographic signal," *J Rehabil Res Dev* **34**, 405–414 (1997).

127. Shafik, A., "A new concept of the anatomy of the anal sphincter mechanism and the physiology of defecation: Mass contraction of the pelvic floor muscles," *Int Urogynecol J Pelvic Floor Dysfunct* **9**, 28–32 (1998).

128. Shephard, R., W. Montelpare, M. Plyley, D. McCracken, and R. Goode, "Handgrip dynamometer, Cybex measurements and lean mass as markers of the aging of muscle function," *Br J Sports Med* **25**, 204–208 (1991).

129. Siroky, M. B., "Electromyography of the perineal floor," *Urol Clin North Am* **23**, 299–307 (1996).

130. Soderberg, G. L., S. D. Minor, and R. M. Nelson, "A comparison of motor unit behavior in young and aged subjects," *Age Ageing* **20**, 8–15 (1991).

131. Solomonow, M., B. H. Zhou, R. V. Baratta, Y. Lu, and M. Harris, "Biomechanics of increased exposure to lumbar injury caused by cyclic loading. Part 1: Loss of reflexive muscular stabilization," *Spine* **24**, 2426–2434 (1999).

132. Stalberg, E., O. Borges, M. Ericsson, B. Essen-Gustavsson, P. Fawcet, P. Nordesjo, B. Nordgren, and R. Uhlin, "The quadriceps femoris muscle in 20–70 year old subjects: Relationship between knee extension torque, electrophysiological parameters and muscle fiber characteristics," *Muscle Nerve* **12**, 382–389 (1989).

133. Staron, R. S., W. J. Kraemer, R. S. Hikida, D. W. Reed, J. D. Murray, G. E. Campos, and S. E. Gordon, "Comparison of soleus muscles from rats exposed to microgravity for 10 versus 14 days," *Histochem Cell Biol* **110**, 73–80 (1998).

134. Thompson, L. V., "Effects of age and training on skeletal muscle physiology and performance," *Phys Ther* **74**, 71–81 (1994).

135. Tischler, M. E., E. J. Henriksen, K. A. Munoz, C. S. Stump, C. R. Woodman, and C. R. Kirby, "Spaceflight on STS-48 and earth-based unweighting produce similar effects on skeletal muscle of young rats," *J Appl Physiol* **74**, 2161–2165 (1993).

136. Vahtera, T., M. Haaranen, A. L. Viramo-Koskela, and J. Ruutiainen, "Pelvic floor rehabilitation is effective in patients with multiple sclerosis," *Clin Rehabil* **11**, 211–219 (1997).

137. Vandervoort, A. A., and A. J. McComas, "Contractile changes in opposing muscles of the human ankle joint with aging," *J Appl Physiol* **61**, 361–367 (1986).

138. Vereecken, R. L., B. Puers, and J. Van Mulders, "Spectral analysis of perineal muscles EMG," *Electromyogr Clin Neurophysiol* **22**, 321–326 (1982).

139. Watson, D. H., and P. H. Trott, "Cervical headache: an investigation of natural head posture and upper cervical flexor muscle performance," *Cephalalgia* **13**, 272–284 (1993).

140. Welsh, L., and O. Rutherford, "Effects of isometric strength training on quadriceps muscle properties in over 55 year olds," *Eur J Appl Physiol* **72**, 219–223 (1996).

141. Wennergren, H., L. E. Larsson, and P. Sandstedt, "Surface electromyography of pelvic floor muscles in healthy children: Methodological study," *Scand J Caring Sci* **3**, 63–69 (1989).

142. Wertsch, J. J., "Anatomical and technical issues in electrodiagnosis," *Phys Med Rehabil Clin N Am* **9**, 1047–9651 (1998).

143. West, J. B., "Climbing Mt. Everest without oxygen: an analysis of maximal exercise during extreme hypoxia," *Respir Physiol* **52**, 265–279 (1983).

144. West, J. B., S. J. Boyer, D. J. Graber, P. H. Hackett, K. H. Maret, J. S. Milledge, R. M. Peters Jr, C. J. Pizzo, M. Samaja, F. H. Sarnquist, et al., "Maximal exercise at extreme altitudes on Mount Everest," *J Appl Physiol* **55**, 688–698 (1983).

145. Widrick, J. J., J. G. Romatowski, J. L. Bain, S. W. Trappe, T. A. Trappe, J. L. Thompson, D. L. Costill, D. A. Riley, and R. H. Fitts, "Effect of 17 days of bed rest on peak isometric force and unloaded shortening velocity of human soleus fibers," *Am J Physiol* **273**, 1690–1699 (1997).

146. Wilder, D. G., A. R. Aleksiev, M. L. Magnusson, M. H. Pope, K. F. Spratt, and V. K. Goel, "Muscular response to sudden load: A tool to evaluate fatigue and rehabilitation," *Spine* **21**, 2628–2639 (1996).

147. Winters, J. M., and J. D. Peles, "Neck muscle activity and 3-D head kinematics during quasi-static and dynamic tracking movements," in J. M. Winters, and S. L. Y. Woo, eds., *Multiple muscle systems: Biomechanics and movement organization*, Springer-Verlag, New York, 1990.

148. Workman, D. E., J. E. Cassisi, and M. C. Dougherty, "Validation of surface EMG as a measure of intravaginal and intra-abdominal activity: Implications for biofeedback-assisted Kegel exercises," *Psychophysiology* **30**, 120–125 (1993).

149. Young, A., M. Stokes, and M. Crowe, "The size and strength of the quadriceps muscles of old and young men," *Clin Physiol* **5**, 145–154 (1985).

150. Young, A., J. Wright, J. Knapik, and A. Cymerman, "Skeletal muscle strength during exposure to hypobaric hypoxia," *Med Sci Sports Exer* **12**, 330–335 (1980).

151. Zedka, M., A. Prochazka, B. Knight, D. Gillard, and M. Gauthier, "Voluntary and reflex control of human back muscles during induced pain," *J Physiol (Lond)* **520**, 591–604 (1999).

152. Zermann, D. H., M. Ishigooka, H. Wunderlich, O. Reichelt, and J. Schubert, "A study of pelvic floor function pre- and postradical prostatectomy using clinical neurourological investigations, urodynamics and electromyography," *Eur Urol* **37**, 72–78 (2000).

153. Zwarts, M. J., T. W. Van Weerden, and H. T. Haenen, "Relationship between average muscle fibre conduction velocity and EMG power spectra during isometric contraction, recovery and applied ischemia," *Eur J Appl Physiol Occup Physiol* **56**, 212–216 (1987).

17

BIOFEEDBACK APPLICATIONS

J. R. Cram

Sierra Health Institute
Nevada City, Ca

17.1 INTRODUCTION

The purpose of this chapter is to provide a basic overview of how surface electromyography (SEMG) may be used in a clinical setting as a biofeedback tool. These applications are commonly found in the fields of clinical psychology and physical medicine. Since the muscles are controlled and regulated by many parts of the brain and nervous system, clinically it is useful to aggregate and consider three domains: the emotions, posture and movement. These frameworks will be briefly reviewed. The chapter will then review a general approach to understanding the clinical assessment and treatment of muscle impairment syndromes. It will expand on how a few of these clinical areas are addressed using biofeedback and will conclude with a brief description of some of the ways SEMG biofeedback is conducted.

Muscle dysfunction can occur for many reasons, some of them having to do with pathology of the muscle tissue, some of them having to do with trauma or injury, and some of them having to do with patterns of use or misuse. George Whatmore [20], one of the founding fathers in the use of SEMG as a clinical feedback tool, referred to muscle dysfunction as *dysponesis*, or bad energy. He noted how problems could emerge when an individual's "performing efforts" (muscle involvement) were too high or too low. The way in which we purposefully move our body may lead the way to our discomfort. Writer's cramp would be an obvious example. He also talked about "bracing efforts," noting how excessive postural tension could effect movement, and overall arousal levels. Finally, he categorized a large number of "representing efforts," to describe the many ways in which the

Electromyography: Physiology, Engineering, and Noninvasive Applications, edited by Roberto Merletti and Philip Parker.
ISBN 0-471-67580-6 Copyright © 2004 Institute for Electrical and Electronics Engineers, Inc.

emotions are reflected within the neuromuscular network. Clinically these are commonly reflected in the patterns of muscle activity associated with our fears and concerns.

All three of these domains—emotions, posture, and movement—interact. The professional athlete knows them well. In the world of golf, how one "addresses" the ball with the posture of the stance, lays down the foundation for the swing. Placement of the feet, for example, will determine the direction of the flight of the ball. How the actual swing comes about needs to be natural, in the "groove." The club head should not go back too far, the elbow needs to stay close to the torso, the wrists are held in a certain way, and the weight of the swing transfers through the hips and feet in a natural way. Learning how to move correctly is essential. And there are video-oriented biofeedback systems that can show the golfer how his stance and movement unfold. With this information, the golfer can then learn to place the backswing at the right distance, hold his arms in the correct way, and learn how to shift his or her weight so that it is easy and natural. But the neuromuscular element that separates the professional golfers from the recreational ones is the professional's ability to control their emotions. Hubbard and Berkoff [11], along with McNutty et al. [17], have suggested that the autonomic nervous system plays a significant role in the calibration of the gamma motor system and the stretch receptor. Thus even a minor increase in autonomic arousal associated with the fear of missing the shot or the thrill of doing so well can unwittingly change the timing of the swing, or the strength of a putt. Along with teaching the golfer how to stand and swing, SEMG biofeedback could be used to teach the golfer how to better control his or her emotional tone.

While the paragraph above speaks more to the role of biofeedback in teaching "peak performance" to a golfer, biofeedback can be used to treat a number of clinical impairment syndromes. The rest of this chapter will focus on these.

17.2 BIOFEEDBACK APPLICATION TO IMPAIRMENT SYNDROMES

The concepts presented below may be studied in greater depth in *Clinical Applications in Surface EMG* by Kasman et al. [13] The seven clinical frameworks presented in Figure 17.1 all feed into a self-perpetuating "faulty motor schema." Just like elevations in blood pressure do not suggest the origin or etiology of the problem, alterations in SEMG activity need to be understood in a broader context. Consider elevations in SEMG levels at rest in the upper trapezius following a motor vehicle accident, for example. These may be due to phobic responses to driving (psychophysiological stress), head forward position (postural dysfunction), fear of movement (learned guarding), herniated cervical discs (reflex spasm), facet blocks (hypermobility of the cervical joints), or tightness of the upper trapezius relative to lower trapezius (muscle imbalances). Note also that clinical syndromes presented in this framework are not mutually exclusive and that patients may exhibit qualities of one or more of the syndromes. Once the etiology is better understood, biofeedback may be one of the treatment elements. Perhaps it can play a role in restoring the normal motor program that has been altered.

17.2.1 Psychophysiological, Stress-Related Hyperactivity

SEMG activity at rest or during movement is elevated either due to general maladaptive coping to stressful situations or a conditioned emotional response to a traumatic event (post traumatic stress syndrome). A patient four weeks post–motor vehicle accident (MVA) who sustained a flexion extension injury to the neck and shoulder region provides a con-

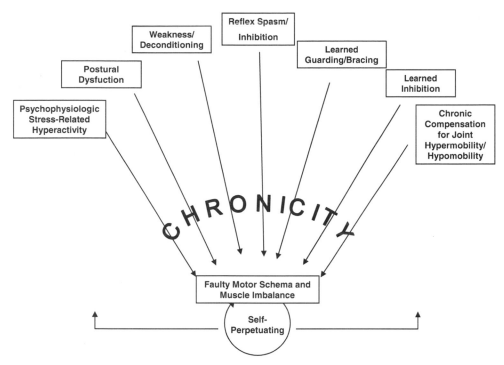

Figure 17.1. Factors contributing to the development of faulty motor schema and muscle imbalances. (From [13] with permission)

venient example. Due to the emotional feature (PTSD) associated with the accident, traditional medical and physical therapies offered to the patient do not produce long-term gains. Clinically a "stress profiling" procedure may be conducted in which the offending muscle is monitored as the patient recalls in detail the events of the MVA. SEMG recruitment patterns noted during recall of the MVA scene (dysponesis) provide evidence as to this emotional component. When identified, a combination of SEMG "downtraining" and psychological therapy to treat the posttraumatic stress disorder that precedes or is done concurrently with the more physically based therapeutic regimes would greatly facilitate the clinical outcome.

17.2.2 Simple Postural Dysfunction

Aberrant motor activity is shown to be a direct function of posture. An example of this may be seen in Figure 17.2. The increased paraspinal muscle activity seen in this tracing is reduced as the head is moved from a head forward position to one in which the head is well positioned over its center of gravity. The initial SEMG elevation is associated with a head forward position and is likely due to the increased load placed on them from the head being forward of its center of gravity. This increased load would also place untoward loading of the articular structures and chronically place the ligaments on stretch. According to McKenzie [16], this chronic physical stress on the soft tissue creates the foundations for pain. In addition muscle length–tension relationships become inefficient. The load moments are increased by lengthening the lever arm through which gravity acts, and the normal torque values are disrupted as some muscles recruitment is at an increased level

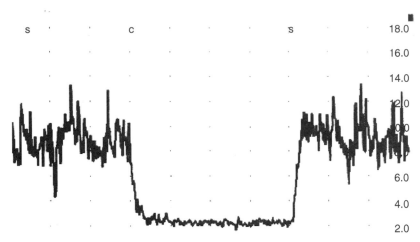

Figure 17.2. SEMG activity of the upper cervical paraspinal muscle recorded during a patient's spontaneous forward head posture (S) and corrected postural alignment (C). (From [13] with permission)

while the antagonists takes on a lessor role. A typical clinical example would be of a patient with headache and tension myalgia of the upper quarter and neck associated with work as a keyboard entry person. Postural examination of the patient shows that the patient sits with a head forward position and with the arms extended slightly while typing.

17.2.3 Weakness/Deconditioning

While muscle weakness may be associated with radiculopathies, plexopathies, neuropathies, or myopathies, the weakness/deconditioning syndrome referred to here involves simple muscle disuse. This may be caused by immobilization after injury or surgery, or as the cumulative effect of poor motor habits and decreased activity. The condition may include atrophic loss of muscle cross-sectional area, inefficient vascularization, and compromised biochemical and physiological function [6]. There may also be a change or diminution in neural drive that accompanies changes in muscle tissue. The patterns of dysponetic SEMG activity may include decrements in peak torque, power deficits (i.e., inability to sustain force through ROM arcs), and impaired fatigue resistance.

Consider the case of a patient who has had his knee immobilized and weight-bearing restricted for six weeks after sustaining a leg fracture. The quadriceps will undergo disuse atrophy during the period of immobilization. Range of motion, strength, endurance, and functional mobility become impaired during the rehabilitation period. When the knee patient is cleared for active strengthening and is examined with SEMG, quadriceps SEMG activity will likely differ between the involved and uninvolved lower extremities. Maximal effort SEMG activity will probably be decreased on the involved side. However, one should be aware that submaximal contractions may show increased activity, presumably reflecting decreased neuromuscular efficiency. In addition the asymmetrical SEMG during maximal effort will be associated with obvious findings of weakness and deconditioning on physical examination.

TABLE 17.1. Reflexive Muscle Spasm in a Muscle Scanning Protocol of a Low-Back Pain Patient with Herniated Disk

Muscle Site	Sit Left	Right	Stand Left	Stand
Cervical	1.2	1.5	1.5	1.7
Trapezius	2.2	3.3	3.2	3.5
T1 paraspinals	1.2	1.3	1.2	1.2
T6 paraspinals	3.7	3.8	5.2	5.4
T10 paraspinals	10.3*	17.1*	4.4	5.9
L3 paraspinals	10.4*	15.2*	4.2	5.1
Abdominals	1.3	1.5	1.6	1.8

Source: From J. R. Cram (ed.), *Clinical EMG for Surface Recording*, Vol. 2 (Nevada City, CA, 1990).

Note: Pain is worse upon sitting. Values are in μV_{RMS}.

* indicates values are outside the normal and expected limits (Cram and Engstrom [4]).

17.2.4 Acute Reflexive Spasm/Inhibition

Spasm is defined as an involuntary hypertonicity induced by the spinal reflex system [14]. Spasm is commonly triggered by noxious mechanical or chemical stimulation of the pain receptors within the muscle or the associated joint. Inhibition is a neurological suppression of muscle activity induced by pain and/or effusion [7]. It is driven, in part, by the golgi tendon organ and designed to protect the tendonous attachments and the joint.

Take an example of a low-back pain patient with a bulging or herniated disc with pain in the right aspect of the low back, radiating down into the right hip and leg. A back patient with a known bulging or herniated disc has decreased range of motion of the torso and poor sitting tolerance. The patient might present with a flexed and laterally shifted trunk posture, visibly and palpably elevated lumbar paraspinal tone. Pain is on the same side as the bulging disc. The pain is increased with sitting (supported or not), and the SEMG activation level also increases. Any movements, active or passive, of the lower extremity lead to an activation of the right erector spinae muscles. The pattern of spasm/activation during rest in the seated posture may be easily seen in Table 17.1, reflected in the "muscle scanning" procedure developed by Cram [4].

An example of acute reflexive inhibition might be seen in a patient with a recent history of trauma and physical examination findings of swelling, tenderness, and inability to tolerate vigorous manual muscle testing of the lower extremity. SEMG monitoring would show a discrete focal drop in SEMG amplitude recorded from the quadriceps during a painful portion of the knee range of motion arc. The focal drop in SEMG activity in this case would be as a consequence of neurophysiologic inhibition.

17.2.5 Learned Guarding/Bracing

This pattern of neuromuscular activity differs from the reflex spasm model, in that the pattern of muscle activity is "learned" or operantly conditioned rather than being strictly mandated by a reflex. The increased muscle activity usually occurs upon movement or postural loading and is performed in an attempt to avoid pain and the "possibility" of further injury. Many times the activation patterns are seen on the side opposite the pain

TABLE 17.2. Protective Guarding Pattern in a Muscle Scanning of a Low-Back Pain Patient

Muscle Site	Sit Left	Right	Stand Left	Right
Frontal	3.5	3.8	1.2	1.3
Temporal	3.5	3.5	4.6	3.6
Masseter	1.3	1.5	1.5	1.6
SCM	1.2	1.5	0.8	0.9
Cervical	1.6	1.7	1.9	1.9
Trapezius	2.2	2.3	3.0	2.5
T1 paraspinals	1.5	1.5	1.3	1.1
T6 paraspinals	3.7	3.5	4.7	4.4
T10 paraspinals	9.6*	19.2*	44.8*	59.2*
L3 paraspinals	2.2	11.2*	19.2*	30.4*
Abdominals	1.3	1.5	1.6	1.8

Source: From J. R. Cram (ed.), *Clinical EMG for Surface Recording*, Vol. 2 (Nevada City, CA, 1990).

Note: The patient's pain is perceived on the left side of the back and hip. Values are in μV_{RMS}.

* indicates values are outside the normal and expected limits (Cram and Engstrom [4]).

(ipsilateral) as the patient exhibits a "learned disuse" of the painful side and a hyperactivity of the nonpainful side.

Consider the pattern of muscle activity for a patient with low back pain shown in Table 17.2. This SEMG data were collected using the muscle "scanning procedure" mentioned above, in which the right and left aspects of multiple muscle groups were quickly sampled in the seated and standing postures. The patient has left-sided back pain that has radiated down into the left hip and leg. Disc herniation is known not to exist. The pattern of activity shows increased SEMG activation on the side opposite of the pain, suggesting a protective guarding pattern. Here the patient has learned to "weight-shift" away from the pain (antalgic posture).

17.2.6 Learned Inhibition/Weakness

This syndrome is similar to the protective guarding and bracing model presented above. It differs in that it focuses on the "inhibition" side of the perspective. It is not uncommon, for example, for the patient to "learn" not to move an injured or painful site. The less they move the muscle or joint, the less pain they feel. And through an operant process of negative reinforcement, they "learn disuse." It is actually quite common in a unilateral neck/shoulder injury to see the injured to have "normal" resting tone while the uninvolved side to be hyperactive at rest (protective guarding). But when the shoulders are elevated, for example, the injured side shows a hypoactive recruitment pattern not reaching the same level of activation as the uninvolved side. It is this hypoactivity during strong recruitment that is referred to here as learned inhibition.

Another example would be an otherwise healthy patient who sustains recurrent strains of the hip adductor muscles while playing racquetball. The pain becomes severe and exacerbated whenever the adductor muscles vigorously contract during functional activities. To avoid the contraction-induced pain, the patient learns to reduce firing of the adductors while performing stressful physical activities. Over a period of time the altered patterns become unconsciously incorporated into the patient's selection of motor programs. Interestingly enough, the adductor SEMG amplitude of this patient appears symmetrical on

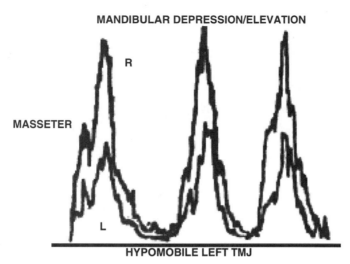

MANDIBULAR DEPRESSION/ELEVATION

Figure 17.3. SEMG activity of the left (*L*) and right (*R*) masseter muscles recorded from a patient with left TMJ pain during three repetitions of jaw opening and closing. Decreased left side muscle activity was associated with hypomobility of the left joint. (From [13] with permission)

both the involved and uninvolved side during walking and low level activities. In fact, when the patient is subjected to an unanticipated postural perturbation, the adductors are noted to recruit normally. The muscles are normally recruited with postural reactions to help prevent a fall. However, during higher velocity and loading conditions such as sustained unilateral stance, lunging, or formal manual muscle testing, activity on the involved side appears to be markedly decreased and impaired.

17.2.7 Direct Compensation for Joint Hypermobility or Hypomobility

In this syndrome the neuromuscular system compensates by attempting to stabilize lax joint structures, by affecting movement against joint stiffness, or by subserving linked compensatory movements over kinetic chains [10]. Although SEMG activity is aberrant, the primary problem is a biomechanical articular fault. The articular fault is causal to a compensatory motor control pattern, which may spontaneously resolve upon improvement in joint mechanics. Chronic joint dysfunction may lead to motor control problems that themselves contribute to deterioration of the kinetic segment and persist even after joint mobility improves. The distinction is made because if aberrant motor activity is felt to be directly compensatory to articular dysfunction, then biofeedback is not a first choice of treatment. The joint dysfunction should be addressed, and then SEMG activity re-assessed.

An example of this may be seen in Figure 17.3. Here a patient with jaw pain is found on physical examination to display hypomobility at the left temporomandibular joint (TMJ). There is a deviation of the midline of the jaw during opening and closing, and a palpable difference between the motions of the left and right mandibular condyles. As opening is initiated (or closing completed) the condyles are felt to spin in place. The condyles are then felt to translate forward as opening continues. This rolling/gliding relationship is necessary for normal jaw range of motion and is expected to be symmetrical at the left and right TMJs. In our case example, SEMG activity shows greater recruitment at the right masseter during jaw opening/closing range of motion. The right mandibular

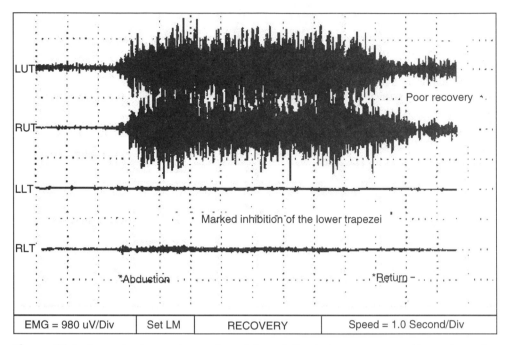

Figure 17.4. A grossly abnormal recording of the relation between upper and lower trapezius during abduction to 90 degrees, in which the upper trapezius dominated over lower trapezius. (From [5] with permission)

condyle translates a greater distance along the articular surface of the zygomatic process. The right masseter is activated to a greater degree to subserve the greater range of movement than the right TMJ. The fundamental problem, however, is not one of the right greater than left masseter sEMG activity, but one of the left less than right joint mobility. SEMG spontaneously becomes symmetrical once the left TMJ is mobilized with manual techniques or exercises.

17.2.8 Chronic Faulty Motor Programs

The eighth and final syndrome is a amalgamation and perpetuation of all of the syndromes above. Here we assume that the central nervous system learns to cope with pain, muscle weakness, joint instabilities, trigger points, myofascial extensibility issues, and so on. As a result there is a learned disruption of the normal agonist–antagonist–synergist relationships. The assessment (and treatment) of this broad syndrome requires SEMG monitoring along with assessment of coincident joint segment dysfunction, soft tissue dysfunction, and behavioral analysis.

Consider Figure 17.4. Here a patient with chronic cervical paraspinal and suprascapular pain is examined following lifting activities at work. Motion takes place throughout the shoulder girdle to elevate the arms to the side (abduction). This includes upward rotation of the scapula, achieved by the coordinated actions of the upper trapezius, the lower trapezius, the lower fibers of the serratus anterior, and numerous other muscles with direct and indirect stabilizing roles. A motor program is a planned set of commands from the central nervous system that serves to coordinate the actions of muscles so that a specific goal is achieved, in this case shoulder abduction. If an inefficient motor program is selected, then one muscle might contract with excessive or reduced tension relative to its synergist, resulting in abnormal loading patterns of both myofascial and articular tissue. With our patient example, it is observed that the SEMG activity of the upper trapezius is considered hyperactive, whereas the activity of the lower trapezius is inhibited. In addition the patient has a poor ability to recognize these patterns of activation and tension and is unable to voluntarily activate the lower trapezius. Biofeedback is used to re-educate the patient about his muscle function, and to develop a more appropriate motor program, one that includes the scapular stabilizers (lower trapezius) in arm movement patterns.

In summary, the preceding impairment syndromes may be seen in a variety of clinical disorders. Some of these have been described and explored above. A partial list of disorders with neuromuscular or musculoskeletal components that are treatable using SEMG biofeedback include tension headache, TMD and myofacial pain disorders, posttraumatic stress disorder, cervical dysfunction, shoulder girdle and upper extremity dysfunctions, low-back dysfunction, hip dysfunction, knee dysfunction, stroke, and urinary incontinence. Application guides that describe clinical protocols for assessment and treatment of many of these disorders are available in [13].

17.3 SEMG BIOFEEDBACK TECHNIQUES

Biofeedback techniques fall roughly into three clinical entities: downtraining (systemic relaxation), uptraining, and coordination training. In the broadest sense, the downtraining techniques are used to facilitate a reduction in muscles that are overactive. As noted above, they may be overactive for many reasons. If the etiology includes an element of emotions, then the down training usually falls under the rubric of systemic relaxation. But there are many other reasons to downtrain the neuromuscular system. Uptraining is actually easier to train, because the patient is being asked to do something, to learn how to turn on a particular muscle or muscle group. This type of training is commonly done when working with inhibited muscles, or muscles that have been weakened due to disuse or injury. Coordination training is considered an advanced level of training and usually follows successful up- or downtraining. It entails trying to teach the patient how to obtain the correct balance of agonists or antagonists. This is more difficult, since the cooperation of muscles involves all three domains of posture, movement, and emotions.

Below I will describe some of the more common SEMG biofeedback techniques. Further examples and explanations of these biofeeedback strategies can be found in [5,13].

17.3.1 Isolation of Target Muscle Activity

Most SEMG biofeedback training strategies begin with assisting the patient to locate, proprioceptively, the dysfunctional muscle. The goal is to learn to isolate it from other muscles. This means learning to contract it alone, and not in concert with surrounding or synergistic muscles. Visual feedback using time series display of either a raw or processed SEMG signal is usually used to guide the patient's efforts. Figure 17.5 shows an example of training a patient to selectively contract the left lower trapezius (LLT) without creating a co-contraction in the right lower trapezius. Here the first four contractions are effective, but the fifth one begins to over generalize to the ipsalateral muscle. Through successive training attempts, the patient is taught to produce more and more effort to activate the muscle of interest, while simultaneously inhibiting associated or surrounding muscles.

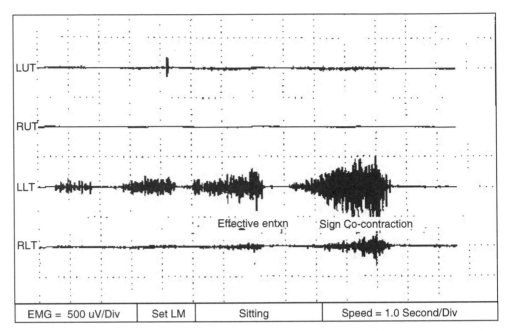

Figure 17.5. Surface EMG recording from the right and left upper and lower trapezius during increasing efforts to recruit and isolate the left lower trapezius. Note that by the fourth attempted recruitment, the patient begins to contract the right lower trapezius along with the left. (From [5] with permission)

This task may involve open isometric contractions, resisted isometric contractions, open movement, and postural adjustments. Once the patient has learned to isolate a particular muscle, then isolation training will proceed to other muscles. In the example above, once the left lower trapezius has been isolated, then the right lower trapezius is attempted, followed by the left upper trapezius, and finally the right upper trapezius. Once isolation is possible, then more advanced coordination training strategies can be employed.

17.3.2 Relaxation-Based Downtraining

Historically SEMG biofeedback training was primarily used as feedback to assist in the cultivation of low arousal and relaxation [2,9,19]. Here it is primarily used to treat the emotional layer of neuromuscular dysfunction. In general, the patient is taught a general or systemic relaxation technique that is assisted by SEMG biofeedback. Usually muscles that are involved in emotional displays (i.e., widely spaced frontal placements or trapezius placements) are monitored. However, it is not uncommon to also monitor the injured site to assess and treat the emotional aspects of trauma. The patient is commonly provided with auditory feedback to guide them on a moment to moment basis, while simultaneously using a time series graph display, usually with a very long sweep time, to see how the relaxation technique has worked over time. The goal is to see a general diminution of resting tone as the patient "relaxes."

The selection of the specific relaxation technique depends on the practitioner's training and the patient's tendencies. A patient who understands best the physical, concrete world might enjoy the systematic exploration of relaxation using a progressive relaxation introduced by Jacobson [12]. Here the patient is taught to tense a release muscles in a systematic step-by-step, muscle-by-muscle process. A patient who is more interested in how the mind effects the body might better enjoy an autogenics training program by Luthe [15]. Here the patient is taught a series of very specific phases or "formulas" to repeat that are known to specifically alter physiological functions. Or a patient whose strongest attribute is visual representation might best enjoy a "guided imagery," such as those suggested by Peper and Holt [18] Here the patient is asked to close their eyes and the therapist guides them through a relaxing experience, such as laying on the beach on a sunny day soaking in the warmth of the sun and the sand.

17.3.3 Threshold-Based Uptraining or Downtraining

Threshold-oriented SEMG biofeedback training utilizes a goal attainment model of training. This technique may be used to teach the patient to either turn on or off a particular muscle or general region. Here the patient is presented with a highly smoothed and processed SEMG signal along with a visual and/or auditory marker set by the therapist. The marker or "threshold" is either a colorful line strategically placed on top of a time series scroll, or an audio event (midi-tone or music), which is turned on or off when the patient meets the specified threshold.

The patient is instructed in strategies on how to exceed or fall below this marker. If the patient succeeds in the attempts to meet or exceed the threshold, the audio event is played, or the color of the SEMG tracing changes, or the therapist who is attending the training says "good" to reinforce the performance. Once the patient can meet or exceed the criterion set by the therapist 80% or 90% of the time, the therapist raises the threshold in the case of uptraining, so that the patient must go to an even higher level of SEMG activity to obtain the threshold oriented feedback reward. The opposite is true, of course, for a downtraining protocol. If the patient is having a difficult time meeting the threshold at least 50% of the time, the therapist can change the threshold to afford a higher level of success.

As one can see, the threshold is systematically set and changed to shape and reinforce the patient's SEMG activity to the desired levels. Many computer-based biofeedback systems are programmed to do this shaping strategy automatically.

17.3.4 Threshold-Based Tension Recognition Training

Threshold markers may be used in many ways to shape a variety of desired SEMG behaviors. One of these behaviors is the ability to accurately perceive how tense or active the patient's muscle really is. As it turns out, patients who have problems with muscle-oriented pain and tension also commonly have lost touch with the proprioceptive nature of what muscle tension feels like. In other words, they commonly don't know whether or not a muscle is tense.

The tension recognition technique differs from the uptraining and downtraining approach, in that the threshold line now represents the "targeted level" of SEMG activity the patient is trying to match, rather than the point to exceed or fall below. Here the patient is systematically trained to go to 5 microvolts, say, over and over and over again. The goal is to train the patient in proprioceptively to know where 5 microvolts is from various resting levels and in a variety of postures. Such threshold training is thought to act like an anchor, allowing the patient to more accurately perceive whether his or her muscle activity is elevated above or relaxed below this threshold. Such sensory-based discrimi-

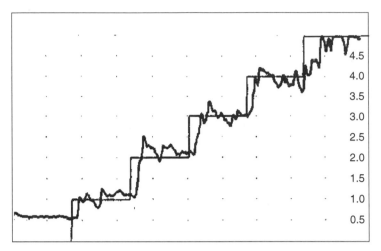

Figure 17.6. Tension discrimination training with use of sEMG feedback from the upper trapezius, with an increase in task complexity from the single-threshold model. The "staircase" template shown here is overlaid onto the graphic display. The patient is instructed to match the template as closely as possible. (From [5] with permission)

nation is very powerful. In one study on headaches [3] this discrimination training approach was demonstrated to promote longer term therapeutic gains than did the general relaxation approach described above.

17.3.5 Tension Discrimination Training

This procedure is a more advanced form of the threshold-based tension recognition training procedure. It differs only in that the patient is trained to discriminate multiple levels of SEMG activity, rather than one. Figure 17.6 demonstrates the use of a visual template in the form of a staircase. Here, the patient is trained using a "step template" to systematically recognize five equally graded steps of SEMG activity. From a systems model, such proprioceptive knowledge of different levels of SEMG activity should allow the muscle to be regulated over a wider range of amplitudes.

17.3.6 Deactivation Training

Deactivation training is basic to all SEMG training procedures. In essence, it trains the patient to "turn off" the SEMG activity following activation. Figure 17.7 shows a time series graph in which the patient exhibits spontaneous SEMG discharge for two seconds following the cessation of a movement. Figure 17.8 shows how the patient was trained to turn off the postmovement discharge immediately following the cessation of the movement using visually guided feedback. A broader discussion of the use of this type of procedure in clinical work may be seen in the efforts of Ettare [8].

17.3.7 Generalization to Progressively Dynamic Movement

Training in muscle control procedures usually begins in a "static" state. Isometric contractions are easier to produce and replicate than fully dynamic ones. But once they have

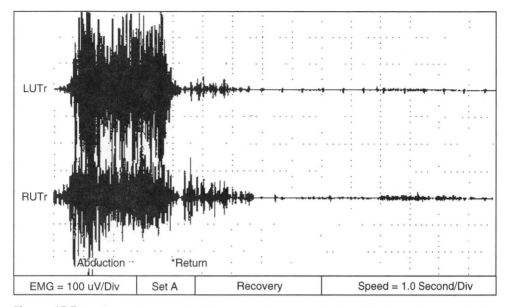

Figure 17.7. Surface EMG recording from upper trapezius during abduction of the arm to 90 degrees and return. Note the presence of activity for two seconds following the return of the arm to the side. (From [5] with permission)

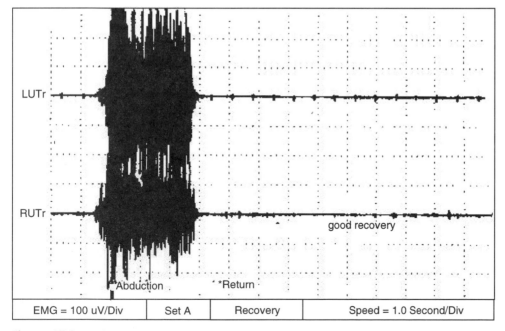

Figure 17.8. Surface EMG recording from upper trapezius during abduction of the arm to 90 degrees and return. Here the patient has been trained to quiet the muscles as quickly as possible following the cessation of movement. (From [5] with permission)

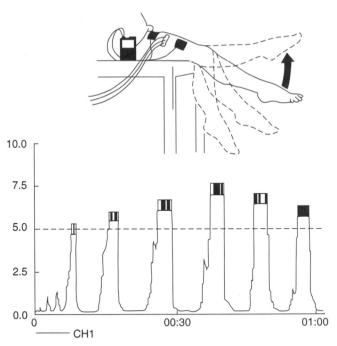

10.0

7.5

5.0 -

2.5

0.0

0   00:30   01:00

——— CH1

Figure 17.9. Standard SEMG-triggered NMES setup. EMG electrodes detect a voltage sum derived from muscle action potentials. The EMG signal is processed in a routine manner, and when the display magnitude exceeds a predetermined value, a relay causes a neuromuscular electrical stimulator to deliver current to the same muscle. The EMG display is nullified during the electrical stimulation because the device would record voltage associated with injection of artificial current, masking the small muscle action potentials. (Courtesy of Verimed International, Inc. Coral Springs, FL; from [13] with permission)

been mastered, it is important to attempt to generalize them. This generalization typically involves larger and larger contractions, and faster and faster movements.

17.3.8 SEMG-Triggered Neuromuscular Electrical Stimulation (NMES)

This is a very sophisticated and powerful procedure that involves not only an SEMG instrument to monitor the electrical activity of the muscle but also a microcurrent device to activate the muscle indirectly through the stimulation of the peripheral motor nerves [1]. The procedure is very similar to the threshold based uptraining one. The only difference is that when the patient reaches a predetermined level of activity, the stimulation device is activated and full muscle recruitment is thereby electrically mandated. Usually this technique is used in very weak muscles. The threshold is initially set at a level that the patient can easily reach but that is nonfunctional. When that threshold is met, a fully functional muscle contraction is created with the microcurrent device. Figure 17.9 shows an example of NMES training of the quadriceps muscle.

17.3.9 Left/Right Equilibration Training

Equilibration training is performed when the left and right aspects of homologous muscle pairs are observed to act differently during symmetrical tasks. In essence, a symmetrical movement, such as forward flexion of the head, should bring about a symmetrical recruit-ment of the sternocliedomastoid muscles. When this does not happen, there is something wrong with the pattern of recruitment.

Equilibration training typically involves simultaneous recording from the right and left aspect of the homologous muscle pair. Typically a processed SEMG signal is displayed on a time series display with the two channels of SEMG aggregated on the same screen using the same sensitivities and sweep times. This way the symmetry of the activity of the two muscles can be easily seen. A skilled therapist will use a variety of therapeutic skills, such as uptraining the lower weaker side to create greater symmetry of recruitment.

17.3.10 Motor Copy Training

This is a more advanced coordination training protocol. Here an SEMG recruitment pattern is generated, stored, and placed on the screen as a background template. The patient is then asked to use his or her own muscle effort during a live SEMG display to "copy" or follow the template.

Typically these types of procedures are done in patients who have unilateral weak-nesses, such as stroke. The uninvolved side is then used during a desired movement to generate the template. The template from the uninvolved side is used by the patient to teach how to use the involved side in a "normalized" fashion.

17.3.11 Postural Training with SEMG Feedback

This is a fairly straightforward procedure. Typically a time series scroll is provided of the postural muscles of interest. The therapist then assists the patient in obtaining a more natural or correct posture, and the SEMG feedback display demonstrates how the improved posture provides less work and stress on the musculoskeletal activity.

17.3.12 Body Mechanics Instruction

This procedure is very similar to the postural training procedure, except this time the patient is doing more vigorous work. It usually involves a time series scroll of a processed signal from muscles that are either prime movers or stabilizer. Commonly the patient is asked to do a task, such as lifting an object from the ground, and the therapist teaches standard body mechanics and observes how well the muscles recruit during the movement. Corrections and suggestions about the lifting technique might be given, and the lift tried again. The SEMG recording may be used to demonstrate how the new technique for lifting will reduce the probability of injury to the involved muscles.

17.3.13 Therapeutic Exercise with SEMG Feedback

In physical medicine, patients are commonly given exercises by the therapist to strengthen a given muscle or muscle grouping. A time series scroll that shows the recruitment pattern of the prime movers and stabilizers of the exercise may be used to verify that the exer-cise is working on the desired muscles. Occasionally, when given a therapeutic exercise, the patient inadvertently uses a muscle substitution pattern to create the desired movement pattern. When SEMG monitoring is used, if a muscle substitution pattern is noted, the patient may be trained using the SEMG feedback to recruit the desired muscles during the exercise.

17.3.14 Functional Activity Performance with SEMG Feedback

All of the prior training techniques led to this procedure. It is essential to take an isolated training procedure and introduce it into real and varied life activities. Refinements of movement, posture, and emotional tone may be done during this time.

17.4 SUMMARY

SEMG biofeedback may be used to assess and treat a variety of musculoskeletal or neuromuscular disorders. Seven muscle dysfunctions were described and reviewed. Emotional, movement, and postural aspects were considered. In addition a variety of biofeedback techniques were described.

REFERENCES

1. Baker, L. L., "Clinical uses of neuromuscular electrical stimulation," in R. P. Nelson, and D. P. Currier, eds., *Clinical electrotherapy*, 2nd ed., Appleton and Lange, Norwalt, CT, 1991.

2. Basmajian, J. V., ed, *Biofeedback: Practice and principles,* 3rd ed., Williams and Wilkins, Baltimore, 1989.

3. Cram, J. R., *EMG biofeedback and the treatment of tension Headaches: A systematic analysis of treatment components.* Behavior Therapy, •• 1980.

4. Cram, J. R., and D. Engstrom "Patterns of neuromuscular activity in pain and non-pain Patients," *Clin Biofeedback Health* **9**, 106–116 (1986).

5. Cram, J. R., and G. S. Kasman, *Introduction to surface EMG*, Aspen Publishing, Gathersburg, PA, 1998.

6. Davies, C. T. M., and A. G. Sargeant, "Effects of exercise therapy on total and component tissue leg volumes or patients undergoing rehabilitation from lower limb injury," *An Hum Biol* **2**, 327–335 (1975).

7. deAndrane, J. R., C. Grant, and A. Dixon, "Joint distention and reflex muscle inhibition in the knee," *J Bone Joint Surg Am* **47**, 313–322 (1965).

8. Ettare, D., and R. Ettare, "Muscle learning therapy: A treatment protocol," in J. R. Cram, ed., *Clinical EMG for surface recordings*, Vol. 2, Clinical Resources, Nevada City, 1990.

9. Gaarder, K. R., and P. S. Montgomery, *Clinical biofeedback: A procedural manual.* Williams and Wilkins, Baltimore, 1977.

10. Hertling, D., and R. M. Kessler, *Management of common musculoskeletal disorders*, 2nd ed., Lippincott, Philidelphia, 1990.

11. Hubbard, D., and G. Berkoff, "Myofascial trigger points show spontaneous needle EMG activity," *Spine* **18**, 1803–1807 (1993).

12. Jacobson, E., *You must relax*, McGraw-Hill, New York, 1976.

13. Kasman, G., J. R. Cram, and S. Wolf, *Clinical applications in surface EMG*, Aspen Publishers, Gaithersburg, MD, 1998.

14. Kraus, H., "Muscle spasm," in H. Kraus, ed., *Diagnosis and treatment of muscle pain*, Quintessence Publishing, Chicago, 1988.

15. Luthe, W., ed., *Autogenics therapy*, Vols. 1–4, Grune and Stratton, New York, 1969.

16. McKenzie, R., *The lumbar spine: Mechanical diagnosis and therapy.* Spinal Publications, Waikanae, New Zealand, 1981.

17. McNutty, W., R. Gervitz, G. Berkoff, and D. Hubbard, "Needle EMG evaluation of trigger point reponses to psycholophysiological stressors," *Psychophysiology* **31**, 313–316 (1994).

18. Peper, E., and C. Holt, *Creating wholeness: A self healing workbook using dynamic relaxation, imagery, and thoughts*, Plenum Publishing, New York, 1993.

19. Schwartz, M., ed., *Biofeedback: A practitioners guide*, Guilford Press, New York, 1995.

20. Whatmore, G., and D. Kohli, *The physiopathology and treatment of functional disorders*, Grune and Stratton, New York, 1974.

18

CONTROL OF POWERED UPPER LIMB PROSTHESES

P. A. Parker, K. B. Englehart, and B. S. Hudgins

Department of Electrical and Computer Engineering and Institute of Biomedical Engineering
University of New Brunswick, Fredericton
New Brunswick, Canada

18.1 INTRODUCTION

The surface myoelectric signal is an effective and important system input for the control of powered prostheses. This control approach, referred to as myoelectric control, has found widespread use for individuals with amputations or congenitally deficient upper limbs. In these systems voluntarily controlled parameters of myoelectric signals from a muscle or muscle group are used to select and modulate a function of a multifunction prosthesis. Figure 18.1 shows a block diagram schematic of the essential elements of a myoelectric control system. The feedforward path is the myoelectric channel, which is a replacement in part for the motor control system.

The concept of myoelectric control was introduced in the 1940s. However, the technology of the day was not adequate to make clinical application viable. It was with the development of semiconductor device technology, and the associated decrease in device size and power requirements, that clinical application saw promise, and research and development increased dramatically. During the 1960s significant progress was made in myoelectric control development, mainly through work in the USSR, Sweden, Britain, Japan, the United State and Canada.

The first commercial system, controlling a powered hand, was produced in the USSR [44], with export to Europe and North America. In Japan research was directed at the development of a multifunction myoelectric hand [42]. Some of the earlier work on pattern

Electromyography: Physiology, Engineering, and Noninvasive Applications, edited by Roberto Merletti and Philip Parker.
ISBN 0-471-67580-6   Copyright © 2004 Institute for Electrical and Electronics Engineers, Inc.

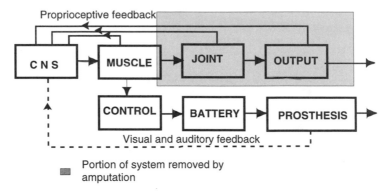

Figure 18.1. Block diagram illustrating relationship between normal and myoelectric control systems (Shaded area is removed by amputation). (Reprinted with permission from [59])

recognition based myoelectric control was carried out in Sweden [34]. In England, Bottomley addressed the issue of control performance improvement through appropriate signal processing techniques [6]. In the United States the development of self-suspended and self-contained myoelectric prostheses advanced [9], and the first myoelectrically controlled elbow prosthesis was developed [52]. Scott and Dorcas [14], in Canada, developed what came to be known as the three-state myoelectric controller, which allowed for the control of a three function device with a single muscle. The early myoelectric controllers operated in an on–off mode to control electrically powered hands with open–close functions. A myoelectric channel was required for each of the open and close functions as a two-muscle controller. Technical advances proceeded during the 1960s and 1970s with small-scale production until a few major commercial suppliers, including Otto Bock and Viennatone, entered the field. (For detailed reviews of the literature in the area and historical perspective, see [59,66].)

The clinical application of myoelectric control of powered prostheses was making a significant impact by the 1970s. Powered prostheses with myoelectric controllers were routinely fitted to upper limb deficient clients, and clinical evaluations of the functional benefits were carried out [26]. Electrically powered prostheses with myoelectric control have several advantages over other types of prostheses: the user is freed of straps and harnesses required of body-powered and mechanical switch control, the myoelectric signal is noninvasively detected on the surface of the skin, the electric battery is possibly the most convenient form of power supply that can be incorporated into a prosthesis, the controller can be adapted to proportional control with relative ease, the electronic circuits (whether analog or digital) are continuously improved and miniaturized, and the muscle activity required to provide control signals is relatively small and can resemble the effort required of an intact limb.

Clinical evaluations of myoelectrically controlled prostheses show in general good acceptance rates by the users [56]. However, the results depend on the type of prosthesis, control strategy, and user training, with low acceptance rates [3] occurring when the user perceives an inadequate controllability—specifically a lack of intuitive and dexterous control. For myoelectric prostheses the control limitations involve the lack of robustness in the processing of the input (the myoelectric signal) to specify the output (joint-space kinematics) and the disparity of the means of manipulation from natural motor control and learning. As discussed in the following section, the random nature of the myoelectric signal and the limits on prosthesis activation delay make fine control difficult. Also, as the level of amputation increases, the disparity between natural control channels and the myoelectric control channel increase. As a result, for unilateral amputees, the prosthetic terminal device is used mostly as an assistive device, and tasks requiring fine manipulation or sensory feedback are performed by the natural hand. If the amputation is bilateral, there is a much more urgent need for multifunction myoelectric control capability.

The efforts put forth to meet the intrinsic technical challenge of this complex control problem have produced anatomical and physiological models and have drawn upon statistical signal processing methods to maximally extract information from the myoelectric signal. The next section of this chapter will discuss the myoelectric signal characteristics, models, and processing in the context of the control of powered upper limb prostheses.

18.2 MYOELECTRIC SIGNAL AS A CONTROL INPUT

The source of control signal for myoelectric controllers is viable residual muscle remaining following amputation or available muscle in the case of a congenital limb deficiency. Given a large superficial muscle and a surface closely spaced bipolar electrode pair, it is possible to acquire myoelectric signal from this muscle alone and achieve a single muscle control channel. Indeed, with a fine wire intramuscular bipolar electrode it is possible to isolate a small muscle segment (a few motor units) and use the motor unit action potential trains as control signal sources. Clinically, however, this latter signal source is not practical due to the invasive transcutaneous nature of the electrode. For surface electrodes the limitations of the single muscle source include the requirement for a superficial muscle, the small interelectrode spacing, and in the case of the congenital amputee the uncertainty of muscle position.

In contrast to the single muscle myoelectric channel referred to above, a widely spaced electrode pair appropriately placed on the limb will acquire signal from a muscle group. Such a multimuscle control signal source is the temporal and spatial sum of the electrical activity of the muscles of the group. The practical limitations of achieving a single muscle source mean that the multimuscle source is the more common control source. It is simply easier to put a widely spaced electrode pair on the limb and use all available signals, rather than searching for critical positions on individual muscles. It is also the case, from a control information point of view, that the temporal and spatial sum of the signals from muscles of a group has certain advantages over a single muscle. This follows from the observation that the contribution made by each muscle of the group to the sum is a function of the intended limb action. Thus the contribution pattern can be voluntarily controlled and the pattern used for control purposes.

A myoelectric channel includes the muscle group, the electrode, and the volume conductor between each muscle of the group and the electrode. In the remainder of this section the control information potential of a single channel and a multichannel will be presented.

18.2.1 Single Myoelectric Channel Model

The block diagram for a linear system model for a myoelectric channel is given in Figure 18.2, where for the ith muscle, $i = 1, \ldots, m$, S_i is the muscle electrical source, $U(t, \lambda_i, p_i)$ is the pooled innervation point process with pooled firing rate λ_i and pattern p_i, $P(t, r_i)$ is the average motor unit action potential seen at the electrode with distance r_i from the source, and $m_i(t)$ is the ith muscle signal. The channel myoelectric signal, $M(t)$, can be expressed as

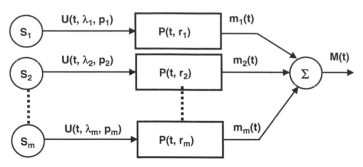

Figure 18.2. Linear system model for a myoelectric channel consisting of muscle sources, S_i, innervation processes, $U(t, \lambda_i, p_i)$ and volume conductor filters, $P(t, r_i)$, $i = 1, 2, \ldots, m$.

$$M(t) = \sum_{i=1}^{m} m_i(t)$$

$$= \sum_{i=1}^{m} U(t, \lambda_i, p_i) * P(t, r_i) \tag{1}$$

where $U(t, \lambda_i, p_i) = \sum_{n=0}^{\infty} \delta(t - T_{ni})$, $\delta(t)$ is the Dirac delta function, and T_{ni} is the nth firing of muscle unit i.

The parameters of $M(t)$ that are under voluntary control are the recruitment parameters λ_i through muscle contraction level. The parameters m and p_i are to a certain extent under voluntary control through the selection of a limb function but are not otherwise considered as control parameters. The observed characteristic of $M(t)$, which is determined by the λ_i, $i = 1, \ldots, m$, is the signal variance σ_m^2 (ac power). This can be shown by first recognizing that with uncorrelated innervation point processes, $U(t, \lambda_i, p_i)$, $i = 1, \ldots, m$, the variance σ_m^2 is the sum of the variances, σ_i^2, of the individual muscle signals. Now considering the variance of $m_i(t)$,

$$\sigma_i^2 = \theta_{ii}(\tau)|_{\tau=0}$$

$$= \theta_{uu}(\tau) * P(\tau, r_i) * P(-\tau, r_i)|_{\tau=0} \tag{2}$$

where $\theta_{ii}(\tau)$ and $\theta_{uu}(\tau)$ are the autocorrelation functions of $m_i(t)$ and $U(t, \lambda_i, p_i)$, respectively. Given that motor unit innervation processes of a muscle are uncorrelated, $\theta_{uu}(\tau)$ is the sum of the motor unit innervation process autocorrelation functions and hence $\theta_{uu}(\tau) = \lambda_i \delta(\tau) + \lambda_i h(\tau)$, where $h(\tau)$ is the renewal density function of the innervation point process [11]. With this substitution (Eq. 2) becomes

$$\sigma_i^2 = (\lambda_i \delta(\tau) + \lambda_i h(\tau)) * P(\tau, r_i) * P(-\tau, r_i)|_{\tau=0}$$

$$= \lambda_i \int_{-\infty}^{\infty} P^2(t, r_i)dt + \lambda_i \int_{-\infty}^{\infty} h(t)\Phi(t, r_i)dt$$

$$= k_i \lambda_i \tag{3}$$

where k_i is a constant resulting from the first integral. The second integral results in a value of zero due to the Poisson statistics of the pooled process, $U(t, r_i, P_i)$. Thus from Eq. (3), $\sigma_i^2 = k_i \lambda_i$, and finally

$$\sigma_m^2 = \sum_{i=1}^{m} \lambda_i k_i \tag{4}$$

From the previous results it is seen that voluntary control of the λ_i through muscle contraction level provides voluntary control over the variance, σ_m^2, of $M(t)$. Voluntary control of the variance, or equivalently the mean absolute value of $M(t)$, is the basis, as discussed later, of the first myoelectric control systems such as three-state controllers. The pattern of contributions by λ_i, $i = 1, \ldots, m$, to $M(t)$, and muscle motor unit firing patterns, p_i, are also to some extent under voluntary control through muscle group function selection. This voluntary pattern control is the basis, as discussed later, for more recent multifunction myoelectric controllers.

While it was implicit in the foregoing that the λ_i were not time varying (i.e., the contraction was static), the model can be extended to include the dynamic case. For time-varying or nonstationary innervation processes, $U(t, \lambda_i(t), p_i)$, the autocorrelation function becomes

$$\theta_{uu}(t, t+\tau) = \lambda_i(t)\delta(\tau) + \lambda_i(t)h(t, \tau) \tag{5}$$

and from Eq. (3)

$$\sigma_i^2(t) = \int_{-\infty}^{\infty} \lambda_i(t)P^2(t, r_i)dt + \int_{-\infty}^{\infty} \lambda_i(t)h(t)\Phi(t, r_i)dt \tag{6}$$

Now $\lambda_i(t)$ is a slowly varying signal compared to $P(t, r_i)$ and is thus approximately constant over the duration of $P(t, r_i)$. Following the same procedure as in the development of Eqs. (3) and (4), Eq. (6) becomes, to a good approximation,

$$\sigma_i^2(t) \approx \lambda_i(t)\int_{-\infty}^{\infty} P^2(t, r_i)dt + \lambda_i(t)\int_{-\infty}^{\infty} h(t)\Phi(t, r_i)dt$$
$$= k_i\lambda_i(t) \tag{7}$$

and

$$\sigma_m^2(t) \approx \sum_{i=1}^{m} k_i\lambda_i(t) \tag{8}$$

In this time-varying case voluntary control of $\sigma_m^2(t)$ through $\lambda_i(t)$ makes possible continuous and proportional prosthesis control.

18.2.2 Single-Channel Control Information

Since, as demonstrated in Section 18.2.1, the parameters under voluntary control (λ_i and p_i), are recruitment related, the information to be extracted from the myoelectric signal for control purposes should be derived from these parameters. Indeed, the combination of firing rate λ_i and pattern p_i will impart a time sequence and parameter pattern to the myoelectric signal that can also be extracted as information. Myoelectric controllers are therefore based on information derived from estimation of the signal variance and/or signal pattern.

It was demonstrated that the signal variance σ_m^2 is a function of firing rate, and hence is an information parameter. Most clinical single channel controllers use this information source. The first issue in this case is the method of choice for the estimation of variance, or another parameter related to the variance. The second is the prosthesis function selection and control algorithm based on the estimate. With respect to the first, the typical and reasonable approach is to estimate the signal mean absolute value (MAV) by means of a rectifier and low-pass filter. The MAV is related to the variance by the signal form factor. With respect to the second, a number of different function selection approaches have been used. They typically involve a decision algorithm to select and turn on–off or control proportionally one of a number of possible prosthesis functions.

Two particularly common selection approaches are level coding and rate coding. In level coding, each prosthesis function is assigned a signal level or more specifically a unique MAV. To select a function, the controller estimates the MAV from the generated myoelectric signal, compares that value with a set of decision boundaries, and chooses the most probable function. The assignment of MAV and decision boundaries values to give minimum decision error is determined by the additive system noise variance and the maximum useable myoelectric variance. Clearly, the performance will depend on the number of prosthesis functions to be controlled. In practice, clinical systems can reliably control three functions, which is a limiation of the approach. For rate coding, the rate of change of the myoelectric signal variance is estimated and the rate used to select a function. In other respects rate coding is similar to level coding. For a review of variance information based controllers, see [59,66].

Repeatable recruitment patterns accociated with individual limb motions/functions have the potential of being richer in information than the variance. It is then possible to provide control over more prosthesis functions for a given decision error. This becomes a pattern classification problem in which the control information is in the features of the pattern. The features, which form the pattern feature vector, are chosen to optimize feature vector discrimination. For the single myoelectric channel the features can be, for example, the coefficients of the autoregressive-moving-average (ARMA) time series model of the signal. That myoelectric signal patterns for different movements exist is demonstrated by Hudgins, [37]. With reference to Figure 18.2, the mechanisms by which the patterns are unique for different movements can be seen. First the spectral characteristics of the individual signals, $m_i(t)$, are strongly dependant on the $P(t, r_i)$. Second, the movement-dependent recruitment patterns, p_i, $i = 1, \ldots, m$, determine the contributions from the individual muscles of the group. Together these factors cause the feature vector pattern to be movement dependant. Pattern information based controllers are discussed in Section 18.4.1. A detailed review of pattern recognition based systems is given in [20,38].

18.2.3 Limitations of the Single-Channel Myoelectric Signal as Control Input

As noted in Section 18.2, control information extraction from the myoelectric signal requires estimation of signal parameters such as variance and ARMA coefficients. Estimates are, by nature, subject to error and thus give "noisy" control information. This noise limits the information content, or more specifically the information transfer rate between the central nervous system and the prosthesis. Consider the time-average estimate $\hat{\sigma}_m^2$ of the signal variance for a static contraction where

$$\hat{\sigma}_m^2 = \frac{1}{T}\int_0^T M^2(t)dt \tag{9}$$

and T is the signal observation window. The noise of this estimate can be found from the variance, $\text{Var}[\hat{\sigma}_m^2]$, given by

$$\begin{aligned}\text{Var}[\hat{\sigma}_m^2] &= E[\hat{\sigma}_m^2] - E^2[\hat{\sigma}_m^2]\\ &= E[\hat{\sigma}_m^2]^2 - \sigma_m^4\end{aligned} \tag{10}$$

where $E[X]$ is the expected value of the random variable X. While $E[\hat{\sigma}_m^2]^2$ can be evaluated for the general case [81], the case in which $M(t)$ is assumed Gaussian gives a simpler result which closely approximates in practice. This follows from the observation that the central limit theorem predicts a Gaussian $M(t)$ at high recruitment values, and for even small m and λ values, $M(t)$ is well approximated as Gaussian. For this case it can be shown [81] that

$$E[\hat{\sigma}_m^2]^2 = \frac{\sigma_m^4}{BT} + \sigma_m^4 \tag{11}$$

and from Eq. (10),

$$\text{Var}[\hat{\sigma}_m^2] = \frac{\sigma_m^4}{BT} \tag{12}$$

where $\Phi_{mm}(f)$ is the power spectral density of $M(t)$ and B is the myoelectric signal *equivalent statistical bandwidth*, given by [4]:

$$B = \frac{\left[\int_0^\infty \Phi_{mm}(f)df\right]^2}{\int_0^\infty \Phi_{mm}^2(f)df} \tag{13}$$

In the case of level-coded systems, the term $E^2[\hat{\sigma}_m^2] = \sigma_m^4$ is the control signal power, and the term $\text{Var}[\hat{\sigma}_m^2]^2 = \sigma_m^4/BT$ is additive noise power. Hence the signal-to-noise ratio (*SNR*) is

$$SNR = \frac{\sigma_m^4}{\text{Var}[\hat{\sigma}_m^2]} = BT \tag{14}$$

The *SNR* will in large part determine the control performance, and from Eq. (14) it can be seen that T is the only design parameter through which the *SNR* can be affected. The value of B can be increased to a limited extent through "whitening" of the myoelectric signal. Whitening is a process by which the myoeletric signal is filtered to give its power spectrum a flat or "white" characteristic. The effectiveness of this process is limited by additive noise and nonstationarities of the myoelectric signal. (A discussion of the whitening process is given in Chapter 6, Section 6.4.)

A typical value for B is 100 Hz, and a typical value of 100 ms for T gives an *SNR* of only 10. This clearly demonstrates one of the limitations of single-channel level-coded myoelectric controllers. Increasing T will improve the *SNR*, but at he expense of the prosthesis dynamic response. Thus there is a trade-off between control performance and dynamic response. It is generally accepted that T should in practice not exceed 200 ms.

A second limitation is the difficulty presented to the prosthesis user of generating the appropriate signal level for the function desired. For the amputee this will be without the benefit of full proprioceptive feedback, and thus level generation errors occur. The error rate will depend on the number of levels and functions to control. This rate has been found to be acceptable for three or fewer levels (as seen by the success of three-state controllers) but grows rapidly beyond this value.

Extraction of information for pattern-based controllers is subject to similar limitations. The estimation of pattern vector feature values has errors that again are determined by the dynamics and signal properties, namely T and B. Operator-induced control error is also present through the generation of inappropriate patterns for the function desired. However, patter-based controllers can be more robust than level-coded controllers in regard to control error. In the case of estimation, the higher dimension of the feature space can better tolerate estimation error. In the case of the operator, movements are more reliably produced than signal levels.

18.2.4 Multiple Myoelectric Channels

If the single channel of Figure 18.2 is repeated, by the addition of more electrode pairs, multi-myoelectric channels are created. The sources, S_i, and the innervation processes, $U(t, \lambda_i, p_i)$, $i = 1, \ldots, m$, are identical for each channel. However, the average motor unit potential for the jth channel, $P(t, r_{ij})$, $j = 1, \ldots, N$, is channel dependent. As seen in Section 18.2.3, the control information of the single myoelectric channel is limited by the estimation error (*SNR*) and can be improved only at the expense of information rate, namely by increasing time averaging via T. However, the control information can be increased, for the a given value of T, by adding additional channels. This allows ensemble averaging for a reduction in parameter estimation error and/or an increase in the pattern feature vector dimension. This approach will give an increase in *SNR* of up to N, and has been demonstrated experimentally by Hogan and Mann [35]. The amount of increase depends on the placement of electrodes for the N channels and the correlation between channels. The gain in information and control performance from multi channels in the case of pattern based controllers has been demonstrated by Kuraganti et al. [46]. For a current review of multifunction myoelectric control, see Englehart et al. [20].

18.3 CONVENTIONAL MYOELECTRIC CONTROL

As mentioned in the previous section, to enable the MES to be used as a control input, some feature of the signal related to user intent must be extracted. Most commercial systems use an estimate of the signal variance as the control input. Considerable research has been focused on the optimum estimator of the signal variance, and several authors [58,35] have shown independently that a square law device followed by averaging is optimum based on error probability. However, a practical implementation of a square law

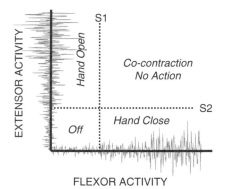

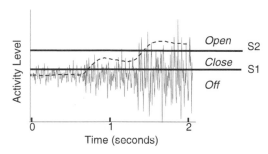

Figure 18.3. Two-channel amplitude-coded myoelectric control.

Figure 18.4. One-channel amplitude-coded myoelectric control.

device is somewhat involved, so most myoelectric control systems approximate the optimum square law by way of a full wave rectifier. By introducing this nonlinearity, the signal can be smoothed or averaged using a low-pass filter to provide a signal amplitude level that reflects the variance of the raw MES. The level of the processed signal can be controlled voluntarily by varying the level of the muscle contraction.

Most commercially available myoelectric control systems are based on amplitude or level coding of the processed MES obtained from an active control muscle of the user. Amplitude coding can be accomplished in several ways. A common approach is to assign each prosthetic limb function to a separate control muscle, as in the Otto Bock two-state system [66]. This way, if a muscle is active above a cutoff threshold (S1 for muscle 1 or S2 for muscle 2), the associated limb function is selected, as shown in Figure 18.3. Appropriate logic circuitry allows only one function to be active at a time.

This approach allows the selection of control muscles based on physiological functions but has the disadvantage of requiring two control muscles for each prosthetic degree of freedom. This limitation often makes the approach impractical for higher level amputees.

In theory, the total range of myoelectric signal produced from a relaxed state to full contraction can be divided into many domains (as shown in Fig. 18.4), each with an associated threshold and terminal device function. The user of this multi-state system produces a fixed level of contraction to match a set threshold (S1 for function 1 or S2 for function 2). The function assigned to this level is then activated.

In practice, however, the number of functions an individual can control with acceptable accuracy using this approach is limited to two per control muscle [71]. Typically these

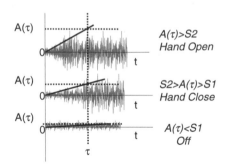

Figure 18.5. Rate-coded myoelectric control.

are hand opening and closing, wrist rotation, or elbow flexion and extension (Fig. 18.5). It is this approach that is used in the UNB three-state myoelectric control system [14]. Attempts to increase the number of control functions per control muscle have been minimally successful and have required the use of level feedback to inform the user of the state of the system [61].

An alternative strategy is used in proportionally controlled devices. Again, each function is assigned a control muscle; however, the speed at which the function is performed is proportional to the amplitude of the activity in the control muscle. Although the initial speed is difficult to control, the visual feedback provided by the active device enables coarse speed control by modifying the strength of the contraction. The Boston elbow is representative of this type of control [74]. In other systems, proportional control has been combined with rate sensitivity to provide three-state proportional control. In these systems the selection of the function is determined by the initial rate of increase in the myoelectric signal amplitude. This rate is computed by integrating the myoelectric activity over a time τ and comparing this value, $A(\tau)$, to the thresholds S1 and S2 to determine the desired prosthetic limb function. The amplitude of the myoelectric signal is then used to control the velocity at which the function is performed [8]. Rate coding and level coding have much in common and in practice may be controlled with similar strategies on the part of the user.

The principal deficiency of amplitude coding is that the number of functions per control channel is limited to usually at most two. To increase the number of functions in the prosthetic device requires an increase in the number of control muscles. This increases the complexity of the system in two ways. Training becomes difficult, as the user must now learn to operate several independent control channels in a coordinated manner. Also the control system circuitry contains several channels of hardware. This leads to problems in control unit size and power consumption. A further problem is encountered when the system is fitted to high-level amputees. These are the individuals who would benefit most from a multifunction prosthesis, but many of them do not have enough control sites to allow a multichannel implementation.

Despite these problems multifunction prostheses have been developed using several channels of amplitude coding. The Boston elbow has been used with some success in combination with an electric hand, but this has required the use of a mechanical switching arrangement to select which device is to be controlled [75]. Another strategy is used in the control of the Utah arm [40]. In this system a quick co-contraction is used to switch between hand and elbow control. Others have provided multifunction control to high-level amputees by combining myoelectric and mechanical 11 systems to meet the individual's requirements [64,49]. More elaborate multifunction prostheses have been attempted but the result is that training the user to isolate the required number of control muscles is impractical. The reason for this lengthy training period is that muscles seldom contract independently. The natural way of activating muscle is within a complex pattern in which many muscles are recruited not only to produce the desired movement but also to stabilize joints and provide resistance [76]. To isolate these muscles requires much concentration. The objective in the design of any prosthetic control system must be to allow the amputee to concentrate on things other than the contraction of specific muscles; the control is then done on a subconscious level, increasing the degree of acceptance of the prosthesis and the efficiency of its use [15].

18.4 EMERGING MEC STRATEGIES

As advances are made in computing devices, signal processing techniques, and instrumentation, new possibilities for advancing the state-of-the-art in myoelectric control have arisen. The goals of current research in myoelectric control are twofold: (1) to provide better accuracy in state selection and (2) to present a more natural means of effecting control. Both of these factors contribute to greater dexterity and therefore better usability (and presumably, acceptance) of prosthetic devices.

As explained in the previous sections, many myoelectric control systems are currently available that are capable of controlling a single device in a prosthetic limb, such as a hand, an elbow, or a wrist. These systems utilize control information from the MES based on an estimate of the amplitude or the rate of change of the MES. Although these systems have been very successful, they do not provide sufficient information to reliably control more than one function (or device) [71]; the extension to controlling multiple functions, as required in high-level limb deficiencies, is a much more difficult problem.

The following sections describe control strategies under development that promise to deliver multifunction control. These techniques still await a full-fledged clinical implementation, but they serve to characterize the capabilities of the next generation of multichannel, multifunction, microprocessor-driven devices.

18.4.1 Pattern Recognition Based Control

If one is to increase the number of devices under the control of the myoelectric signal, it is clear that a more sophisticated means of discriminating different muscle states is needed. Two things are needed for this to be possible:

1. *More information must be extracted from the MES about the active muscle state.* The manner in which one might extract more information from the MES could involve one or both of the following approaches:

   - Use multiple channels of MES, providing localized information at a number of muscles sites.
   - Develop a *feature set* that extracts as much information as possible from the MES and serves to discriminate different classes of movement.

2. *A classifier capable of exploiting this information must be constructed.* The role of the classifier is to assimilate and exploit the information it receives, and decide from which class the information originated.

This suggests a pattern recognition based approach to MES control, as depicted in Figure 18.6. The details of each stage of a pattern recognition based ocntrol system are discussed in the following sections.

Figure 18.6. Pattern recognition problem.

MES Measurement Strategies. When using the surface myoelectric signal, the primary concern regarding the placement of recording electrodes is to capture as much novel information about the muscle activity as possible. To accomplish this, when placing electrodes on the upper limb, one is faced with two possibilities:

1. *A single bipolar channel, with the bipolar electrodes spaced widely apart.* This technique, used by Hudgins [37], involves placing one electrode on the biceps and one on the triceps. This approach captures the activity of a large volume of muscle, all superimposed in to a single gross myoelectric channel. The drawback of this approach is that there is no spatial discrimination in the activity of different muscles, and novel information from different muscles may exhibit destructive interference.

2. *Multiple bipolar channels, with closely spaced electrode pairs.* Because the pickup region under closely space pairs is more local, multiple channels are needed to capture the activity of different muscle groups. The advantages address the drawbacks of the single channel; spatial discrimination is now possible, and no destructive cancellation occurs.[1]

It has been shown in several studies that multiple MES, channels provide much better discrimination among control states than do single channels [46,21]. The question that remains is: How many channels can be added, and still provide an improvement in control accuracy?

The nature of the myoelectric signal also has an influence on its capabilities as a control signal. Almost all approaches in the past have used the myoelectric signal collected during constant, somewhat steady-state contractions. This mostly random signal has been characterized by its statistical properties, as discussed earlier. The steady-state MES has very little temporal structure due to the active modification of recruitment and firing patterns needed to sustain a contraction [12]. The reason is the presence of feedback paths, both intrinsic (the afferent neuromuscular pathways) and extrinsic (the visual system).

While investigating the properties of the MES coincident with the onset of rapid contractions, Hudgins [37] observed a substantial degree of structure in the transient waveforms. Examples of these data are shown in Figure 18.7; these were acquired during small but distinct contractions, using a single bipolar electrode pair placed over the biceps and triceps muscle groups. This arrangement was intended to allow a large volume of musculature to influence the measured activity. Figure 18.7 shows typical patterns corresponding to flexion/extension of the elbow, and pronation/supination of the forearm.

---

[1] Potential disadvantages of multiple-channel systems include the need for more instrumentation and more data to process. Whereas the number of channels was practically limited in the past by the sheer bulk of the instrumentation and the difficulty in maintaining good contact, advances in electrode array miniaturization, fabrication, and interfacing techniques have made multichannel systems more feasible. Processing power is no long a concern, given the capabilities of today's microprocessors, so data bandwidth is not an issue, either.

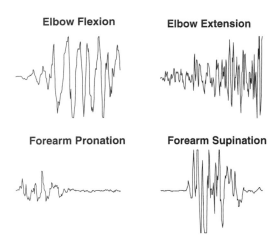

Figure 18.7. Patterns of transient MES activity recorded using a single bipolar electrode pair, placed over the biceps and triceps. (Reprinted with permission from [20])

These patterns exhibit distinct differences in their temporal waveforms. Within a set of patterns derived from the same contraction, the structure that characterizes the patterns is sufficiently consistent to maintain a visual distinction among different types of contraction. This observable structure has been reported by others [77], and it indicates an orderly recruitment of motor units. This may result from a "motter plan" in the central nervous system [31], the absence of sensory feedback paths in such a rapid burst of activity [65], or a combination of both.

The presence of determinism in the transient MES accompanying the onset of contraction suggests that these data should provide a powerful means of discriminating MES patterns corresponding to different movement types. This has been demonstrated by Hudgins et al. [37] in a prosthetic control system, which will be described below, and by Farry et al. [25], with application to teleoperation of a robotic hand.

Feature Extraction. The most intuitive, and most widely used, feature that describes the MES has been the index of gross activity, which be the variance, mean absolute value, or some other similar measure. With the motivation of providing more information about the MES in each channel, multivariate features sets have been proposed and used successfully. At first, limited by the available computational power of the day, the features were based on time domain statistics, such as variance, zero crossings, and the "length" of the waveform locus. With increasing computational resources came systems based on autocorrelation coefficients, spectral measures, and time series models. Current approaches seek to exploit the temporal structure in MES patterns by using the short-time Fourier transform, wavelet and wavelet packet transforms, and higher order spectral analysis.

Classifiers. The practical methodologies that exist for pattern classification may be loosely grouped into three categories. Historically the two classical methods are the *statistical* [70,16] and the *syntactic* [69] approaches. The third, and most recently established type of pattern classifier is the *learning* (or *neural*) approach. Learning algorithms have their origins in perceptrons [62] and adaptive linear elements [73], and have matured into the diverse field of artificial neural networks [39].

In terms of the use of classifiers in MES control systems, statistical classifiers were used almost exclusively until about the mid-1980s, at which point the first applications of artificial neural networks began to appear. A variety of artificial neural networks architecture and learning algorithms have been investigated in the context of MES pattern recognition, including simple feedforward multilayer perceptrons [43], dynamic networks [23], and self-organizing feature maps [30,10]. Recent investigations have seen the application of genetic algorithms [24] and fuzzy logic classifiers [78,50].

If a consensus can be drawn from these investigations, it is that although powerful classifiers may marginally improve the classification of the MES, it is the *feature set* that is crucial to overall performance. This will be elucidated in the following section.

Evolution of Pattern Recognition Based Systems. The first pattern recognition based approach to MES control was developed by a group of researchers at Temple University in Philadelphia in the late 1960s [27]. Their system extracted a simple amplitude measure from six channels of the MES placed at various locations about the shoulder. They used a statistical classifier to discriminate six classes of motion: elbow flexion, elbow extension, humeral rotation (out), humeral rotation (in) forearm supination, and forearm pronation. The system was accurate about 75% of the time.

Other pattern recognition based approaches to myoelectric control were developed at the Chalmers University in Göoteborg, Sweden [48] (the control of a three degrees of freedom hand prosthesis), and the University of California, Los Angeles [51] (the control of multi-joint robotic manipulators). Both of these approaches used multiple channels of amplitude information and statistical pattern classifiers.

Thwarted by the difficulty associated with instrumenting many channels of MES, and maintaining good contact in each of these channels, others sought to enhance MES control by maximizing the information content in each channel. Saridis et al. [63] were among the first to suggest the use of statistical properties (other than integrated amplitude) as a means of conveying the information content of the myoelectric signal from a given control site. The features of the myoelectric signals were the mean absolute value, the variance, the magnitude of the third moment, zero crossings, the autocorrelation function, and the power spectral density. A statistical analysis of these features determined those that contained the maximum information for pattern separation to be identified. Zero crossings and signal variance was found to provide the best classification of the remaining motions. Two channels of MES (biceps and triceps) were used to classify three types of motion with roughly 85% accuracy. Graupe et al. [33] focused on the feature set as well, performing a fourth-order autoregressive analysis on a single channel of MES. Doershuk et al. [13] adopted Graupe's time series approach, and extended it to a multiple-site control system.

As explained previously, Hudgins [37] described the presence of a deterministic pattern in the MES activity accompanying the onset of a sudden contraction. A MES control system was developed to exploit this determinism. A block diagram of this system is shown in Figure 18.8.

A feature set comprising a set of time domain statistics (mean absolute value, waveform locus, zero crossings, turns) is computed on a series of six 40 ms segments. These features capture various aspects of the temporal structure in the 240 ms burst of MES activity recorded from the biceps and triceps, commencing when the signal exceeds an amplitude threshold. A multilayer perceptron artificial neural network is trained upon exemplars of four classes of limb motion. As indicated in the figure, a typical prosthetic control scenario is the actuation of elbow flexion, elbow flexion, 16 forearm pronation, and forearm supination. This system was shown to be roughly 90% accurate on an ensemble of 15 nor-

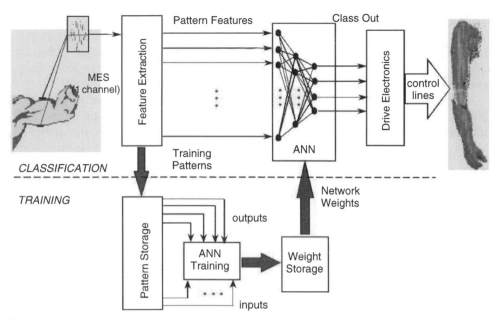

Figure 18.8. A multifunction control scheme based on the transient MES, using an artificial neural network. (Reprinted with permission from [20])

mally limbed and limb-deficient subjects. The entire system has been implemented as an embedded controller [36].

Improvements to Hudgins's control scheme have focused on the feature extraction stage. Specifically, the efficacy of time-frequency features have been investigated as a means of improving classification accuracy. The short-time Fourier transform (STFT) is an obvious candidate for capturing both time and frequency information, due to its simplicity and computational efficiency. Farry [25] and Gallant [30] have demonstrated good results using variants of the STFT. Englehart [18,22] provides a comparison of Hudgins's time-domain feature set, the STFT, the wavelet transform, and the wavelet packet transform. Each of the time-domain feature sets are shown to outperform Hudgins's time-domain features; the wavelet packet transform offered the best overall performance. From a roster of 16 subjects, a classification rate of 96% was observed on a four class problem when using two MES channels, and virtually no error when using four MES channels. Using four channels on the six-class problem, about 96% accuracy was obtained. This approach has been recently extended to operate in a *continuous mode* [19], processing and classifying the MES elicited by continuous muscle activity in real time while still retaining the accuracy of burst mode classification. This approach promises a more natural control interface, allowing greater dexterity and interactivity when performing a sequence of manipulations.

Fundamental to the success of the work of Gallant and Englehart was the use of *dimensionality reduction*, a scheme of reducing the number of features in the feature set so as not to overwhelm the classifier. Gallant used an iterative technique called *projection pursuit*, which seeks to map a higher dimensional space (the STFT features) onto a lower dimensional space using a metric that tends to cluster the data. Englehart used a simpler projection method, *principal components analysis*, which projects the higher dimensional space onto a lower dimensional space such that the reconstruction error is minimized. This method has the advantage of having a well understood, easily computed analytic solution. Both projection pursuit and principal components analysis dramatically improve the classification results when using features based on the STFT, the wavelet transform, or the wavelet packet transform.

There is much left to be done in the field of feature extraction for classification. Quadratic time-frequency methods hold promise if their computational complexity can be sufficiently reduced so as to be feasible for real-time implementation. Nonlinear projection methods have recently been introduced [45]; these may offer greater class discrimination capabilities than projection pursuit or principle components analysis.

18.4.2 Intelligent Subsystems

The goal of pattern recognition based systems is to discriminate the intended degrees of freedom from the MES as accurately as possible, in a sequential manner. This modality of control can be referred to as *sequential control*.

> *Sequential Control.* The degrees of freedom to be functionally replaced are actuated one at a time, in a sequence, as one executes a multifunction task.

Clearly, it would be awkward to accomplish most active daily living tasks using sequential control. In order to replicate the kinematics and dynamics of normal human upper limb movement, one must have simultaneous and coordinated control of each degree of freedom.

> *Simultaneous, Coordinated Control.* The degrees of freedom are actuated simultaneously, in a coordinated fashion, as one executes a multifunction task.

We are capable of this type of control at a subconscious level. The central nervous system manages coordinated limb motion by organizing control in a hierarchical manner. The premotor cortex is responsible for high-level planning of coordinated tasks, and instructs the motor cortex to issue the necessary stimuli to the appropriate motor nerves. The motor nerves innervate muscles in an intricate pattern of recruitment so as to produce smooth and dexterous motion. Feedback paths of tactile force, muscle tension, and muscle length allow precise regulation of force output and limb position.

Clearly, it is unreasonable to expect that the exquisite control of the central nervous system might be extracted from the MES. It is an unavoidable reality that the surface MES conveys the control information elicited by the neuromuscular system as a second-order effect. What we measure as the surface MES is the superposition of complex neuromuscular activity, and inevitably, spatial and temporal information is lost.

As a consequence the amount of information that may be conveyed by the surface MES is a subset of the delivered to the muscles by the motor neurons. The ability to control an upper limb with the dexterity of a normally limbed individual, however, requires all of the information delivered (and retrieved) by the nervous system.

This is not to say that an acceptable compromise can not be found. Although the dimensionality of the control problem is beyond that of the information that may be derived from the surface MES, the controllability may be augmented by the presence of an intelligent control subsystem. The purpose of this subsystem should be to assume the responsibility of certain aspects of control. It must be understood that there is a trade-off between the degree to which the subsystem assumes control and the degree to which the system is under the intentional control of the user. Some compromise as to the degree of intervening control must be determined.

There have been numerous attempts to augment myoelectric control with some form of intelligent subsystem, with the primary goal of allowing multiple degrees of freedom to be controlled, while reducing the mental burden of the user.

Kinematic Coupling and Trajectories. Perhaps the simplest approach to managing coordinated control is to assign a kinematic coupling between degrees of freedom. In this strategy the position of at least one joint (e.g., a wrist) follows another joint (e.g., an elbow) in some predetermined manner (a ratio, perhaps). This strategy can be used to determine a useful trajectory of a limb for specific purposes, such as feeding [60]. The use of predefined trajectories based on the limb kinematics of useful tasks has been successfully used to reduce the dimension of the control problem [7,29,67]. Though fixed trajectories serve a useful dedicated purpose, interactive control of individual degrees of freedom is hindered; tasks that require manipulation that does not match the preassigned functions may be difficult to accomplish.

Heuristic Rules. Another strategy that has been used to relieve the mental burden of MES control is the use of heuristic rules to coordinate multi-joint movement. Maric et al. [53] used heuristic rules to govern four basic coordinated actions: *reach, move, orient,* and *grasp-release*. Using a joystick as a control input, Zadaca et al. [80] demonstrated how coordinated multi-joint activities can be accomplished with a relatively low-dimensional control input. They combined resolved endpoint control (see next section) with heuristic rules to control orientation in three degrees of freedom; these were humeral rotation, forearm rotation (pronation-supination), and terminal device prehension. The inputs to the heuristic algorithm were direction, position, and speed of the prosthetic joints. The heuristic rules may be summarized as:

Prehension. When reaching out to grasp an object, open the terminal device. If a reaching motion is sensed and the terminal device is not loaded, the terminal device will be opened during positioning. This saves time and emulates normal reaching movements.

Humeral Rotation. The idea behind this rule is to use humeral rotation instead of shoulder rotation when the upper limb is nearly vertical. When a shoulder rotation command is issued, the program checks if the current upper limb position is nearly vertical. If so, shoulder rotation is canceled, and a proper humeral rotation command is issued.

Wrist Rotation. The wrist rotation rule attempts to orient the terminal device in the direction of motion.

Stop Motors to Prevent Collision Damage. This rule provides protection from possible collision damage to the arm by checking current positions to previous positions when motion was intended.

Although a rule-based approach does reduce the dimensionality of the control problem, movement is constrained to that specified in the heuristic program; this was discovered by the group as they attempted active daily living tasks. Neither of these investigations developed to the point of using a MES interface for control.

Perhaps the best contemporary use of heuristic rules is exemplified by the Oxford intelligent hand [47]. A microprocessor-controlled device, the hand has control inputs consisting of two MES channels, slip sensors and force sensors at the fingertips, and position feedback of the fingers. The hand relives the conscious effort of the user by "sensing" which of two grip types is most appropriate, grasping an object with an appropriate level of force upon contact, and, via slip detection, securing a grip by automatic tightening of grasp.

The Oxford intelligent hand is an integral part of an ambitious project entitled TOMPAW (Totally Modular Prosthetic Arm with high Workability). This is a collaborative effort among several centers of excellence in England, Scotland, and Sweden, funded by the European Union. The intended outcome of the project is a complete modular arm system where every joint has an independent microprocessor controller. The separate controllers will communicate under the supervision of a master controller responsible for co-ordinating the actions of multiple joints in response to a simple input form the user.

Endpoint Control. There is evidence to suggest that humans plan upper limb activities in three-dimensional Cartesian coordinates [5], and that this target activity is transformed into the joint space forces (muscle activity) required to produce the desired limb dynamics [68]. Therefore the conscious command input consists of little more than an *endpoint*, and the neuromuscular and skeletal system take care of the kinematics and dynamics of the joints.

This physiological analogue suggests a desirable strategy for multifunctional control of prostheses. It is unreasonable to expect that sufficient information can be extracted from the MES to control many degrees of freedom in a coordinated manner; the dimensionality of the problem is simply too high, given the limited amount of information in the MES. If we reduce the problem to that of specifying an endpoint, however, the dimensionality of the problem is significantly reduced.

There have been some attempts at using an endpoint control strategy to manipulate multi-joint artificial limbs [80], but these did not use the myoelectric signal as the control input. A simple implementation of MES-driven endpoint control has been developed [1] that determines elbow position from the biceps/triceps MES and an assumed model of upper limb dynamics.

The extension of MES-driven endpoint control to the multi-joint case is much more complex. In this case an intelligent subsystem must augment the endpoint command to infer the kinematic and dynamic behaviour of the prosthetic limb. The tasks involved in multi-joint control include trajectory generation and motion control. *Kinematic control* is the assignment of resolving joint positions that correspond to the endpoint trajectory, given initial and goal coordinates, possibly with some intermediate (via) points and motion constraints. The *dynamic control* problem consists of finding the joint torques that will cause the arm to follow a specified trajectory while satisfying some measure of accuracy.

Most of the research in kinematic and dynamic control has been in the field of robotics. A successful resolution of these control methods requires efficient and stable matrix inversion and accurate knowledge of manipulator kinematic parameters. Neural network approaches are model free (they do not need a priori knowledge of parameters) and estimate the inverse solution in a lookup table fashion rather than by solving the control equations. For these reasons they are attractive alternatives to analytical methods. Neural network based approaches to kinematic control include [2,17,41,79]. Perhaps the most complete solution to dymanic control has been described by Kawato et al. [32,55]. These methods may be applied to the prosthetic control problem in the context of an intelligent subsystem.

18.5 SUMMARY

The myoelectric signal serves as an important control input for powered limb prostheses. These myoelectric control systems rely on the myoelectric signal to convey information regarding intent from the user to the prosthesis controller. The control information required of the myoelectric signal can vary from simple switch on–off commands in single muscle systems to complex multifunction commands from patterns generated by muscle groups.

The limitation in all cases is the accuracy and rate of information that can be achieved via the myoelectric channel. Present commercial controllers place reasonably low demands on information, and can be accommodated by one or two channels. More sophisticated prostheses will require much higher rates and completely new control strategies. To these ends considerable work is directed to investigation of control strategies based on classification of myoelectric signal patterns for continuous/simultaneous control of devices in multifunction prostheses.

REFERENCES

1. Abul-Haj, C. J., and N. Hogan, "Functional assessment of control systems for cybernetic elbow prostheses. Part I: Description of the technique," *IEEE Trans BME* **37**, 1025–1036 (1990).
2. Albus, J., "A new approach to manipulator control: The cerebellar model articulation controller," *J Dyn Syst Meas Contr* **97**, 270–277 (1975).
3. Atkins, D., W. H. Donovan, and A. Muilenberg, "Retrospective analysis of 87 children and adults fitted with electric prosthetic componentry," Assoc Children's Prosthetic-Orthotic Clinics Conf, St. Petersburg, FL, 1993, p. 4.
4. Bendat, J. S., and A. G. Piersol, *Measurement and analysis of random data*, J Wiley, New York, 1966.
5. Bizzi, E., N. Hogan, F. A. Mussa-Ivaldi, and S. Giszter, "Does the nervous system use equilibrium-point control to guide single and multiple joint movements?" *Beh Brain Sci* **15**, 603–613 (1992).
6. Bottomley, A. H., "Myoelectric control of powered prostheses," *J Bone Joint Surg* **37**, 506 (1955).
7. Carlson, L. E., and C. W. Radcliffe, "A multi-mode approach to coodinated prosthesis control," *Proc 4th Int Symp External Control of Human Extremities*, Dubrovnik, Yugoslavia, 185–196 (1972).
8. Childress, D., "A myoelectric three state controller using rate sensitivity," in *Proc ACEMB*, Chicago, 4–5 1969.
9. Childress, D. S., and J. N. Billock, "Self-containment and self-suspension of externally powered prostheses for the forearm," *Bull Prosthet Res* **10**, 4–21 (1970).
10. Christodoulou, C., and C. S. Pattichis, "A new technique for the classification and decomposition of EMG signals," in *1995 IEEE Int Conf on Neural Networks Proc*, Vol. 5, IEEE, New York, 1995, pp. 2303–2308.
11. Cox D. R., and N. L. Smith, "Om the superposition of renewal processes," *Biometrica* **41**, 91 (1954).

12. DeLuca, C. J., "Physiology and mathematics of myoelectric signals," *IEEE Trans BME* **26**, 313–325 (1979).

13. Doershuk, P. C., D. E. Gustafson, and A. S. Willsky, "Upper extremity limb function discrimination using EMG signal analysis," *IEEE Trans BME* **30**, 18–28 (1983).

14. Dorcas, D. S., and R. N. Scott, "A three-state myoelectric control" *Med Biol Eng* **4**, 367 (1966).

15. Doubler, J. A., and D. S. Childress, "An analysis of extended physiological proprioception as a prosthesis control technique," *J Rehab Res Dev* **21**, 5–18 (1984).

16. Duda, R. O., and P. E. Hart, *Pattern classification and scene analysis*, Wiley, New York, 1973.

17. Elsley, R. K., "A learning architecture for control based on back-propagation neural networks," *Proc IEEE Conf Neural Networks*, 1988, pp. 587–594.

18. Englehart, K., "Signal representation for classification of the transient myoelectric signal," PhD thesis, University of New Brunswick, Fredericton, New Brunswick, 1998. (unpublished)

19. Englehart, K., and B. Hudgins, "A robust real time control scheme for multifunction myoelectric control," *IEEE Trans BME* **40**, 848–854 (2003).

20. Englehart, K., B. Hudgins, and P. A. Parker, "Multifunction control of powered prostheses using the myoelectric signal," in Teodorescu and Jain, eds., *Intelligent technologies for rehabilitation*, CRC Press, Boca Raton, FL, 2001, pp. 1–61.

21. Englehart, K., B. Hudgins, and P. A. Parker, "A Wavelet Based Continuous Classification Scheme for Multifunction Myoelectric Control," *IEEE Trans BME* **48**, 302–311 (2001).

22. Englehart, K., B. Hudgins, P. A. Parker, and M. Stevenson, "Classification of the Myoelectric Signal using Time-Frequency Based Representations," *Med Eng Phys* **21**, 431–438 (1999).

23. Englehart, K., B. Hudgins, M. Stevenson, and P. A. Parker, "Classification of transient myoelectric signals using a dynamic feedforward neural network," World Congr Neural Net, Washington, DC, 1995.

24. Farry, K. A., J. J. Fernandez, R. Abramczyk, M. Novy, and D. Atkins, "Applying genetic programming to control of an artificial arm," *Myoelectric Control '97 (MEC'97) Conf*, Institute of Biomedical Engineering Fredericton, New Brunswick, Canada, July 23–25, 1997, pp. 50–55.

25. Farry, K., I. D. Walker, and R. G. Baraniuk, "Myoelectric teleoperation of a complex robotic hand," *IEEE Trans Robot Autom* **12**, 775–788 (1996).

26. Feeny, R. J., and I. Hagaeus, "Evaluation of the EMG-controlled hand prosthesis," *Proc 3rd Int Symp External Control of Human Extremities*, ETAN, Dubrovnik, Yugoslavia, 1970.

27. Finley, R. R., and R. W. Wirta, "Myocoder studies of multiple myocoder response," *Arch Phys Med Rehabil* **48**, 598 (1967).

28. Flash, T., and N. Hogan, "The coordination of arm movements: An experimentally confirmed mathematical model," *J Neurosci* **5**, 1688–1703 (1985).

29. Freedy, A., F. Hull, and J. Lyman, "Adaptive aiding for artificial limb control," *Bull Prosthe Res* (Fall) 3–15 (1971).

30. Gallant, P. J., "An approach to myoelectric control using a self-organizing neural network for feature extraction," MS thesis, Queens University, Kingston, Ontario, 1993.

31. Georgopoulos, A., et al. "Cognitive neurophysiology of the motor cortex," *Science* **260**, 47–52 (1993).

32. Gomi, H., and M. Kawato, "Neural Network for a closed-loop system using feedback-error learning," *Neural Net* **6**, 993–946 (1993).

33. Graupe, D., and W. K. Cline, "Functional separation of EMG signal via ARMA identification methods for prosthetic control purposes," *IEEE Trans Syst Man Cybern* **5**, 252–259 (1975).

34. Herberts, P., "Myoelectric signals in control of prostheses," *Acta Orthop Scand* (suppl) **40**, 124 (1969).

35. Hogan, N., and R. Mann, "Myoelectric signal processing: Optimal estimation applied to electromyography. Part I: Derivation of the optimal myoprocessor," *IEEE Trans BME* **27**, 382–395 (1980).

36. Hudgins, B., K. Englehart, P. A. Parker, and R. N. Scott, "A microprocessor-based multifunction myoelectric control system," 23rd Canadian Med Biol Eng Soc Conf, Toronto, May, 1997.

37. Hudgins, B. S., P. A. Parker, and R. N. Scott, "A new strategy for multifunction myoelectric control," *IEEE Trans BME* **40**, 82–94 (1993).

38. Hudgins, B. S., P. A. Parker, and R. N. Scott, "A pattern recognition approach to multifunction myoelectrc control," *Med Life Sci Eng* **13**, 21–38 (1994).

39. Hush, D. R., and B. G. Horne, "Progress in supervised neural networks: What's new since Lippmann," *IEEE Sig Process Mag* (January), 8–39 (1993).

40. Jacobson, S. C., D. F. Knutti, R. T. Johnson, and H. H. Sears, "Development of the Utah Artificial Arm," *IEEE Trans BME* **29**, 249–269 (1982).

41. Josin, G., D. Charney, and D. White, "Robot control using neural networks," *Proc Int Conf Neural Net*, 1988, pp. 625–631.

42. Kato, I., et al., "Multifunctinal myoelectric hand prosthesis with pressure sensory feedback system—WASEDA Hand-4P," Proc 3rd Int Symp External Control of Human Extremities, Dubrovnik, Yugoslavia, 1969, pp. 155–170.

43. Kelly, M., and P. A. Parker, "The application of neural networks to myoelectric signal analysis: A preliminary study," *IEEE Trans BME* **37**, 221–230 (1990).

44. Kobrinski, A. E., et al., "Problems of bioelectric control," in J. F. Coles, ed., *Automatic and remote control, Proc 1st IFAC Int Conf*, Vol. 2, Butterworths, London, 1960.

45. Kramer, M. A., "Nonlinear principal component analysis using autoassociative neural networks," *AIChe J* **37**, 233–243 (1991).

46. Kuruganti, U., B. Hudgins, and R. N. Scott, "Two-channel enhancement of a multifunction control scheme," *IEEE Trans BME* **42**, 109–111 (1995).

47. Kyberd, P., "The intelligent hand," *IEE Rev*, (September) 31–35 (2000).

48. Lawrence, P., P. Herberts, and R. Kadefors, "Experiences with a multifunctional hand prosthesis controlled by myoelectric patterns," in M. M. Gavrilovic and A. B. Wilson, eds., *Advances in external control of human extremities*, Etan, Belgrade, 1973, pp. 47–65.

49. Leal, J. M., T. W. Williams, and J. M. Malone, "Adaptation of the VA/USMC electric hand to the Liberty Mutual Boston Elbow for above-elbow amputees," *Bull Prosthet Res* **18**, 56–60, (1981).

50. Leowinata, S., B. Hudgins, and P. A. Parker, "A multifunction myoelctric control strategy using an array of electrodes," 16th An Congr Int Society Electrophysiology and Kinesiology, Montreal, PQ, Canada, 1998.

51. Lyman, J. H., A. Freedy, and R. Prior, "Fundamental and applied research related to the design and devolopment of upper-limb externally powered prostheses," *Bull Prosthet Res* **13**, 184–195 (1976).

52. Mann, R. W., "Cybernetic limb prosthesis: The ALZA distinguished lecture," *An Biomed Eng* **9**, 1–43 (1981).

53. Maric, M. R., and M. M. Gavrilovic, "An evaluation of synergistic control for the rehabilitation manipulator," *Proc 4th Int Symp External Control of Human Extremities*, Etan, Dubrovnik, Belgrade, Yugoslavia, 1972, pp. 277–287.

54. McKerrow, P. J., *Introduction to robotics*, Addison-Wesley, Reading, MA, 1991.

55. Miyamoto, H., M. Kawato, T. Setoyana, and R. Suzuki, "Feedback-error-learning neural network for trajectory control of a robotic manipulator," *Neural Net* **1**, 251–265 (1988).

56. Northmore-Ball, M. D., H. Heger, and G. A. Hunter, "The below-elbow prosthesis," *J Bone Joint Surg* **62**, 363 (1980).

57. Parker, P. A., "Optimal signal processing for multifuntion myoelectric communication channel," PhD thesis, University of New Brunswick, Fredericton, New Brunswick, Canada, 1975.

58. Parker, P. A., J. Stuller, and R. N. Scott, "Signal processing for the multistate myoelectric channel," *Proc IEEE* **65**, 662–674 (1977).

59. Parker, P. A., and R. N. Scott, "Myoelectric control of prostheses," *CRC Cri Rev Biomed Eng* **13**, 283–310 (1986).

60. Reswick, J. B., H. W. Merler, W. H. Do, C. K. Taft, J. H. Brunell, R. W. Corell, D. H. Thomas, K. M. Griggs, and D. J. Gawlowicz, "Conscious control of programmed motions in orthotics and prosthetics," Report EDC4-62-2, Case Institute of Technology, Cleveland, OH, 1962.

61. Richard, P., R. Gander, P. Parker, and R. N. Scott, "Multistate myoelectric control: The feasibility of 5-state control," *J Rehab Res Dev* **20**, 84–86 (1983).

62. Rosenblatt, F., "The perceptron: A probabilistic model for information storage and organization in the brain," *Psycho Rev* **65**, 386–408 (1958).

63. Saridis, G. N., and T. Gootee, "EMG pattern analysis and classification for a prosthetic arm," *IEEE Trans BME* **20**, 403–409 (1982).

64. Schmidl, H., "The I. N. A. I. L. experience fitting upper-limb dysmelia patients with myoelectric control," *Bull Prosthet Res* **10–27**, 17–42 (1977).

65. Schmidt, R. A., *Motor control and learning: A behavioral emphasis*, Human Kinetics, Champaign, 1988.

66. Scott, R. N., and P. A. Parker, "Myoelectric prostheses: state of the art," *J Med Eng Technol* **12**, 143–151 (1988).

67. Simpson, D. C., and G. Kenworthy, "The design of a complete arm prosthesis," *Biomed Eng* **8**, 56–59 (1973).

68. Srinivasan, S., "A rhythmic movement control scheme using artificial neural networks," PhD thesis, University of Saskatchewan, Sasketoon, Saskatchewan, 1992.

69. Fu, K. S., *Syntactic pattern recognition and applications*, Prentice-Hall, Englewood Cliffs, NJ, 1982.

70. Tou, J. T., and R. C. Gonzalez, *Pattern recognition principles*, Addison-Wesley, Reading, MA, 1974.

71. Vodovnik, L., J. Kreifeldt, R. Caldwell, L. Green, E. Silgalis, and P. Craig, "Some topics on myoelectric control of orthotic/prosthetic systems," Rep EDC 4-67-17, Case Western Reserve University, Cleveland, OH, 1967.

72. Wada, Y., and M. Kawato, "A neural network model for arm trajectory formation using forward and inverse dynamics models," *Neural Nets* **6**, 919–932 (1993).

73. Widrow, B., and M. A. Lehr, "30 years of adaptive neural networks: Perceptron, madaline, and backpropagation," *Proc IEEE* **78**, 1415–1442 (1990).

74. Williams, T. W., "The Boston Elbow," in *Proc 4th Congress ISEK (International Society of Electromyography and Kinesiology)*, Boston, 1979, pp. 66–67.

75. Williams, T. W., "Practical methods for controlling powered upper-extremity prostheses," *Assist Technol* **2**, 3–18 (1990).

76. Wirta, R. W., D. R. Taylor, and F. R. Findley, "Pattern recognition arm prosthesis: A historical perspective—Final report," *Bull Prosthet Res* **10–30**, 9–35 (1978).

77. Yamazaki, Y., M. Suzuki, and T. Mano, "An electromyographic volley at initiation of rapid contractions of the elbow," *Brain Res Bull* **30**, 181–187 (1993).

78. Yang, Y. S., F. K. Lam, F. Chan, and P. A. Parker, "A new fuzzy approach for pattern recognition with application to EMG classification," Int Conf Neural Networks, Washington, DC, June 2–6, 1996.

79. Yeung, D. T., and G. A. Bekey, "Using a context-sensitive learning network for robot arm control," Proc IEEE Int Conf Robotics and Automation 1989, pp. 1441–1447.

80. Zadaca, H., J. Lyman, and A. Freedy, "Studies and development of heuristic endpoint control for artificial upper limbs," University of California, Los Angeles, Rep UCLA-ENG-74-79; Biotechnology Lab Techn Rep 54, 1974.

81. Zhang, Y. T., P. A. Parker, and R. N. Scott, "Study of the effects of motor unit recruitment and firing statistics on the signal-to-noise ratio of a myoelectric channel," *Med Biol Eng Comput* **28**, 225–231 (1990).

Electromyography: Physiology, Engineering, and Noninvasive Applications, edited by Roberto Merletti and Philip Parker.
ISBN 0-471-67580-6 Copyright © 2004 Institute for Electrical and Electronics Engineers, Inc.

Prostheses. *See also* Powered upper limb
    prostheses
  multifunction, 462–463
  multifunctional control of, 470
  myoelectric, 145, 454–455
PSD estimation, 135, 142. *See also* Power
    spectral density (PSD)
  during voluntary contractions, 146
  using wavelet packets, 279
  using wavelet shrinkage, 279–280
Pseudorandom electrical activity, 42
Psychological stress, 354
  musculoskeletal disorders and, 355
  neuroendocrine systems sensitive to, 355
  trapezius muscle response to, 356–357
Psychophysiological stress-related
    hyperactivity, 436–437
Psychosocial stress, 353

Quadratic time-frequency methods, 468
Quadrifillar needle electrode (QNE), 41
"Quasi-stationary" signal, 144, 293

Race walking, 371–372
Radial conductivity, 88
RADIUS value, selecting, 158–159
Rainoldi, A., 233, 403, 417, 420, 423
Ramp contractions, 154
Random processes, 264
  cumulants of, 266
Raster plots, MUAP, 62
Rate coding, 458, 462
  motor unit, 6–9
Reciprocal inhibition, 14
Recorded signal, frequency content of, 42
Recording equipment, EMG, 43–45
Recordings, 72–73
  high-density multichannel, 338–341
Recruitment (REC), 101
  influence on MMG, 316–318
Recruitment threshold, 9
Recruitment threshold force, 71
Recurrence histograms, 66
Recurrence map (RM), 157, 158
Recurrence quantification analysis (RQA), 134,
    154–162, 373–374
  application to surface EMG signal analysis,
    159–162
  mathematical basis of, 155–159
Recurrence representation (RM), 159
Reference decomposition, 67, 70
Reference MUs, association with detected
    MUs, 67–68
Reference signal, 191

Reflexes, 11–12
Reflexive spasm, acute, 439
Reflex overflow, 14
Reflex responses, 15
Rehabilitation medicine, 403–425
  EMG in back and neck pain, 404–411
  EMG of the pelvic floor, 411–417
  hypobaric hypoxia, 420–423
  microgravity effects on neuromuscular
    system, 423–425
Reinnervation, collateral, 33, 34
Relaxation-based downtraining, 444–445
Relinearization, 143–144
Renewal point process, 65
Renshaw cell mediated inhibition, 14
Repetitive stimulation, 334
"Representing efforts," 435–436
Ring electrode, 177, 178
Romberg test, 423
Root mean square (RMS), 346
  indicators of, 140
  value of, 236
Running, 372

Satellite potential, 66
Scaling function, 276–277
Scanning EMG, 41–42
Screen quality, 45
SD signal, 115. *See also* Single differential
    (SD)
Segmentation phase performance, 68–69
"Segmentation" strategy, 408
Selective muscle stimulation, 93, 94
Selective recruitment, 9
SEMG amplifier, input impedance of, 116–118
SEMG amplitude normalization procedures,
    368
SEMG biofeedback. *See also* SEMG feedback;
    Surface electromyography (SEMG)
  in ergonomics, 352
  techniques of, 443–450
SEMG decomposition, 301
SEMG feedback. *See also* SEMG biofeedback;
    Surface electromyography (SEMG)
  functional activity performance with, 450
  postural training with, 449
  therapeutic exercise with, 449
SEMG measurements, 338–341
SEMG models, structure-based, 207–209
SEMG normalization/calibration, 346–347
SEMG sensors, recommendations for, 123–127
SEMG signal conditioning system, 120–121
SEMG signals. *See also* Surface
    electromyography (SEMG)